Progress in
Biochemical Pharmacology
Vol. 4

Progress in Biochemical Pharmacology

Vol. 4

Recent Advances in Atherosclerosis

Epidemiology - Intact Organisms - Lipoproteins - Plasma Lipids
Whole Artery - Tissue Cultures - Hormones - Primates - Non Primates
Platelets - Clinical

Edited by
C.J. MIRAS, Athens; A.N. HOWARD, Cambridge; R. PAOLETTI, Milan

With 2 color plates, 236 figures and 151 tables

BASEL (Switzerland)   S. KARGER   NEW YORK

International Symposium on Atherosclerosis, Athens 1966. Organised by the European Society for Biochemical Pharmacology. All papers reviewed at the end of 1967 by the authors

Progress in Biochemical Pharmacology

Editor: R. PAOLETTI, Milan

Vol. 1: First International Symposium on Radiosensitizers and Radioprotective Drugs, Milan, May 1964. X + 750 p., 287 fig., 158 tab., 1965.

Vol. 2: Second International Symposium on Drugs Affecting Lipid Metabolism, Milan, September 1965. Part I: Cholesterol and Atherosclerosis - Plasma Triglycerides and Lipoproteins - Drugs Affecting Plasma Lipids. XII + 516 p., 2 cpl., 168 fig., 133 tab., 1967.

Vol. 3: Second International Symposium on Drugs Affecting Lipid Metabolism, Milan, September 1965. Part II: Fatty Acid - Prostaglandins - FFA Mobilization - FFA and Triglyceride Transport - Liposoluble Vitamins. XII + 528 p., 183 fig., 160 tab., 1967.

Vol. 4: International Symposium on Atherosclerosis, Athens, May/June 1966. Epidemiology - Intact Organisms - Lipoproteins - Plasma Lipids - Wole Artery - Tissue Cultures - Hormones - Primates - Non Primates - Platelets - Clinical. VIII + 636 p., 2 cpl., 236 fig., 151 tab., 1968.

S. Karger AG, Arnold-Böcklin-Strasse 25, CH-4000 Basel 11 (Switzerland)

All rights, including that of translation into other languages, reserved.
Photomechanic reproduction (photocopy, microcopy) of this book or parts thereof without special permission of the publishers is prohibited.

©

Copyright 1968 by S. Karger AG, Basel
Printed in Switzerland by National-Zeitung AG, Basel

Contents

Epidemiology

GROEN, J.J. (Jerusalem): Recent Advances in the Epidemiology of Atherosclerosis 1
ARAVANIS, C.; LEKOS, D.; DONTAS, A.; MICHAELIDES, G. and KEYS, A. (Athens): On the Study of Ischemic Heart Diseases on the Islands of Crete and Corfu, Greece 12
BRUNNER, D. (Jaffa): The Influence of Physical Occupational Activity on the Incidence of Ischemic Heart Disease 19
GROEN, J.J.; DREYFUSS, F. and GUTMANN, L. (Jerusalem): Epidemiological, Nutritional and Sociological Studies of Atherosclerotic (Coronary) Heart Disease Among Different Ethnic Groups in Israel 20
McGILL, H.C. Jr. (New Orleans): Geographic Pathology of Atherosclerosis . . 26
STAMLER, J.; BERKSON, D.M.; MOJONNIER, LOUISE; LINDBERG, H.A.; HALL, YOLANDA; LEVINSON, M.; BURKEY, FRANCES; MILLER, WILDA; EPSTEIN, M.B. and ANDELMAN, S.L. (Chicago, Ill.): Epidemiological Studies on Atherosclerotic Coronary Heart Disease: Causative Factors and Consequent Preventive Approaches 30
Discussion Session 1 50

Intact Organisms

AHRENS, E.H. Jr. (New York, N.Y.): Studies of Cholesterol Metabolism in Intact Organisms 54
KINSELL, L.W.; WOOD, P.D.S.; SHIODA, R.; SCHLIERF, G. and LEE, Y.L. (Oakland, Cal.): Effects of Diet upon Plasma, Bile and Fecal Steroids . . 59
MIETTINEN, T.A. (Helsinki): Effect of Dietary Cholesterol and Cholic Acid on Cholesterol Synthesis in Rat and Man 68
TAYLOR, C. BRUCE; MIKKELSON, B.; ANDERSON, J.A. and FORMAN, D.T. (Evanston, Ill.): Human Serum Cholesterol Synthesis 71
WAGENER, H. and SCHWARTZKOPFF, W. (Heidelberg): Distribution and Turnover of Arachidonic Acid in Normals and Hyperlipemic Patients 81
Discussion Session 2 88

Lipoproteins

ALAUPOVIC, P. (Oklahoma City, Okla.): Recent Advances in Metabolism of Plasma Lipoproteins: Chemical Aspects 91
BEAUMONT, J.L. (Paris): Hyperlipidemia with Circulating Anti β Lipoprotein Auto-Antibody in Man. Auto-Immune Hyperlipidemia, its Possible Role in Atherosclerosis 110
BLATON, V.; PEETERS, H. (Brugge); GRESHAM, G.A. and HOWARD, A.N. (Cambridge): Differential Fatty Acid Composition of Alpha and Beta Lipoproteins in Baboons 122

BOBERG, J. and NYE, E.R. (Stockholm): Recirculation of H3-Labelled Palmitate in Plasma Lipoproteins 126
BOYD, G.S.; ONAJOBI, FUNMILAYO D. and PERCY-ROBB, I.W. (Edinburgh): Studies on Plasma Lipoproteins in the Rat 132
BRUNNER, D. and LOEBL, J. (Jaffa): Alpha-Cholesterol Percentage in Coronary Patients and in Healthy Controls 140
COHEN, L. and DJORDJEVICH, Juliana (Chicago, Ill.): Variations in Human Serum Alpha-1-Lipoproteins 141
GUSTAFSON, A. (Göteborg): A Hypothetical Transformation of Chylomicrons and VLD Lipoproteins into Lipoproteins of Higher Density 143
PEETERS, H. and BLATON, V. (Brugge): Lipid Fatty Acid Relationships in Electrochromatographic Lipoprotein Fractions 144
SCANU, A. and GRANDA, J.L. (Chicago, Ill.): Comparative Optical Properties of Human Serum Low- and High-Density Lipoproteins before and after Delipidation 153
WALTON, K.W. (Birmingham): The Role of Low-Density Lipoproteins in the Pathogenesis of Human Atherosclerosis 159
SKOŘEPA, J.; HRABÁK, P.; NOVAK, Š.; TODOROVIČOVÁ, H. and MARČAN, K. (Prague): Ultracentrifugal Fractionation of Post Heparin Lipolytic Acticity in Human Plasma 161
Discussion Session 3 163

Plasma Lipids

CARLSON, L.A. (Stockholm): Recent Advances in the Metabolism of Plasma Lipids 170
FROSCH, B. and WAGENER, H. (Heidelberg): Conjugated Bile Acids in Serum of Patients with Essential Hyperlipemia and Hypercholesterolemia 179
GERÖ, S.; BIHARI-VARGA, M. and SZÉKELY, J. (Budapest): Gastric Mucosubstances and Serum Lipid Pattern 183
GOULD, R.G. and SWYRYD, E.A. (Palo Alto, Cal.): Metabolism of Plasma Lipids 191
GROSS, R.C.; REAVEN, G.M. and FARQUAHR, J.W. (Palo Alto, Cal.): Turnover of Endogenous Plasma and Liver Triglyceride in Man and the Dog . . . 197
KURIEN, V.A. and OLIVER, M.F. (Edinburgh): Changes in Serum Free Fatty Acids Levels after Acute Vascular Occlusion 204
RIFKIND, B.M,; BEGG, T.; JACKSON, I. and GALE, M. (Glasgow): Blood Lipid Levels as Related to Adiposity and Obesity 206
NIKKILÄ, E.A.; MIETTINEN, T.A.; PELKONEN, R. and TASKINEN, M.-R. (Helsinki): Plasma Insulin Response to Glucose in Endogenous and Alimentary Hyperglyceridemia 208
Discussion Session 4 209

Whole Artery

LOFLAND, H.B. (Winston-Salem, N.C.): Recent Advances in Arterial Metabolism: The Whole Artery 211
ADAMS, C.W.M.; ABDULLA, Y.H.; BAYLISS, O.B.; MAHLER, R.F. and ROOT, M.A. (London): Quantitative Histochemical Observations on Certain Oxidative and Lipolytic Enzymes in Human Aortic Wall 218

BILLIMORIA, J.D. and ROTHWELL, T.J. (London): Factors Affecting the Synthesis of Individual Phospholipids in the Rat Aorta 225
BÖTTCHER, C.J.F. (Leiden): Origin and Development of Atherosclerotic Lesions 231
BOWYER, D.E.; HOWARD, A.N.; GRESHAM, G.A.; BATES, D. and PALMER, B.V. (Cambridge): Aortic Perfusion in Experimental Animals. A System for the Study of Lipid Synthesis and Accumulation 235
CHRISTENSEN, S. (Aarhus): Intimal Uptake of Plasma Lipoprotein and Atherosclerosis 244
KRUT, L.H. and WILKENS, J.A. (Cape): The Filtration of Plasma Constituents into the Wall of the Aorta 249
STEIN, Y.; EISENBERG, S. and STEIN, O. (Jerusalem): Metabolism of Lysolecithin by Human Umbilical and Dog Carotid Arteries 253
Discussion Session 5 260

Tissue Cultures

POLLAK, O.J. (Dover, Del.): Recent Advances in the Metabolism of Arteries. Tissue Cultures, Homogenates and Slices 265
HOLLANDER, W.; KRAMSCH, D.M. and INOUE, G. (Boston, Mass.): The Metabolism of Cholesterol, Lipoproteins, and Acid Mucopolysaccharides in Normal and Atherosclerotic Vessels 270
McMILLAN, G.C. and STARY, H.C. (Montreal, Que.): Radioautographic Observations on DNA Synthesis in the Cells of Arteriosclerotic Lesions of Cholesterol-Fed Rabbits 280
MEAD, J.F.; HAGGERTY, D.F.; GERSCHENSON, L.E. and HARARY, I. (Los Angeles, Cal.): Recent Advances in Polyunsaturated Fatty Acid Metabolism . . . 282
PATELSKI, J.; WALIGÓRA, Z. and SZULC, S. (Poznań); BOWYER, D.E.; HOWARD, A.N. and GRESHAM, G.A. (Cambridge): Lipolytic Enzymes of the Aortic Wall 287
POLLAK, O.J. and ADACHI, Minoru (Dover, Del.): Human and Rabbit Arterial Cells Compared in Tissue Cultures 294
LAZZARINI ROBERTSON, A. Jr. (Cleveland, Ohio): Oxygen Requirements of the Human Arterial Intima in Atherogenesis 305
ROTHBLAT, G.H.; HARTZELL, R.; MIALHE, H. and KRITCHEVSKY, D. (Philadelphia, Pa.): Cholesterol Ester Metabolism in Tissue Culture Cells . . . 317
ZEMPLÉNYI, T.; MRHOVÁ, O.; URBANOVÁ, D. and KOHOUT, M. (Prague-Krč): Study of Factors Affecting Arterial Enzyme Activities in Man and Some Animals 325
Discussion Session 6 330

Hormones

FURMAN, R.H.; ALAUPOVIC, P.; BRADFORD, R.H. and HOWARD, R.P. (Oklahoma City, Okla.): Gonadal Hormones, Blood Lipids and Ischemic Heart Disease 334
HIRSCH, R.L. (New York, N.Y.): Biologic Half-Life of Lipoprotein Triglyceride in the Normal and Diabetic Rabbit 351
MUKHERJEE, S.; GUPTA, S. and BHOSE, A. (Calcutta): Effects of Gonadal Hormones on Cholesterol Metabolism in Rat 352
PICK, RUTH; CLARKE, G.B. and KATZ, L.N. (Chicago, Ill.): Estrogens and Atherosclerosis 354

RENOLD, A.E.; GONET, A.E.; STAUFFACHER, W. and JEANRENAUD, B. (Geneva): Laboratory Animals with Spontaneous Diabetes and/or Obesity: Suggested Suitability for the Study of Spontaneous Atherosclerosis 363
ROBINSON, R.W. (Worcester, Mass.): Survival after Cerebral Thrombosis . . . 370
SÓLYOM, A. and PUGLISI, L. (Milan): On the Effect of Norepinephrine and Nicotinic Acid on Free Fatty Acid Transport and Incorporation in Tissue Lipids 371
Discussion Session 7 373

Primates

WISSLER, R.W. (Chicago, Ill.): Recent Progress in Studies of Experimental Primate Atherosclerosis 378
ANDRUS, S.B. (Boston, Mass.); PORTMAN, O.W. (Beaverton, Oregon) and RIOPELLE, A.J. (Covington, La.): Comparative Studies of Spontaneous and Experimental Atherosclerosis in Primates. II. Lesions in Chimpanzees Including Myocardial Infarction and Cerebral Aneurysms (with colour plate I) 393
CLARKSON, T.B.; BULLOCK, B.C. and LEHNER, N.D.M. (Winston-Salem, N.C.): Pathologic Characteristics of Atherosclerosis in New World Monkeys . . . 420
HAUST, M. DARIA (London, Ont.): Electron Microscopic and Immuno-Histochemical Studies of Fatty Streaks in Human Aorta (with colour plate II) 429
HOWARD, A.N.; GRESHAM, G.A. and BOWYER, D.E. (Cambridge) and LINDGREN, F.T. (Berkeley, Cal.): Aortic and Coronary Atherosclerosis in Baboons . . 438
THOMAS, W.A.; LEE, K.T. and KIM, D.N. (Albany, N.Y.): Metabolic Studies of Protein Synthesis in Aortas of Monkeys Fed Atherogenic Diets 445
Discussion Session 8 452

Non Primates

GRESHAM, G.A. and HOWARD, A.N. (Cambridge): Recent Advances in the Pathology of Atherosclerosis in Animals other than Primates 460
HESS, R. and STÄUBLI, W. (Basel): Mechanism of Atherogenesis in the Parakeet 470
KRITCHEVSKY, D. and TEPPER, SHIRLEY A. (Philadelphia, Pa.): Influence of Special Fats on Experimental Atherosclerosis in Rabbits 474
KRITCHEVSKY, D. and TEPPER, SHIRLEY A. (Philadelphia, Pa.): Influence of Linolexamid on Experimental Atherosclerosis in Rabbits 480
MALMROS, H. (Lund) and STERNBY, N.H. (Malmö): Induction of Atherosclerosis in Dogs by a Thiouracil-Free Semisynthetic Diet Containing Cholesterol and Hydrogenated Coconut Oil 482
THOMAS, W.A.; LEE, K.T.; KIM, D.N.; SISSON, J.A. and SCOTT, R.F. (Albany, N.Y.): Aspects of Protein Metabolism in Rats Fed Atherogenic Diets . . . 488
Discussion Session 9 501

Platelets

MUSTARD, J.F.; GLYNN, M.F.; JØRGENSEN, L.; NISHIZAWA, E.E.; PACKHAM, M.A. and ROSWELL, H.C. (Hamilton and Toronto, Ont.): Recent Advances in Platelets, Blood Coagulation Factors and Thrombosis 508

CONSTANTINIDES, P. (Vancouver, B.C.): The Cause of Cerebral Artery Thrombosis in Man 533
ENGELBERG, H. and FUTTERMAN, M. (Beverley Hills, Cal.): The Effect of Catecholamines upon the *in vitro* Thrombotic Coagulation of Human Blood . . 542
FRENCH, J.E. and BARCAT, J.A. (Oxford): The Fine Structure of Platelets and Platelet Aggregates *in vivo* 550
HAWKEY, CHRISTINE (London): The Breakdown of Artificial Platelet Thrombi *in vitro* 556
WALTON, K.W. and WOLF, P. (Birmingham): The Nature and Significance of Platelet-Products in Plasma 559
Discussion Session 10 561

Clinical

SCHETTLER, G. (Heidelberg): Current Trends in the Treatment of Atherosclerosis from the Clinical View Point 565
BIZZI, A. (Milan): Inhibition of Fatty Acid Release by Pyrazole Derivatives . 573
CHRUŚCIEL, T.L.; HERMAN, Z.S.; BRUS, R.; PLECH, A.; HABCZYŃSKA, D. and WAŻNA-BOGUŃSKA, C. (Zabrze): Influence of Clofibrate and Vatensol on the Development of Experimental Atherosclerosis in White Rats 578
GEY, K.F. and LORCH, E. (Basel): Comparison of Nicotinic Acid and B-Pyridylcarbinol with Respect to Lipid Metabolism in the Rat *in vivo* 585
LIJÓ-PAVÍA, J.J. (Buenos Aires): Blindness Caused by Atherosclerosis of the Choroid and Retina Treatment with Enzymes of Prostatic Origin 591
SHIMAMOTO, T.; ATSUMI, T.; NUMANO, F. and FUJITA, T. (Tokyo): Treatment of Atherosclerosis with Pyridylcarbamate 597
THORP, J.M. and COTTON, R.C. (Macclesfield) and OLIVER, M.F. (Edinburgh): Role of Endocrine System in the Regulation of Plasma Lipids and Fibrinogen, with Particular Reference to the Effects of 'Atromid'-S 611
Discussion Session 11 618

Authors' Index 627
Subject Index 629

From the Department of Medicine A, Hadassah-Hebrew University Hospital and Medical School, Jerusalem, Israel

Recent Advances in the Epidemiology of Atherosclerosis*

J.J. GROEN, Jerusalem

Introduction

From prevalence and incidence surveys, hospital admission, autopsy and mortality studies, carried out in many places of the world, a more and more complete picture now begins to emerge, from which our present knowledge of the epidemiology of the atherosclerotic diseases can be synthetized. Most of this knowledge concerns atherosclerotic heart disease, but data are now also accumulating on the epidemiology of the other manifestations and complications of atherosclerosis and I shall have the opportunity to refer to these later. During the past years several surveys of the progress of knowledge have appeared, and I would like to pay a special tribute to the excellent symposia published in the Journal of Chronic Diseases [1–3].

Table I summarizes the factors which have now been identified as playing a role in the pathogenesis of coronary heart disease, and I would like to comment shortly on each of them.

We are accustomed to regard *age* as an important factor in the development of the disease. Autopsy studies have revealed that the early stages, macroscopically visible as fatty streaks and soon followed by fibrous plaques occur already towards the end of the first, and increase during the second decade of life.

* These researches were supported by U.S.P.H.S. grant No. H-4427, National Heart Institute, National Institutes of Health, Bethesda, Md., USA.

It takes two or more decades before the lesions become so advanced that they give rise to clinical manifestations in an appreciable number of cases. We tend to forget, however, that this picture does not represent the natural history of the disease in the biological sense of the word. It only holds for populations living the type of life which is commonly associated with the modern western culture. In other population groups, like the South African Bantu, the natives of Guatemala and the Yemenites and Bedouin in Israel, the disease is so rare in all age groups that age can hardly be called a risk factor per se. The same holds for the influence of *sex*. The preponderance of its incidence among the male over the female sex is well established. However, this male over female preponderance is so much more marked in the countries of the so-

Table I. Pathogenetic ('increased risk') factors in coronary heart disease, identified by epidemiological studies

1. Age and sex (in 'western' countries).
2. Social group (a) ethnic, (b) economic.
3. Diet (saturated fats; cholesterol; sugar?).
4. Lack of physical activity.
5. Hypercholesterolaemia (-betalipoproteinaemia; triglyceridaemia).
6. Obesity?
7. Cigarette smoking.
8. Certain diseases: Diabetes; hereditary hyperlipaemias; hypertension.
9. Psychological factors? ('Personality'; Work problems; Marital difficulties; Excessive selfcontrol.)
10. Unknown genetic factors?

called Western culture, that it appears doubtful whether the difference in susceptibility between the sexes is caused by endogenous factors. The difference is certainly not due to different levels of serum cholesterol because these are either equal in both sexes, or, if anything, slightly higher in females. Differences in serum alpha-beta lipoprotein ratios and values have been reported, but these are not confirmed by all authors, and the differences are small only. It might well be that there are other harmful factors in the ways of life of Western culture, which affect men more than women, but these are still hypothetical.

Impressive evidence is now available that coronary heart disease is unequally distributed over different *ethnic groups*, and it has been shown convincingly, that certain populations like the Navahoes and Bantu, the inhabitants of Crete and Corfou, the Yemenites and Bedouin in Israel and certain Japanese groups are almost free from this condition. There are also some indications that there is unequal distribution of the disease among socio-economic groups within Western populations. In this connection, it should be remembered that the groups in which the disease is rare differ not only in their ethnic composition from the Western populations, but also in their socio-economic level.

Most of the evidence indicates that these ethnic and socio-economic differences are not genetic in nature, but environmental. However, the environment and the ways of life of these almost 'coronary free' groups differ from those of the populations of the Western culture in so many respects that it has not yet been possible to designate one single factor as the cause. Indeed, most investigators regard coronary heart disease as a result of a multi-factorial constellation [4].

A lot of work has been done during the past twenty years on animals, human volunteers, patients and controls, to elucidate the role of hypercholesterolemia in the production of the disease. Indeed, the international epidemiological studies of Keys and his coworkers and many others have shown a highly significant correlation between average serum cholesterol levels and the mortality from and prevalence of the coronary heart disease. This correlation, added to the experimental evidence and the results of incidence studies, like those of Framingham, Albany and Chicago, leave little doubt about the fact that high serum cholesterol is one of the most important pathogenetic factors and indicators of an increased risk to acquire the disease. Although it is proven that cholesterol circulates in the form of lipoproteins, of which the light density beta-lipoproteins probably are the causative agents, there is no convincing evidence that for epidemiological studies the serum concentrations of beta-lipoprotein (or triglycerides) are more useful than the cholesterol level.

Work on animals and volunteers has taught us that the main factors in the diet which raise the serum cholesterol are the quantities of saturated fatty acids (among which myristic and palmitic acids are probably the most important), of cholesterol, and probably of sugar. These nutrients are present in high amounts in the diets of Western populations; they are consumed in much smaller amounts by so-called underdeveloped groups, who have both a low serum cholesterol, and a low incidence of coronary heart disease.

An important question which has not yet been answered satisfactorily concerns the influence of other nutrients on the serum cholesterol. In particular, we do not know whether polyunsaturated fatty acids, like linoleic acids, *specifically* lower the serum cholesterol, or whether they share this effect with protein and polysaccharides, in other words, with all other nutrients which can replace the saturated fats in the diet. So long as this problem has not been solved, the present recommendations to the public to replace the dietary saturated-, by polyunsaturated fatty acids [5], although justified, appear somewhat one-sided. It is too often forgotten that the Middle Eastern and other populations which have a conspicuously low incidence of myocardial infarction do *not* eat high amounts of polyunsaturated fats; they do *not* use maize-, soya-or safflower oil. They subsist on diets which are low in total and saturated fat, in cholesterol and in sugar, but high in bread and thereby in starch and vegetable protein (Table II). Such diets may therefore be recommended with at least as much justification [6].

Another problem, as yet unsolved, is the contradiction between the dietary studies in animals and human volunteers, which showed an important effect of diet on the serum cholesterol in the same

Table II. Comparison of the composition of an average 'Western' with a Trappist monk, Yemenite and Bedouin diet (in grams per day)*

	Calories	Protein Animal	Vege-table	Carbohydrate Mono/Disacch.	Starch	Sat. F.A.	Fat Oleic A.	Linol. A.	Cholesterol mg %
'Western'	2250	85 (15%)		250 (45%)			100 (40%)		550
			55		115	135	50	32	9.5
Trappist	3200	110 (14%)		480 (65%)			78 (21%)		160
			40	70	40	440	18	40	5
Yemenite	2200	88 (15%)		355 (65%)			48 (19%)		400
			30	58	40	315	18	20	6.5
Bedouin	2250	63 (11%)		410 (74%)			38 (15%)		10
			5	58	20	390	8	12	9.0

* Figures in brackets: Caloric %.

individual, and the apparent lack of correlation between dietary history and serum cholesterol in different subjects, which has recently emerged from the work of several English and American epidemiological research groups. Whereas there is unanimity in the finding that individuals with a high serum cholesterol are predisposed for coronary artery disease, it could not be proven that such predisposed individuals had eaten differently than others. This might mean that the methods used for appraisal of the diets were not accurate enough, but it is also possible that in a Western culture all examined subjects consume so much saturated fat that it is already maximal in its cholesterol raising effect, so that the differences in serum cholesterol among them are due to other factors than the diet.

Among these it seems at present very likely that *lack of regular physical activity* plays an important role. No direct methods to measure energy expenditure by epidemiological methods are available, but indirect methods point to an effect of exercise on serum cholesterol. Thus we are confronted with the fact that two important causes of hypercholesterolemia and thereby of coronary heart disease namely, type of diet and lack of physical activity are deeply engrained in the Western ways of life. This would mean that a prevention of coronary heart disease by way of lowering the serum cholesterol in the masses of the population would not only be a medical, but also a socio-cultural challenge.

The role of *obesity* is gradually becoming more clear. It appears that a high amount of fat in the body does not per se mean hypercholesterolemia or increased risk of coronary heart disease. It is the type of nutrition which produces the obesity, which appears to be of importance. If, like it is usually the case in the Western world, obesity is the result of an overconsumption of saturated fat and sugar, it is generally accompanied by hypercholesterolemia, and thereby with an increased risk. When, like we find it in the Middle East, the obesity is the result of the over-eating of bread, or, as it is the case for the Trappist monks, of peas and beans, it is accompanied by a low serum cholesterol and it carries no extra risk for the development of coronary heart disease.

Cigarette smoking has been shown to increase the chances for development of coronary heart disease, especially in young males. The mechanism is unknown; it does not seem to work via an increase in serum cholesterol.

Of the diseases which have been shown to predipose for coronary heart disease, some of the so-called *hereditary hyperlipideamias* are of considerable biochemical interest, because they may help us to arrive at an understanding of the pathogenetic mechanisms. However, the frequency of these conditions in the general population is not so great as to make them important from an epidemiological point of view. This is the case, however, with diabetes and hypertension. As far as *diabetes* is concerned, it is quite possible that part of the increased risk which patients from this disease run, is due to the diets, which the majority of doctors traditionally still prescribe for this condition. These diets are so high in saturated fat and so low in bread, peas and beans, rice and other staple foods, that we need not to be surprised about the high blood cholesterol and the frequency of atherosclerotic cardio-vascular complications in diabetic patients. The rarity of such complications in Trappist monks and in Yemenite Jews (among both of whom diabetes is not rare) [8] seems to indicate that diabetes need not per se lead to coronary heart disease, if not treated by high saturated fat, i.e. atherogenic diets. Whether the diabetic state itself predisposes to atherosclerosis and coronary heart disease is still not quite proven in my opinion. (Only one thing is certain, viz., that the risk of acquiring coronary heart disease by a high saturated fat diet cannot be compensated by even high amounts of insulin.) It follows that those physicians who are responsible for the treatment of diabetic patients should use as yard-sticks for adequate control not only the blood sugar and the excretion of keto-bodies, but also the serum cholesterol level and adjust the diet and exercise of these patients accordingly.

The promotion of atherosclerosis and coronary heart disease by sustained high blood pressure is now well proven by clinical and epidemiological observations. To ascribe it to the effect of an increased strain on the arterial wall, causing an increased risk of mechanical damage to the endothelium and media, seems a very likely explanation. An interesting finding among the Bedouin in Israel was the observation that although they had an about equal prevalence of hypertension as a Jewish control group, they had a conspicuously lower prevalence of hypertensive heart disease. As the Bedouin, as already mentioned, have a very low serum cholesterol and very low prevalence of coronary heart disease, it might be surmized that in a population group whose arteries are not affected by atherosclerosis, the heart is much less affected by a sustained high blood pressure. If the coronary vessels are wide open, the left ventricle can compensate the burden of the increased pressure by hypertrophy, but when these vessels are narrowed or even occluded by atherosclerosis, the heart of the hypertensive will decompensate sooner. It may sound like a speculative heresy, but it could be doubted whether 'pure' hypertensive heart disease (i.e. without arteriosclerosis) exists!

This leads us to another finding of Israeli investigators. In contrast to the very different distribution of coronary heart disease among the different ethnic groups, now proven by hospital admission and mortality statistics, by prevalence and autopsy studies, the distribution of the *cerebral complications* of atherosclerosis, viz., apoplexia and cerebral thrombosis (taken together) is the same in these groups (Fig. 1) [9]. This difference in distribution means that whatever the two main manifestations of atherosclerosis have in common, there must also exist important differences in their pathogenesis. It seems a likely hypothesis to connect these findings with the different ethnic distribution of serum cholesterol, which is so highly correlated with coronary heart disease and with the equal ethnic distribution of hypertension, as the main pathogenetic factor in the production of cerebral vascular accidents.

In 1964, YATES [10] drew attention to the discrepancy between the rise in death rate from coronary heart disease, registered during the years 1932–1961 in the United Kingdom, and the constancy of the mortality from cerebral vascular accidents during this period. However, when he separated the cerebral vascular deaths into those caused by a cerebral haemorrhage and by cerebral thrombosis, it appeared that death from hemorrhage had declined and from thombosis had increased twofold. He concluded that whereas similar pathogenic factors seem to play a role in thrombosis of the cerebral and coronary arteries, cerebral hemorrhage differed from these two in its etiology.

Similar findings were reported from South Africa. Whereas the death rates from coronary heart disease among the Bantu are conspicuously low, their mortality from cerebral vascular disease is about the same as among South African Whites. In two recent papers STALLONES [3] and BERGSON and STAMLER [11] have summarized the present status of knowledge of the epidemiology of cerebral vascular disease. Especially these last authors have sug-

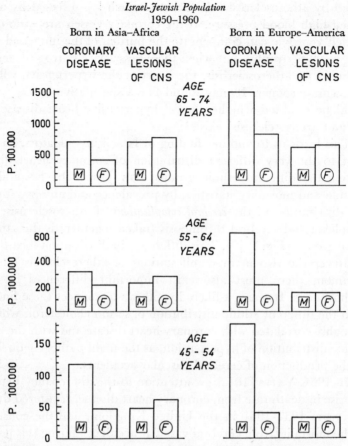

Fig. 1. Average annual mortality rates (per 100,000 inhabitants) in three age groups by continent of birth and sex, for coronary disease (No. 420 of ICD) and for vascular lesions of CNS (Nos. 330–334 of ICD).

gested that the equal mortality from cerebral vascular disease among so-called 'primitive' and Western populations can be explained by the fact that hypertension is equally frequent in almost all underdeveloped and developed populations examined so far. BERGSON and STAMLER base this opinion on comparisons between mortality from cerebral vascular and coronary disease in the United States. The White population of the United States has a much higher mortality from coronary heart disease than the Coloured. But in cerebral vascular mortality the Non-Whites are slightly higher than the Whites. Whereas coronary heart disease mortality is much higher among men than among women (of both groups), cerebral vascular mortality is slightly higher among females than among males. The slight preponderance of Coloured over Whites in their cerebral vascular mortality correlates with a higher prevalence of hypertension among them. As hypertension, moreover, is equally distributed among men and women (or in some countries even more frequent among women), this also correlates with the sex distribution of cerebro-vascular mortality. The latest communications from Japan seem to fit into the same picture.

We may even extend the pathogenetic hypothesis one step further. The present available epidemiological data are compatible with the concept that atherosclerosis can be distinguished into at least two diseases of different etiology: (a) Hypercholesterolemia, in animals and in man, is an important factor in the production of *atherosis of the intima* of the arteries. It is specifically this atherosis and its complications which are more frequent in Western than in Eastern populations. (b) The *sclerotic changes in the tunica media* of the arteries however, are hardly different in frequency in Western and Eastern population groups. They seem to be in the first place the result of mechanical (hemodynamic) damage to the vascular wall, in which hypertension plays the main role as an increased pathogenetic risk factor. In the Western countries atherosclerosis is a combination (to a varying degree in different individuals and in different regions of the body) of intima-atherosis and media-sclerosis and its incidence is increased by both nutritional and hemodynamic damage to the large arteries, leading to coronary- and cerebral-thrombosis. In the so-called primitive or Eastern groups, only the hemodynamic strains operate and they produce predominantly (although not exclusively) the media sclerosis of the large and small vessels, leading to cerebral hemorrhage as the most frequent clinical sequence.

A few words must be said about the present status of knowledge of the role of so-called *psycho-social factors* in the production of coronary heart disease. Whereas such a role seems to be undeniable for the clinician and the family doctor who has an opportunity to watch the personality and social and family setting in which a myocardial infarct develops, we have to acknowledge that too little scientific proof for the role of such factors in the pathogenesis of the disease is still available. All we have are hypotheses about certain predisposing personality features (or behaviour patterns) and about inter-human conflicts connected with the work and family situation in the Western culture and the excessive way in which certain individuals try to inhibit their emotional reactions to such situations. We have also some significant differences in the answers given by individuals with coronary heart disease to certain questions in epidemiological questionnaires, compared to healthy persons of the same age. The work of Friedman and Rosemann has even shown that certain answers and behaviour patterns are predictive indicators of the future incidence of coronary heart disease. It is obvious however, that we need a great deal more evidence before we can regard the correlation as proven. If proven, it seems most likely that the hemodynamic changes which strong emotions produce in cardiac output and stroke volume, increase the strain on the walls of the coronary arteries. Thus, certain inter-human conflicts and the specific meaning which they have for certain individuals, may ultimately contribute to coronary heart disease and myocardial infarction.

Epidemiological research in coronary heart disease started by the use of merely descriptive statistical methods and the accumulation of data on its distribution by age, sex, ethnic group etc. In a second phase the researches were aimed at the testing of certain etiological hypotheses and the detection of those factors by which groups of special risk could be identified. It is only very recently that the results of a few preventive studies have been published. The favourable results, now well documented [12, 13, 14], obtained in the so-called anti-coronary groups by a combined control of diet, physical activity and smoking (to which possibly an enhanced optimism of the volunteers may be added) fill epidemiologists and the public with new hope. After having identified the role of environmental conditions and some individual factors in the etiology, the task of epidemiology in the future will be more and more to device well controlled *preventive* trials.

The greatest physician of all times, Hippocrates the Greek, did not describe coronary heart disease. This may seem strange at first sight if we consider his extraordinary capacities for recognizing most of the major clinical syndromes as we still see them today. He certainly knew and described apoplexia. I find it therefore difficult to assume that such striking clinical pictures as angina pectoris or of death from acute myocardial infarction could have escaped his observation. Could it be that the disease was less frequent in his time? If we look again at Table I which summarizes the main pathogenetic factors, as our modern researches have identified them, it would seem that his contemporaries were exceptionally well protected: Their diet did not contain any butter; olive oil was the only fat used by the Ancient Greeks. We know that they consumed high amounts of bread and that sugar and chocolate were unknown in their days. They had no cars and their main past-time was physical exercise, training for the Olympic games. Cigarettes had not yet been invented. Indeed, from the interesting studies of our Greek colleagues in the Island of Crete, we may derive further support for our assumption that coronary heart disease was rare in Hippocrates' time, and that this is the reason why he did not describe the condition!

Hippocrates was not only a clinician, he was also a pioneer epidemiologist. It is only fitting that, as clinicians and epidemiologists, meeting in this ancient Greek city, we remember the Great Master to whom we are so much indebted.

References

1. EPSTEIN, F.H.: Epidemiology of coronary heart disease. J. Chron. Dis. *18:* 735 (1965).
2. SACHETT, D.L. and WINKELSTEIN, W.: Epidemiology of aortic and periferal atherosclerosis. Id. 775.
3. STALLONES, R.A.: Epidemiology of cerebrovascular disease. Id. 859.
4. ROSE, G.A. and BLACKBURN, H.: Cardiovascular population studies. Methods (W.H.O., Geneva 1966).
5. AMERICAN HEART ASSOCIATION: Dietary fat and its relation to heart attacks and strokes. Circulation *23:* 133 (1961); News Release June 5 (1964).
6. GROEN, J.J.: Effect of bread in the diet on serum cholesterol. Amer. J. clin. Nutr. (In press).
7. GROEN, J.J.; TIJONG, B.; KOSTER, M.; WILLEBRANDS, A.F.; VERDONCK, G. and PIERLOOT, M.: Hypertension and coronary heart disease among Trappist and Benedictine monks. Amer. J. clin. Nutr. *10:* 456 (1962).
8. COHEN, A.M.: Prevalence of diabetes among different ethnic groups in Israel. Metabolism *10:* 50 (1961).
9. KALLNER, G. and GROEN, J.J.: Mortality and hospitalization in relation to coronary and cerebral vascular disease in Israel. J. Atheroscler. Res. *6:* 419 (1966).
10. YATES, P.A.: A change in the pattern of cerebrovascular disease. Lancet *i:* 65 (1965).
11. BERGSON, D.M. and STAMLER, J.: Epidemiological findings in cerebrovascular diseases and their implications. J. Atheroscler. Res. *5:* 189 (1965).
12. CHRISTAKIS, G.; RINZELER, F.J. and MOSLOVSKY, A.: Summary report of research activities of the Anti-Coronary Club. Publ. Health Rep. *81:* 1 (1966).
13. STAMLER, J.: Present Symposium.
14. LEREN, P.: Effect of plasma cholesterol lowering diet in male survivors of myocardial infarction. Acta Med. Scand. Suppl. 466 (1966).

Author's address: J.J. Groen, M.D., Hadassah Hebrew University Hospital and Medical School, P.O.P. 499, *Jerusalem* (Israel).

Department of Cardiology, University of Athens, and Department of Physiological Hygiene, University of Minnesota

On the Study of Ischemic Heart Diseases on the Islands of Crete and Corfu, Greece*

C. Aravanis, D. Lekos, A. Dontas, G. Michaelides, Athens, and A. Keys, Minneapolis, Minn.

Introduction

This report presents data on two groups of men aged 40–59 who were fully examined in 1960 and 1961 and have been followed since. These subjects number 686 on the island of Crete and 529 on the island of Corfu, comprised 97.6% and 95.3% respectively of all men of those ages. The project is a part of an international co-operative program concerned with the epidemiology of cornary heart disease (CHD) in middle-aged men in defined geographical areas.

The primary purposes of the program include coverage on (1) ischemic heart disease, (2) comparison of the men with and without these disease, based on items such as body weight, body fatness, anthropometric characteristics, blood pressure, smoking habits, physical activity, serum cholesterol, electrocardiogram, dietetic studies, pulse waves and respiratory function tests, (3) distribution, by five year ages, of characteristics that may be related to etiology or development of these diseases in the samples and comparison with data from other similar or different samples, (4) analysis of interrelationships among the measured variables, (5) characterization and comparison of the samples regarding diet, (6) incidence of CHD in the follow-up and search for characteristics that may differentiate the sick from the healthy men, (7) age trends in characteristics that may be relevant to those diseases and to the aging process, and (8) to obtain informations – in parallel with programs

* This work is supported by U.S.P.H.S. grants HE-06090-05.

elsewhere – on the etiology and natural history of cardiovascular diseases with attention to the posibilities for prevention.

The characteristics of the examined smaples are: co-operation, stability in residence, occupation and habits, no marked contrasts in economic circumstances, and no sharp retirement and reduction in physical activity with age. There is adequate distribution of moderate to heavy work and activity patterns are relatively fixed. No problems of infectious diseases exist.

Results

It was found that: (1) The *height* is normally distributed in both areas. On the average, men are almost identical in height. Regarding the *weight*, the prevalence of overweight tends to be less than in other co-operative studies. The men in Crete are less often grossly overweight (20% above the standard average for height and age) than the men in Corfu. Among men aged 40–59 the prevalence of overweight was 10% in Crete and 14% in Corfu; this difference although small is significant (chi-square 4.09 and p is less than 0.05) (Table I).

Table I. Prevalence of overweight (110 or more and 120 or more per cent of 'standard' average for height and age)

Sample	40–44 110%	120%	45–49 110%	120%	50–54 110%	120%	55–59 110%	120%
Crete	13.1	2.5	10.9	3.0	10.8	4.5	4.1	1.4
Corfu	13.3	7.5	20.7	4.4	10.7	6.0	11.9	6.3

(2) Skinfold thickness: departs most notably from a normal distribution. In both areas men are similar in body fatness (sum of skinfolds) though men in Crete tend to be thinner. The 'average' thickness of subcutaneous fat (if allowance is made for the true skin) is barely 5 mm (Table II). Using a standard classification criteria, 13% of

Table II. Median values for relative body weight and the sum of skinfolds

	40–44	45–49	50–54	55–59	40–44	45–49	50–54	55–59
RBW	94	91	92	88	94	93	92	90
SF	14	14	15	14	16	15	14	14

the men in Crete have some degree of obesity and only 2% were classed as extremely obese. The corresponding figures for Corfu are 16 and 4%.

(3) Blood pressure: The men in Crete and Corfu do not differ importantly in arterial blood pressure. Systolic blood pressure tends to rise with age, but the average yearly increase is only 0.5 mm Hg in Crete and 0.3 mm Hg in Corfu. Diastolic blood pressure shows even less trend, with an average rise of 0.2 mm Hg per year. Over the age range 40–59 the prevalence of hypertension shows no consistent age trend in Corfu, and in Crete the prevalence actually tends to fall from age 40–44 to 55-59. If hypertension is judged to be present when systolic blood pressure is 160 mm or more, 12.8% of the men 40–59 were hypertensives in both areas. For a diastolic blood pressure of 95 mm or more, 11.8% were hypertensives (Table III).

Table III. Median values and prevalence of systolic and diastolic hypertension

Age	Crete				Corfu			
	Systolic	≥160	Diastolic	≥95	Systolic	≥160	Diastolic	≥95
40–44	131	6.3	80	13.9	130	6.7	81	8.3
45–49	132	8.0	80	13.4	130	13.2	80	14.9
50–54	135	15.6	81	11.0	134	13.0	81	13.0
55–59	138	21.2	83	9.6	135	18.2	81	11.1

Though the picture of blood pressure of this group is so stable, individual changes from 1960 to 1965 were many and large. Shifts in the diastolic blood pressure were even more numerous. More striking were the shifts in men at the extremes of the distribution of blood pressure.

(4) Smoking habits: Almost a fourth of the men in both areas never smoked cigarettes and at the time of the examination 42.8% in Crete and 36.5% in Corfu were non-smokers (Table IV). The smokers in

Table IV. Cigarette smoking habits

Sample	Age	Never	Quit	1–9	10–19	20 or more
Crete	40–59	23.8	19.0	10.2	17.2	29.8
Corfu	40–59	24.6	11.9	13.4	28.4	21.7

Crete tended to smoke more heavily than the smokers in Corfu and this difference is statistically highly significant. The difference of the smokers (20 cigarettes or more daily) in Crete and Corfu has a chi-square value of 9.55 and p is less than 0.01.

(5) Activity: In Crete over 60% of the men are habitually engaged in heavy physical work and less than 7% were sedentary or did light work. Corfu stands in sharp contrast, with less than a third in heavy physical activity and an equal number in light work. Three-fourths of the Cretans and over half of the men in Corfu are farmers. For both areas combined 49% did heavy physical work and 17% were sedentary or lightly active (Table V).

Table V. Habitual physical activity (1 sedentary and light, 2 moderate, 3 very active)

Age	Crete				Corfu			
	Total men	% in activity 1	2	3	Total men	% in activity 1	2	3
40–59	686	6.9	31.0	62.1	529	31.6	37.2	31.2

(6) Blood serum cholesterol: In regard to serum cholesterol there is a significant tendency for values to be lower in Corfu than in Crete, but the difference is trifling, only 6 mg in the middle of the distribution. The median cholesterol values are shown in Table VI. The

Table VI. Serum cholesterol (Mean values: mg/100 ml)

Age class	40–44	45–49	50–54	55–59
Corfu (1961)	202.0	204.0	207.0	200.2
Crete (1960)	201.0	200.0	214.2	213.1
Crete (1965)	217.6	213.2	224.8	218.4

Mean 5 year rise for all 548 men in Crete 11.89 mg%.

prevalence of hypercholesterolemia, using a standard criteria, is 27% in Crete and 24% in Corfu. Grade 4 hypercholesterolemia characterized 7% of the men in Crete and 6% of the men in Corfu. Average serum cholesterol values on the same men show small seasonal variations. Though reliable comparisons can be made among populations, it is difficult to distinguish sharply between individuals within these relatively homogeneous populations. Data on serum cholesterol values for the five year follow-up period show that the average value for all ages (40–59 in 1960) rose 11.90 mg per 100 ml, but this change is strictly related, falling from 16.6 for men initially 40–44 to only 5.1 mg per 100 ml for men initially 55–59, this trend being statistically highly significant. These differences between 1960 and 1965 are made up of two factors, a general shift independent of age, which may be related to a richer diet, plus an age trend *per se*.

(7) EKG findings: In both samples, but particularly in Crete, the EKG findings are notable for the relatively small number of abnormalities. In Crete one man showed a definite myocardial infarction, and in Corfu three men were in this category. In both samples the rate of prevalence of this finding is only 3.3 per thousand. A striking feature of the EKG's in Crete is the high frequency of *sinus bradycardia*, more than 10%. This accounts for the relatively higher frequency of long P–R interval. In Crete 11 men showed depression of the S–T at rest and 4 inverted T waves. Nine men showed S–T depression in the post-exercise EKG. Significant EKG abnormalities were more numerous in Corfu; twenty showed S–T depression at rest and 11 negative T waves. The exercise test added 14 cases of S–T depression and one negative T. Left axis deviation was more common in Crete than in Corfu (Table VII).

Table VII. Frequency of resting EKG findings

Item	EKG Code	40–44 Crete	40–44 Corfu	45–49 Crete	45–49 Corfu	50–54 Crete	50–54 Corfu	55–59 Crete	55–59 Corfu
Q Waves I	1	0	0	0	0	1	1	0	2
	2	1	0	0	0	1	2	0	2
	3	1	2	2	3	2	4	1	1
S-T Depres. IV									
S-T-J 1 mm	1	0	0	0	1	2	3	0	3
S-T-J 0.5–1 mm	2	2	3	4	3	2	4	3	1
T-Negative V									
–5 mm or more	1	0	0	0	0	0	0	0	2
–1 mm to –5 mm	2	1	1	1	1	1	4	1	3
A-V Block VI									
Complete	1	0	0	0	0	0	0	0	0
Partial	2	0	1	0	0	0	1	0	0
Ventric. Blocks									
L.B.B.B.	1	0	1	0	0	0	1	0	0
R.B.B.B.	2	1	1	1	1	0	3	3	3

(8) Dietary studies: Repeated in different seasons dietary studies showed no significant differences between Crete and Corfu or between seasons in total calories. The average is 2792 calories per day. Alcohol provided a small but of significance amount of calories. Average total protein intake varied little between places or seasons and only ¼ of the protein was from animal sources. Total fat intake was high at all times in both areas, notably from olive oil, and covers more than 30% of total calories. The intake of fats of animal origin is small (Table VIII). These diets are low in both saturated (S) and poly-

Table VIII. Means for nutrient intakes

Item	Crete	Corfu
Total cal./day	2748 ± 91	2836 ± 75
Alcohol cal.	128 ± 20	228 ± 28
% cal. total protein	10.5 ± 0.3	11.3 ± 0.3
% cal. animal	3.1 ± 0.4	3.2 ± 0.3
% cal. total fats	40.3 ± 1.0	32.7 ± 0.9
% cal. animal	6.0 ± 0.6	3.9 ± 0.4
% cal. olive ol	30.9 ± 1.2	23.8 ± 0.9
% cal. alcohol	4.3 ± 0.9	7.8 ± 0.9

Table IX. Mortality

Crete (5 year, 686 men)	Corfu (4 ½ year, 529 men)
Total: 10 cases ($14.0^0/_{00}$)	Total: 16 cases ($30.2^0/_{00}$)
CHD: 1 case ($1.47^0/_{00}$)	CHD: 3 cases ($5.6^0/_{00}$)

unsaturated fatty acids (P). The average percentage of total calories supplied by P is 2.2 in Crete and 2.6 in Corfu. The corresponding figures for S are 9.7 for Crete and 6.6 for Corfu.

(9) Mortality: The total mortality (all causes) in both areas has been remarkably low. Among 686 men in Crete the total 5 year mortality was 10 deaths ($14.7^0/_{00}$). In Corfu among 529 men of the same age the total 4 and a half year mortality was 16 men ($30.2^0/_{00}$). Mortality from CHD was $1.47^0/_{00}$ for Crete and $5.6^0/_{00}$ for Corfu.

(10) Follow-up studies: Five year re-examination in Crete. The collected up to-day data clearly show the remarkable relative rarity of EKG abnormalities in this population, aged 45–64 (average 54.2 years). Besides absence of any sign of new myocardial infarction, all major abnormalities (S–T depression, negative T waves, A–V block,

bundle branch blocks, atrial fibrillation) are unusually rare. The EKG records together with the mortality data and the absence of angina pectoris allow the conclusion that CHD is indeed uncommon in these populations.

Summary

Data of two groups of men aged 40–59, fully examined in 1960 and 1961 and followed since in regard to cardiovascular diseases are presented.

Variables taken into consideration were: food, blood pressure, anthropometric characteristics, obesity, activity, smoking, EKG, cholesterol, respirometry.

Prevalence of CHD appears to be remarkably low in both areas, particularly in Crete. Total mortality is low as well as mortality from CHD.

Data of the examined variables are presented.

Authors' addresses: Christ Aravanis, M.D., 47 Vasilissis Sophias Ave.; Lekos, D., M. D., 52 Patriarchou Ioakim St.; Dontas, A., M. D., 10 Alopekis St.; Michaelides, G., M. D., 5 Pindarou St., *Athens* (Greece); and Keys, A., Ph. D., Department of Physiological Hygiene, Stadium, Gate 27, University of Minnesota, *Minneapolis, Minn.* (USA).

Government Hospital Donolo, Jaffa

The Influence of Physical Occupational Activity on the Incidence of Ischemic Heart Disease

D. Brunner, Jaffa

The frequency of IHD in a 15-year period was retrospectively surveyed in 5,279 men and 5,229 women, aged 40 to 69, living in Israeli collective settlements (Kibbutzim) under uniform environmental conditions. (Profession and job position has no influence on life standard. Food is prepared in one kitchen and served in a communal dining room.)

The incidence of the anginal syndrome, myocardial infarction and fatalities due to IHD was 2.5 to 4 times higher in sedentary workers than in non-sedentary workers. Women, who in the Kibbutz society do a full day's work, show a very low incidence rate of fatal IHD.

The male/female ratio for angina pectoris for sedentary as well as physically active workers was between 1.1 to 3.1, but myocardial infarction occurred 16 to 20 times more frequently in men than in women.

No difference in serum total cholesterol, triglycerides, alpha-cholesterol percentage and in body weight was found between sedentary and non-sedentary workers of both sexes.

It is concluded that physical occupational activity should be considered a valuable principle in the prevention of ischemic heart disease.

Author's address: D. Brunner, M.D., Government Hospital, Donolo, *Jaffa* (Israel).

From the Department of Medicine A, Hadassah Hebrew University Hospital and Medical School and the Israel Institute for Applied Social Research, Jerusalem, Israel

Epidemiological, Nutritional and Sociological Studies of Atherosclerotic (Coronary) Heart Disease Among Different Ethnic Groups in Israel*

J.J. GROEN, F. DREYFUSS and L. GUTTMAN, Jerusalem

In 1953 DREYFUSS [1] drew attention to the fact that myocardial infarction was a more frequent cause for admission to hospital among Jews who had immigrated to Israel from Western than from Eastern countries. These observations were confirmed by DREYFUSS, TOOR, AGMON and ZLOTNICK [2], and by KALLNER and GROEN [3]. The same trend appeared to be present in the mortality figures [4, 3]. Recently these last authors have demonstrated that the lower mortality from coronary heart disease is evident specifically in Jews stemming from North Africa, Iraq and Yemen. Among the last group the disease is positively rare.

UNGAR and his coworkers [5] found the same difference in distribution between the Jews of different origin in standardized autopsy studies. The present authors demonstrated the existence

Table I. Myocardial infarct, angina pectoris, coronary heart disease and country of origin among Jewish port workers (Males only)

Country of origin	Myocardial infarct Abs. No.	(Exp. No.)	%	Angina pectoris Abs. No.	(Exp. No.)	%	Coronary h. dis. Abs. No.	(Exp. No.)	%	Total
Israel	1	(0.8)	3.2	2	(1.3)	6.4	3	(2.0)	9.7	31
Western Europe	2	(1.6)	3.3	4	(2.5)	6.6	6	(3.9)	10.0	60
Eastern Europe	13	(12.8)	2.4	23	(22.2)	4.3	36	(35.3)	6.8	533
Asia and Africa	2	(2.8)	1.7	2	(5.0)	1.7	4	(7.8)	3.4	118
Others										2
All males	18			31			49			744

* These researches were supported by Grant H-4427 of the U.S.P.H.S., National Heart Institute, National Institutes of Health, Bethesda, Md., USA.

of a difference in ethnic distribution in a prevalence study among the workers in the port of Haifa (Table I), and this was recently confirmed in an extensive study of Government and Municipal employees [6]. In another part of our prevalence study it was shown that the Arab Bedouin of the Negev Desert had a conspicuously lower prevalence of the disease than a Jewish control group; the prevalence among Arab villagers was somewhat higher, but still lower than mostly found in Western populations (Table II).

Table II. Coronary and hypertensive heart disease among Nomadic Arabs (Bedouin), Village Arabs and Jewish portworkers in Israel (Males age 30–70)

	Total	Myocard. inf. Abs. No.	%	Angina pect. Abs. No.	%	Cor. h. dis. Abs. No.	%	Hyper. h. dis. Abs. No.	%
Bedouin	510	1	0.2	?*	?*	1+	0.2+	9	1.7
Arab villagers	254	3	1.2	9	3.5	12	4.7	6	2.3
Jewish portworkers	744	18	2.4	31	4.2	49	6.6	13	1.8

* The number of cases of angina pectoris among the Bedouin is unavailable because of the language difficulties between the subjects and the interpreters.

Thus there can be no doubt that there is indeed a conspicuous difference in the distribution of coronary heart disease between these different ethnic groups living in the same country. This difference is not due to differences in age distribution, hospital admission

Table III. Comparison of average blood cholesterol by age and sex between Negev Bedouin, Arab villagers and Jewish port workers in Haifa

Age groups		Males	Females	Both sexes
30–39	N.B.	149	148	149
	A.V.	156	165	161
	J.H.	195	194	195
40–49	N.B.	154	161	156
	A.V.	170	178	173
	J.H.	207	215	209
50–59	N.B.	154	184	160
	A.V.	156	180	170
	J.H.	206	249	210
60–69	N.B.	150	177	154
	A.V.	169	183	175
	J.H.	196	235	199
Total	N.B.	151	162	159
	A.V.	164	174	169
	J.H.	199	209	204

bias, or diagnostic accuracy; it is a real difference in occurrence rate of the disease.

Toor et al. [7, 8] and Brunner et al. [9, 10] were the first to demonstrate that Yemenite Jews had a very low serum cholesterol. Our investigations also showed that the differences in the prevalence of coronary heart disease observed among the three ethnic groups run parallel to those in serum cholesterol (Table III and IV). It was

Table IV. Myocardial infarct and coronary heart disease by serum cholesterol level in three ethnic groups (Males only)

Serum cholesterol mg%	Bedouin			Arab villagers			Jewish portworkers		
	Total	M.I. %	C.H.D. %*	Total	M.I. %	C.H.D. %	Total	M.I. %	C.H.D. %
–150	270	0.4	–	91	1.2	2.2	56	–	–
150–199	167	–		108	0.9	6.5	339	2.1	6.0
200–269	36	–		33	–	6.4	311	3.0	8.1
270+	–	–		1	–	–	35	5.7	11.3
No blood obtained	37	–		21	–	–	3	–	–
Total	510	0.2	?*	254	1.1	4.7	744	2.4	6.6

* No figures available because the numbers of angina pectoris among the Bedouin are unreliable.

found moreover, that Yemenites who had been in the country for eight years or more had higher serum cholesterol levels than new immigrants [7]. That these serum cholesterol levels are probably not a genetic but an acquired characteristic was shown by Groen and his coworkers [11]: When a group of non-Yemenite healthy volunteers ate a Yemenite diet, their blood cholesterol dropped to the same low level as found in the Yemenites themselves. Brunner and Loebl showed the reverse to be also true: When Yemenites ate Western diets, their serum cholesterol increased [12]. Further studies have revealed that the factors in the diet responsible for this low serum cholesterol in Yemenites or Bedouin were a low content of total and saturated fat, of cholesterol and sugar; the quantities of vegetable protein and poly-saccharides in these diets are high, mainly due to a high consumption of bread which is a characteristic of the nutrition of all population groups in the Middle East. The diets do *not* contain specially high amounts of oils and thereby are *not* particularly rich in polyunsaturated fatty acids [11, 13, 14].

As other differences which may play a role in the low frequency of coronary heart disease, it was found that the Bedouin spend more physicial exercise in their work and ways of transportation than the Jewish control group (Table V). There were also differences in

Table V. Physical activity among the Bedouin (0.2% M.I.) and the Jewish portworkers (2.4% M.I.)

| | What is your usual means of transportation? (in %) | | | |
| | Bedouin | | Portworkers | |
	Males	Females	Males	Females
By cart or car	1	–	3	2
By bus or tender	17	6	80	82
Motor cycle or scooter	–	–	3	2
Bicycle	0.2	–	6	2
Horse, donkey or camel	38	20	–	–
Walking	44	74	8	12

economic level and in so called psychosocial ways of life as revealed by different answers to a questionnaire between patients with myocardial infarction and others. The accompanying tables show some of these differences (Tables VI, VII, and VIII). Similar

Table VI. Myocardial infarct and coronary heart disease and work problems in the past (Haifa portworkers)

| | Total | Myocardial infarct | | | Coronary heart disease | | |
		Abs. No.	Exp. No.	%	Abs. No.	Exp. No.	%
Very many or many	171	10	4.1	5.8	22	11.3	12.9
Some or none	573	8	13.9	1.4	27	37.7	4.8
All	744	18		2.4	49		6.6

Table VII. Myocardial infarction and coronary heart disease and attitude of coworkers (Haifa male portworkers)

| Liked by coworkers | Total | Myocardial infarction | | | Coronary heart discase | | |
		Abs. No.	(Exp. No.)	%	Abs. No.	(Exp. No.)	%
Very much and fairly	666	15	(16.3)	2.2	37	(43.8)	5.5
Indifferent or not liked	74	3	(1.7)	4.1	12	(5.2)	16.2
No coworkers	4						
All males	744	18		2.4	49		6.6

Table VIII. Myocardial infarct and coronary heart disease among Jewish portworkers by age, work problems and attitude of coworkers (Males only)

Age Group	Work problems + unliked			Work problems + liked			No work problems + unliked			No work problems + liked		
	M.I. (%)	CHD (%)	Total	M.I. (%)	CHD (%)	Total	M.I. (%)	CHD (%)	Total	M.I. (%)	CHD (%)	Total
Under 49	–	2 (9.1)	22	3 (1.7)	7 (4.0)	176	–	1 (4.3)	23	3 (1.1)	6 (2.3)	262
50–69	3 (15.8)	8 (42.1)	19	9 (9.4)	15 (15.6)	96	–	1 (10)	10	–	9 (6.7)	136
All males	3 (7.3)	10 (24.4)	41	12 (4.4)	22 (8.1)	272	–	2 (6.1)	33	3 (0.8)	15 (3.7)	397

differences in answers to this questionnaire, pertaining to the ways of life of the deceased, were given by first degree relatives of patients who had died from myocardial infarction, compared to those given by relatives of patients who died from other causes [15].

It will be the task of further investigations to measure the influence of each of these different factors in the multifactorial causal constellation of atherosclerosis and coronary heart disease. It is interesting that no ethnic or sex differential distribution was found in the mortality or hospitalization figures for cerebral vascular disease [3, 4].

References

1. DREYFUSS, F.: Incidence of myocardial infarction in various communities in Israel. Amer. Heart J. *45:* 749 (1953).
2. DREYFUSS, F.; TOOR, M.; AGMON, J. and ZLOTNICK, A.: Myocardial infarction in Israel. Cardiologia *30:* 387 (1957).
3. KALLNER, G. and GROEN, J.J.: Mortality and hospitalization in relation to coronary and cerebral vascular disease in Israel. J. Atheroscler. Res. *6:* 419 (1966).
4. KALLNER, G.: Epidemiology of arteriosclerosis in Israel. Lancet *i:* 1155 (1958).
5. UNGAR, H. and LAUFER, A.: Necropsy survey of atherosclerosis in the Jewish population of Israel. Path. et Microbiol. *24:* 711 (Basel 1961).
6. DUBLIN, T.; GROEN, J.J.; MEDALIE, J.; NEUFELD, H. and RISS, E.: Hypertension and ischaemic heart disease in government and municipal employees in Israel. Amer. J. publ. Hlth. (In print).
7. TOOR, M.; KATCHALSKY, A.; AGMON, J. and ALALLOUF, D.: Serum lipids and atherosclerosis among Yemenite immigrants in Israel. Lancet *i:* 1270 (1957).
8. TOOR, M.: Atherosclerosis and related factors in immigrants to Israel. Circulation *22:* 265 (1960).
9. BRUNNER, D.; LOEBL, K.; FISCHER, M. and SHICK, J.: Cholesterol and phospholipids in healthy males of various origins in Israel. Harefuah *48:* 1 (1955).
10. BRUNNER, D. and LOEBL, K.: Serum cholesterol, electrophoretic lipid pattern, diet and coronary heart disease. Ann. Int. Med. *49:* 732 (1958).
11. GROEN, J.J.; BALOGH, M. and YARON, E.: Effect of the Yemenite diet on the serum cholesterol of healthy non-Yemenites. Israel. J. Med. Sciences *2:* 196 (1966).
12. BRUNNER, D. and LOEBL, K.: Effect of Western diet on serum cholesterol of Yemenite Jews. Discussion at the present Symposium (Karger, Basel/New York 1967).
13. COHEN, A.M.; BAVLY, S. and POZNANSKI, R.: Change of diet of Yemenite Jews in relation to diabetes and coronary heart disease. Lancet *ii:* 1399 (1961).
14. GROEN, J.J.: Effect of bread in the diet on serum cholesterol. Amer. J. clin. Nutr. (In print).
15. GROEN, J.J.; DRORY, S.: Psychosocial factors in coronary heart disease; a combined sociological and autopsy study. Congress of Geographical Pathology (Leiden 1966) (S. Karger, Basel/New York 1967).

Authors' address: Prof. J.J. Groen, F. Dreyfuss. M.D., and L. Guttman, M.D., Dept. of Medicine A. Hadassah Hebrew University Hospital, P. O. Box 499, *Jerusalem* (Israel).

From the Department of Pathology Louisiana State University School of Medicine, New Orleans, La., USA

Geographic Pathology of Atherosclerosis*

H. C. McGill, Jr., New Orleans, La.

Introduction

Many pathologists have attempted to reconstruct the natural history of atherosclerosis by examining the arteries of patients dying of causes other than atherosclerosis, particularly those of young individuals. In the middle 1950's, pathologists began to report comparative studies of atherosclerosis in autopsied patients in different geographic locations. With few exceptions, these workers concluded that the severity of coronary and aortic atherosclerosis was correlated with the incidence of coronary heart disease; and that the severity of atherosclerosis in its preclinical stages was correlated with economic development and with fat and cholesterol content of the diet. One exception to this tendency was that aortic fatty streaks in children and adolescents did not parallel advanced lesions and clinical disease in older persons.

Methods

Previous geographic studies usually compared only two or three different groups at a time; numbers of cases were limited; and

* This paper reports partial results of the International Atherosclerosis Project, an international cooperative research project supported by grants HE-04152 and HE-07913 from the National Heart Institute, United States Public Health Service; and by many institutions and individuals in the countries represented. The project was sponsored by the Department of Pathology of Louisiana State University School of Medicine and the Institute of Nutrition of Central America and Panama, Guatemala.

methods of grading varied so that results from one study could not be compared with those from another. In 1959, a group of pathologists organized the International Atherosclerosis Project to collect better information on this subject. They developed a standard operating protocol and central laboratory facilities in Guatemala and New Orleans. Participating pathologists met in Guatemala in February, 1960, at the Institute of Nutrition of Central America and Panama and began to collect arteries in May, 1960. By September, 1965, the participants (Table I) had submitted to the central laboratory 23,205 sets of coronary arteries and aortas from autopsied patients 10 to 69 years of age in 19 localities.

Table I. Participants in the International Atherosclerosis Project, 1960–65

Location		Participating pathologist	Races
Bogotá	Colombia	Egon Lichtenberger	Mestizo
Cali	Colombia	Pelayo Correa	Mestizo, Negro
Caracas	Venezuela	Luis Carbonell	White, Negro
San José	Costa Rica	Jorge Salas	White
Durban	South Africa	John Wainwright	Indian, Bantu
Guatemala	Guatemala	Carlos Tejada and Carlos Restrepo	Mestizo
Kingston	Jamaica	William B. Robertson	Negro
Lima	Peru	Javier Arias-Stella	Mestizo
Manila	Philippines	Benjamin Barrera	Filipino
Mexico	Mexico	Ruy Perez Tamayo	Mestizo
New Orleans	United States	Henry C. McGill, Jr., and Jack P. Strong	White, Negro
New Delhi	India	V. Ramalingaswami	Indian
Osaka	Japan	Toru Miyaji	Japanese
Oslo	Norway	Lars A. Solberg and Aagot C. Loken	White
Recife	Brasil	Humberto Menezes	White, Negro
San Juan	Puerto Rico	Lorenzo Galindo	White, Negro
Santiago	Chile	Sergio Donoso	White
São Paulo	Brasil	Mario Montenegro	White, Negro
Turin	Italy	Aldo Stramignoni	White

The central laboratory stained the arteries with Sudan IV and packed them in plastic bags (Fig. 1). The specimens were identified only by randomly assigned accession numbers. A team of five pathologists estimated the percent of intimal surface in each arterial segment involved by atherosclerosis, and then estimated the percent of involved intimal surface covered by each of four types of lesions –

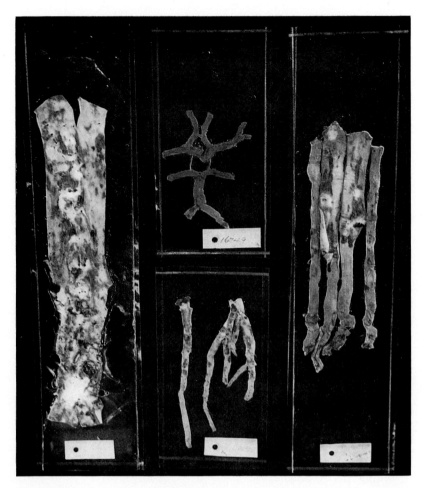

Fig. 1. Aorta, coronary arteries, and cerebral arteries from a case included in the International Atherosclerosis Project. The arteries have been stained in Sudan IV and are packed in clear plastic bags.

fatty streaks, fibrous plaques, complicated lesions (haemorrhage, ulceration, thrombosis), and calcification.

The statistical laboratory computed the absolute percent of intimal surface of each segment involved by each type of lesion, and computed mean values of percent surface involved by lesion, artery, age, sex, location-race group, and cause of death. For most analyses, we combined fibrous plaques, complicated lesions, and calcification into one measure which we called 'raised lesions'.

Analyses of Data

Cause of death effects. There are many sources of bias in the autopsy population that limit its usefulness as a sample of the living population. Age, sex, and race are obvious characteristics by which the cases must be stratified for intergroup comparisons. Differing causes of death may also affect the degree of atherosclerosis in autopsied cases. We conducted preliminary analyses of the material in order to select the best possible group of autopsied cases for geographic and racial comparisons of atherosclerosis. The results indicated that certain cause-of-death groups could be pooled without introducing serious bias. Subsequent geographic comparisons were based on this group, which comprised about 65% of the total pool of cases.

Ranking populations by advanced lesions. We computed an unweighted mean of raised lesions in the five arterial segments, four age groups from 25 through 64 years, and two sexes for each location-race group. There was a continuous range in extent of raised lesions from a high of 18.3% in the New Orleans White to a low of 6.2% in the Durban Bantu. Oslo, Durban Indian, New Orleans Negro, Manila, and Caracas also ranked relatively high and just below New Orleans White. All other populations were low in extent of raised lesions, ranging from 6.2 to 10.8%. In most location-race groups the extent of raised lesions paralleled the prevailing estimate of mortality from coronary heart disease in the corresponding living populations.

Other analyses. Other analyses of the data included a detailed study of the distribution of fatty streaks among the different groups of young persons; the relationship of coronary atherosclerosis to coronary heart disease; the extent and severity of atherosclerosis in persons dying of selected diseases as compared to others; the topography of atherosclerotic lesions in the coronary arteries and the correlation between coronary atherosclerosis and heart weight. These analyses, together with detailed reference tables and discussion of the results, will be published as a group of papers in a journal supplement.

Author's address: Henry C. McGill, Jr., M.D., Dept. of Pathology, Louisiana State University School of Medicine, *New Orleans, La.* (USA).

From the Heart Disease Control Program, Division of Adult Health and Aging, Chicago Board of Health; the Chicago Health Research Foundation and the Department of Medicine, Northwestern University Medical School, Chicago, Illinois

Epidemiological Studies on Atherosclerotic Coronary Heart Disease: Causative Factors and Consequent Preventive Approaches

J. Stamler, D. M. Berkson, Louise Mojonnier, H.A. Lindberg, Yolanda Hall, M. Levinson, Frances Burkey, Wilda Miller, M. B. Epstein and S. L. Andelman, Chicago, Ill.

Since the mid 1950's, major prospective epidemiological studies in large living population groups have shown unequivocal relationships between several traits and habits (singly and in combination), and risk of premature clinical atherosclerotic coronary heart disease [1-19]. The findings of our group in a sizeable cohort of men originally aged 40–59, employed by the Peoples Gas Light and Coke Company in Chicago, are representative [7]. Of the 1,594 male employees in this age group on January 1, 1958, 1,465 were examined for the first time for the purposes of this study in 1958. Of these 1,465 men, 1,329 were free of evidence of definite clinical coronary heart disease, were followed without loss over the subsequent years, and underwent no systematic intervention by the research group. Data on the relationship between the major coronary risk factors and incidence from new coronary heart disease and from all causes – for the period 1958–62 – are presented in Figs. 1–6 [1, 7]. Clearly, the men with original systolic blood pressures under 130 mm Hg had incidence rates of coronary heart disease approximately one third as great as the men with pressures of 150 mm Hg or greater (Fig. 1). These findings are typical of those reported by others. The data are similar on serum cholesterol and cigarette smoking respectively as risk factors (Figs. 2 and 3). The cumulative impact of combined risk factors is well illustrated

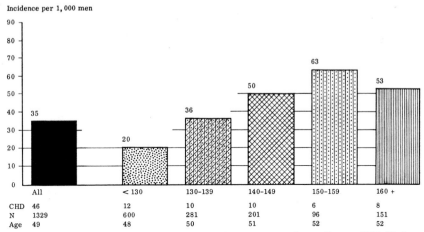

Fig. 1. Systolic blood pressure and incidence of new coronary heart disease, 1958–62, in 1,329 men originally age 40–59, free of definite clinical coronary heart disease and followed longterm without systematic intervention. Peoples Gas Light and Coke Company study.

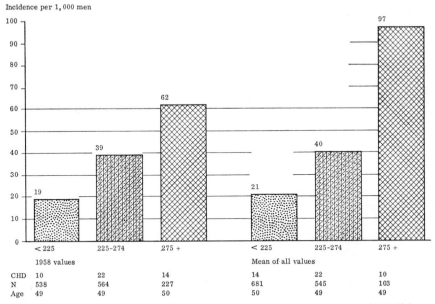

Fig. 2. Serum cholesterol level and incidence of new coronary heart disease, 1958–62 in 1,329 men originally age 40–59, free of clinical coronary heart disease, and followed longterm without systematic intervention. Peoples Gas Light and Coke Company study. The first set of bars deals with the men classified by their first cholesterol determination only, in 1958. The second set of bars deals with the men classified based on the mean of all serum cholesterol determinations available on each man for the period of 1958–62.

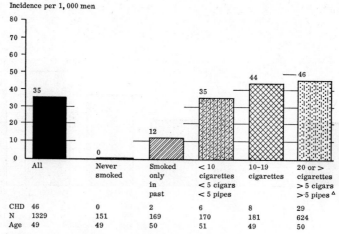

Fig. 3. Smoking and incidence of new coronary heart disease, 1958–62 in 1,329 men originally age 40–59, free of clinical coronary heart disease and followed longterm without systematic intervention. Peoples Gas Light and Coke Company study.

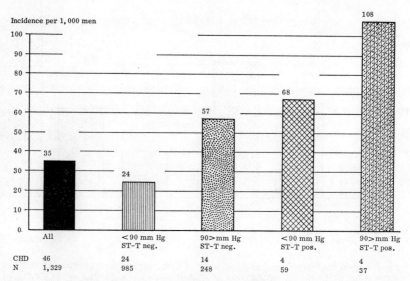

Fig. 4. Diastolic pressure and so-called minor non-specific T-wave abnormalities, and incidence of new coronary heart disease, 1958–62 in 1,329 men originally age 40–59, free of clinical coronary heart disease and followed longterm without systematic intervention. Peoples Gas Light and Coke Company study.

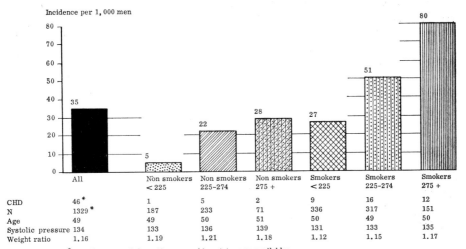

Fig. 5. Smoking and serum cholesterol (1958 value), and incidence of new coronary heart disease, 1958–62 in 1,329 men originally age 40–59, free of clinical coronary heart disease and followed longterm without systematic intervention. Peoples Gas Light and Coke Company study.

in Figs. 4–6. Diabetes (symptomatic or asymptomatic) is another major coronary risk factor, as is an habitual diet high in calories, total fat, saturated fat, cholesterol, total carbohydrate, sugar and salt. Obesity, physical inactivity and tension-generating personality-behavior pattern are probably important as well.

As already indicated, significant evidence on the role of several risk factors was already available by the mid 1950's. In 1957, therefore, our research group concluded that the time had come to go beyond descriptive-analytical epidemiologic studies, and proceed to experimental epidemiologic investigation. The further basis for this conclusion was the additional fact that the major coronary risk factors are amenable to control and correction by nutritional-hygienic-pharmacologic means. Therefore, the possibility arises of converting high-risk persons into medium or low-risk, i.e. achieving the primary prevention of the disease.

Based on these concepts our group undertook in 1957 to begin to accrue practical research experience in the preventive management of Chicago men age 40–59 free of clinical atherosclerotic disease, but designated coronary-prone because of their findings with regard to hypercholesterolemia, overweight, hypertension, and fixed minor non-specific T-wave abnormalities in the electrocardio-

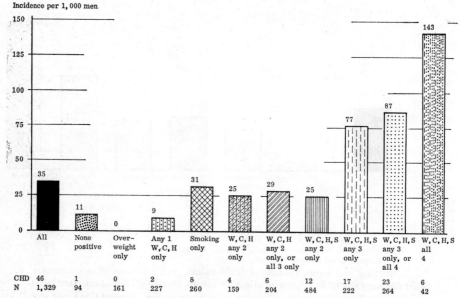

Fig. 6. Relationship between status with respect to four coronary risk factors (hypercholesterolemia, hypertension, overweight and cigarette smoking) and incidence of new coronary heart disease, 1958–62 in 1,329 men originally age 40–59, free of clinical coronary heart disease and followed longterm without systematic intervention. Peoples Gas Light and Coke Company study. W. is overweight, C. is hypercholesterolemia, H. is hypertension, S. is cigarette smoking. The criteria are: Overweight – a ratio of observed weight to desirable weight of 1.15 or greater; hypercholesterolemia – a serum cholesterol level of 250 mg/100 ml or greater; hypertension – a diastolic blood pressure of 90 mm Hg or greater; smoking – ten or more cigarettes per day.

gram (i.e. low voltage, diphasic or flat T-waves) [1, 20, 21]. The original high-risk criteria utilized in the beginning of this Coronary Prevention Evaluation Program (CPEP) are indicated in the first four points of Fig. 7. The fifth combination of abnormalities was incorporated in the study subsequently, based on the unequivocal demonstration of the role of cigarette smoking as a major coronary risk factor.

Participants for this study have been recruited over years as volunteers, largely from several major industrial organizations in Chicago. All volunteers were invited to become active participants. Since the study began as an effort to accrue experience with management of the major coronary risk factors, and since it drew its participants largely from industry, no double blind assignment of men to experimental and control groups was accomplished. However, matching groups – with similar medical findings and combinations of coronary risk factors – were identified from three

1. Absence of organic heart disease, clinical atherosclerotic disease in other major arterial beds, and diagnosed diabetes mellitus requiring drug treatment.
2. Presence of severe hypercholesterolemia – 325 mg/100 ml or greater – as a single risk factor.
3. Presence of hypercholesterolemia (260 mg/100 ml or greater), overweight (15% or more above desirable weight), hypertension (95 mm Hg or greater diastolic pressure) – any 2 or all 3.
4. Presence of fixed minor non-specific T wave abnormalities in the ECG, plus at least one other risk factor.
5. Presence of cigarette smoking – 10 or more/day – plus at least one other risk factor.

Fig.7. Criteria for the higher risk designation in the Coronary Prevention Evaluation Program.

major prospective epidemiological studies (the Peoples Gas Company, the Western Electric Company and the Railroad Employees studies) [7, 10, 22]. Thus, from the beginning the perspective was to carry the study through longterm, gradually building up an adequate cohort of men to permit evaluation of the decisive endpoints, i.e. mortality from coronary heart disease and from all causes.

As of our most recent data analysis through March 31, 1966, 335 high risk men meeting these criteria had been enrolled in the study and placed on the preventive regimen. Of the 335 participants, 117 had the combination – hypercholesterolemia, hypertension, overweight, cigarette smoking, any three or all four, with or without concomitant abnormality of the T-wave. Men with this combination of risk factors have been designated very high risk, based on experiences of the prospective studies [1, 7, 8].

The Coronary Prevention Evaluation Program aims to correct and control five coronary risk factors, i.e. hypercholesterolemia, obesity, hypertension, cigarette smoking and physical inactivity. Its methods – elaborated in the initial years of endeavor – have been previously described in detail [1, 20, 21]. In brief, major attention is given to accomplishing a change in eating habit in order to achieve and maintain a correction of hypercholesterolemia and overweight. The recommended diets can be briefly categorized as moderate in calories, total fat, total carbohydrate, sugars, and alcohol; low in saturated fat and cholesterol; high in protein and other essential nutrients (Fig.8). These are *not* low fat diets, but rather moderate fat diets, e.g. 25 to 30% of calories from total fat rather than the usual American 40–45%. They *are* diets low in

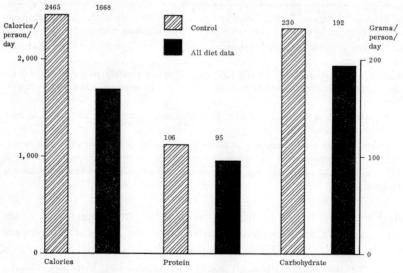

Fig. 8a.

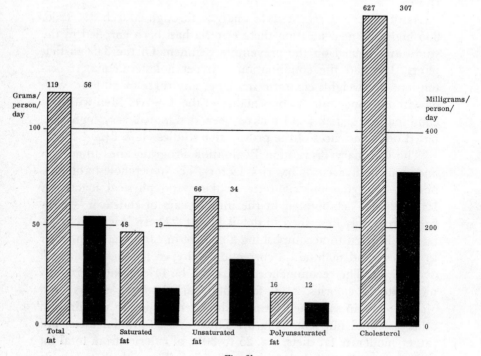

Fig. 8b.

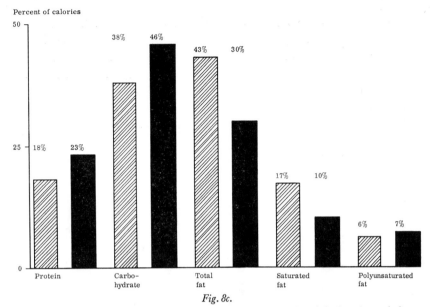

Fig. 8c.

Fig. 8a–c. Mean daily nutrient intake during the control period and during the period on the recommended diet for the first 99 participants in the Coronary Prevention Evaluation Program.

total fat and cholesterol. This is their decisive characteristic, of key importance in effecting and maintaining reduced serum cholesterol levels in the usual type of 'moderate' mixed hypercholesterolemic and hypertriglyceridemic diet-induced hyperlipidemia commonly seen in persons eating American diets. The CPEP diets are moderate, not high, in polyunsaturated fatty acids, and contain no special supplements of oils, since such measures are generally not necessary as long as effective control is achieved of calorie, saturated fat and cholesterol intake [1].

This change in nutrient pattern was achieved on a diet of mixed ordinary foodstuffs, without formulae or 'crash' or pharmacologic approaches. The recommendations entailed an emphasis on low-fat meats eaten in moderation, and a de-emphasis of high fat meats; an emphasis on low fat dairy products and a de-emphasis of high fat dairy products; ample use of fish, seafood and lean poultry (white meat of broiler and turkey); making egg yolk a once-in-a-while food; emphasis on low-saturated-fat, high-polyunsaturated-fat salad and cooking oils, and on the newer margarines of similar composition, and de-emphasis of bacon, lard, butter, suet and the older margarines as fat sources; emphasis on grains, starchy vegetables and fruits as such, without the preparation of 'rich' dishes with quantities of added saturated fats; moderation in intake of sugars, candy, etc. while favoring carbohydrate spreads (honey, jam, jelly, marmalade) rather than hard fats; moderation in intake of alcoholic beverages (Fig. 9).

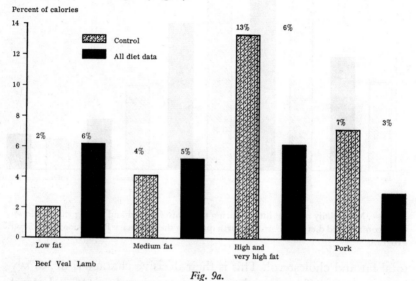

Fig. 9a.

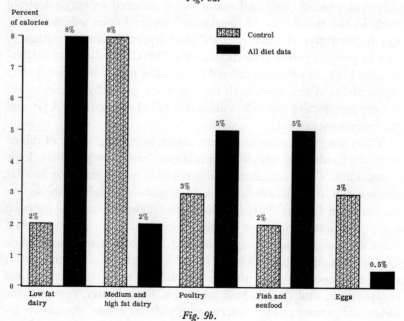

Fig. 9b.

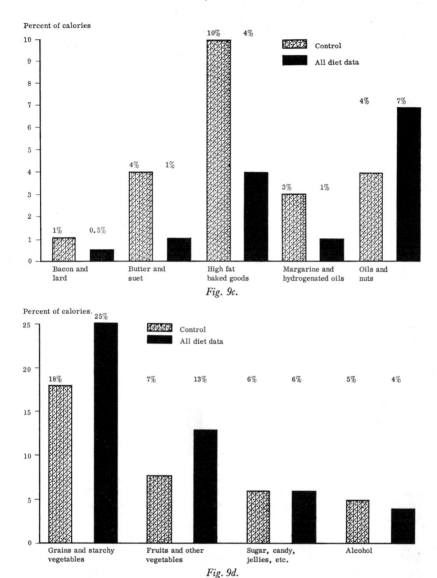

Fig. 9a–d. Mean daily percent of calories from various food groups during the control period and during the period on the recommended diet for the first 99 participants in the Coronary Prevention Evaluation Program.

The program also focuses on longterm control of the other risk factors, when present, since its basic objective is to accrue experience in the testing of the general hypothesis that concurrent control of the major coronary risk factors induces a significant reduction in incidence of premature clinical atherosclerotic coronary heart disease.

In the beginning, one of the major uncertainties concerned the ability to maintain free living men in such a study for the four or five years needed to achieve a definitive assessment of effect on CHD incidence and mortality. The goal set initially was to keep four-year drop-out rate under 50%. This objective has been more than achieved. Thus, for the 335 high risk men, 6 year drop-out rate is 337/1,000 (Fig. 10). The drop-out rate is highest during the

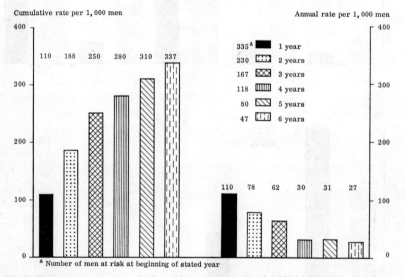

Fig. 10. Drop-out rates – cumulative and per annum – for the 335 higher risk participants in the Coronary Prevention Evaluation Program as of March 31, 1966. These rates were calculated by the life table method, with age correction by five year age groups to the U.S. male population, 1960.

first year after diet recommendation and is insignificant (approximately 30/1,000/year) during the fourth, fifth and sixth year. The six year drop-out rates were generally lower in the men age 50–59, compared with the men age 40–49, the lowest rate being in the men age 55–59 (Fig. 11). The drop-out rate was higher in the 117 very high risk men, being 459/1,000 at six years, compared with a rate of 337/1,000 for the entire group of 335 high risk men. This differential in drop-out rate is an important finding in relation to evaluation of the mortality endpoint (see below).

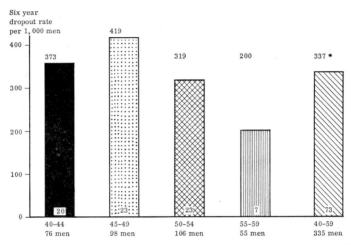

Fig. 11. Six year cumulative drop-out rate by age for the 335 higher risk men in the Coronary Prevention Evaluation Program as of March 31, 1966. These rates were calculated by the life table method. * Rate age-corrected by 5-year age groups to U.S. male population, 1960.

The most recent evaluation of the effects of the program on major coronary risk factors was carried out in January, 1966, and involved the 208 high risk men in the program as of April 30, 1965. All available data through September 30, 1965 were utilized in this data analysis, which focused on the findings in the continuing participants. The data analysis involved comparison of mean control values with mean for all values on each man throughout the period of his participation in the study. Data with regard to mean serum cholesterol changes are presented in Fig. 12. For the 156 non-dropout participants, a 16.1% average decline in serum cholesterol was recorded. Of these 156 men, 105 had control serum cholesterol levels of 260 mg/100 ml or greater. Their serum cholesterol fall averaged 19.4%. For the 22 men with control serum cholesterol levels of 325 mg/100 ml or greater, the fall averaged 32.2%. For the 105 hypercholesterolemic men (260 mg/100 ml or greater), an analysis was also accomplished of the distribution of serum cholesterol responses. Of these 105 men, 82 experienced a serum cholesterol fall of 10% or greater, 55 experienced a fall of 15% or greater, 28 experienced a fall of 20% or greater (Fig. 13). This wide range of responses reflects several factors, e.g. initial level, adherence to dietary recommendation, and inherent responsiveness. The relative role of the latter two of these cannot readily be assessed

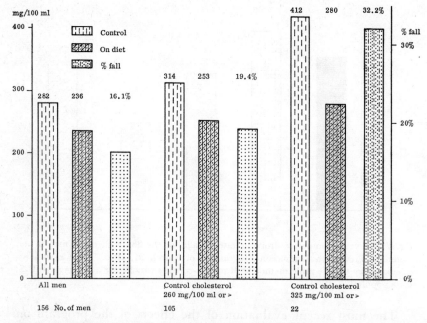

Fig. 12. Effects of the Coronary Prevention Evaluation Program on serum cholesterol level in the non-drop-out participants. The data are for the 156 non-drop-outs from the cohort of 208 men in the program as of April 30, 1965, with use of all data available on these men through September 30, 1965. Of these 156 continuing participants, 105 had control serum cholesterol levels of 260 mg/100 ml or greater, 22 had control serum cholesterol levels of 325 or greater.

% Fall	Number of men	Rate/1,000
<10.0	23	219.5
10.0–14.9	27	257.1
15.0–19.9	27	257.1
20.0–24.9	8	76.2
25.0–29.9	7	66.6
30.0–34.9	4	38.0
35.0 > △	9	85.7

Fig. 13. Distribution of serum cholesterol responses in the 105 higher risk non-drop-out participants in the Coronary Prevention Evaluation Program with control serum cholesterol levels of 260 mg/100 ml or greater (see previous figure). The rate per thousand represents the number of men per thousand experiencing the cited fall in serum cholesterol level; e.g. 27 of the 105 men had a fall of 15.0–19.9 % in serum cholesterol level during their entire period of follow-up after diet recommendation. As in the previous figure, the analysis included all data available on these men through September 30, 1965.

in a nonmetabolic ward study. In men with associated hypertriglyceridemia, serum triglyceride levels tended to fall concomitant with the correction of hypercholesterolemia. No tendency was recorded for serum triglycerides to rise.

The mean reductions in weight for all of the non-drop-out participants, and for those 15% or more, and 25% or more above desirable weight are recorded in Fig. 14. The distribution of the

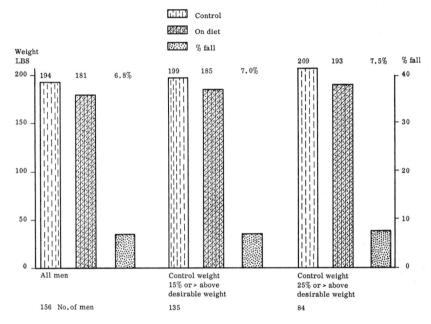

Fig. 14. Effects of the Coronary Prevention Evaluation Program on weight in the 156 non-drop-out participants from the cohort of 208 men in the program as of April 30, 1965. The analysis included all data on these men through September 30, 1965. Of the 156 continuing participants, 135 had a control ratio of observed weight to desirable weight of 1.15 or greater, 84 had a control ratio of 1.25 or greater.

weight response for the 135 men 15% or more above desirable weight is presented in Fig. 15. In general, moderate weight reduction was achieved, with a significant percent of the men losing sufficient weight so that they no longer fell into the defined overweight category. However, success was rarely achieved in reducing the men to the desirable weight level for their height, as indicated by the life insurance statistical tables, i.e. a significant percent of them remained moderately overweight and obese.

Fall (lbs.)	Number of men	Rate/1,000
<5	22	163.0
5–9	55	407.6
10–14	35	259.3
15–19	13	96.3
20–24	7	51.9
25>	3	22.2

Fig. 15. Distribution of weight response in the 135 higher risk non-drop-out participants in the Coronary Prevention Evaluation Program with a control ratio of observed weight to desirable weight of 1.15 or greater (see the previous figure).

Analyses of the data on serum cholesterol and weight year by year for the men active in the study at least four years revealed sustained maintenance of the change in serum cholesterol level and weight, with only slight tendency to regression toward control levels, due to a degree of 'slipping' of dietary control.

With regard to hypertensive men, when weight reduction and moderate restriction of dietary salt intake did not achieve a normalization of blood pressure, antihypertensive drugs were recommended. By these procedures, elevated blood pressure has been reduced and controlled in a majority of men with that risk factor. With respect to the cigarette smokers, approximately 30% have achieved an abandonment of this habit for at least one year or longer, either by giving up all smoking, or by switching to pipe or cigars in moderation, without inhaling. These are the alternative recommendations of the Program to the cigarette smoking participants. Although detailed objective assessment has not been readily feasible, the general impression is that a majority of the sedentary men have changed their habits of living sufficiently to permit their classification as light activity.

As of March 31, 1966, the median follow-up time for the men in this study was 3 years; 131 men had been in the Program for more than 4 years, 95 for more than 5 years. During this period of follow-up, a sufficient number of deaths occurred to permit preliminary evaluation of the trend, at least on the basis of a comparison of the non-drop-out and drop-out men. (Data are currently being analyzed from the three prospective studies with groups of men comparable to the CPEP participants.) The cumulative mortality from coronary heart disease is presented in Fig. 16. For the 262 men remaining active in the Program, two CHD deaths were recorded,

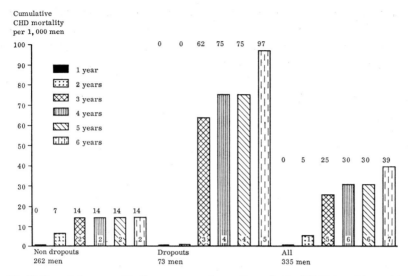

Fig. 16. Cumulative mortality from coronary heart disease in the 335 higher risk participants in the Coronary Prevention Evaluation Program as of March 31, 1966. The rates were calculated by the life table method, with age correction by five year age groups to the U.S. male population, 1960. Cumulative mortality rates are given for the 262 non-drop-out participants, the 73 drop-outs and the total group of 335 men.

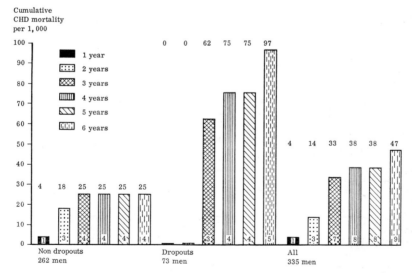

Fig. 17. Cumulative mortality from all causes in the 335 higher risk participants in the Coronary Prevention Evaluation Program as of March 31, 1966. The rates were calculated by the life table method, with age correction by five year age groups to the U.S. male population, 1960.

a death rate of 14/1,000 in six years. This rate was calculated by the life table method, with age correction by five year age groups to the U.S. male population in 1960. Among the 73 drop-outs, five CHD deaths were recorded, i.e. a six year rate of 97/1,000. The mortality rates from all causes for these two groups were 25 and 97/1,000 respectively (Fig. 17). Two of the non-drop-outs died from neoplastic disease, of the lung and gastrointestinal tract respectively. Although the differences in mortality rates between the non-drop-outs and the drop-outs are sizeable, they are not statistically significant at the five percent level of probability. Moreover, as noted earlier, the drop-outs are more heavily weighted with very high risk men, and therefore the two groups are not strictly comparable in terms of their original risk factor status. This fact is particularly important, since 8 of the 9 deaths were in men from the very high risk group.

In order to evaluate this matter further, a life table analysis was made for the 117 very high risk men, divided into non-drop-outs and drop-outs. Their six year death rates from coronary heart disease and from all causes are summarized in Fig. 18 and 19. Of the 82 non-drop-outs, only 1 died of coronary heart disease, and 3

Group	Number of men	Number of deaths	Death rate	S.E.
All	117	6	108.9	45.3
Non drop-outs	82	1	21.7	19.8
Drop-outs	35	5	228.7	90.0

Fig. 18. Six year mortality rates from coronary heart disease in the 117 very high risk participants in the Coronary Prevention Evaluation Program, as of March 31, 1966. The rates were calculated by the life table method, with age correction by five year age groups to the U.S. male population, 1960. S.E. is a standard error of the six year mortality rate.

Group	Number of men	Number of deaths	Death rate	S.E.
All	117	8	147.7	50.6
Non drop-outs	82	3	79.8	46.2
Drop-outs	35	5	228.7	90.0

Fig. 19. Six year mortality rates from all causes in the 117 very high risk participants in the Coronary Prevention Evaluation Program, as of March 31, 1966. The rates were calculated by the life table method with age correction by five year age groups to the U.S. male population, 1960.

died of all causes. Of the 35 drop-outs, 5 died, all of coronary heart disease. The coronary heart disease six-year mortality rates were 21.7 and 228.7/1,000 respectively, a difference which is statistically significant at the 5% level of probability. However, the death rates from all causes – although sizeably different – are not significantly different.

At this juncture these results can be characterized as encouraging, in terms of ability to hold free-living high-risk middle-aged men in such a program longterm, and to achieve a significant change in coronary risk factor status by nutritional-hygienic means in a high percent of such men. The preliminary findings with respect to mortality have encouraged our research group to continue this study, and to expand the number of its participants sizeably. Viewed overall, our experience supports the concept that control of the major coronary risk factors offers the possibility of a breakthrough against the widespread current epidemic of premature atherosclerotic coronary heart disease.

References

1. STAMLER, J.: Lectures on preventive cardiology (Grune and Stratton, New York 1967).
2. GOFMAN, J.W. and ANDRUS, E.C.: Evaluation of serum lipoprotein and cholesterol measurements as predictors of clinical complications of atherosclerosis. Report of a cooperative study of lipoproteins and atherosclerosis. Circulation 14: 691 (1956).
3. DAWBER, T.R.; MOORE, F.E. and MANN, G.V.: Coronary heart disease in the Framingham study. Amer.J.Publ.Hlth. 47: 4, Part 2, p.4 (1957).
4. DOYLE, J.T.; HESLIN, A.S.; HILLEBOE, H.E.; FORMEL, P.F. and KORNS, R.F.: A prospective study of degenerative cardiovascular disease in Albany: Report of three years' experience – I. Ischemic heart disease. Amer.J.Publ.Hlth. 47: 4, Part 2, p.25 (1957).
5. CHAPMAN, J.M.; GOERKE, L.S.; DIXON, W.; LOVELAND, D.B. and PHILLIPS, E.: The clinical status of a population group in Los Angeles under observation for two to three years. Amer.J.Publ.Hlth. 47: 4, Part 2, p.33 (1957).
6. STAMLER, J.; LINDBERG, H.A.; BERKSON, D.M.; SHAFFER, A.; MILLER, W. and POINDEXTER, A. with the assistance of COLWELL, M. and HALL, Y.: Prevalence and incidence of coronary heart disease in strata of the labor force of a Chicago industrial corporation. J.chron.Dis. 11: 405 (1960).
7. STAMLER, J.: Atherosclerotic coronary heart disease – The major challenge to contemporary public health and preventive medicine. Conn.Med. 28: 675 (1964).
8. DAWBER, T.R.; KANNEL, W.B. and McNAMARA, P.M.: The prediction of coronary heart disease. Trans. Assoc. Life Ins. Med. Dir. Amer. 47: 70 (1964).
9. ZUKEL, W.J.; LEWIS, R.H.; ENTERLINE, P.E.; PAINTER, R.C.; RALSTON, L.S.; FAWCETT, R.M.; MEREDITH, A.P. and PETERSON, B.: A short-term community study of the epidemiology of coronary heart disease – A preliminary report on the North Dakota study. Amer.J.Publ.Hlth. 49: 1630 (1959).

10. PAUL, O.; LEPPER, M.H.; PHELAN, W.H.; DUPERTUIS, G.W.; MACMILLAN, A.; MCKEAN, H. and PARK, H.: A longitudinal study of coronary heart disease. Circulation *28:* 20 (1963).
11. PELL, S. and D'ALONZO, C.A.: Acute myocardial infarction in a large industrial population. Report of a 7-year study of 1,356 cases. J.amer.med.Ass. *185:* 117 (1963).
12. DOYLE, J.T.: Etiology of coronary disease: risk factors influencing coronary disease. Mod.Conc.Cardiovasc.Dis. *35:* 81 (1966).
13. DOYLE, J.T.; DAWBER, T.R.; KANNEL, W.B.; KINCH, S.H. and KAHN, H.A.: The relationship of cigarette smoking to coronary heart disease. J.amer.med.Ass. *190:* 886 (1964).
14. CHAPMAN, J.M. and MASSEY, F.J.: The interrelationship of serum cholesterol, hypertension, body weight, and risk of coronary disease. Results of the first ten years' follow-up in the Los Angeles heart study. J.chron.Dis. *17:* 933 (1964).
15. KEYS, A.; TAYLOR, H.L.; BLACKBURN, H.; BROZEK, J.; ANDERSON, J.T. and SIMONSON, E.: Coronary heart disease among Minnesota business and professional men followed fifteen years. Circulation *28:* 381 (1963).
16. EPSTEIN, F.H.: The epidemiology of coronary heart disease – A review. J.chron. Dis. *18:* 735 (1965).
17. OSTRANDER, L.D., Jr.; FRANCIS, T., Jr.; HAYNER, N.S.; KJELSBERG, M.O. and EPSTEIN, F.H.: The relationship of cardiovascular disease to hyperglycemia. Ann. intern.Med. *62:* 1188 (1965).
18. BORHANI, N.O.; HECHTER, H.H. and BRESLOW, L.: Report of a ten-year follow-up study of the San Francisco longshoremen – mortality from coronary heart disease and from all causes. J.chron.Dis. *16:* 1251 (1962).
19. HAMMOND, E.C.: Smoking in relation to mortality and morbidity. Findings in first 34 months of follow-up in a prospective study started in 1959. J.nat.Cancer Inst. *32:* 1161 (1964).
20. STAMLER, J.: Current status of the dietary prevention and treatment of atherosclerotic coronary heart disease. Prog.Cardiovasc.Dis. *3:* 56 (1960).
21. STAMLER, J.; BERKSON, D.M.; YOUNG, Q.D.; LINDBERG, H.A.; HALL, Y.; MOJONNIER, L. and ANDELMAN, S.L.: Diet and serum lipids in atherosclerotic coronary heart disease – Etiologic and preventive considerations. Med.Clin.N.Amer. *47:* 3 (1963).
22. TAYLOR, H.L.; KLEPETAR, E.; KEYS, A.; PARLIN, W.; BLACKBURN, H. and PUCHNER, T.: Death rates among physically active and sedentary employees of the railroad industry. Amer.J.Publ.Hlth. *52:* 1697 (1962).

Acknowledgements

It is a pleasure to acknowledge the cooperation and support of Eric Oldberg, M.D., President, Chicago Board of Health and Chairman, Chicago Health Research Foundation. It is also gratifying to express appreciation to all the members of the Foundation Board of Directors, to the entire Chicago Board of Health and its several Divisions cooperating in the research undertakings of our group summarized in this paper. It is also a pleasure to pay tribute to the entire staff of the Heart Disease Control Program, Division of Adult Health and Aging, Chicago Board of Health aiding in this research.

We are also grateful to Paul Meier, Ph.D., of the Department of Statistics and the Biological Sciences Computation Center, University of Chicago. It is a further pleasure to express appreciation to the several Chicago organizations giving invaluable cooperation in this research effort, particularly the Peoples Gas Light and Coke Company, its Chairman, Remick McDowell and its President, Leslie A. Brandt; the American Oil

Company and its Medical Directors, Gilbeart H. Collings, Jr., M.D. and John Malia, M.D.; the Newspaper Division of the Field Enterprises, its Vice-President and General Manager, Mr. Wilbur C. Munnecke and its Medical Officer, Jacques M. Smith, M.D.; Armour and Company and its Chairman of the Board, William Wood Prince; the Stop and Shop Organization and its President, Gardner H. Stern, Sr., and its Vice-President, Gardner H. Stern, Jr.

The research presented in this paper was made possible by grants from the American Heart Association; the Chicago Heart Association; the National Heart Institute, the National Institutes of Health, U.S. Public Health Service (H.E. 04197 and H.E. 09426); the Division of Community Health Services, U.S. Public Health Services (C.H. 00075); the Corn Products Institute of Nutrition; the National Dairy Council and the Wesson Fund for Medical Research.

Authors' address: J. Stamler, M. D.; D. M. Berkson, M. D.; Louise Mojonnier, Ph. D.; H. A. Lindberg, M.D.; Yolanda Hall, M.S.; M. Levinson, M.D.; Frances Burkey, R.N.; Wilda Miller, M.P.H.; M.R. Epstein, Ph.D. and S.L. Andelman, M.D., M.P.H., Heart Disease Control Program, Chicago Board of Health, Chicago Civic Center, *Chicago, Ill. 60602* (USA).

Discussion Session 1

H. Malmros (Lund): Could you tell us the composition of the fatty acids of the bread used by the different populations you have studied?

J.J. Groen (Jerusalem): The bread eaten by the Yemenites in Israël is mostly the regular Israël bread and is baked from wheat flour of 70–75% extraction. The Bedouins eat bread prepared from 100% extracted wheat flour. The diets of both groups contain only very small amounts of saturated fats, because they eat practically no margarine, butter or cream, little milk, cheese, oil or eggs and only lean meat. As a result, almost all the linoleic acid of the diets are derived from bread and, as the content in bread is not very high, the total intake is about 5–10 g/day. This is not more than the amount in the diet of 'western' Jews in Israël who use butter, oil (mostly cotton seed but also some maize and soya oil) and consume linoleic acid also in their margarine (which contains 20% linoleic acid in Israël). Therefore, the low blood cholesterol in some eastern groups cannot be explained by a large quantity of linoleic acid in their diet; it is caused by a low intake of saturated fat and cholesterol. You will find the particulars in our papers in the Amer.J.clin.Nutrition, 1964 and 1966 and the Israël J. of med. Sciences, 1966.

H. Engelberg (Beverly Hills, Calif.): In view of the marked increase (2–3 times) in mortality of heavy smokers in this age group of men, were the data divided into non-smokers and smokers?

J. Stamler (Chicago, Ill.): In all our studies, including the Coronary Prevention Evaluation Program, detailed attention has been paid to smoking as a risk factor. As already indicated, it is one of the five abnormalities that the Program attempts to correct. Of the 335 men in the Program as of March 31, 1966, 169, just about half, were initially heavy smokers (e.g. 10 or more cigarettes per day). Of the 117 very high risk men, with three or more abnormalities, 107 were heavy smokers. Of the 9 men who died, 8 were smokers, including one man who died from a primary carcinoma of the lung. Of the 7 men whose deaths were attributed to coronary heart disease, 6 were smokers, three of them had 3 risk factors, three had all 4 risk factors (hypercholesterolemia, hypertension, smoking, overweight).

O.J. Pollak (Dover, Del.): Were there familial hypercholesterolemics among those with very severe hypercholesterolemia? If so, how many? Did these differ from others in response to hypocholesterolemic measures?

J. Stamler (Chicago, Ill.): Our latest analysis of serum cholesterol responses involved the 208 higher risk men in the study as of April 30, 1965. Of these, 156 were continuing participants when the data analysis was accomplished late in 1965. Of the 156, 22 were severely hypercholesterolemic, i.e. had levels of 325 mg/100 ml or greater, with a mean level for that group in excess of 400 mg/100 ml. Some of these men reported a positive family history of premature coronary disease, probably due to a familial tendency to hyperlipidemia. However, no systematic effort was made to study all members of the families of these men. Therefore it is not possible to give a precise statistic on the number with familial hypercholesterolemia.

As indicated in the text of our presentation, the serum cholesterol response to dietary recommendations varied in these hyperlipidemic men. It was marked in some, moderate in others and slight in a few. Unfortunately, it is not possible in a study of free-living men to determine precisely whether a poor response is a function of unsatisfactory adherence to the recommended diet, or whether it is due to endogenous metabolic non-responsiveness, or a combination of these. During the last year or two, our group has begun to utilize the paper electrophoretic technique of FREDRICKSON and LEES, together with determination of both serum cholesterol and triglyceride levels, to accomplish a more precise characterization of the type of hyperlipidemia present in these men. Glucose tolerance testing is also done on each of them, as well as evaluation for hypothyroidism and other abnormalities making for secondary hyperlipidemia. No detailed analysis has as yet been made of our findings along these lines. However, my preliminary impression is that men with Type II electrophoretic patterns, i.e. men with familial hyper-beta-lipoproteinemia – with clear plasma, markedly elevated serum cholesterol, normal triglycerides, markedly elevated low density beta-lipoproteins, without elevated pre-beta or very low density lipoproteins – tend to be the least likely to respond well to dietary change. This is an initial impression, and it will require further evaluation of the data, as well as the more precise control that is possible only in a metabolic ward study, to check its validity.

T. ZEMPLÉNYI (Prague): Dr. STAMLER, you referred to diabetes as a risk factor and the same was noted in the contribution of Dr. GROEN. It seems from recent data that the correlation between diabetes and atherosclerosis (ischemic heart disease) is not as clearcut as assumed before (see the discussion on this topic a few days ago at the International meeting in Sliac, Czechoslovakia). What is your experience from your studies?

J. STAMLER (Chicago, Ill.): Unfortunately, I was not able to attend the Sliac meeting, and I am not sure I have seen the data to which you refer. In our prospective epidemiologic studies at the Peoples Gas, Light & Coke Company in Chicago, the findings to date are consistent with the longstanding clinical observations, i.e. people with diagnosed diabetes have an increased risk of developing premature clinical atherosclerotic disease of the coronary, cerebral and/or peripheral major arteries. During 1965, our group was for the first time able to administer a 50 gm load of glucose, with collection of blood for plasma glucose determination at one hour, in all participants in the long-term study at the Gas Company. A preliminary analysis of the data revealed – consistent with findings from several recent reports – that the men with clinical diagnoses of atherosclerotic disease had higher mean plasma glucose levels at one hour than those diagnosed as clinically normal. It also demonstrated that men with other coronary risk factors, e.g. hypercholesterolemia, overweight, hypertension, singly or in combination, tended to have higher mean plasma glucose levels (as a group) than men free of these abnormalities. Similar findings were recently reported from the Tecumseh, Michigan study by Drs. EPSTEIN, TH. FRANCIS and their colleagues. Moreover, Drs. TH. R. DAWBER and W. KANNEL of the Framingham study have presented prospective data, involving 10 or 12 years of follow-up, demonstrating that high random blood sugar levels are associated with increased risks of developing clinical atherosclerotic disease. Thus, our direct experience and that of our colleagues engaged in prospective epidemiologic research in the United States reinforce the long-standing conclusion that diabetes is a major risk factor for atherosclerotic disease. The validity of this conclusion is further underscored by the type of data that Dr. FREDRICKSON has been compiling, demonstrating that at least three of the major types of familial hyperlipidemia are associated with abnormalities of glucose tolerance. Whatever the complex metabolic mechanisms involved, these findings of an association between abnormalities of glucose tolerance and hyperlipidemia – together with the extensive documentation on the role of hyperlipidemia as a coronary risk factor – point out the likelihood that abnormality of carbohydrate metabolism is truly associated with increased risk of atherosclerotic disease, at least under conditions of life in the United States.

Against this background, I would be most curious to see the data presented at Sliac, suggesting a contrary conclusion. I am, of course, aware that limited data are available, for example on Yemenite Jews, indicating that diabetics living under conditions of economic underdevelopment, i.e. habitually eating diets low in calories, total fat, saturated fat, cholesterol, sugar, apparently do not manifest increased risk of premature clinical atherosclerotic disease. Provided further research confirms this important observation, it may be concluded that it reaffirms a basic theoretical concept, i.e. that diabetes – like hypertension and cigarette smoking – enhances risk of atherosclerotic disease only among peoples exposed to a potentially atherogenic diet, i.e. a diet high in calories, fat, saturated fat, cholesterol, sugar. A significant corollary follows: the diets widely used for the treatment of diabetics from the preinsulin era down to the present – diets high in total fat, saturated fat, cholesterol (and all too often excessive in calories as well, with tolerance by some physicians) – are wrong, and wrong not only quantitatively (i.e. in calories), but qualitatively (i.e. in composition) as well. There is reason to believe that exogenous insulin may have metabolic effects intensifying atherogenesis, in the presence of a high-fat, high-cholesterol diet. Thus Western medicine – or at least a significant sector of it – may have unwittingly been adding insult to injury, in terms of the atherogenic threat to the diabetic.

D. Brunner (Jaffa): There is no obligatory association between diabetes mellitus and atherosclerosis. We have reported the findings in 76 Yemenite diabetic Jews.

34 of them had clinically diagnosed diabetes for more than 5 years. At the time of the study, 56 of them received insulin, 20 of them less than 30 units per day, 24 from 31–60 units and 12 patients more than 60 units per day. 7 patients were treated with hypoglycemic drugs.

In those 76 Yemenite agricultural workers only one case of myocardial infarction in a 76 year old man occured. No case of ECG findings, indicating coronary artery disease, no case of peripheral vascular disease and no case of diabetic nephropathie were found.

The average cholesterol levels of the 76 Yemenite diabetics was 233 ± 47 mg % which is higher than the cholesterol level of healthy Yemenites (160 ± 25 mg %) but lower than the average cholesterol values of non-Yemenite diabetics, matched to the Yemenite group in respect to age, sex, obesity, duration of diabetes and amount of daily insulin. On the other hand the cholesterol levels of 50 non-Yemenite diabetics suffering from diabetes for more than 15 years and inflicted with many atherosclerosis and angiopathic complications were lower than the cholesterol of the diabetic Yemenite.

S. Gerö (Budapest): Have you any data on the behavior pattern of the two groups, dropout and non-dropout?

J. Stamler (Chicago, Ill.): In recent years, new participants in the Coronary Prevention Evaluation Program have been given a psychological test, developed by Dr. M. Zaks of Northwestern University Medical School, called the Personal Opinion Inventory (P.O.I.). Like the Minnesota Multiphasic Personality Inventory, considerable work has been done on the development of several scales for scoring the P.O.I., and on the verification of these scales. This test was also given by our group to the Chicago participants in the National Diet-Heart Study. Analyses of these data have not yet been completed. Therefore, it is not possible at this juncture to make any objective statements. All of us hope that such tests will eventually accomplish the prospective identification of those likely to drop out of such a study.

P. Constantinides (Vancouver, B.C.): Were the five deaths among the dropouts cardiovascular deaths? Were autopsies done on these deaths?

J. Stamler (Chicago, Ill.): As indicated in the presentation, all five deaths thus far recorded among the 63 dropouts have been cardiovascular deaths. Autopsies were done on 2 of these, clinical data and death certificates are available on all five. I do not believe there is any question about the cause of death.

S. MANCINI (Milan): What is your feeling about non-responsive non-dropout subjects that you have followed?

J. STAMLER (Chicago, Ill.): In response first to Dr. MANCINI, the continuing participants, i.e. the non-dropouts, of course exhibit varying degrees of adherence longterm to the recommended diet, ranging from excellent to good to fair to poor. Unfortunately, continuing participation – in the form of coming in for periodic visits and examinations – may not be synonymous with optimal adherence. Objective measures of adherence in such a longterm study involving free living men are hard to come by. Our general impression is that once a fall in weight and serum lipids occurs in the early part of the study, effectiveness of continued adherence can be gauged by sustained maintenance of those reductions.

B.M. RIFKIND (Glasgow): Have you found any correlation between the degree of weight reduction achieved and the falls obtained in serum lipid levels?

J. STAMLER (Chicago, Ill.): Most of the men in the Coronary Prevention Evaluation Program – 309 of 335 as of March 31, 1966 – were 15% or more above desirable weight. Therefore their recommended diets aimed at reduction of both weight and serum cholesterol levels. During the initial months after diet recommendation, both weight and serum cholesterol usually fell. Eventually, an isocaloric diet was established, moderate in calories and total fat, low in saturated fat and cholesterol, moderate in polyunsaturated fat. With longterm adherence to this diet, both weight and serum cholesterol remained reduced.

No detailed statistical analysis has as yet been done on the precise relationship between degree of weight reduction and serum cholesterol response. To speculate for a moment, it is likely that the data will show such a relationship, if for no other reason than the fact that the men experiencing the best weight reduction were generally the best adherers to the recommended diet, both quantitatively and qualitatively.

In view of the question and one of its possible implications, it is worth emphasizing that hypercholesterolemia can be corrected on an isocaloric diet, without any weight reduction. This has been our experience with hypercholesterolemic non-obese men for whom a diet deficient in calories was not prescribed. This has also been the experience of the National Diet-Heart Study, as well as many other investigations, including those under metabolic ward conditions carried out by Drs. AHRENS, BEVERIDGE, CONNOR, KEYS, KINSELL, and other colleagues.

J.J. GROEN (Jerusalem): It remains for me to close this session and to thank the participants and those who have taken part in the discussion. I hope you will forgive me when I thank only one of you personally, i.e. Dr. Paul D. WHITE. There are many reasons for this but one I may mention particularly: Within a few days from today, Dr. WHITE hopes to celebrate his 80th birthday and this will be a day of rejoicing not only for him but also for all his friends and admirers assembled here. Hippocrates, the great Greek physician, and still an example to all of us, lived to be over a hundred years. Likewise, we wish you, Dr. WHITE, the Hippocrates of modern cardiology, that you may continue in full facility of your capacities, as a doctor, researcher and humanitarian for many more years, and that you may live to see, as a reward for your efforts, the conquest of atherosclerosis!

From the Rockefeller University, New York, N.Y.

Studies of Cholesterol Metabolism in Intact Organisms

E. H. AHRENS, Jr., New York, N.Y.

The Section on Whole Body Metabolism included six reports dealing with sterol metabolism in the whole mammalian organism. KINSELL, TAYLOR, BORGSTRÖM and the author discussed their studies in man; MIETTINEN and MALMROS their observations in rats and rabbits, respectively. Certain major difficulties were not entirely resolved in the course of the discussion; these centered around the methods used in measuring fecal steroids and around the interpretation of isotopic studies.

In obtaining the balance data which must underlie all calculations of synthesis, absorption, feedback control mechanisms, transfer between body pools, and flux, AHRENS emphasized the need for accurate measurement of neutral and acidic fecal steroids. The methods already reported from his laboratory [1, 2] have not required modification in the last two years, and have been successfully adopted in other laboratories. The original publications included various proofs of accuracy and precision; additional evidence, now in print [3], confirms that essentially the same results can be obtained by the isotopic balance method introduced by HELLMAN, ROSENFELD, INSULL and AHRENS in 1957 [4] and by the newer chemical isolation procedure referred to above, which depends on measurement by gas-liquid chromatography (GLC).

The concept of a sterol balance depends on the investigator's ability to recover quantitatively all the sterols excreted by the organism, as well as all conversion products. Clearly, if the sterol ring structure were degraded in the body to forms no longer recognizable as steroids, balance studies would appear to be out of the question, due to losses on the output side. It came as an unwelcome surprise, then, that AHRENS and his colleagues had evidence that in many patients sterols undergo important degrees of ring degradation during their transit through the body. However, since plant sterols suffer the same degree of degradation but are not absorbed to any significant degree, it becomes possible to correct for the losses of cholesterol (and of cholesterol conversion products) by measuring the recovery of plant sterols (and of their neutral steroid conversion products); the percentage recovery of plant sterols proves to be the same as that of cholesterol (and its conversion products) and can be used as a convenient correction factor.

The evidence for these statements was presented. It consisted of two approaches: first, the raw data on recovery in the stools of plant sterols during long-term feeding experiments in which measured amounts of plant sterols were fed in each feeding for many weeks, and second, the recovery of ^{14}C in neutral and acidic fecal steroids when 4-^{14}C-cholesterol was fed in the diet for many weeks until the isotopic steady state was achieved. After that time, by definition, the output of ^{14}C must equal the intake. However, in four patients the recovery of ^{14}C varied from 38 to 88 % during the isotopic steady state (from seven to nine 4-day stool collections were analyzed in each of the four patients, 39–181 days after isotope administration). But, simultaneously the recovery of plant sterols was also measured in each of these samples, and when the ^{14}C recovery was corrected according to the % recovery of plant sterols, the corrected ^{14}C recovery became respectively 102, 93, 101, and 99%.

Thus, the ability to achieve a nearly perfect ^{14}C recovery during the isotopic steady state was taken as clear proof that the plant sterols and cholesterol (and their steroidal conversion products) were degraded to the same extent, as if the enzymatic mechanism responsible for the ring degradation 'saw' no distinction between any of these sterols. Since the three major plant sterols in the diet (campesterol, stigmasterol and β-sitosterol) differ from cholesterol only in the side chain, it would appear that the active site of the enzyme must operate at the A-ring end of the four sterol prototypes.

The measurement of plant sterols in the feces also serves another purpose: that of calculation of fecal flow rates, with the plant sterols serving as internal markers, similarly to the well-known use of chromic oxide. However, the plant sterols are preferable to chromic oxide in studies of sterol balance, for they probably pass through the intestine in the same non-aqueous phases of the intestinal juices. Furthermore, they are routinely measured by GLC whenever analyses are made of fecal neutral steroids derived from cholesterol, so that no additional laboratory work is required. Thus, in the four studies referred to above, the coefficients of variation (S.D. ÷ mean fecal neutral steroid excretion) were reduced from 42, 31, 39, and 22% to 10, 10, 14, and 11%. The same degree of 'smoothing' of the excretion data was achieved by chromic oxide measurements made simultaneously; thus, 4-day calendar collections of stools could be equally as well corrected to 4-day pools in a physiological sense by analysis of plant sterol recoveries, assuming equal degrees of plant sterol degradation in all collections.

Finally it was emphasized that all previously reported studies of sterol balance which had not taken sterol degradation into account must be viewed with suspicion, including reports from the author's group. Any future studies of cholesterol metabolism must measure the degree of sterol degradation in each patient studied, since there is great variation from one patient to another in this regard, and in any one patient from one regimen to another. This injunction applies equally, whether balances are carried out by the isotopic balance method of HELLMAN and associates, or by the GLC method of the Rockefeller group – the only two reported methods which are adequate for accomplishing sterol balances in intact organisms.

KINSELL reported the results of 11 sterol balance studies, based on the isotopic balance concept and on an as yet unpublished method for fractionation into acidic and neutral steroids. According to his calculations, all 11 patients showed a larger excretion of fecal steroids when fed unsaturated fat; he concluded that an increased steroid excretion explained the decreases in serum cholesterol concentrations which occurred on the unsaturated fat intake. The following criticisms of his conclusions can be listed: (1) Mean excretions were compared without standard deviations, even though these means were derived from series of stool collections; thus, the statistical significance of the paired means could not be assessed; (2) the studies were not carried long enough to determine whether the purported excretion increase was temporary or sustained; thus, one could not guess whether the increase was due to greater cholesterol synthesis or to flux from the tissues; (3) no data were given on stool weights on the two regimens; thus, it was not possible to judge whether the purported flux on unsaturated fat was due merely to increased peristalsis and more rapid transit time; (4) no corrections were made for fecal flow; thus, the results obtained for individual stool pools would have been needlessly large, diminishing the opportunity to show significant differences in excretion from one regimen to the next; (5) no corrections for possible sterol degradation were made; indeed, the method used requires feeding a diet completely free of plant sterols, so that any possibility of calculating and correcting for sterol degradation is ruled out.

TAYLOR reviewed the studies he and his colleagues have published since 1960, dealing with isotopic steady state experiments in man patterned after the experimental model in rats introduced by CHAIKOFF and associates in 1957. The results TAYLOR obtained showed that in man dietary cholesterol is relatively less well absorbed than in dogs, rats and rabbits, and that considerable cholesterol synthesis persists despite high cholesterol intakes. TAYLOR's important pioneering studies in this area have recently been confirmed by WILSON, FRANTZ and the author.

BORGSTRÖM reported studies on cholesterol absorption in man, based on 5-day stool collections after a single peroral dose of $4\text{-}^{14}C\text{-}$cholesterol mixed with varying amounts of unlabeled cholesterol mixed in butter. He reported a uniform absorption of 30% of dietary cholesterol up to the largest dose of 6 gm. These results, which do not concur with recent studies of absorption in man carried out by TAYLOR, WILSON and by the author, are not easily interpretable. First, no assessment was made of sterol degradation; since the conclusion is based on labeled cholesterol *not* recovered in the 5-day stool collection period, a mean sterol degradation of only 30% would lead to the quite different conclusion that no cholesterol at all had been absorbed. Second, the possibility of isotope loss through simple exchange with mucosal cholesterol – yet with no *net* absorption – cannot be ruled out.

MIETTINEN reported an experiment in rats fed either cholesterol or taurocholate or both, in which flash labeling with ^{3}H-mevalonate plus ^{14}C-acetate was carried out. The ratio of isotopes in serum cholesterol was then measured at intervals up to 4 days, in order to see the possible inhibition of synthesis of cholesterol by cholesterol itself or by taurocholate at the HMG-reductase step. He failed to find evidence for bile acid feedback control, whereas with cholesterol feeding the incorporation of acetate into cholesterol actually increased while that from mevalonate remained unchanged. TAYLOR had also found that cholesterol feeding in man led to greater incorporation of labeled acetate into serum cholesterol. Conceivably the findings of these two groups could be explained by changes in precursor pool sizes, rather than by changes in synthesis rates.

MALMROS reported that rabbits fed hard fat diets (but free of dietary cholesterol) over many months not only had much higher serum cholesterol concentrations, but that total carcass cholesterol was also increased significantly. Sterol balance studies were not performed, but it seemed clear that cholesterol synthesis and storage must have been greater on the hard fat diets.

References

1. GRUNDY, S.M.; AHRENS, E.H., Jr. and MIETTINEN, T.A.: Quantitative isolation and gas-liquid chromatographic analysis of total fecal bile acids. J.Lipid Res. 6: 397–410 (1965).
2. MIETTINEN, T.A.; AHRENS, E.H., Jr. and GRUNDY, S.M.: Quantitative isolation and gas-liquid chromatographic analysis of total dietary and fecal neutral steroids. J.Lipid Res. 6: 411–424 (1965).
3. GRUNDY, S.M. and AHRENS, E.H., Jr.: An evaluation of the relative merits of two methods for measuring the balance of sterols in man: isotopic balance versus chromatographic analysis. J.clin.Invest. 45: 1503 (1966).
4. HELLMAN, L.; ROSENFELD, R.S.; INSULL, W., Jr. and AHRENS, E.H., Jr.: Intestinal excretion of cholesterol: a mechanism for regulation of plasma levels. J.clin.Invest. (Abstract) 36: 898 (1957).

Author's address: Prof. E. H. Ahrens, Jr., The Rockefeller University, York Avenue and 66th Str., New York, N.Y. 10021 (USA).

From the Institute for Metabolic Research, Highland General Hospital, Oakland, California

Effects of Diet upon Plasma, Bile and Fecal Steroids

L.W. Kinsell, P.D.S. Wood, R. Shioda, G. Schlierf and Y.L. Lee, Oakland, Cal.

More than 15 years have elapsed since work was begun in our laboratory which resulted in the original report indicating that substitution of a variety of vegetable (unsaturated) fats for animal fat and subsequently for saturated vegetable fat would result in a very significant and maintained fall in the level of plasma cholesterol. These observations have been abundantly confirmed [1]. Since that time, work has been carried out in a number of laboratories designed to clarify the mechanism of this effect. This has remained a controversial subject [2–7].

During the past three years my colleagues and I have addressed ourselves intensively to this problem. Dr. Wood has been primarily responsible for the development of precise methodology which gives us confidence in the accuracy of our findings.

In the course of this three-year period, steroid (neutral sterol and bile acid) excretion in the stool has been studied in 11 subjects during the intake of saturated and unsaturated fats, respectively. In the evolutionary process technics have been modified to some degree, but perhaps the most striking fact is that, despite considerable variation in absolute amounts of daily fecal steroid excretion, as well as in the magnitude of differences in relation to diet, greater steroid excretion has been observed during the intake of unsaturated as compared to saturated fats *in every subject*, regardless of the order in which such fats were given. In Figure 1 the mean changes are observed in five consecutive studies carried out under reasonably comparable conditions, in which the statistical significance of the excretion changes is apparent. All eleven studies have

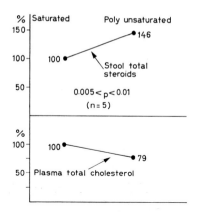

Fig. 1. Replacement of predominantly saturated (nearly sterol-free) fat with equal amounts of polyunsaturated fat results in significant increase in stool steroid excretion in association with decrease in plasma cholesterol.

been carried out on the metabolic ward using formula diets which, with two exceptions, have been essentially sterol-free.

Since another matter of major clinical interest, namely, effects of dietary cholesterol and dietary fat, respectively, on plasma cholesterol, has also been controversial, I shall devote some of my remarks to a study which has been carried out (and which is still in progress) in one subject (a 60-year-old lady with psychoneurotic problems, but no evidence of organic disease) in whom data have been obtained during the intake of essentially sterol-free quantitatively constant formula diets containing saturated and unsaturated fat, respectively, and also during periods of supplementation of such diets by the addition of 750 mg of tritium-labeled cholesterol daily. Prior to the institution of these studies, the patient's miscible cholesterol pools were labeled by intravenous administration of 4-C^{14}-labeled cholesterol.

In Figure 2 three phases are shown of the study in this subject: (a) during the intake of sterol-free partially saturated fat, (b) with the addition of 750 mg of tritium-labeled cholesterol daily, and (c) with the substitution of trilinolein for the partially saturated fat, cholesterol intake remaining unchanged. From this figure certain things are apparent, namely:

(1) When cholesterol was added to the diet, there was a modest but significant increase in the level of plasma cholesterol and in the excretion of total stool steroids. This increase could be accounted for by the excretion of exogenous cholesterol.

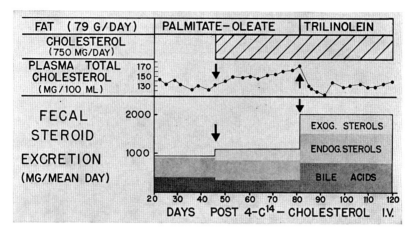

Fig.2. Replacement of palmitate-oleate triglyceride with trilinolein (in the presence of 750 mg of dietary cholesterol daily) results in (a) prompt and maintained decrease in plasma cholesterol, (b) increase in stool steroids (exogenous and endogenous).

(2) When polyunsaturated replaced saturated fat in the diet, there was an immediate, profound and maintained increase in the total fecal steroid excretion at the same time that an abrupt and maintained fall in plasma cholesterol occurred. The mean daily increase in total steroid excretion amounted to 914 mg. Calculations of change in absorption of exogenous cholesterol indicated a decrease from 59% to 33%. Presumably the same percentage change occurred in the reabsorption of endogenous cholesterol. How much of the increased bile acid excretion was attributable to decreased absorption as compared to increased production is impossible to tell from these data.

Figure 3 shows the increase in plasma chemical cholesterol in comparison with the percent increase in tritiated cholesterol*. The excellent agreement in these two figures indicates, at least in this patient under these conditions, that even in the presence of a relatively high cholesterol intake there is quite limited contribution to *plasma* cholesterol by dietary cholesterol.

Your moderator Dr. AHRENS and his colleagues have published data [5] which they feel support certain concepts, namely:

---

* The H^3-cholesterol used was uniformly labeled. As would be anticipated, bile data indicate that H^3 is lost to a very appreciable degree in the process of conversion of cholesterol to bile acids. It is probable that relatively little H^3 is exchanged for H^1 in plasma and bile free cholesterol.

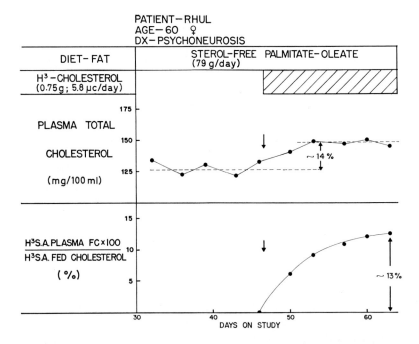

Fig. 3. Addition of 750 mg of H^3-labeled cholesterol to partially saturated fat results in moderate increase in plasma cholesterol (14% by chemical determination; 13% by isotopic measurement).

(1) A quite low level of bile acid excretion under essentially all conditions;

(2) no significant increase in steroid excretion in response to change from saturated to unsaturated dietary fat (although Dr. AHRENS, in the paper which you have just heard, apparently feels that he wishes to reconsider this latter statement in view of their more recent observations);

(3) no change in the rate of decrease of plasma cholesterol specific activity when the dietary fat is changed from saturated to unsaturated; hence, he feels that the fall in plasma cholesterol level is probably attributable to redistribution of cholesterol within other portions of the miscible pool [7].

In an attempt to resolve some of these disagreements, Dr. AHRENS' laboratory and our laboratory have exchanged stool specimens obtained from patients studied in the Rockefeller Institute and the Institute for Metabolic Research, respectively. At a previous meeting [8] Dr. AHRENS, in the course of comments upon some of the data obtained by each of us, indicated that he felt that our methods were in error because, on one occasion, the stool bile acid specific activity was slightly lower than the plasma *total* cholesterol specific activity. He felt, therefore, that our relatively high bile acid excretion figures were attributable to measurement of non-bile acid material, thus giving falsely low specific activity figures and consequently falsely high calculated chemical figures.

Before proceeding with further data, it may be well to refer briefly to the literature with regard to bile acid excretion. The data from the Rockefeller group provide a range from 120 to 225 mg per day.

Following are figures reported over the past few years from laboratories using 'dependable' methodologies:

MOORE, FRANTZ ET AL., 1962 [9]	mean 515 mg/day
AVIGAN and STEINBERG, 1965 [6]	mean 665; max. 2320
WILSON and LINDSEY, 1965 [10]	range 230–1070
BERGSTRÖM, 1962 [11]	~ 800

In the studies which we have carried out to date, the range of bile acid excretion has varied from 132 to 1138 mg/day, with a mean of 498 (n = 35). It would appear, therefore, that although considerable individual variation is observed by all workers, our range is essentially in accord with that obtained by the majority.

Since basically the existing disagreements appear to center around the methodologies involved, it may be well to look at the data in a patient currently under study in whom, during the course of the study, we have determined specific activity of (a) plasma free cholesterol, (b) bile free cholesterol and bile acid, and (c) stool neutral sterol and bile acid. As will be apparent from a subsequent figure, the specific activities of free sterols in plasma and bile obtained on the same day have been essentially identical throughout.

Figure 4 shows the bile acid/cholesterol specific activity ratios in stool and bile at comparable times. It is apparent that at all times a normal precursor-product relationship is present, that is, the bile acid specific activity is greater than the cholesterol specific activity in both bile and feces and the magnitude of this ratio is essentially what one might expect. As the study progresses the ratio tends to approach unity. This is in accord with the findings of others [12]. Since in bile the problems associated with bile acid

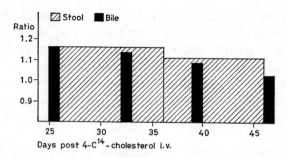

Fig.4. Bile acid/cholesterol specific activity ratios in bile and in stool at comparable times are similar (consistently > 1.0).

measurement (of mass as well as of radioactivity) are inherently much less than in stool, we find such correlation most reassuring.

In more than 90% of all subjects whom we have studied, a fecal bile acid/cholesterol specific activity ratio of more than 1.0 and less than 1.3 has been present except in the late stages of the study, at which time (due perhaps to a combination of the physiological processes involved as well as to some degree of loss of accuracy as the total countable activity diminishes) ratios approaching 1, or on occasion slightly less than 1, are observed.

Figure 5 shows comparative data resulting from measurements in two stool specimens carried out by the Rockefeller group and our group. In stool specimen No. 1 the Rockefeller bile acid/

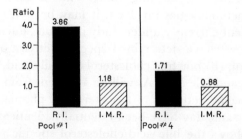

Fig.5. Stool bile acid/cholesterol specific activity ratios in two stool specimens as determined in two laboratories using quite different methodologies (see text). (R.I. = Rockefeller Institute; I.M.R. = Institute for Metabolic Research.)

cholesterol specific activity ratio approached 4. Ours was approximately 1.2. In the second stool specimen (in the same patient at a considerably later time), the Rockefeller figure was 1.7 and ours approximately 0.9. We believe that the latter figure from our laboratory is slightly in error, probably for the reasons indicated above. We believe quite equally that the Rockefeller figure in the first pool is 'physiologically improbable' as very probably is the second value also. We suspect that such figures arise from presently undetectable errors in the GLC procedure for determination of bile acids with resultant considerably elevated specific activity figures for the latter.

Finally, one comes to the matter of decay curves of plasma free cholesterol following preliminary labeling of the body pools by intravenously injected C^{14}-cholesterol. Recalling the paper of CHOBANIAN and HOLLANDER [13], one would anticipate that equilibration of all of the miscible pools in the body would be achieved at approximately 35–50 days, diet and other conditions remaining reasonably constant. In Figure 6 a portion of the same study is shown in which a 'truly exponential slope' was obtained by the fiftieth day under conditions of intake of relatively saturated fat supplemented with 750 mg of cholesterol per day. When the fat

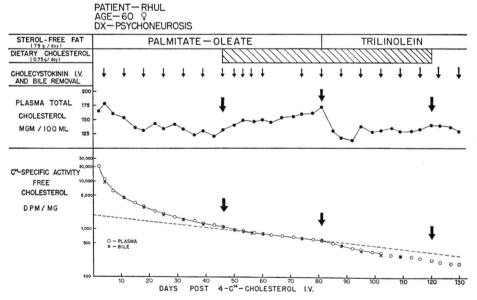

Fig.6. Changes in rate of fall of plasma free cholesterol specific activity (and in plasma free cholesterol level) in response to substitution of polyunsaturated for partially saturated fat. The findings are compatible with increase in mobilization of cholesterol from partially labeled portions of the miscible pools, with or without changes in synthesis rate (see Fig.2 and text) in response to ingestion of unsaturated fat.

was changed to trilinolein, all other factors remaining constant, there was an abrupt increase in the rate of fall of the specific activity, followed by a gradual return toward the original exponential slope. If one compares this figure with other data obtained during the same periods in this patient (Fig. 2), it becomes clear that, however one explains the changes in the slope, they are not compatible with the concept of decrease in plasma cholesterol as the result of redistribution among other portions of the body miscible pool. Extensive consideration of the interpretation of these changes will be the subject of a more detailed paper, but it may be appropriate at this time to indicate that they are quite compatible with the interpretation of (a) initial change due to decrease in plasma pool size (with or without increased synthesis), (b) subsequent change due, at least in part, to increased mobilization from some portion(s) of the body miscible pools. Hopefully, such source(s) could include atheromata.

To summarize, the following points would seem to be reasonably supported by the available data:

(1) The hypocholesterolemic effect of dietary unsaturated fat is attributable, in whole or part, to increased excretion of cholesterol and its metabolites. It seems probable that at least a portion of this increased excretion is referable to decrease in absorption and reabsorption.

(2) The increased excretion rate continues for a considerable period of time (at least 80 days on the basis of current data). Since the increased excretion is associated with a new plateau of plasma cholesterol, it is obvious that there must be increased rate of addition to the plasma pool. Interpretation of the changes in specific activity of plasma and bile cholesterol in the light of the foregoing provides support for the concept of increased mobilization of cholesterol from one or more portions of the body miscible pools in response to unsaturated fat intake. The data available do not preclude some change in rates of synthesis.

(3) It is obvious that the data from the Rockefeller group and from ours are not in agreement. It is our belief that our own data are substantially correct, but together with all responsible workers in the field we believe that a great amount of work lies ahead before anyone can be totally certain of the absolute interpretation of his findings. We feel that all of us owe a debt of gratitude to Dr. AHRENS and his associates for their pioneer work in the field of

gas-liquid chromatography of steroids and look forward to the day when this type of technic will supplant the extremely tedious isotopic balance procedure presently employed in our laboratory.

Acknowledgements

Supported in part by research grants HE-00955 and HE-09346 from the National Institutes of Health, grants from The Corn Products Company and the Safflower Council of the National Institute of Oilseed Products. We wish to thank Dr. F.H. Mattson and colleagues of The Procter & Gamble Company for generous supplies of sterol-free synthetic triglycerides, and Dr. Maurice E. Stansby and associates of the U.S. Bureau of Commercial Fisheries for generous supplies of molecularly distilled fats.

The excellent technical assistance of Miss Marjorie Coelho, Mrs. Gloria Bivins and Mrs. Virginia Pomeroy is acknowledged.

References

1. Kinsell, L.W.: Relationship of dietary fats to atherosclerosis, in Progress in the chemistry of fats and other lipids, editors: Holman, R.T., Lundberg, W.O. and Malkin, T. pp. 138–170 (Pergamon Press, New York 1963).
2. Danielson, H.: Present status of research on catabolism and excretion of cholesterol, in Advances in lipid research, vol. 1. pp. 335–385 (Academic Press, New York 1963).
3. Hellman, L.; Rosenfeld, R.S.; Insull, W., Jr. and Ahrens, E.H., Jr.: Intestinal excretion of cholesterol: A mechanism for regulation of plasma levels. J. clin. Invest. 36: 898 (1957).
4. Grundy, S.M.; Ahrens, E.H., Jr. and Miettinen, T.A.: Quantitative isolation and gas-liquid chromatographic analysis of total fecal bile acids. J. Lipid Res. 6: 397–398 (1965).
5. Spritz, N.; Ahrens, E.H., Jr. and Grundy, S.: Sterol balance in man as plasma cholesterol concentrations are altered by exchange of dietary fats. J. clin. Invest. 44: 1482–1493 (1965).
6. Avigan, J. and Steinberg, D.: Sterol and bile acid excretion in man and the effects of dietary fat. J. clin. Invest. 44: 1845–1856 (1965).
7. Spritz, N.; Grundy, S. and Ahrens, E.H., Jr.: Studies on the mechanism of diet-induced alterations of plasma cholesterol. J. clin. Invest. 42: 981–982 (1963).
8. Wood, P. and Kinsell, L.: Int. Symp. on Drugs Affecting Lipid Metabolism, Milan, Sept. 1965. Progress in Biochemical Pharmacology, II (1966) in press.
9. Moore, R.B.; Anderson, J.T.; Keys, A. and Frantz, I.D., Jr.: Effects of dietary fat on the fecal excretion of cholesterol and its degradation products in human subjects. J. Lab. clin. Med. 60: 1000 (1962).
10. Wilson, J.D. and Lindsey, C.A., Jr.: Studies on the influence of dietary cholesterol on cholesterol metabolism in the isotopic steady state in man. J. clin. Invest. 44: 1805–1814 (1965).
11. Bergstrom, S.: Metabolism of bile acids. Fed. Proc. 21/4: part 2, 28–32 (1962).
12. Lindstedt, S.: Equilibration of dietary cholesterol and bile acids in man. Bile acids and steroids. Clin. chim. Acta 7: 1–4 (1962).
13. Chobanian, A.V. and Hollander, W.: Body cholesterol metabolism in man. The equilibration of serum and tissue cholesterol. J. clin. Invest. 41: 1732–1737 (1962).

Authors' addresses: Dr. L. W. Kinsell, Dr. P. D. S. Wood, Dr. R. Shioda, Dr. G. Schlierf and Y. L. Lee, Institute for Metabolic Research, Highland General Hospital, *Oakland, Calif.* (USA).

From the III. Department of Medicine and Department of Medical Chemistry,
University of Helsinki

Effect of Dietary Cholesterol and Cholic Acid on Cholesterol Synthesis in Rat and Man*

T.A. Miettinen, Helsinki

The effectiveness of the liver to suppress its cholesterol synthesis as a response to cholesterol feeding (negative feed back mechanism) has been studied *in vivo* only in a small number of human subjects mainly because the adequate methods available require either feeding of cholesterol or heavy water [5, 6] over a period of several weeks, carefully controlled sterol balance techniques [1, 3] or combination of these procedures. During our investigations on the mechanism of hypercholesterolemia, the effectiveness of the feed back mechanism was studied by injecting a mixture of acetate-C^{14} and mevalonate-H^3 successively during low and high cholesterol (or bile acid) feeding periods (keeping the basic diet constant) followed by the determination of radioactivity in serum cholesterol. Activation of the feed back system by dietary sterol will inhibit the incorporation of C^{14} into cholesterol. A possibly expanded pool size of readily exchangeable cholesterol is corrected by the uninfluenced formation of H^3-cholesterol to which C^{14}-cholesterol is related. Consequently, an increase of the ratio H^3/C^{14} of serum cholesterol reflects effectiveness of the feed back mechanism. Preliminary results of the application of the procedure has been reported [4] and will be extended in the present paper.

Table I shows that the ratio H^3/C^{14} of serum cholesterol increases three-fold in rats (five in each group) when the sterol-free diet fed is supplemented with crystalline cholesterol, indicating that the con-

* This study was supported by grants from the State Medical Research Council of Finland, Yrjö Jahnsson Foundation and U.S. Public Health Service.

Table I. Effect of dietary cholesterol and taurocholate on the incorporation of simultaneously administred C^{14}-acetate and H^3-mevalonate to serum cholesterol in rats on sterol-free diet

Dietary addition	DPM/Mg of cholesterol		H^3/C^{14} *
	H^3	C^{14}	
None	8570	456	20±1.2
Cholesterol (0.5%, 5 days)	6317	98	67±6.4
Cholesterol (0.5%, 14 days)	5747	111	60±5.0
Taurocholate (0.4%, 10 days)	8213	337	28±3.0

* Mean ± SE.

version of acetate to cholesterol was diminished due to activation of the feed back system. The low specific activity of tritium in the serum cholesterol of cholesterol fed groups may be ascribed to a larger pool size of exchangeable cholesterol rather than to inhibited mevalonate incorporation. That bile acids seems to play an unimportant role in the regulation of the feed back system is evidenced in Table I by the finding that the ratio H^3/C^{14} increased only slightly as response to taurocholate feeding. The effect of the supplementation of a solid food diet with cholesterol (2 g/day for 5 days) and taurocholate (800 mg/day for 5 days) on the incorporation of C^{14}-acetate and H^3-mevalonate into serum cholesterol was studied in three human subjects. Table II shows that the ratio H^3/C^{14} re-

Table II. Effect of dietary cholesterol and taurocholate on the incorporation of acetate-C^{14} and mevalonate-H^3 to serum cholesterol in three human subjects on solid food diet

Dietary addition	Case 1 H^3/C^{14}	Case 2 H^3/C^{14}	Case 3 H^3/C^{14}
None	15.4	16.0	17.8
Cholesterol	13.2	17.7	47.1
Taurocholate	14.0	20.9	16.3

mains almost constant in all three studies of the normocholesterolemic case 1, increased somewhat in the hypercholesterolemic case 2 as a response to taurocholate feeding and showed almost three-fold increase (about 60% inhibition of synthesis) in the normocholesterolemic case 3 during cholesterol feeding. Results confirm earlier findings, obtained with other methods, that dietary cholesterol clearly inhibits endogenous synthesis of cholesterol in rat. In man this inhibition is detectable only occasionally. Dietary bile acids, on the other hand, seems to be less active or ineffective inhibitors of cholesterol synthesis in rat and man. However, sterol balance and the disappearance of plasma radioactive cholesterol have indicated that bile acid feeding inhibits cholesterol synthesis in man [2].

References

1. GRUNDY, S.M.; AHRENS, E.H., Jr. and MIETTINEN, T.A.: Quantitative isolation and gas-liquid chromatographic analysis of total fecal bile acids. J. Lipid Res. 6: 397–410 (1965).
2. GRUNDY, S.M.; HOFMANN, A.F.; DAVIGNON, J. and AHRENS, E.H., Jr.: Human cholesterol synthesis is regulated by bile acids. J. clin. Invest. 45: 1018 (1966).
3. MIETTINEN, T.A.; AHRENS, E.H., Jr. and GRUNDY, S.M.: Quantitative isolation and gas-liquid chromatographic analysis of total dietary and fecal sterols. J. Lipid Res. 6: 411–424 (1965).
4. MIETTINEN, T.A.: Effect of dietary cholesterol on endogenous synthesis of cholesterol in rat and human. 3rd FEBS Meeting, Warsaw 1966, Meeting edition, p. 184 (Academic Press, London/New York 1966).
5. TAYLOR, C.B.; PATTON, D.; YOGI, N. and COX, G.E.: Diet as source of serum cholesterol in man. Proc. Soc. exp. Biol., N.Y. 103: 768–772 (1960).
6. TAYLOR, C.B.; MIKKELSON, B.; ANDERSON, J.A. and FORMAN, D.T.: Human serum cholesterol synthesis measured with the deuterium label. Arch. Path. 81: 213–231 (1966).

Author's address: Dr. T. A. Miettinen, III. Department of Medicine, University of Helsinki, *Helsinki* (Finland).

Department of Pathology, Evanston Hospital, Evanston, Illinois; Northwestern University Medical School, Chicago, Illinois; and Mayo Clinic, Rochester, Minnesota

Human Serum Cholesterol Synthesis*

C. BRUCE TAYLOR, B. MIKKELSON, J. A. ANDERSON and D. T. FORMAN, Evanston, Ill.

GOULD and TAYLOR (1950) reported a marked suppression of newly synthesized cholesterol in the liver and plasma of dogs ingesting a high cholesterole diet; TOMKINS et al. (1953) demonstrated a similar homeostatic mechanism in rats. The mechanism of this inhibition of cholesterol synthesis by dietary cholesterol has been studied extensively and shown to possess the characteristics of a negative 'feed-back' mechanism or control system. Thus, in this control system, a single reaction, the reduction of B-OH B methyl glutaryl CO A to mevalonic acid was shown by SIPERSTEIN and GUEST (1960) to be the primary site of depression of cholesterologenesis produced by exogenous (dietary) cholesterol. Studies by COX et al. (1963) in our laboratories have shown an almost complete lack of change in the rate of appearance of newly synthesized plasma cholesterol in man regardless of the type of diet (cholesterol-rich or cholesterol-low) employed. These findings corroborated earlier studies by DAVIS et al. (1959) which indicated that man (unlike the dog and rat) has little demonstrable 'feed-back' control of *in vitro* hepatic cholesterol synthesis.

Employing an indirect method reported by MORRIS et al. (1957) for estimating relative contributions of dietary and endog-

---

* Supported by the National Institutes of Health, Department of Health, Education, and Welfare (Graduate Research Training Grant 2 GM-697); the Chicago, Illinois and American Heart Associations; the Glenview Area United Fund, Glenview, Ill.; the Life Insurance Medical Research Fund; the Thomas J. Dee and the George C. Moody Memorial Research Funds of Evanston Hospital, and the Dr. Gladys Henry Dick Memorial Pathology Fund of Evanston Hospital.

enously synthesized cholesterol to plasma cholesterol in studies on rats, TAYLOR et al. (1960) and KAPLAN et al. (1963) fed ring-labeled 4-C^{14} cholesterol to dogs and humans. Our studies in the dog were comparable to those reported in the rat by MORRIS et al. (1957); these animals both demonstrated nearly complete replacement of endogenous sources of plasma cholesterol by ring-labeled dietary cholesterol (Fig. 1). The indirect studies were in good agreement with earlier studies demonstrating the dominant homeostatic role played by the liver in the cholesterol metabolism of the rat and dog. The results from the ring-labeled cholesterol feeding

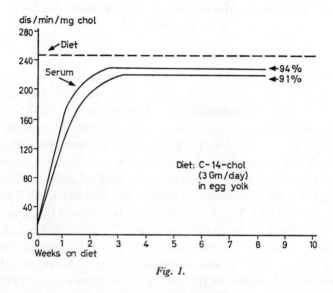

Fig. 1.

studies in man were very different from those observed in dogs and rats; only about a quarter to a third of labeled dietary cholesterol appeared in plasma with the remaining large fraction of ¾ to ⅔ of plasma cholesterol apparently being derived from endogenous synthesis (Fig. 2). As reported earlier by KAPLAN et al. (1963), since most subjects showed an appreciable rise in serum cholesterol while on the ring-labeled cholesterol regimen, there is suggestive evidence that, in the human, dietary cholesterol is additive to an essentially quantitatively unaltered endogenous supply of plasma cholesterol.

In order to better understand cholesterol metabolism in man and determine the effects of dietary cholesterol on the rate of

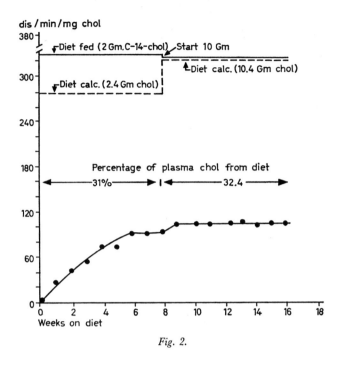

Fig. 2.

synthesis of serum cholesterol, a deuterium-labeling technique was employed. The rationale and methodology were based on the work of LONDON and RITTENBERG (1950) who showed that a single adult male subject on a regular diet (probably containing 0.5 to 1 g of cholesterol per day) had about 0.5 g of newly synthesized cholesterol appear in his plasma each day. Since much of the quantitative estimation of human cholesterol synthesis is based on this single study, we felt the need for additional similar studies with subjects on cholesterol-free and cholesterol-rich diets.

Materials and Methods

TAYLOR et al. (1965) carried out eight endogenous serum cholesterol synthesis studies on five healthy human subjects (male and female). The subjects' age, race, sex, and national origin varied considerably. Each subject served as his or her own control and they were alternately placed on a cholesterol-low diet and a cholesterol-rich diet. The cholesterol-low diet consisted mainly of egg whites, beans, lentils, powdered fat-free milk, and dry cottage cheese. This regimen excluded meat, egg yolks, cream, and butter fat. All subjects were placed on this diet six weeks prior to the initiation of the studies.

Baseline levels of normally occurring deuterium in urine and in serum cholesterol were obtained from all subjects. Subsequently, subjects were administered an oral priming dose of heavy water (D_2O) in concentrations high enough to raise the body water D_2O content to 0.5 moles per cent in excess of normal baseline D_2O. Thereafter, subjects received oral doses of D_2O daily, approximating 10% of their original priming dose in order to compensate for normal turnover of body water. All studies were run for at least 40 to 50 days' duration, and almost daily urine and serum samples were collected from subjects.

After completion of this initial study, subjects were allowed 7–10 months for the elimination of D_2O of body water, cholesterol and other endogenously synthesized deuterated organic molecules. Mass spectrometric analysis verified the return to baseline levels of D_2O contents of body water and serum cholesterol.

The cholesterol-rich study was then instituted. This study provided an estimated daily dietary intake of 2.5 to 3.0 g of cholesterol. The egg was the major cholesterol-carrying vehicle since an egg yolk contains approximately 250 mg cholesterol. Again, urine and serum samples were collected almost daily for determination of deuterated body water levels and deuterium concentration in newly synthesized cholesterol.

The isolation, combustion and reduction of serum cholesterol were carried out in the following manner. Serum specimens were hydrolyzed with alcoholic KOH and extracted with petroleum ether and one portion of the extract analyzed for total serum cholesterol using the procedure reported by ABELL et al. (1952), with a Lieberman-Burchand color reaction (20–1, acetic anhydride-sulfuric acid). The remainder was converted to cholesterol digitonide. Pyridine was then added in order to split off the digitonide and the cholesterol was extracted with acidified ether, purified and weighed. Specimens so isolated usually yielded from 2 to 4 mg of pure cholesterol.

The isolated cholesterol was then combusted with copper oxide at 600°C under high vacuum conditions in a sealed ampule of Corning No. 1720 ignition glass. The ampule containing the

CO_2 and H_2O of combustion was then placed in liquid nitrogen in order to condense and freeze the gases formed. Carbon dioxide was removed and water driven over into a reduction ampule containing 0.5 g zinc. The reduction tube was then heated to 500°C in order to reduce the H_2O and D_2O to H_2 and D_2 gas.

The deuterium assays were carried out in a Nier Model, RCA isotope ratio mass spectrometer at the Mayo Clinic. The mass spectrometric analyses consisted of measuring the mass 3 to mass 2 ratio. This ratio was then converted to heavy water concentration by reference to a calibration derived from five standards of various concentrations of D_2O and H_2O. The magnitude of error in the reduction and mass spectrometer analyses was of the order of $\pm 4\%$.

Results

The principal advantage of using deuterium is its negligible radioactivity which makes it possible to carry out long-term studies on humans without radiation as a prime hazard.

By enriching each subject's body water deuterium with the initial priming dose and daily doses of deuterium oxide, we were able to elevate and maintain the subject's labeled body water between 0.5 and 0.6 moles per cent of D_2O. The labeling of newly synthesized cholesterol was adequate to provide meaningful data when D_2O enrichment was at 0.5 moles per cent of D_2O in body water. Maximum enrichment of 0.1% of D_2O in body water was shown to be experimentally unsatisfactory.

The maximal deutero-cholesterol concentrations were reached by 38–44 days in all eight studies. (Fig. 3 is a graph of data from two studies on one subject.) These times are comparable to the study reported by London and Rittenberg (1950) in which maximum deutero-cholesterol-digitonide concentrations were reached on the 36th day.

Exogenous cholesterol feeding markedly increased the serum cholesterol levels. There was a marked increase in serum cholesterol of from 20 to 56%. Since our studies showed simultaneous marked reduction of absolute deutero-cholesterol concentrations in the cholesterol-rich subjects, we felt this lower isotope content was due to the dilution effect of exogenous cholesterol on endogenously synthesized deutero-cholesterol. The most striking example

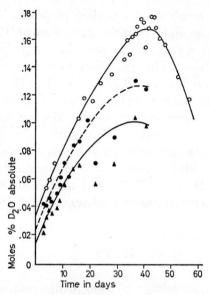

Fig. 3. This figure graphically presents progressive and maximal deuterium enrichment of serum cholesterol of subject 1 (a 43-year-old man) while on cholesterol free (open circles O—O) and cholesterol-rich (solid triangles ▲—▲) diets. Moles/100 cc. D_2O absolute are plotted on the ordinate and time in days is plotted on the abscissa. The middle (corrected) curve (solid circles ●--●) is derived from the same data as the cholesterol-rich curve following correction for dilution of deuterated cholesterol by unlabeled dietary cholesterol.

was Subject 2 (Fig. 4) whose mean serum cholesterol increased 56% while on the cholesterol-rich diet. His absolute deutero-cholesterol concentration, however, fell from 0.1640 moles per cent (low-cholesterol diet) to 0.0650 moles per cent of D_2O on the cholesterol-rich diet. We corrected for the dilution of endogenously synthesized deutero-cholesterol by dietary cholesterol by determining the degree of dilution by exogenous cholesterol and increasing the moles per cent of D_2O absolute by a corresponding factor. The effectiveness of this correction is demonstrated by the corrected curve obtained when the subject was on the cholesterol-rich diet. This effect was seen in all corrected curves.

By plotting the data on semi-log paper in the manner of LONDON and RITTENBERG (1950) one can obtain the half-turnover time of serum cholesterol. This half-turnover time is the time required to reach 50% of maximum deutero-cholesterol concentration. It also represents the time required for half the serum cholesterol molecules present in circulation to be synthesized. After

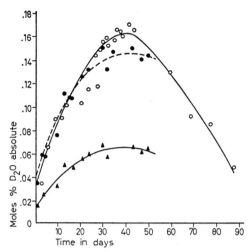

Fig. 4. Graphic presentation of progressive and maximal deuterium enrichment of serum cholesterol of subject 2 (a 25-year-old man) while on cholesterol free (open circles O—O) and cholesterol-rich (solid triangles ▲—▲) diets. Moles/100 cc. D_2O absolute are plotted on the ordinate and time in days is plotted on the abscissa. The third curve (solid circles ●--●) plots deuterium enrichment data taken during the cholestrol rich dietary period and corrected for dilution of deuterium label by unlabeled dietary cholesterol. Note that in this study (as in those shown in Fig. 3) the presence or absence of dietary cholesterol seems to have little effect on the rate of appearance of endogenously synthesized (deuterated) cholesterol.

obtaining the $t\frac{1}{2}$ (estimated half-turnover time in days), one can calculate the turnover rate of endogenously synthesized plasma cholesterol. This turnover is the mean survival time of serum cholesterol molecules and is related to the rate at which serum cholesterol molecules are synthesized. It is interesting to note the relatively unchanged $t\frac{1}{2}$ and T (estimated turnover time in days) in both low-cholesterol and cholesterol-rich groups (Table I).

The estimated quantities of newly synthesized cholesterol appearing in the plasma each day are also reasonably comparable to the quantity reported by LONDON and RITTENBERG (Table II). Again, the essentially complete lack of an effect of dietary cholesterol on estimated daily plasma synthesis is striking.

Discussion

The above studies and our earlier studies indicate an essential absence of a 'feed-back' mechanism in man for compensation for dietary cholesterol. These studies suggest a continuous unaltered rate of delivery of endogenous cholesterol to the plasma regardless

Table I. Estimated turnover and half-turnover time values for subjects on cholesterol-free and cholesterol-rich diets

Subject[a]	Diet	t½ (days)[b]	T (days)[b]
Subject of London and Rittenberg	(?) chol.-rich	8	12
No. 1	chol.-free	10.5	15.0
No. 1	chol.-rich	10.0	14.4
No. 2	chol.-free	8.0	11.6
No. 2	chol.-rich	8.5	12.3
No. 3	chol.-free	12.5	18.0
No. 3	chol.-rich	12.0	17.3
No. 4	chol.-free	9.5	13.7

[a] A fifth subject receiving a much lower labeling dose of D_2O is not listed.
[b] t½ = estimated half-turnover time in days; T = estimated turnover in days.

A compilation of the data on all subjects and comparison with the data of London and Rittenberg indicate that the absence of cholesterol or the presence of 2.5 to 3.0 g, per day in the diet results in no significant change in the t½ or T of plasma cholesterol.

Table II. Data used for estimation of grams of plasma cholesterol synthesized per day

Subject	Diet	T (days)[a]	Plasma vol. (ml)	Mean serum cholesterol (mg percent)	M^b = total plasma chol. (g)	m^c = g plasma cholesterol syn./day
Subject of London and Rittenberg	(?) chol.-rich	12	3750	175	6.56	0.546
No. 1[d]	chol.-free	15.0	2990	218	6.5	0.43
	chol.-rich	14.4	2990	261	(7.8–1.3)	0.45
No. 2[d]	chol.-free	11.6	2330	175	4.1	0.35
	chol.-rich	12.3	2330	273	(6.4–2.3)	0.33
No. 3[d]	chol.-free	18.0	2891	150	4.3	0.24
	chol.-rich	17.3	2891	193	(5.6–1.3)	0.25
No. 4	chol.-free	13.7	3180	110	3.5	0.26

[a] T = estimated turnover time in days.
[b] M = total quantity of plasma cholesterol derived from plasma volume and mean serum cholesterol.
[c] m = quantity of plasma cholesterol synthesized per day derived from dividing M by T, in gram.
[d] In subjects 1, 2, and 3 the total plasma cholesterol in grams during cholesterol-rich studies has been corrected for estimated dilution by unlabeled dietary cholesterol.

of diet. They further suggest that absorbed dietary cholesterol is added to endogenous serum cholesterol, resulting in a net higher serum cholesterol concentration. There still remains another important facet of human cholesterol metabolism requiring further study; this concerns the ability of man to alter his rate of conversion of cholesterol to bile acids and alter fecal excretion of both bile acids and cholesterol to compensate for variable quantities of dietary cholesterol. Studies of this nature have been reported in dogs by ABELL et al. (1956), in rats by WILSON (1964), and are being actively studied in man by GRUNDY et al. (1965), MIETTINEN et al. (1965) and WILSON and LINDSEY (1965). On the basis of our present knowledge it would appear that man's principal protection from hypercholesteremia of dietary origin is his rather limited capacity for intestinal absorption of cholesterol as reported by KAPLAN et al. (1963). Enhancement of fecal steroid excretion has been achieved by exchanging polyunsaturated fats for saturated fats as reported by WOOD et al. (1966), and by ingestion of cholestyramine resin, a bile acid sequestering agent, reported by HASKINS and VAN ITALLIE (1965). Hydrophilic colloids have been reported to lower serum cholesterol levels and may also enhance fecal sterol excretion. These studies have been reported by GARVIN et al. (1965). The role played by dietary constituents on the enterohepatic circulation of bile acids and cholesterol, and in turn on the fecal output of sterols is a very new area of study which may prove to be highly significant in human cholesterol metabolism.

Summary

Studies outlined above indicate that man has an essential absence of an hepatic 'feed-back' mechanism for compensation for dietary cholesterol. In man there seems to be a continuous unaltered rate of delivery of endogenous cholesterol to plasma regardless of dietary cholesterol absorption. Our present knowledge suggests that man's principal protection from hypercholesteremia of dietary origin is his limited capacity for intestinal absorption of cholesterol.

References

ABELL, L. L.; LEVY, B. B.; BRODIE, B. B. and KENDALL, F. E.: A simplified method for the estimation of total cholesterol in serum and demonstration of its specificity. J. biol. Chem. *195:* 357 (1952).

ABELL, L. L.; MOSBACH, E. H. and KENDALL, F. E.: Cholesterol metabolism in the dog. J. biol. Chem. *220:* 527–536 (1956).

COX, G. E.; TAYLOR, C. B.; PATTON, D.; DAVIS, C. B., Jr. and BLANDIN, N.: Origin of plasma cholesterol in man. Arch. path. *76:* 60–88 (1963).

DAVIS, C. B., Jr.; COX, G. E.; TAYLOR, C. B., and CROSS, S. L.: Cholesterol synthesis in human liver. Surg. Forum *9:* 486–489 (1959).

GARVIN, J. D.; FORMAN, D. T.; EISEMAN, W. R. and PHILLIPS, C. R.: Lowering of human serum cholesterol by an oral hydrophilic colloid. Proc. Soc. exp. Biol. N. Y. *120:* 744–746 (1965).

GOULD, R. G., and TAYLOR, C. B.: Effect of dietary cholesterol on hepatic cholesterol synthesis. Fed. Proc. *9:* 179 (1950).

GRUNDY, S. M.; AHRENS, Jr., E. H. and MIETTINEN, T. A.: Quantitative isolation and gas-liquid chromatographic analysis of total fecal bile acids. J. Lipid Res. *6:* 397–410 (1965).

HASKINS, S. A. and VAN ITALLIE, T. B.: Cholestyramine resin therapy for hypercholesteremia. J. amer. med. Ass. *192:* 289–293 (1965).

KAPLAN, J. A.; COX, G. E. and TAYLOR, C. B.: Cholesterol metabolism in man. Arch. Path. *76:* 359–368 (1963).

LONDON, I. M., and RITTENBERG, D.: Deuterium studies in normal man. I. The rate of synthesis of serum cholesterol. II. Measurement of total body water and water absorption. J. biol. Chem. *184:* 687–691 (1950).

MIETTINEN, T. A.; AHRENS, E. H., Jr. and GRUNDY, S. M.: Quantitative isolation and gas-liquid chromatographic analysis of total dietary and fecal neutral steroids. J. Lipid Res. *6:* 411–424 (1965).

MORRIS, M. D.; CHAIKOFF, I. L.; FELTS, J. M.; ABRAHAM, S. and FANSAH, N. D.: The origin of serum cholesterol in the rat: diet versus synthesis. J. biol. Chem. *224:* 1039 (1957).

SIPERSTEIN, M. D. and GUEST, M. J.: Studies on the site of feedback control of cholesterol synthesis. J. clin. Invest. *39:* 462 (1960).

TAYLOR, C. B., MIKKELSON, B.; ANDERSON, J. A. and FORMAN, D. T.: Human serum cholesterol synthesis measured with the deuterium label. Arch. Path. *81:* 213–231 (1966).

TAYLOR, C. B.; PATTON, D.; YOGI, N. and COX, G. E.: Diet as source of serum cholesterol in man. Proc. Soc. exp. Biol. *103:* 768–772 (1960).

TOMKINS, G. M.; SHEPPARD, H. and CHAIKOFF, I. L.: Cholesterol synthesis by liver. III. Its regulation by ingested cholesterol. J. biol. Chem. *201:* 137 (1953).

WILSON, J. D.: The quantification of cholesterol excretion and degradation in the isotopic steady state in rat: The influence of dietary cholesterol. J. Lipid Res. *5:* 409–417 (1964).

WILSON, J. D. and LINDSEY, C. A., Jr.: Studies on the influence of dietary cholesterol on cholesterol metabolism in the isotopic steady state in man. J. clin. Invest. *44:* 1805–1814 (1965).

WOOD, P. D. S. and KINSELL, L. W.: Dietary fat in relation to plasma cholesterol, fecal cholesterol and bile acids in man. Progr. biochem. Pharm. 2 (Karger, Basel/New York 1966) in Press.

Authors' addresses: C. Bruce Taylor, M. D.; B. Mikkelson and D. T. Forman, Ph. D., Department of Pathology, Evanston Hospital, 2650 Ridge Avenue, *Evanston, Ill. 60201* (USA). J. A. Anderson, Section of Biophysics, Mayo Clinic, *Rochester, Minn.* (USA).

From the Department of Medicine (Ludolf Krehl Clinic), University of Heidelberg, Germany (Director: Prof. Dr. G. SCHETTLER)

Distribution and Turnover of Arachidonic Acid in Normals and Hyperlipemic Patients

H. WAGENER and W. SCHWARTZKOPFF, Heidelberg

As yet only a few investigations on the kinetics of individual human plasma lipid fractions were performed. The metabolism of albumin-bound C^{14}-labeled unesterified fatty acids in normals was studied by LAURELL (1957) and FREDRICKSON and GORDON (1958), yielding information on the short half-life of plasma free fatty acids. Similar investigations on the kinetics of cholesterol were performed by CHOBANIAN et al. (1962). In order to study the dynamics of esterified fatty acids in humans we used radioactive arachidonic acid. This unsaturated fatty acid was chosen in view of the altered fatty acid composition of lipid fractions in different forms of hyperlipemia characterized by decreased relative amounts of polyunsaturated fatty acids (HAMMOND and LUNDBERG, 1955; BJÖRNTORP and HOOD, 1960; SCHRADE et al., 1961; KINGSBURY et al., 1962; BÖHLE and HARMUTH, 1963).

Methods

Labeled arachidonic acid was prepared by the tritium gas exchange method of WILZBACH (1957) using the mercury acetate complexes of arachidonic acid in order to prevent addition of tritium to the double bonds. These complexes may then be split without isomerization. The purified acid had a specific activity of 27.7 $\mu C/g$. Amounts corresponding to 10 μC were complexed with albumin in isotonic saline as the potassium salt and intravenously injected into 9 normolipemic and 2 hyperlipemic subjects. Plasma samples were obtained before and at different times up to 3 weeks following the injection. Total lipids were extracted according to FOLCH et al. (1957). Aliquots of the extracts were used for determination of plasma lipids, preparation of fatty acid methyl esters, and liquid-scintillation counting. Radioactivity values were calculated as percentage of total injected dose per liter plasma (%D/L). Plotting of radioactivity values on a semilogarithmic scale yielded exponential radioactivity-time curves which could be resolved into several components with different slopes. From these individual curves half-life times, rate constants, and distribution volumes were derived. Total pools of exchangeable arachidonic acid were calculated as products of distribution volume and arachidonic acid concentration in plasma the latter being obtained from plasma total fatty acids and their percentage composition. The distribution volumes were compartmentated according to the percentage portions from the sum of zero time radioactivity values. In the same way the total miscible pools were divided. Amounts of exchangeable arachidonic acid which participate in transfer reactions in the various compartments were obtained by multiplication of the aliquots of miscible arachidonic acid with the corresponding rate constants. Elimination of arachidonic acid was calculated from total miscible pools and the rate constants of the original radioactivity-time curves.

Results

Fig. 1 shows the radioactivity-time curve of a normolipemic subject which could be resolved into 5 individual exponential curves. The first part of the original curve with a fast decrease of radioactivity besides distribution of the injected fatty acid corresponds to incorporation into various plasma lipid fractions which may also comprise transesterification processes. Therefore all these reactions are summarized as transfer. The last part of the curve showing a steady decrease of radioactivity represents elimination of the administered arachidonic acid mainly due to catabolic reactions.

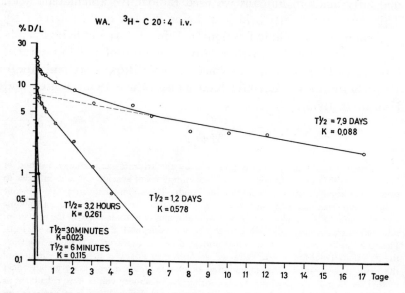

Fig. 1. Tritium-labeled arachidonic acid kinetics after intravenous injection in a normolipemic subject.

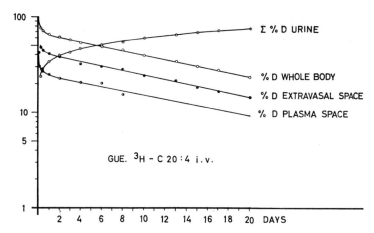

Fig. 2. Radioactivity in total body, extravascular space, plasma space, and urine after intravenous injection of tritium-labeled arachidonic acid in a normolipemic subject.

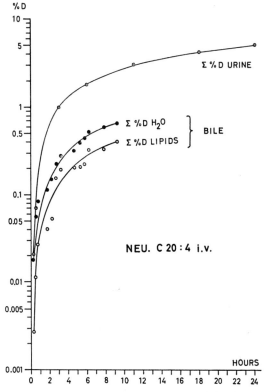

Fig. 3. Radioactivity in bile and urine after intravenous injection of tritium-labeled arachidonic acid in a normolipemic subject.

Determination of radioactivity in various urine samples obtained during the experimental period permitted calculation of total radioactivity in the extraplasmatic space. Fig. 2 illustrates that alterations of total body and extravascular radioactivities are proportional to plasma radioactivity values. Radioactivity excreted in urine increases during the experiment. From the intersection of total body and urine radioactivity curves a biological half-life time of 5.8 days is obvious in this normolipemic patient. Biological half-life in all subjects ranged between 5.8 and 13.0 days with no difference between normals and hyperlipemics.

In one normal subject bile was collected during the first nine hours after arachidonic acid injection. As is evident from Fig. 3 there was a smaller excretion of radioactivity in this fluid than with urine consisting of 60% tritiated water and 40% radioactive lipids.

The main data of arachidonic acid cinetics obtained in normo- and hyperlipemic subjects are summarized in Table I. The distribution volume of arachidonic acid was smaller in hyperlipemic subjects than in normals. Whereas no difference in the absolute miscible pool of arachidonic acid between the two groups could be dem-

Table I. Arachidonic acid kinetics in normo- and hyperlipemic subjects (s. = significant, n.s. = not significant)

Arachidonic acid kinetics	Normolipemic subjects	Hyperlipemic subjects	Significancy
Plasma concentration (% of total fatty acids)	2.6±1.3	1.3±1.1	n.s.
Distribution volume (L)	9.3±2.2	3.8±0.9	s.
Exchangeable pool (g)	1.3±0.8	0.7±0.2	n.s.
Exchangeable pool (% of total fatty acids)	25.9±16.1	4.8±3.8	s.
T ½ plasma (days)	8.7±2.7	5.6±2.4	n.s.
Exchangeable extraplasmatic pool (g)	0.82±0.43	0.22±0.12	s.
Exchangeable pool in plasma (g)	0.48±0.38	0.46±0.10	n.s.
Exchangeable pool in plasma (% of total exchangeable pool)	34.6±8.1	69.2±8.4	s.
Exchangeable extraplasmatic pool: exchangeable pool in plasma	1.99±0.57	0.46±0.17	s.
Total transfer (g/day)	110.9±218.3	1.4±1.1	n.s.
Elimination (g/day)	0.10±0.05	0.11±0.10	n.s.
Elimination (% of exchangeable pool/day)	8.4±2.1	14.4±8.2	n.s.
Total transfer: elimination	1108.3±1160.3	12.7±5.2	s.

onstrated the relative amounts of exchangeable arachidonic acid were significantly smaller in hyperlipemics. In the hyperlipemic subjects considerably smaller amounts of arachidonic acid were present in the extraplasmatic compartments than in normals. The plasma of both normo- and hyperlipemic persons contained similar absolute amounts of arachidonic acid. Related to the total miscible pool the patients possessed twice as much arachidonic acid in plasma than normals. The ratios of extravascular exchangeable arachidonic acid to arachidonic acid present in plasma were significantly lower in the hyperlipemic persons. Neither total transfer nor absolute and relative elimination values were different in both groups. But the transfer to elimination ratios in patients were significantly lower than in normals indicating a diminished transfer in relation to elimination in the hyperlipemic group.

Discussion

These results indicate disturbances of distribution, transfer, and elimination of arachidonic acid in hyperlipemic persons which may be due to lower total body amounts of this acid, smaller compartments or inhibition of distribution. There are clues for all these possibilities. Whereas the relative concentrations and the absolute amounts of total miscible arachidonic acid were not different in both groups the relative exchangeable amounts of this polyunsaturated fatty acid were considerably lower in hyperlipemics. Furthermore the total distribution volumes of arachidonic acid were smaller in subjects with hyperlipemia. This difference seems to be mainly due to smaller extraplasmatic compartments in which early transfer reactions occur. The diminution of this compartment then causes impaired distribution of arachidonic acid which leads to an accumulation in plasma, an abnormality responsible for reduced amounts of arachidonic acid taking part in transfer processes. The accumulation of lipids in plasma of hyperlipemic patients was characterized as 'retention hyperlipemia' by THANNHAUSER (1958) who attributed this disturbance to a dysfunction of capillaries in lipid absorbing organs. Later on impaired splitting of chylomicrons by lipoprotein lipase could be excluded. But in fat-induced hyperglyceridemia abnormally low activities of this enzyme were demonstrated (FREDRICKSON et al., 1963). Our results in addition reveal inhibited distribution of an unsaturated fatty acid. By treatment with chlorophenoxyisobutyrate elevated plasma lipid concentrations in one hyperlipemic patient could be considerably decreased. Repetition of the injection of labeled arachidonic acid then showed that arachidonic acid kinetics were also changed comprising increases of relative arachidonic acid concentration, arachidonic acid miscible pool, exchangeable amounts in extraplasmatic compartments, transfer, and elimination.

Summary

Investigations on the kinetics of intravenously injected tritium-labeled arachidonic acid revealed smaller distribution volumes and lower amounts of exchangeable arachidonic acid in hyperlipemic patients than in normals. In the patients more miscible arachidonic acid was present in plasma. There were no differences in total transfer and elimination between both groups. In individual patients transfer to elimination ratios were considerably lower than in individual normals. These results indicate a relative deficiency, differences in compartmentation, and disturbances in extraplasmatic distribution of arachidonic acid in hyperlipemic patients.

References

BJÖRNTORP, P. and HOOD, B.: Pattern of cholesterol ester fatty acids in serum in normal subjects and in patients with essential hypercholesterolemia. Circulat. Res. *8:* 319–323 (1960).

BÖHLE, E. und HARMUTH, E.: Die Fettsäurenzusammensetzung der Lipidfraktionen des Blutserums bei verschiedenen Hyperlipidämien. Z. klin. Chem. *1:* 83–91 (1963).

CHOBANIAN, A.V.; BURROWS, B.A.; HOLLANDER, W.; SULLIVAN, M. and COLOMBO, M.: Body cholesterol metabolism in man. II. Measurement of the body cholesterol miscible pool and turnover rate. J. clin. Invest. *41:* 1738–1744 (1962).

FOLCH, J.; LEES, M. and SLOANE-STANLEY, G.H.: A simple method for the isolation and purification of total lipides from animal tissues. J. biol. Chem. *226:* 497–509 (1957).

FREDRICKSON, D.S. and GORDON, R.S.: The metabolism of albumin-bound C^{14}-labeled unesterified fatty acids in normal human subjects. J. clin. Invest. *37:* 1504–1515 (1958).

FREDRICKSON, D.S.; ONO, K. and DAVIS, L.L.: Lipolytic activity of post-heparin plasma in hyperglyceridemia. J. Lipid Res. *4:* 24–33 (1963).

HAMMOND, E.G. and LUNDBERG, W.O.: The effect of low-fat diet and atherosclerosis on the polyunsaturated fatty acids of human blood plasma. Arch. Biochem. *57:* 517–519 (1955).

KINGSBURY, K.J.; MORGAN, D.M.; AYLOTT, C.; BURTON, B.; EMMERSON, R. and ROBINSON, D.J.: A comparison of the polyunsaturated fatty acids of the plasma cholesteryl esters and subcutaneous depot fats of atheromatous and normal people. Clin. Sci. *22:* 161–170 (1962).

LAURELL, S.: Turnover rate of unesterified fatty acids in human plasma. Acta physiol. scand. *41:* 158–167 (1957).
SCHRADE, W.; BIEGLER, R. and BÖHLE, E.: Fatty acid distribution in the lipid fractions of healthy persons of different age, patients with atherosclerosis and patients with idiopathic hyperlipidaemia. J. Atheroscl. Res. *1:* 47–61 (1961).
THANNHAUSER, S.J.: Lipidoses. Diseases of the intracellular lipid metabolism. 3rd Edition. (Grune & Stratton, New York/London 1958).
WILZBACH, K.E.: Tritium-labeling by exposure of organic compounds to tritium gas. J. amer. chem. Soc. *79:* 1013 (1957).

Author's address: Priv.-Doz. Dr. med. H. Wagener, *69 Heidelberg*, Bergheimerstr. 58 (Germany).

Discussion Session 2

B.C. TAYLOR (Evanston, Ill.): I wish to comment on Dr. BORGSTRÖM's report on cholesterol absorption measured by recovery of fecal radioactive cholesterol (originally labelled dietary cholesterol).

The possible shifting of plasma cholesterol to extrahepatic compartments needs intensive study. In preliminary studies we have found no evidence of increased accumulation of cholesterol in extrahepatic tissues (such as muscles, adipose tissue and skin) in intact dogs and rats. Man, rabbit, monkey and the hypothyroid dog may behave differently and accumulate cholesterol in extrahepatic tissues, since they develop xanthomatosis and other animals do not.

H. MALMROS (Lund): It is of course very important to know what really happens, when saturated is replaced by polyunsaturated fat. We all know that the serum cholesterol will decrease but the mechanism of this cholesterol-lowering effect is still obscure. It has not been possible to demonstrate without any doubt a change in the cholesterol excretion, degradation or synthesis or in the bile acid production. Therefore it has been suggested that unsaturated fats effect a redistribution of cholesterol between serum and tissue for instance the muscles. There are numerous difficulties in the methodology. In animal experiments it is however possible to determine the total amount of cholesterol in the whole body. In rabbits, fed pellets or corn oil, we have found about 2 g of cholesterol in the total carcass. After feeding saturated fat – palm oil, butter or coconut oil – we have recovered 4–6 g cholesterol in the body. In 4 cases we have first produced a marked hypercholesterolemia by feeding the animals 20% hydrogenated coconut oil and then after 15 weeks we have replaced the saturated fat by the same amount of corn oil. This substitution was followed by a rapid decrease in the serum cholesterol. At the end of the experiment, after 57 weeks, the rabbits were killed and the total amount of cholesterol in the carcass was determined. We found the following values: 2.05, 2.49, 2.77, 2.81 g. It is also interesting to know that in 3 cases of 4 we did not find any macroscopic lesions in the aorta at the autopsy. I think these results suggest, that in rabbits, an isocaloric fat substitution is followed by a real increase in the excretion or degradation of cholesterol or bile acids.

E.H. AHRENS (New York, N.Y.): Professor MALMROS, I wonder if it is possible in your experiments, which I really consider to be most important, to know, on a balance basis, whether or not you could account for this accumulation of cholesterol in the tissues of the animals. You know how much you feed, how much is excreted presumably in the faeces and how much is in the rabbit. Does it all add up?

H. MALMROS (Lund): No, because we have not fed cholesterol; we fed coconut fat and do not know how much cholesterol has been synthesized from the fat.

E.H. AHRENS (New York, N.Y.): You do not know how much is excreted?

H. MALMROS (Lund): No.

E.H. AHRENS (New York, N.Y.): What you do know is that these is more in the rabbit at the end of the experiment after feeding coconut fat.

H. MALMROS (Lund): Yes, and less if the rabbits were first fed hydrogenated coconut oil and then corn oil for some months.

R.G. GOULD (Palo Alto, Cal.): I would like to suggest a possible explanation of why the human appears to differ from all other species in not having a feed-back inhibition of cholesterol synthesis in the liver. When cholesterol is fed to rats the level of cholesterol in the liver is increased. As Dr. FRANTZ first suggested, it may be the concentration in liver of cholesterol that inhibits the synthesis rather than the fact that it was fed. If the human subjects being tested already have a high liver cholesterol, a feed-back inhibition would be already at its maximum capacity. Human livers generally, have higher cholesterol levels than other species. Until we can do these studies in humans, in whom there is reason to believe the liver cholesterol levels are normal, we cannot reach a decision. It might take a long, perhaps an indefinite period of time on a cholesterol free diet to reduce the human liver to such levels in many areas of the world. I would like to ask if anyone knows of studies similar to those Dr. TAYLOR and Dr. MIETTINEN reported today that have been done on humans in which liver cholesterol levels were determined or in which plasma cholesterol levels were 150 mg% or less? One hesitates to believe that the human is so different from so many experimental species in such a fundamental regard. Except for the results of BHATTATHIRY and SIPERSTEIN on liver biopsies, there is very little evidence to support the idea of feed-back inhibition in humans.

C. B. TAYLOR (Evanston, Ill.): Our 3 subjects on the deuterium incorporation study were on a cholesterol free diet for 6 weeks before the deuterium incorporation was started. Our laboratory found essentially complete recovery of cholesterol synthesis in rat liver slices from rats on cholesterol rich diets for 2 months, if they were subsequently on cholesterol free diets for 15 days (Fed. Proc. *15*: Abstract No. 1737, 1956). As I recall our original studies on dogs, Dr. GOULD, we were able to get maximal rates of hepatic cholesterol synthesis after we had fed dogs a cholesterol free diet for only 3 or 4 weeks (GOULD and TAYLOR, Fed. Proc. *9*: 179, 1950).

Although I agree that it would be well to get human synthesis data on an habitual vegetarian eating no cholesterol I have the feeling that 6 weeks on a cholesterol free diet should be an adequate period of time before doing control studies to determine cholesterol synthesis.

M. PASCAUD (Paris): I should like to point out a remark on what concerns the incorporation of arachidonic acid investigated by Dr. WAGENER. The specific activity refers to the total plasma lipids and not to its constituent. In our experiments with labelled linoleic acid in rat, we observed that, a few days after administration of C^{14} linoleic acid, the radioactivity retained in the body lipids is not restricted to the linoleic acid containing components; it is found, for a great part, in other newly synthesized fatty acids (e.g. palmitic) through degradation of linoleic acid to acetyl CoA. This observation is certainly valid for arachidonic acid. Then, it appears necessary, for an accurate determination of arachidonic acid incorporation, to isolate the acid by gas chromatography.

H. WAGENER (Heidelberg): In our studies, kinetics of intravenously administered ^{3}H-arachidonic acid were determined by measuring radioactivity of purified plasma total lipids. In this way radioactivity was evaluated in neither different lipid fractions nor various fatty acids. Alterations, as mentioned by Dr. PASCAUD, may of course, occur. But as we were interested in differences of kinetics between normals and hyperlipemic patients, such procedures were not absolutely necessary. Nevertheless, investigations including differentiations as outlined before are at present being undertaken.

D. E. BOWYER (Cambridge): Is there any information on whether esterified cholesterol rather than free cholesterol causes 'feed-back' inhibition of hepatic cholesterol synthesis, for example in the rabbit?

G. GOULD (Palo Alto, Cal.): The mechanism of the inhibition of cholesterol biosynthesis in the liver of cholesterol fed animals is not known. We have found increases in both the free and esterified cholesterol fractions of whole liver whenever synthesis is decreased by cholesterol feeding. However, most of the inhibition is located in the microsomes and our preliminary results on analysis of microsomes indicate very little increase in free cholesterol, whereas the esterified cholesterol concentration, which is extremely small in normal rat livers, increases several fold in cholesterol-fed animals. Unfortunately, the addition of either cholesterol or its esters has not been reported to produce the same type of inhibition as is found in cholesterol-fed animals, so the mechanism of this effect is yet to be elucidated.

C. B. Taylor (Evanston, Ill.): Agrees with Dr. Gould's answer.

R. Paoletti (Milan): Dr. Miettinen, you mentioned that blood sterols are probably modified in human intestinal tract as far as the sterol ring is concerned. I wonder if this is true also for normal and germ-free rats?

T. Miettinen (Helsinki): Actually I cannot give the answer to this question because we have studied conventional and not germ-free rats with no special attention to the degradation of cholesterol ring. Plant sterols are converted by intestinal bacterias to secondary products (coprostanol and coprostanone derivatives) analogously in the rat and human. Studies with rats on regular chow diet demonstrated only slightly positive plant sterol balance (less than 10%) indicating that degradation was negligible if any.

E. H. Ahrens (New York, N.Y.): We do not yet know what products are produced when cholesterol and the plant sterols are degraded. What we do know is that the tritium and carbon-14, which previously were fixed to the sterol ring structure, are not present in the feces to the extent expected from consideration of the isotopic steady state; to be still more specific, the isotope is not in the aqueous phase after extraction of neutral and acidic steroids. This is the clearest evidence that the losses are not methodologic. Thus we must seek the lost isotopes in body water (^3H) or in respired $^{14}CO_2$, and this we are planning to do.

L. W. Kinsell (London): We have no quarrel with the recovery of conventional bile acids, but as you indicated intestinal flora do some strange things to the steroids. Consequently none of us really knows what he is looking for. Our guess would be that this is where the difficulty comes in terms of recovery in stools by gas-liquid chromatography.

From the Cardiovascular Section, Oklahoma Medical Research Foundation, and Department of Biochemistry, University of Oklahoma School of Medicine, Oklahoma City, Oklahoma

Recent Advances in Metabolism of Plasma Lipoproteins: Chemical Aspects

P. ALAUPOVIC, Oklahoma City, Okla.

I. Introduction

The importance of plasma lipids in the genesis of atherosclerosis and the rapid development of new, simple and efficient analytical tools for their separation and determination have provided important stimuli for many qualitative and quantitative studies of the various lipid components of plasma lipoprotein classes [1-6]. While these studies have contributed significantly to our knowledge of the lipid composition of lipoproteins in health and disease, commensurate information regarding the composition and nature of protein moieties ('apolipoproteins') has not been available. However, the recognition of the importance of protein moieties for structural stability and, possibly, specificity of plasma lipoproteins, the paucity of data concerning their chemical and physical-chemical properties, and their distribution within the individual lipoprotein classes, have emphasized the need for better understanding of this aspect of lipoprotein chemistry.

This review of progress in the chemistry of human plasma lipoproteins is limited to a brief discussion of chemical and physical-chemical properties of immunochemically-homogeneous apolipoproteins. New data on the distribution of apolipoproteins in all major lipoprotein classes are presented. In view of the disclosures of a marked heterogeneity of proteins in various ultracentrifugally-defined lipoproteins, a new classification system is suggested which is based on apolipoproteins as the major distinguishing components and criteria.

II. Chemical and Physical-Chemical Properties of Apolipoproteins

It has now been generally accepted that the ultracentrifugally-defined spectrum of lipoproteins consists of five major classes: (1) chylomicrons ($S_f > 5,000$), (2) very low-density lipoproteins (VLDL, d <1.006 g/ml, S_f 20–5,000), (3) low-density lipoproteins (LDL, d 1.006–1.063 g/ml, S_f 0–20), (4) high-density lipoproteins (HDL, d 1.063–1.210 g/ml), and (5) very high-density lipoproteins (VHDL, d >1.210 g/ml). The VHDL class can be divided, according to Scanu [5], into a protein-phospholipid and a free fatty acid-albumin complex. Oncley and co-workers [7] and Ewing et al. [8] have shown, on the basis of studies on macromolecular distributions, that at least the first four major classes of lipoproteins represent polydisperse systems, heterogeneous in respect to both particle size and hydrated density. Analyses of N-terminal amino acids [9–13] have indicated (Table I) that LDL are characterized by glutamic acid

Table I. N-terminal amino acids of major classes of human serum lipoproteins

Lipoprotein classes	Aspartic acid	Glutamic acid	Serine	Threonine
Chylomicrons	+	+	+	+
Very low-density lipoproteins	+	+	+	+
Low-density lipoproteins	traces	+	traces	traces
High-density lipoproteins	+	traces	traces	traces

and HDL by aspartic acid as N-terminal amino acids and contain single, distinct protein moieties. Traces of other amino acids in these two classes have been considered merely as contaminations resulting from incomplete centrifugal separation of lipoproteins. However, high yields of serine and threonine, in addition to the above N-terminal amino acids, in chylomicrons and VLDL [11–13] have suggested that both of these lipoprotein classes may contain more than one apolipoprotein. These findings could be interpreted as indicating that at least some lipoprotein classes consist of particles differing not only in size and hydrated density but also with respect to the type of protein moiety. Such an interpretation immediately poses several obvious questions. How many different proteins? Are some apolipoproteins of chylomicrons and VLDL the same as those of LDL and/or HDL? Do all N-terminal amino acids represent apolipoproteins or, perhaps, contaminating globular proteins? After all, is the evidence provided by terminal amino acid analyses sufficient proof of protein heterogeneity of lipoproteins? In order to answer these or similar questions we decided to approach this problem in the following manner. Because of the apparent protein homogeneity of LDL and HDL, the first requirement would consist of total delipidization of carefully separated and washed LDL and HDL lipoproteins followed by the isolation and chemical and immunochemical characterization of the corresponding apolipoproteins. The results of these characterization studies, in concert with those available from literature, would be indispensable as references for comparison.

Assuming that VLDL consist of a mixture of polydisperse classes of lipoproteins, each of which is characterized by a specific apolipoprotein, the next requirement was the development of a procedure which would permit separation of chylomicrons and VLDL into subfractions of narrow density ranges. Once this was accomplished the chemical characteristics of the protein moieties of VLDL subfractions could be compared with those of LDL and HDL apolipoproteins. A standardized procedure for the preparative fractionation of VLDL into five subfractions has been developed in our laboratory [14]. Unfortunately, the results of terminal amino acid and immunochemical analyses (to be presented in more detail later in this discussion) showed that the subfractionation of VLDL which we had achieved with this procedure had not yielded lipoproteins characterized by single distinct protein moieties. On the other hand, it was established that the heterogeneity of VLDL with respect to protein moieties involves almost the entire density range. Under these circumstances, the obvious alternative approach to this problem consisted of delipidization of the entire range of chylomicrons and VLDL and separation of corresponding apolipoproteins. Since the usual methods for total or partial delipidization of VLDL either failed to yield water-soluble proteins [5, 15] or resulted in relatively poor yields of lipid-protein residues [16-19], a new method of partial delipidization of serum lipoproteins had to be developed [20] for this purpose.

Before proceeding further with the discussion of the characteristics of LDL and HDL apolipoproteins and the presentation of the results obtained by applying the new method for partial delipidization to VLDL, it seems appropriate to describe briefly two techniques which, in the author's opinion, have, so far, been of utmost importance in studies of serum apolipoproteins: the technique for total delipidization of lipoproteins developed by Scanu et al. [21] and the method for partial delipidization introduced by Gustafson [20].

A. *Delipidization techniques.* The extraction of lipoproteins with ethanol-diethyl ether mixtures results in complete or nearly complete delipidization of lipoproteins. The modification introduced by Scanu et al. [21] involves the extraction of lipoproteins with a mixture of ethanol-diethyl ether (3:1, v/v), followed by a continuous extraction of sample with diethyl ether at —20°. The application of this procedure to HDL yields a water-soluble protein essentially free of lipids [21, 22]. The importance of this accomplishment lies in the fact that, so far, this is the only structural or non-structural apolipoprotein soluble in its lipid-free form without addition of any solubilizing agents [23]. It is thus uniquely suited as a model compound for lipid binding studies [24] and determination of physical-chemical properties.

Total delipidization of LDL and VLDL, on the other hand, yields water-insoluble apolipoproteins. The application of the Gustafson method of partial delipidization [20] to these two classes of lipoproteins succeeded where the method of total delipidization failed: it provided soluble phospholipid-protein residues. The method permits removal of neutral lipids preferentially, without use of detergents. Briefly, the method consists of lyophilization of lipoproteins in the presence of insoluble starch, followed by several consecutive extractions of the dried mixtures of lipoprotein and starch with n-heptane at —12° and recovery of the corresponding phospholipid-protein residues by extraction with buffer solutions. All neutral lipids and some phospholipids, mainly lecithin, are removed. The recovery of soluble residues is relatively high for most VLDL subfractions (40–90%) and HDL (100%) but not LDL (10%). The residues obtained from VLDL have higher phospholipid-protein ratios than those obtained from LDL and HDL. Since differences in the amount of firmly bound phospholipid are probably determined by the binding properties of proteins, the higher phospholipid-protein ratios of VLDL are interpreted as indicating the presence of protein(s) which are different from those of LDL and HDL.

B. Protein moiety of HDL (Apolipoprotein A). The similarity in the amino acid composition of the apolipoproteins A isolated from various subfractions of HDL [10, 25, 26] seems to indicate identity of the protein moieties of the entire HDL class. The content of nonpolar amino acids is relatively small (37%). The presence of cystine has now been established [26, 27]. Aspartic acid is the N-terminal and threonine is the C-terminal amino acid [10, 28]. Peptide patterns of combined tryptic and chymotryptic [29] and tryptic and peptic [30] hydrolysates have been published. Several laboratories [5] have reported the presence of small amount of carbohydrates. A recent analysis [23] indicated the following sugar composition: glucosamine (1.4%), fucose (0.5%), sialic acid (0.4%) and a mixture of hexoses, presumably glucose and mannose in equal amounts, (0.7%).

It has been reported independently by several groups of investigators [23, 26, 31, 32] that ultracentrifugally isolated HDL and HDL_3 (1.125–1.210 g/ml) exist in two forms which differ in electrophoretic mobility in agar gel and cross-react only partially as antigens. LEVY and FREDRICKSON [26] considered the slower-moving or 'lipid-poor' form as a possible transformation artifact of the faster-moving or 'lipid-rich' form from which it could be obtained by prolonged ultracentrifugation, storage or partial delipidization. However, they found a small quantity of 'lipid-poor' form of HDL in fresh plasma. Very recently NEFF and BLOCK detected two antigenic components of HDL which were separated from LDL by gel filtration on Sephadex G-200 [33]. AYRAULT-JARRIER *et al.* [31], and SCANU [23] demonstrated immunochemically the presence of two antigenic forms in apolipoprotein A. Studies in this laboratory [34] have extended these observations by showing that four subfractions of HDL (1.110–1.210 g/ml) contain two electrophoretically separable apolipoproteins present in varying concentrations in each subfraction.

The determination of physical-chemical properties of apolipoprotein A, particularly sedimentation coefficient and molecular weight, has represented a very difficult task resulting frequently in seemingly contradictory data and reports. SCANU *et al.* [21] suggested first a molecular weight of 75,000 and a sedimentation coefficient of 4.1 S. Shore and Shore [27] observed, however, two peaks with values ranging from 2.3 to 2.6 and from 4.2 to 4.6, respectively. In the presence of 0.08% sodium dodecyl sulfate (SDS) apolipoprotein A showed only one peak (2.6 S) and had a molecular weight of 36,500. Subsequently, it has been observed in this laboratory [22, 28, 35] that apolipoprotein A undergoes a concentration-dependent aggregation producing several starch gel electrophoretic bands (at pH 7–8.5) and distinct schlieren boundaries (4.1 S and 7.7 S). Addition of urea, or increase of pH to 11, results in dissociation of the aggregates into a single major component, $s^{\circ}_{20,w}$ 4.1. Addition of SDS causes further dissociation of apolipoprotein A into a subunit, $s^{\circ}_{20,w}$ 2.5, in agreement with a similar observation of Shore and Shore. The molecular weight of this protein subunit, isolated from a subfraction of HDL (d 1.150–1.185 g/ml) and measured by the Archibald method, was 28,000 [34]. When SDS was removed by prolonged dialysis, apolipoprotein A reassociated into a unit with the sedimentation coefficient of the initial preparation and a molecular weight of 55,900. In a very recent study Scanu [23] confirmed the subunit concept and polymerization phenomenon of apolipoprotein A, but found a value of 21,500 for the molecular weight (in presence of SDS) of protein moieties isolated from either HDL_2 (1.063–1.125 g/ml) or HDL_3 (1.125–1.210 g/ml). In contrast, Shore and Shore [36] reported differences in the amino acid composition and in the molecular weights between apolipoproteins A of HDL_2 and HDL_3; they obtained in the presence of SDS values of 23,500 and 32,000, respectively, and interpreted their data as an indication that the protein moieties of these two HDL are not identical. Although many important details are still missing, the subunit concept of the apolipoprotein A structure has been generally established and already proved very useful for a better understanding of the occurrence of its polymorphic forms.

It seems appropriate to complete the discussion of apolipoprotein A by mentioning studies on its lipid recombining capacity. The initial demonstration [24] of this capacity has been extended by Sanbar and Alaupovic to studies [22, 28] of I^{131}-labeled apolipoprotein A binding with serum lipids in solutions containing urea (1–8 M) or SDS (0.1%) or succinic anhydride; in each instance the ability of labeled apolipoprotein A to bind serum lipids *in vitro* was maintained. It has been concluded from these studies that a high degree of folding is not an important requirement of protein for lipid binding. This conclusion is further supported by recent findings [37] that the high a-helix content (55–65%) of apolipoprotein A is disrupted by the action of urea.

   C. *The protein moiety of LDL (Apolipoprotein B)*. An earlier report [38] claiming delipidization of LDL and water solubility of the recovered protein moiety could not be confirmed [5]. As a matter of fact, all reported procedures for total delipidization of LDL [5, 30, 39, 40] yield water-insoluble products. Therefore, virtually nothing is known about the physical-chemical properties of apolipoprotein B. The complete amino acid composition has been reported recently from several laboratories [26, 40, 41]. The nonpolar amino acids comprise 42% of the total content [40]. Glutamic acid appears to be the only N-terminal [9, 10, 11] and serine the only C-terminal amino acid [10]. Peptide patterns of peptic and tryptic digests of apolipoprotein B have been obtained [30]. The carbohydrate content of apolipoprotein B has been determined independently by Ayrault-Jarrier *et al.* [42] and by Marshall and Kummerow [43]. The average composition includes 2.7% hexoses, 0.3% fucose, 1.2% glucosamine and 1.4% N-acetyl-neuraminic acid [42]. Immunochemically, only a single antigenic form of LDL has been obtained [15]. In the absence of any experimental evidence, it has been hypothesized that apolipoprotein B might consist of subunits of a minimal molecular weight of 100,000 [40] which circulate in plasma in various degrees of polymerization. In contrast, most recent studies from this laboratory [44] on the physical-chemical properties of immunochemically homogeneous subfractions of LDL (S_f 3–50) indicate a rather constant proportion of protein and phospholipid, suggesting only a single type or degree of subunit association of apolipoprotein B. In order to resolve these problems a method of delipidization which would yield lipid-free and water soluble apolipoprotein B is urgently needed.

   D. *Protein moieties of VLDL*. The application of the partial delipidization procedure [20] to LDL and HDL resulted, in each instance, in the isolation of phospholipid-protein residues characterized immunochemically by a single protein moiety, apolipoproteins A and B, respectively [30].

   The VLDL ($S_f > 20$) isolated from pooled sera of hyperlipemic subjects yielded, upon partial delipidization, a mixture of three phospholipid-protein residues [30]. These residues were separated by a combination of Pevikon zone electrophoresis and preparative ultracentrifugation and identified, on the basis of sedimentation rates, as 4S, 14S and 7S phospholipid-protein residues. Each residue was characterized by determination of its protein and phospholipid content, sedimentation coefficient, diffusion coefficient and hydrated density, by immunochemical specificity, peptide patterns of tryptic and peptic hydrolysates, N-terminal amino acid analysis and by the quantitation of individual phospholipids [30]. Some of these results are presented in Table II. The striking similarities in the phospholipid/protein ratios and physical-chemical properties, the identical peptide patterns and N-terminal amino acids, and the immuno-

Table II. Phospholipid-protein residues of very low-density lipoproteins ($S_f > 20$)

Phospholipid-protein residue	Phospholipid/protein ratio	Precipitin Reactions Anti-HDL serum	Anti-LDL serum	Anti-VLDL serum	Major N-terminal amino acid	Peptide pattern	Apolipoprotein
4S	0.9	+	−	+	aspartic acid	identical to HDL	A
14S	0.7	−	+	+	glutamic acid	identical to LDL	B
7S	2.4	−	−	+	serine threonine	different from HDL and LDL	C

chemical identity reactions, proved that the protein moieties of 4S and 14S residues are the same as apolipoprotein A and B, respectively. The 7S residue contained another protein, isolated for the first time and designated apolipoprotein C. The 7S residue was characterized by a high phospholipid/protein ratio (2.4), by threonine and serine as N-terminal amino acids and by peptide patterns different from those of HDL and LDL. Immunochemically, it gave no precipitin line with antibodies to HDL, LDL, albumin or γ-globulin, but showed a single line with antibodies to VLDL [45]. Because of the high phospholipid binding capacity it is suggested that apolipoprotein C plays a major role in maintaining the structural stability of protein-poor VLDL particles.

E. The protein moiety of VHDL. When human serum is subjected to preparative ultracentrifugation in a solvent density 1.21 g/ml at 105,000 xg for 17 h or more, the majority of the lipoproteins undergoes flotation. However, 8–15% of the total serum phospholipid [46] and very small amounts of triglyceride and cholesterol undergo sedimentation. PHILLIPS [47] has shown that the principal phosphatide is lysolecithin which accounts for about half of the total phospholipid in the d>1.21 fraction and about half of the total lysolecithin of serum. The d>1.21 fraction also contains most of the 0.3 to 0.6 millimole of plasma free fatty acids bound to albumin. The nature of the protein moiety of the phospholipid-protein complex and the relationship of the complex to HDL has not been known. Studies by SANBAR and by FURMAN et al. [28, 48] of the distribution of radioactivity following incubation of I^{131}-labeled apolipoprotein A with serum (30% radioactivity remained in the VHDL fraction) and the exchange reaction between I^{131}-labeled d>1.21 fraction and non-radioactive HDL indicated that apolipoprotein A may be the protein moiety of at least some VHDL. This suggestion has been supported recently by immunological studies [26, 5] which showed that the d>1.21 fraction of serum contains a protein identical to the protein moiety of HDL. On the other hand, SWITZER and EDER [49] concluded that as much as 95% of total serum lysolecithin is bound to albumin and that a negligible amount of other phospholipids is carried by an ultracentrifugally-altered HDL. Since a phospholipid-protein complex has not been isolated as a single entity, the controversy regarding the nature of the VHDL protein moiety has not been resolved.

In an attempt to answer this question the d>1.21 fraction was adjusted to a solvent density 1.25 g/ml with cesium chloride and centrifuged at 165,000 xg for 44 hours [50]. The top fraction, $VHDL_1$ (d 1.21–1.25 g/ml), differed considerably from the bottom fraction, $VHDL_2$ (d>1.25 g/ml), in respect to the percent lipid composition and the nature of the protein moieties. The $VHDL_1$ contained 15–25% and the $VHDL_2$ 75–85% of the total phospholipid of the parent d>1.21 fraction. The $VHDL_1$ gave no precipitin lines with antibodies to LDL, albumin or any other globular protein, but gave two precipitin lines with antibodies to HDL. Both fractions showed identity reactions with HDL_2 and HDL_3. The lipid moiety of $VHDL_1$ was characterized by a high content of lecithin (71.6%), negligible amounts of lysolecithin (3.5%) and free fatty acids and a low cholesterol/phospholipid ratio (0.07). It has been concluded on the basis of immunochemical studies and comparison of peptide patterns that $VHDL_1$ contains the same protein moiety as HDL, namely, apolipoprotein A. In contrast, $VHDL_2$ was characterized by a high content of lysolecithin (70.8%) and free fatty acids and less lecithin (24.3%). The protein moieties of $VHDL_2$ consisted of the globular serum proteins, including albumin and traces of apolipoprotein A. Thus the d>1.21 fraction (VHDL) can be separated at the present time into two major groups. The $VHDL_1$ represents the continuation, i.e., the portion having the greatest hydrated density, of a polydisperse system of HDL. The $VHDL_2$, containing only traces of apolipoprotein A-containing lipoproteins, consists of the albumin-fatty acid and the albumin-lysolecithin complexes.

III. Distribution of Apolipoproteins in Major Lipoprotein Classes

Preliminary experiments on the distribution of apolipoproteins, or the corresponding phospholipid-protein residues recovered from Pevikon blocks, indicate some interesting differences in the apolipoprotein content of VLDL in various hyperlipemic states [30]. Subjects with dietary carbohydrate-accentuated lipemia contain principally apolipoprotein B (70–80%) and relatively small amounts of apolipoprotein A (8–10%) and C (10–15%). On the other hand, the lipemia of a 'mixed-type' (i.e., neither fat-induced nor carbohydrate-accentuated) is characterized by a high content of apolipoprotein C (48%) and B (48%) and a very small amount of apolipoprotein A (4%).

To study qualitatively the distribution of apolipoproteins in five subfractions [14] of chylomicrons and VLDL (S_f>5,000, S_f 400–5,000, S_f 100–400, S_f 50–100 and S_f 20–50) utilizing immunochemical techniques, rabbits were immunized against VLDL (S_f>20) isolated from the serum of a patient with 'mixed-type' lipemia [45]. The antibodies produced gave single immunoprecipitin lines with the LDL and HDL, and three lines with albumin-free VLDL. The results of studies on the distributions of proteins in pooled *lipemic* sera [51] are shown in Fig. 1. The central wells contained antibodies as indicated in the photograph. The lower left well in each pattern contained the fraction with S_f>5,000; four other VLDL subfractions with decreasing flotation coefficients were placed clockwise in the remaining wells. The precipitin lines with antibodies to HDL and LDL were obtained close to the wells with antigens. The precipitin line close to the central well in experiment with antibodies to VLDL is characteristic of apolipoproteins C-carrying lipoproteins. It is thus apparent that all three apolipoproteins were

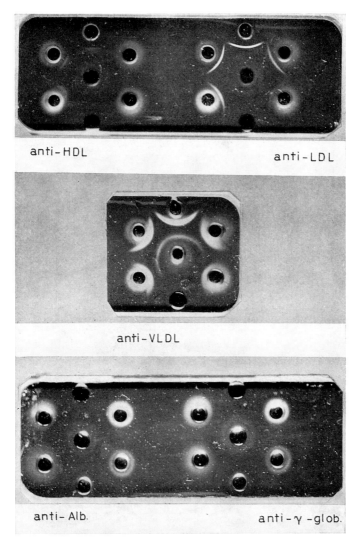

Fig. 1. Gel diffusion patterns of five subfractions of very low-density lipoproteins. The lower left well in each pattern contains the subfraction with $S_f > 5,000$ (chylomicrons), four other subfractions (S_f 400–5,000, S_f 100–400, S_f 50–100, S_f 20–50) are placed clockwise in the remaining wells. The central wells contain the antibodies as indicated on the photograph.

present in first three fractions with highest flotation coefficients ($S_f > 100$) including chylomicrons. Fraction S_f 20–50 contained only apolipoprotein B. None of the subfractions reacted with antibodies to either albumin or γ-globulin. Due to the relatively small amounts of VLDL in the serum of fasting *normal* subjects, studies on the distribution of apolipoproteins are not as yet conclusive. It seems, however, that apolipoproteins A and B represent the minor and major components, respectively. Evidence for the presence of a mixture of apolipoproteins in VLDL has been reported by several laboratories. SCANU and PAGE [17] presented evidence of the occurrence of apolipoproteins A and B in chylomicrons. SHORE and SHORE [12] noted the production of HDL following lipolysis of VLDL with S_f 20–400 by postheparin plasma, and FURMAN ET AL. [52] demonstrated the presence of HDL and LDL in VLDL by preparative ultracentrifugation of lipemic serum before and after partial disruption of VLDL by sonic forces. Recently, LEVY ET AL. [53] reported immunochemical evidence of the presence of apolipoproteins A and B in the electrophoretically defined 'pre-beta' (VLDL) lipoprotein fraction.

In view of the protein heterogeneity of VLDL and several reports [26, 31, 54, 53] providing immunochemical evidence of the presence of apolipoproteins A and B in the lipoprotein fraction isolated between d 1.063–1.110 g/ml, it became obvious that the entire spectrum of lipoproteins should be re-examined as a mixture of apolipoproteins. The LDL were separated by differential ultracentrifugation into five subfractions and HDL and VHDL into six. Each subfraction was tested with antibodies to LDL and HDL by agar gel immunodiffusion. Results are presented in Table III. The LDL contained traces of apolipoprotein A even in subfractions isolated

Table III. Immunochemical properties of human serum low-density and high-density lipoproteins

Lipoprotein fraction density range g/ml	Precipitin reaction Anti-HDL serum	Anti-LDL serum	Apolipoprotein
1.006–1.020	—	+	B
1.020–1.030	(+)	+	(A) B
1.030–1.040	(+)	+	(A) B
1.040–1.050	+	+	A, B
1.050–1.063	++	+	A, B
1.063–1.110	++	+	A, B
1.110–1.130	++	(+)	A
1.130–1.150	++	—	A
1.150–1.185	++	—	A
1.185–1.210	++	—	A
1.210–1.250	++	—	A

between d 1.020–1.040. In contrast to the report by LEVY and FREDRICKSON [26], the subfractions d 1.050–1.063 and d 1.063–1.110 gave two precipitin lines with antibodies to HDL. Previous reports of the presence of apolipoprotein B in the subfraction d 1.063–1.110 have been confirmed. It should be pointed out that the apolipoprotein A-carrying lipoproteins present in the fraction of d <1.063 g/ml could be removed only by precipitation with antibodies to HDL. The presence of small amounts of these lipoproteins seems to depend to some extent on the nutritional state of the donor; sera obtained after 24 h of fasting contained apolipoprotein A-lipoproteins only in d 1.050–1.063 fraction. The importance of a possible dietary effect on the formation of lower density HDL remains to be established.

IV. Proposal of a new Classification System of Human Serum Lipoproteins

The hydrated densities of serum lipoproteins have provided the basis of a widely accepted classification system of lipoproteins comprising five major classes, each of which is characterized by a relatively narrow range of particle sizes and densities. Since these physical-chemical properties are largely a consequence of the neutral lipid composition of lipoprotein particles, this classification system depends on the amounts and relative proportions of various lipid components, primarily triglycerides and cholesterol esters. Fig. 2

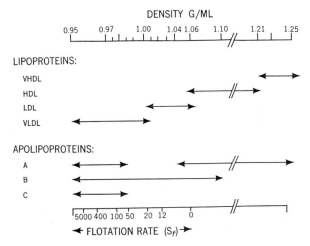

Fig. 2. Classification of serum lipoproteins based either on the particle densities or apolipoproteins.

presents schematically the usually accepted density ranges of the major lipoprotein classes, VLDL (including chylomicrons), LDL, HDL and VHDL, as determined experimentally by studies on the distribution of macromolecular particles [7, 8]. Although each class consists of a heterogeneous polydisperse system of particles differing both in size and hydrated density, the usefulness and popularity of this classification system has been greatly advanced and supported by the development of preparative ultracentrifugal techniques which enable the separation and isolation of the major lipoprotein classes and by the many metabolic and clinical studies which relate certain derangements of fat transport with a particular lipoprotein class.

Chemical, physical-chemical and immunochemical characterizations of various ultracentrifugally-defined serum lipoprotein classes have revealed, however, another type of heterogeneity, namely, heterogeneity in respect to protein moieties. To demonstrate this protein heterogeneity, the pattern of distribution of apolipoproteins corresponding to the distribution of lipoproteins based on densities is shown in Fig. 2. The heterogeneity of the ultracentrifugally-defined lipoprotein families in respect to density and protein is evident. For example, VLDL consist of three families of lipoproteins based on the presence of apolipoproteins A, B and C. Since all three lipoproteins have a similar lipid composition, their hydrated densities and ultracentrifugal behavior are similar. Both LDL and HDL contain two families of lipoproteins based on apolipoproteins A and B as the protein moieties. The VHDL contain apolipoprotein A and albumin. Although albumin is not indicated in Fig. 2, its ability to bind fatty acids and lysolecithin warrants its inclusion as an apolipoprotein in any classification system based on protein moieties. Certain segments of the spectrum such as S_f 16–50 (apolipoprotein B) and d 1.110–1.210 g/ml (apolipoprotein A) consist of single families since they are characterized by a single apolipoprotein. There is some evidence that the range of density over which an apolipoprotein is found may be influenced by diet or a derangement of lipid transport in individual subjects. Thus, traces of apolipoprotein A-carrying lipoproteins have been detected in the d 1.006–1.040 g/ml range in healthy subjects six to eight hours following a standard meal. However, data in Fig. 2 are derived from studies of sera obtained only from fasting subjects (normolipemic and hyperlipemic).

Recent studies of the characterization and distribution of the protein moieties have prompted a revaluation of the current classification system of serum lipoproteins. Is a classification of lipoproteins based on such variable physical-chemical properties as size or density the most useful system from the chemical and metabolic viewpoints? Is there then an alternative approach which would offer greater insight into the formation, metabolism and chemical composition of lipoproteins?

Since the known lipid components occur in all ultracentrifugally-defined lipoprotein classes, with the possible exception of albumin-fatty acid and albumin-lysolecithin complexes, *apolipoproteins* are the only components which offer a distinct *chemical* differentiation of various lipoprotein families. Three distinct protein moieties have been identified thus far in the spectrum of human serum lipoproteins: apolipoproteins A, B and C. Together with albumin, they provide a logical classification system of lipoproteins which is based on chemistry, as proposed previously by ALAUPOVIC et al. [56]. This classification (Fig. 2) recognizes three polydisperse systems of apolipoprotein A-, B-, and C-carrying lipoproteins and simpler albumin-fatty acid and albumin-lysolecithin complexes. Although this formulation of lipoprotein families recognizes the inevitable heterogeneity of particles with respect to size, it provides for protein homogeneity and emphasizes the absolute essentiality of apolipoproteins for the formation and possible change in composition of lipoproteins. Each of these newly formulated lipoprotein families represents a *continuum* of particles within characteristic and specific ranges of density. Apolipoprotein C- and B-carrying lipoproteins occur in the 0.95–0.99 and 0.95–1.110 g/ml ranges, respectively. Apolipoprotein A-carrying lipoproteins have been found in two density ranges: 0.94–0.99 g/ml and 1.04–1.25 g/ml. Apolipoprotein A appears to have the unique capacity to form lipoprotein particles of any hydrated density and thus, under certain circumstances, may be found in any portion of lipoprotein density spectrum. The albumin-lipid complexes are found at densities greater than 1.25 g/ml. Methods are not yet available for the isolation, in preparative amounts, of the entire class of any lipoprotein characterized by a single apolipoprotein. Precipitation techniques utilizing macromolecular compounds [1, 4, 5, 7, 32] or specific antibodies [44] represent the most promising approach to the problem at present.

It is suggested that this newly proposed classification system, based on protein moieties as the essential components of lipoprotein particles, is justified chemically and biologically. *Chemically*, it utilizes unique chemical components, i.e., apolipoproteins, as criteria for differentiation of various lipoprotein families. It recognizes the capacity of each apolipoprotein to bind various amounts of lipid (thus to form lipoproteins of various hydrated densities) but also indicates differences and limitations in binding capacities of individual apolipoproteins. *Biologically*, it points out differences in the metabolic origin and fate of various lipoprotein families. Genetic disorders of lipid transport such as Tangier disease and abetalipoproteinemia [5] are characterized primarily by the absence of an apolipoprotein and only secondarily by decrease of a specific density class. Other differences in the origin and fate of individual apolipoproteins may be revealed to be of significance in the development of atherosclerosis.

Acknowledgements

The author wishes to thank Doctors ROBERT H. FURMAN and REAGAN H. BRADFORD for many stimulating discussions and criticisms, and to Mrs. M. FARMER for assistance in the preparation and typing of the manuscript. The work of author quoted in this review paper has been supported in part by grants from the U.S. Public Health Service (HE-6221 and HE-2528), the American Heart Association, and the Oklahoma Heart Association.

References

1. GURD, F.R.N.: Some naturally occurring lipoprotein systems. In: Lipide chemistry. p. 260 (Wiley, New York 1960).
2. LINDGREN, F.T. and NICHOLS, A.V.: Structure and function of human serum lipoproteins. In: The plasma proteins. p. 2 (Acad. Press, New York 1960).
3. FREEMAN, N.K.; LINDGREN, F.T. and NICHOLS, A.V.: The chemistry of serum lipoproteins. In: Progress in the chemistry of fats and other lipids. *7:* 216 (Pergamon Press, Oxford 1963).
4. ONCLEY, J.L.: Lipid protein interactions. In: Brain lipids and lipoproteins, and the leucodystrophies. p. 1 (Elsevier, Amsterdam 1963).
5. SCANU, A.M.: Factors affecting lipoprotein metabolism. In: Advances in lipid research. *3:* 63 (Acad. Press, New York 1965).
6. GOFMAN, J.W. and YOUNG, W.: The filtration concept of atherosclerosis and serum lipids in the diagnosis of atherosclerosis. In: Atherosclerosis and its origin. p. 197 (Acad. Press, New York 1963).
7. ONCLEY, J.L.: Lipoproteins. In: Proc. of an international symposium on lipid transport. p. 70 (Thomas, Springfield 1964).
8. EWING, A.M.; FREEMAN, N.K. and LINDGREN, F.T.: The analysis of human serum lipoprotein distributions. In: Advances in lipid research *3:* 25 (Acad. Press, New York 1965).
9. AVIGAN, J.; REDFIELD, R. and STEINBERG, D.: N-terminal residues of serum lipoproteins. Biochim. biophys. Acta *20:* 557–558 (1956).
10. SHORE, B.: C- and N-terminal amino acids of human serum lipoproteins. Arch. Biochem. *71:* 1–10 (1957).

11. RODBELL, M.: N-terminal amino acid and lipid composition of lipoproteins from chyle and plasma. Science *127:* 701–702 (1958).
12. SHORE, B. and SHORE, V.: Some physical and chemical properties of the lipoproteins produced by lipolysis of human serum S_f 20–400 lipoproteins by post-heparin plasma. J. Atheroscler. Res. *2:* 104–114 (1962).
13. BOBBITT, J. L. and LEVY, R. S.: Proteins of S_f 10–400 lipoproteins from lipemic human plasma. Biochemistry *4:* 1282–1288 (1965).
14. GUSTAFSON, A.; ALAUPOVIC, P. and FURMAN, R. H.: Studies of the composition and structure of serum lipoproteins: Isolation, purification, and characterization of very low-density lipoproteins of human serum. Biochemistry *4:* 596–605 (1965).
15. SCANU, A.; LEWIS, L. A. and PAGE, I. H.: Studies on the antigenicity of β- and a_1-lipoproteins of human serum. J. exp. Med. *108:* 185–196 (1958).
16. AVIGAN, J.: Modification of human serum lipoprotein fractions by lipide extraction. J. biol. Chem. *226:* 957–964 (1957).
17. SCANU, A. and PAGE, I. H.: Separation and characterization of human serum chylomicrons. J. exp. Med. *109:* 239–256 (1959).
18. HAYASHI, S.; LINDGREN, F. and NICHOLS, A.: Degradation of S_f 20–400 and high density lipoproteins of human sera by ethyl ether. J. amer. chem. Soc. *81:* 3793–3798 (1959).
19. GRUNDY, S. M.; DOBSON, H. L.; KITZMILLER, G. E. and GRIFFIN, A. C.: Lipid transport and lipoprotein interconversions. Amer. J. Physiol. *200:* 1307–1310 (1961).
20. GUSTAFSON, A.: New method for partial delipidization of serum lipoproteins. J. Lipid Res. *6:* 512–517 (1965).
21. SCANU, A.; LEWIS, L. A. and BUMPUS, F. M.: Separation and characterization of the protein moiety of human a_1-lipoprotein. Arch. Biochem. *74:* 390–397 (1958).
22. SANBAR, S. S. and ALAUPOVIC, P.: Effect of urea on behavior of the protein moiety of human serum a-lipoproteins in solution. Biochim. biophys. Acta *71:* 235–236 (1963).
23. SCANU, A.: Forms of human serum high density lipoprotein protein. J. Lipid Res. *7:* 295–306 (1966).
24. SCANU, A. and HUGHES, W. L.: Recombining capacity toward lipids of the protein moiety of human serum a_1-lipoprotein. J. biol. Chem. *235:* 2876–2883 (1960).
25. SCANU, A. and HUGHES, W. L.: Further characterization of the human serum D 1.063–1.21, a_1-lipoprotein. J. clin. Invest. *41:* 1681–1689 (1962).
26. LEVY, R. I. and FREDRICKSON, D. S.: Heterogeneity of plasma high density lipoproteins. J. clin. Invest. *44:* 426–441 (1965).
27. SHORE, V. and SHORE, B.: The protein subunit of human serum lipoproteins of density 1.125–1.200 gram/ml. Biochem. biophys. Res. Comm. *9:* 455–460 (1962).
28. SANBAR, S. S.: Structure and metabolism of serum high density lipoproteins. Ph. D. Diss. Oklahoma City (1963).
29. RODBELL, M. and FREDRICKSON, D. S.: The nature of the proteins associated with dog and human chylomicrons. J. biol. Chem. *234:* 562–566 (1959).
30. GUSTAFSON, A.; ALAUPOVIC, P. and FURMAN, R. H.: Studies of the composition and structure of serum lipoproteins. Separation and characterization of phospholipid-protein residues obtained by partial delipidization of very low-density lipoproteins of human serum. Biochemistry *5:* 632–640 (1966).
31. AYRAULT-JARRIER, M.; LEVY, G. and POLONOVSKI, J.: Etude des a-lipoprotéines sériques humaines par immunoelectrophorèse. Bull. Soc. Chim. biol. *45:* 703–713 (1963).
32. BURSTEIN, M. and FINE, J. M.: Isolement d'une fraction des alpha-1-lipoprotéines sériques après précipitation par le sulfate de dextrane en présence des cations bivalents et de saccharose. Rev. franc. Et. clin. biol. *9:* 105–108 (1964).

33. NEFF, B.J. and BLOCK, W.D.: Characterization of serum lipoproteins separated by gel filtration. Fed. Proc. 25, Part I: 794 (1966).
34. ALAUPOVIC, P.; SULLIVAN, M.L. and WALRAVEN, S.L.: Unpublished experiments.
35. ALAUPOVIC, P.; SANBAR, S.S.; FURMAN, R.H. and KRATOHVIL, J.: Studies on the association-dissociation reaction of the protein moiety of human serum α-lipoproteins. Abstracts of the Intern. Conf. on Biochem. of Lipids (Stockholm 1963).
36. SHORE, B. and SHORE, V.: The protein subunits of high density lipoproteins of human serum. Fed. Proc. 25, Part I: 705 (1966).
37. SCANU, A.: Studies on the conformation of human serum high-density lipoproteins HDL_2 and HDL_3. Proc. nat. Acad. Sci., Wash. 54: 1699–1705 (1965).
38. ONCLEY, J.L.; GURD, F.R.N. and MELIN, M.: Preparation and properties of serum and plasma proteins. XXV. Composition and properties of human serum β-lipoprotein. J. amer. chem. Soc. 72: 458–472 (1950).
39. BANASZAK, L.J. and MCDONALD, H.J.: The proteolysis of human serum β-lipoproteins. Biochemistry 1: 344–349 (1962).
40. MARGOLIS, S. and LANGDON, R.G.: Studies on human serum β_1-lipoprotein. I Amino acid composition. J. biol. Chem. 241: 469–476 (1966).
41. LEVY, R.S. and LYNCH, A.C.: Amino acid analysis of human serum lipoproteins. Fed. Proc. 21: 75 (1962).
42. AYRAULT-JARRIER, M.; CHEFTEL, R.I. and POLONOVSKI, J.: Les glucides de la β-lipoprotéine $S_f 1.063$ 0–12 du sérum sanguin humain. Bull. Soc. Chim. biol. 43: 811–816 (1961).
43. MARSHALL, W.E. and KUMMEROW, F.A.: The carbohydrate constituents of human serum β-lipoprotein: galactose, mannose, glucosamine and sialic acid. Arch. Biochem. 98: 271 (1962).
44. LEE, D.; ALAUPOVIC, P. and FURMAN, R.H.: Physical-chemical characterization and structural aspects of low-density serum lipoproteins. Circulation 34, Suppl. No. II:18 (1966).
45. ALAUPOVIC, P.; LEDFORD, J.; FURMAN, R.H. and OLSON, A.C.: Unpublished experiments.
46. FURMAN, R.H.; NORCIA, L.N.; FRYER, A.W. and WAMACK, B.S.: Lipoprotein recovery following ultracentrifugal fractionation at solvent density 1.21 as determined by cholesterol and lipid phosphorus analyses of supernatant and bottom fractions. J. lab. clin. Med. 47: 730–734 (1956).
47. PHILLIPS, G.B.: Lipid composition of human serum lipoprotein fraction with density greater than 1.210. Proc. Soc. exp. Biol., N.Y. 100: 19–22 (1959).
48. FURMAN, R.H.; SANBAR, S.S.; ALAUPOVIC, P.; BRADFORD, R.H. and HOWARD, R.P.: Studies of the metabolism of radioiodinated human serum alpha lipoprotein in normal and hyperlipemic subjects. J. lab. clin. Med. 63: 193–204 (1964).
49. SWITZER, S. and EDER, H.A.: Transport of lysolecithin by albumin in human and rat plasma. J. Lipid Res. 6: 506–511 (1965).
50. ALAUPOVIC, P.; SANBAR, S.S.; FURMAN, R.H.; SULLIVAN, M.L. and WALRAVEN, S.L.: Biochemistry 5: 4044–4053 (1966).
51. LEDFORD, J.H.; ALAUPOVIC, P. and FURMAN, R.H.: Studies on the protein moieties (apolipoproteins) of very low-density lipoproteins of hyperlipemic human serum. Circulation 34, Suppl. No. II:18 (1966).
52. FURMAN, R.H.; HOWARD, R.P.; LAKSHMI, K. and NORCIA, L.N.: The serum lipids and lipoproteins in normal and hyperlipidemic subjects as determined by preparative ultracentrifugation. Effects of dietary and therapeutic measures. Changes induced by *in vitro* exposure of serum to sonic forces. Am. J. clin. Nutr. 9: 73–102 (1961).

53. LEVY, R.I.; LEES, R.S. and FREDRICKSON, D.S.: The nature of pre-beta (very low density) lipoproteins. J. clin. Invest. *45:* 63–77 (1966).
54. AYRAULT-JARRIER, M.; LEVY, G.; WALD, R. and POLONOVSKI, J.: Separation par ultracentrifugation des a-lipoprotéines du sérum humain normal. Bull. Soc. Chim. biol. *45:* 349–359 (1963).
55. ALADJEM, F.; LIEBERMAN, M. and GOFMAN, J.W.: Immunochemical studies on human plasma lipoproteins. J. exp. Med. *105:* 49–67 (1957).
56. ALAUPOVIC, P.; GUSTAFSON, A.; SANBAR, S.S. and FURMAN, R.H.: The protein moieties of human serum lipoproteins – a basis for classification. Circulation *30*, Suppl. *III:* 1 (1964).

Author's address: P. Alaupovic, Cardiovascular Section, Oklahoma Medical Research Foundation and University of Oklahoma School of Medicine, *Oklahoma City, Okla.* (USA).

(Groupe de Recherches sur l'Athérosclérose [INSERM].
Hôpital Boucicaut, Paris, France)

Hyperlipidemia with Circulating Anti β Lipoprotein Auto-Antibody in Man. Auto-Immune Hyperlipidemia, its Possible Role in Atherosclerosis

J. L. Beaumont, Paris

An anti β lipoprotein auto-antibody and an hyperlipidemia were found together in two men. One of them had also an IgA* myeloma [1, 2, 3, 13] and both had clinical evidence of severe ischaemic diseases. The main data of these two cases suggested the possibility of 'auto immune hyperlipidemia', a pathological entity of perhaps great significance for the pathogenesis of atherosclerosis [4]. This paper summarizes the clinical and biological data collected in these cases and gives additional information on the antibody which was isolated and purified in one of them.

Clinical Reports

Case 1: Mr. G. René, born in 1906, was followed as an out patient from October 1964 to October 1965 for hyperlipidemia with xanthomatosis, multiple myeloma and coronary insufficiency.

The hyperlipidemia (Table I) was first found in 1961 when a milky serum and a total cholesterol of 340 mg/100 ml were detected. A few weeks later the first cutaneous xanthomas appeared rather suddenly and spread to both hands, elbows, knees and buttocks. Since then, the cutaneous xanthomas have extended to almost all the skin, including the face, staying plane at the palms and tuberous elsewhere. In 1964, tendon xanthomas, xanthelasma and corneal arcus were also noticed. From October 1964 to October 1965, 15 blood controls showed constant hyperlipidemia (Fig. 1) with values ranging from 410 to 990 mg/100 ml for the total serum cholesterol; 170 to 1,440 for the triglycerides and 400 to 1,000 for the phospholipids (Table I). Blood sugar, urea, and uric acid were normal. Low fat and hypocaloric diets both reduced slightly the lipidemia but it never returned to normal figures. Triparanol (given in 1961), chlorophenoxyisobutyrate, and melphalan had little or no effect.

* According to the nomenclature of immunoglobulins recommended by WHO.

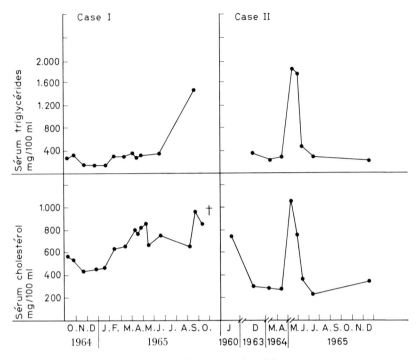

Fig. 1. Serum cholesterol and triglycerides in case 1 and 2.

Table I. Auto immune hyperlipidemia (Fasting serum lipids)

	Serum lipids (mg/100 ml)*	
	Case 1	Case 2
Cholesterol	410–990	210–1,040
Triglycerides	180–1,440	210–1,840
Phospholipids	400–900	310–1,150
NEFA	20–65	20–55

* Range observed during the all survey.

The myeloma could have been suspected in 1961 when an electrophoretic tracing showed a tall and acute peak in the γ_1 globulin area, but it was only proved in 1964–65 on the following findings: paraproteinemia, plasmocytosis and bone impairment. The paraproteinemia amounted 38% of the total serum proteins which were high (9 to 12 g/100 ml). The paraprotein migrated as a γ_1 globulin during electrophoresis in paper, agar and agarose; as a Sa_2 in starch gel. On analysis by double diffusion or immuno-electrophoresis, with specific antiserums, it proved to be an IgA globulin. The IgG and IgM globulins were slightly decreased. The plasmacytosis amounted to 15% in the sternal bone marrow with grossly abnormal cells. There was a diffuse decalcification of the skeleton and a fracture of a dorsal vertebra was found on March 1965 when the patient complained about pains along the intercostal spaces. Related to the myeloma were also a RBSR of 126 mm in one hour; a moderate anemia (3,300,000 RBC pmm^3) with normal platelets and WBC; a slight prolongation of the thrombin and thromboplastin times; and two months before death, an hemorrhagic, cutaneous, renal, and retinal syndrome and some neurologic troubles which were refered to the great elevation of blood viscosity [19] and subsided by repeated plasmapheresis. Melphalan and corticoids did not change the course of the disease.

The coronary insufficiency was strongly marked by typical and severe angina pectoris on effort beginning on January 1965, and increasing until the death. Death occurred suddenly at home three days after an intractable pain lasting several hours followed by electrical evidence of a myocardial infarction. The autopsy could not be made.

Among the other facts which characterize this first case, one can notice that Mr. G. who received no blood transfusion before March 1965, experienced during his life many manifestations of allergy, specially: spasmodic coryza, giant urticaria, allergy to penicillin. In his family, no other case of xanthomatosis, hyperlipidemia, or myeloma was found, but the inquiry was very incomplete. His daughter, 23 years old, had also spasmodic coryza, her blood lipids and proteins were normal.

Case 2: Mr. M. Ladislas, born in 1925, was followed since 1963 till now for an arteriosclerosis obliterans and an hyperlipidemia.

Intermittent claudication of the two legs began in 1959 when he was 34 years old and never subsided. By 1965, the pain came on walking of about 50 yards. The pulsations were absent in both posterior tibial and popliteal arteries, they were decreased in the femoral arteries and an aortography showed diffuse impairment of the arterial wall and almost complete occlusion of the right femoral artery. The ECG suggested that the coronary arteries were also involved, showing signs of postero-lateral subepicardial ischemia.

The hyperlipidemia was first found in 1960 when a milky serum and a total cholesterol of 715 mg/100 ml were detected. Since then, 10 blood controls showed in all instances but one (Fig. 1) high values ranging from 300 to 1,040 mg/100 ml for the total serum cholesterol; 210 to 1,840 for the triglycerides and 310 to 1,150 for the phospholipids (Table I). Once, however, the cholesterol was found at 210 mg within the normal range but the triglycerides were still abnormally high at 270 mg. Both low fat diet and chlorophenoxyisobutyrate given separately seemed to induce a real decrease in the values of the lipidemia.

There are no cutaneous or tendon xanthomas, no corneal arcus. Mr. J. is apparently a strong man weighing 81 kg and measuring 1 m 61 with a blood pressure of 130/80 mm/Hg, no liver or spleen enlargment; RBC, WBC and platelets are normal; the total serum protein stays around 7.5 g % and the electrophoretic and immuno electrophoretic study does not show any abnormal component except for the great augmentation of the β lipoproteins. The RSR was once found slightly accelerated (15 mm in one hour). In 1947, Mr. J. was severely wounded in the face by a bullet and several skin homografts had to be made but, as far as he remembers, he received no blood transfusion and had no infection of his wound.

Typical Features

The hyperlipidemia is made of lipoproteins (Lp) which behave like abnormal β Lp (or low density Lp). Although, they are antigenically β Lp, they migrate badly on paper and in agar (Fig. 2)

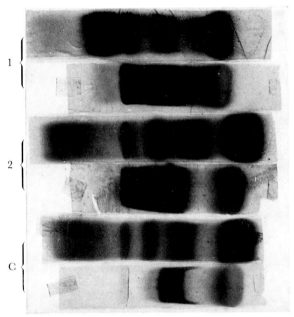

Fig. 2. Electrophoretic patterns in agar of the serum in case 1, case 2 and a control (C). Each upper strip is stained with Amido black and lower with Sudan black.

and do not migrate at all in starch gel. In the ultracentrifuge [12] most of them move as very low density Lp (D 1006–1019), but one finds also in case 1 that 1) the D 1019–1063 class is very poor; 2) a large part of the antigenically β Lp are heavier than normal β Lp and float only at D 1080; 3) there is no Lp in the D <1006 class even when the triglyceride content of the serum is very high or after a fat meal (Table II). In both cases, the oral vitamin A

Table II. Auto immune hyperlipidemia (Fasting cholesterol in the serum lipoproteins)

	Cholesterol in the lipoproteins (mg/100 ml of serum)		Mean values for 20 normal controls
	Case 1	Case 2	
Chylomicra 25,000 g ×1 h	0	12 (3.5)	0
D 1,006–1,019 lipoproteins	206 (34)	95 (27)	32 (17)
D 1,019–1,063 lipoproteins	3 (0.5)	146 (41)	100 (54)
D>1,063 lipoproteins	400 (65.5)	100 (28)	55 (29)

* % of total cholesterol.

tolerance test [5] elicits a great prolongation of the vitaminemia and analysis shows that the vitamin accumulates in those flottation classes which are increased, including the D 1063–1080 class found in case 1 (Table III).

Table III. Auto immune hyperlipidemia. Oral vitamin A tolerance test.

Whole serum vitamin A I.U./100 ml

		Prior to the vitamin A test		During the vitamin A test					
		Before taking any vit. A	After 100,000 I.U. vit. A for 2 days	Hours after taking 500,000 U. of vit. A					
				3	6	9	12	24	48
Case 1		210	710	1,090	2,270	1,960	1,720	1,070	680
Case 2		250	480	840	1,330	1,430	2,770	2,270	860
26 controls (upper limit)		–	260	2,820	3,240	1,950	1,230	390	–
Serum lipo- proteins prepared by U.C. in case 1	D<1006	–	0	0	–	–	0	0	0
	1006–1019	–	350	630	–	–	900	560	360
	1019–1063	–	70	70	–	–	40	30	70
	>1063	–	300	450	–	–	620	460	300

* * *

The presence of a circulating anti β Lp auto antibody was first suspected in case 1 when, in several ways, evidence was achieved of an intimate association between the IgA paraprotein and the Lp which were increased [1, 13] (Fig. 3A). The complexes are soluble *in vivo* but precipitate *in vitro* when the serum is mixed with certain mediums or after their preparation by flottation. After these first statements, it was also found [2, 3] in case 1 that the patient's serum behaves like an anti β Lp antiserum when used against human serum and purified human β Lp in double diffusion [18] immunoelectrophoretic [10] and passive hemagglutination tests [11] (PHA). The results were more positive when the complexed Lp had been removed from the serum before its use. In the best condition, reciprocal dilution titers ranged from 12,500 to 312,000 when a PHA +++ is taken as the end point (Table IV). At last, a method was developed [6] starting from isolated complexes, which allowed the preparation of a very pure and active γ globulin. In case 2 no precipitating or agglutinating activity was found in the

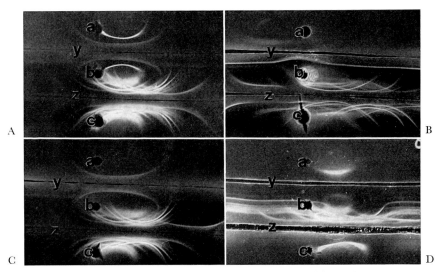

Fig. 3. Immunoelectrophoretic patterns in agar stained with amido black.
A: normal human serum (a, b, and c). Rabbit antiserum against washed low density Lp D 1006–1019 of case 1 (Y); horse anti serum against pooled normal human serum (Z).
B: Rabbit anti human β Lp anti serum (a); serum of case I (b); normal human serum (c); pure human β Lp (Y); horse antiserum against pooled normal human serum (Z).
C: Normal human serum (a, b, and c); rabbit antiserum against pure antibody of case 1 (Y); horse antiserum against pooled normal serum (Z).
D: Pure human β Lp (a and c). Normal human serum (b); serum of case 1 without β Lp (Y); horse antiserum against pooled normal human serum (Z).

Table IV. Auto immune hyperlipidemia. Passive hemagglutination power of serum extracts against benzidinised RBC sensitized with pure β Lp

	Higher dilution giving a +++ positive agglutination (reciprocal)	Amount of Lp giving complete inhibition of about 5 +++ units µg
Case 1		
Whole serum	625	not studied
Serum without its β Lp	12,500 to 312,500*	2.5**
Extract from the complexed Lp of 1 ml serum	72,500**	2.5**
Case 2		
Whole serum	0	not studied
Serum without its β Lp	16	not inhibited
Extract from the complexed Lp of 8 ml serum	625***	Complete inhibition with less than 5***

* The titer depends on the sample studied.
** Figures obtained in recent studies with improved methods.
*** Figures obtained in the first study [4]. Case 2 was not restudied with the improved methods.

serum but starting from the isolated low density lipoproteins and applying the methods already used in case 1, a very active γ globulin could also be prepared which gave typical results in classical precipitation and PHA tests when purified β Lp were used as antigen [4].

All the positive reactions found in case 1 and 2 could be easily inhibited by pure β lipoproteins (see add.) (Table IV).

*   *   *

Additional information could be gathered in case 1 because a plasmapheresis yielded enough plasma to allow further basic work [6].

The antibody is an IgA globulin like the paraprotein. Whole serum analysis shows that the precipitating and agglutinating power is in the γ 1 globulin area after electrophoresis in agar (Fig. 3 A, B), and comes out between 0,1 and 0,2 M during DEAE cellulose chromatography using phosphate buffer pH 6.5 with a molarity gradient from 0.0175 to 0.5 M. Analysis of the extracts prepared from the isolated complexes (washed many times before the extraction procedure) show that they contain only one very active protein. It is a γ_1 globulin on electrophoretic analysis which reacts with anti IgA but not with specific anti IgG and IgM antiserum. The serum of rabbits immunised with the extracts when reacted with pooled whole human serums, develops only the precipitation line of the IgA globulin (Fig. 3 C). More than 90% of this pure protein is precipitable by β Lp and about 15×10^{-9} of it gives still +++ PHA when mixed in a 0.5 ml final volume with 250×10^6 RBC coated with pure β Lp.

The antigen is the β Lp (see add.). Whole serum and active extracts give only one precipitation line (Fig. 3 D). In inhibition experiments, the purer the β Lp used, and the smaller is the quantity necessary for complete inhibitions. In the best conditions available now this end point was achieved with 2.5 μg of β Lp for about 5 agglutinating units (PHA +++). The different classes of low density lipoproteins gave positive inhibition results. Chylomicra were also inhibiting but much larger quantities were necessary, probably because of the numerous other proteins which are associated with chylomicra in the circulating blood. No inhibition was seen with human serum from which β and α Lp had been removed or with many other proteins, including albumin, and γ globulins. On the other hand, all the β Lp studied so far react with the antibody; including the patient's Lp, those of 35 men, and those prepared from rabbit, dog, rat, and guinea pig serum.

PHA tests demonstrated that the antibody gave positive +++ results over a wide T° range and is still very active at 37°C. Pro zones were seen when using the whole serum but not with pure antibody. This one and other results suggest that two different forms of the antibody, 'complete and incomplete' could co-exist in the whole serum [6].

* * *

In case 2, the quantity of antibody recovered by extracting the isolated low density lipoproteins, although it was sufficient to get typical precipitation and agglutinating patterns (at significant dilution levels) (Table IV), was not sufficient for more detailed study. However, it must be pointed out that the active γ globulin extracted in case II reacted strongly with the rabbit antiserum prepared with the pure antibody coming from case 1, suggesting that it could be also an IgA globulin.

Comments

In case 1 and 2 there is an hyperlipidemia, clinical evidence of ischaemic disease and in the serum an immunoglobulin (Ig) which behaves like a circulating anti β lipoprotein auto-antibody. Although nosology is quickly moving in this field it is likely that these Ig are antibodies and indeed auto antibodies. Their specificity for β Lp was demonstrated after high purification and it is obvious that they react with self β Lp. The lack of individual and species specificity can hardly be held against this assumption since the β Lp of vertebrates are known to have common antigenic sites [20]. On that account, one can only suppose that the site (or sites) involved here are different from the group specific sites found on human β Lp after iso immunisation (see add.) [7]. The Ig nature of the precipitin found in the present cases and the precipitin activity of the highly purified IgA globulin extracted from the complexed β Lp in case 1 makes it easy also to distinguish them from the precipitins found in the serum of patients with deficiency of β Lp [17], and also from other precipitins found in tissue extracts [9] because the specificity, namely the Ig nature, of the latter remains open to question.

The frequency of this sort of hyperlipidemia has to be methodically evaluated with appropriate tools. Although hyperlipidemia associated with myeloma and xanthomatosis is very rare, a similar hyperlipidemia to the one seen in case 2 is much more common. Thus it is remarkable that anti β lipoprotein antibodies have not been found already in similar cases. However, signs of an association between the paraproteins and the β Lp were detected in other cases of myeloma with xanthomatosis [14, 15] and in a recent case in which the authors did not find any obvious sign of association they noticed 1) that the β Lp did not enter in starch gel during electrophoresis, and 2) that the patient's serum contained a 'factor' which was able to prevent the β Lp of normal serum from entering into the gel [8]. It is obvious that the presence of an anti β Lp auto antibody would easily explain these facts. One reason for which, if present, it could have been missed, is perhaps that in most cases, there is probably no antibody excess directly detectable in the serum like in case 1. All the antibody is then combined with the circulating anti β Lp like in case 2 and not detectable unless it is extracted from the circulating complexes. Thus it is not impossible that such 'hidden' antibodies are present in many cases of hyperlipidemia.

The part played by the antibody in producing the hyperlipidemia has yet to be explained. The vitamin A tolerance test shows that the clearing of the soluble antibody-β Lp complexes is very slow and this very likely accounts for the accumulation in the circulating blood of cholesterol, triglycerides and other exogenous and perhaps endogenous fats which are carried by β Lp. This impaired clearing of the Lp could depend on its coating by the antibody. However, this hypothesis has to be checked before rejecting the unlikely possibility that the antibody is a result of the hyperlipidemia rather than a cause.

The role of this sort of hyperlipidemia in the pathogenesis of atherosclerosis is well supported by clinical evidence of ischaemic diseases in case 1 and specially in case 2 in which arteriosclerosis obliterans began when the patient was only 34 years old. However, in the lack of statistical data, it can only be remembered that the clearing of post prandial lipemia was found impaired in many patients who have ischaemic diseases related to atherosclerosis [5, 16]. As a cause of delayed clearing it is thus possible for anti β Lp auto antibody to play an important role in the pathogenesis of atherosclerosis at an epidemiological level. Also open to question is the way in which the β Lp antibody complexes can induce the atheroma in the arterial wall.

The origin of the anti β Lp auto-antibody rises an other very important question. The association with a myeloma in case 1 suggests that the antibody is produced by an abnormal clone of plasma cells. If true, its anti β Lp specificity could be only 'fortuitous' and not 'auto-immune'. But there was no detectable sign of myeloma in the anamnesis of case 1 and a severe wound followed by skin homografts in case 2. Because of these reasons, the name 'auto-immune hyperlipidemia' has been proposed [4] for the new pathological entity found in these two patients.

Summary

An anti β lipoprotein auto antibody and an hyperlipidemia were found together in two men. The main data of these two cases lead to the concept of 'auto-immune hyperlipidemia' a new pathological entity of perhaps great significance for the pathogenesis of atherosclerosis.

This entity is characterised by 1) a rather variable hyperlipidemia which can reach very high levels. 2) A circulating antibody which produces soluble complexes with the circulating β Lipoprotein. Most of those complexes move in the ultracentrifuge as very low density lipoproteins: D <1006 and D 1006–1019. However some of them are heavier than normal β lipoprotein and float between D 1063 and 1080. 3) A delayed clearing of the lipoproteins complexed with the antibody which was found with the oral vitamin A tolerance test. 4) Clinical evidence for atherosclerotic induced ischaemic disease.

References

1. BEAUMONT, J. L.; JACOTOT, B.; BEAUMONT, V.; WARNET, J. et VILAIN, C.: Myélome, hyperlipidémie et xanthomatose. Nouv. Rev. franç. Hémat. 5: 507–517(1965).
2. BEAUMONT, J. L.; JACOTOT, B.; VILAIN, C. et BEAUMONT, V.: Myélome, hyperlipidémie et xanthomatose. III. Un syndrome dû à la présence d'un auto-anticorps anti β lipoprotéine. Nouv. Rev. franç. Hémat. 5: 782–792 (1965).

3. BEAUMONT, J.L.; JACOTOT, B.; VILAIN, C. et Mme V. BEAUMONT: Présence d'un auto-anticorps anti β lipoprotéine dans un sérum de myélome. C.R. Acad. Sci. *260:* 5960–5962 (1965).
4. BEAUMONT, J.L.: L'hyperlipidémie par auto-anticorps anti β lipoprotéine. Une nouvelle entité pathologique. C.R. Acad. Sci. *261:* 4563–4566 (1965).
5. BEAUMONT, J.L.; LELLOUCH, J. et SWYNGHEDAUW, B.: Hyperlipémie post-prandiale et hyperlipidémie dans les maladies par athérosclérose. J. Atherioscl. Res. *5:* 125–134 (1965).
6. BEAUMONT, J.L. and LORENZELLI, L.: L'auto anticorps antilipoprotéine (anti-Pg) du γ A myélome avec hyperlipidémie: Methode d'isolement et de purification à partir des complexes circulants. Ann. Biol. clin. *25:* 655–675 (1967).
7. BLUMBERG, B.S.: Iso antibodies in humans against inherited serum low density lipoproteins. The Ag system. Ann. N.Y. Acad. Sci. *103:* 1052–1057 (1963).
8. COHEN, L.; BLAISDELL, R.K.; DJORDJEVICH, J.; ORMISTE, V. and DOBRIOLOVIC, L.: Familial xanthomatosis and hyperlipidemia and myelomatosis. Amer. J. Med. *40:* 299–317 (1966).
9. DUFOUR, D. et TREMBLAY, A.: Réaction de précipitation en gélose, en double diffusion et par immunoélectrophorèse entre une lipoprotéine sérique et un constituant spécifique de certains tissus autologues chez le rat normal. Rev. cand. Biol. *23:* 501–503 (1964).
10. GRABAR, P. et WILLIAMS, C.A., Jr.: Méthode immunoélectrophorétique d'analyse de mélanges de substances antigéniques. Biochim. biophys. Acta *17:* 67–74 (1955).
11. HALPERN, B.N.; JACOB, M.; BINAGHI, R. et PARLEBAS, J.: Mise en évidence et dosage des anticorps allergiques par l'hémagglutination *in vitro* dans les syndromes allergiques humains et expérimentaux. Rev. franç. Allerg. *4:* 201–214 (1961).
12. HAVEL, R.J.; EDER, M.A. and BRAGDON, J.H.: The distribution and chemical composition of ultracentrifugally separated lipoproteins in human serum. J. clin. Invest. *34:* 1345–1353 (1955).
13. JACOTOT, B.; NGUYEN TRONG, T. et BEAUMONT, J.L.: Myélome, hyperlipidémie et xanthomatose. II Recherches complémentaires sur l'association entre la paraprotéine et les lipoprotéines légères. Nouv. Rev. franç. Hémat. *5:* 777–779 (1965).
14. KAYDEN, M.J.; FRANKLIN, E.C. and ROSENBERG, B.: Interaction of myeloma gamma globulin with human beta lipoprotein. Circulation *26:* 659 (1962).
15. LEWIS, L.A. and PAGE, I.M.: An unusual serum lipoprotein globulin complex in a patient with hyperlipemia. Amer. J. Med. *38:* 286–297 (1965).
16. MORETON, J.R.: Chylomicronemia, fat tolerance and atherosclerosis. J. lab. clin. Med., *35:* 373–384 (1950).
17. MURRAY, R.F. and BLUMBERG, B.S.: Precipitin activity in the sera of patients with deficiency of β lipoproteins (acanthocytosis). Nature *208:* 357–359 (1965).
18. OUCHTERLONY, O.: Diffusion-in-gel methods for immunological analysis. Prog. in Allergy *5:* 1–38 (1958).
19. SMITH, E.; KOCWA, S. and WASSERMAN, L.R.: Aggregation of IgG globulin *in vivo.* I. The hyperviscosity syndrome in multiple myeloma. Amer. J. Med. *39:* 35–48 (1965).
20. WALTON, K.W. and DARKE, S.J.: Immunological characteristics of human low density lipoproteins and their relation to lipoproteins of other species. – In Protides of the biological fluids, 146–148 (Elsevier, Amsterdam 1963).

Note added in proof

Since this paper was submitted it was found in case 1 that the anti lipoprotein antibody also reacts with the pure *a* Lp of serum. The corresponding antigenic site, hitherto unknown, is carried by *a* Lp as well as β Lp and the name of Pg antigen was proposed for it (J.L. BEAUMONT: C.R. Acad. Sciences, 264, 1966).

Acknowledgements

We are indebted to Violette Beaumont, M.D.; B. Jacotot, M.D.; C. Vilain, M.D. and Marie Françoise Poullin, Ph.D. for their great contribution, and we wish to acknowledge gratefully the technical assistance of Liliane Lorenzelli, Colette Charmeau, Annie Stora and Mrs. Ferry.

Author's address: J. L. Beaumont, M. D., Groupe de Recherches sur l'Athérosclérose (INSERM), Hôpital Boucicaut, *Paris* (France).

From the Simon Stevin Institute for Scientific Research, Brugge

Differential Fatty Acid Composition of Alpha and Beta Lipoproteins in Baboons

V. Blaton, H. Peeters, Brugge, and G.A. Gresham, A.N. Howard, Cambridge

The use of baboons in the field of atherosclerosis would seem to offer certain advantages as they are in many respects similar to man, and develop the disease in the wild state [1].

Although early attempts to produce atherosclerosis in primates by feeding cholesterol were unsuccessful [2, 3] more recently fibrous lesions were claimed in monkeys deficient in pyridoxine [4]. By feeding a diet containing only moderate amounts of cholesterol it has been possible to induce aortic atherosclerosis in young baboons [5]. This diet contains 15% egg yolk and 15% butter and has a cholesterol content of 0.5% which is about two and a half times the average human intake. This type of diet was used in this study and fed during a period of 4 years. Young control animals were fed plain monkey pellets.

The serum lipids of baboons, submitted to the experimental diet for two years, were separated by thin layer chromatography and their fatty acid composition determined by gas liquid chromatography [6].

In this work the pooled serum of four baboons under diet during four years was examined according to the procedure described for human serum [7] and compared with the pooled serum of two control animals.

Considering the alpha 1 and beta-lipoproteins of control animals and of animals under diet, we observed that the lipoproteins are less saturated as well in the control animals as under the atherogenic diet. Diet however increases the unsaturation both in alpha 1 and in beta lipoproteins with a predominant increase of the 18:1 and 18:2 content (Table I).

Table I. The fatty acid ratios in electrochromatographic alpha 1 and beta-lipoproteins of baboon serum

Baboon	a_1 Control	Ath. diet	β Control	Ath. diet
$\dfrac{\Delta:0}{\Delta:1\ \Delta:2}$	0.90	0.75	0.63	0.49
$\dfrac{\Delta:0}{\Delta:1}$	1.93	1.52	1.67	1.05
$\dfrac{\Delta:0}{\Delta:2}$	1.69	1.49	1.01	0.91
$\dfrac{18:0}{18:2}$	0.68	0.60	0.36	0.38
$\dfrac{18:1}{18:2}$	0.77	0.90	0.55	0.78
$\dfrac{18:2}{20:4}$	2.34	1.36	3.86	2.06

The alpha 1 and beta triglyceride, cholesterol ester and phospholipid ratios have also been determined. For the alpha fraction (Table II) the unsaturation of the cholesterolesters is pronounced in the controls but decreases with diet. For the phospholipids there is more unsaturation with diet. The triglycerides become slightly more

Table II. The fatty acid ratios of the lipid classes extracted from alpha 1 lipoprotein

Baboon	a_1 TG Control	Ath. diet	a_1 CE Control	Ath. diet	a_1 PL Control	Ath. diet
$\dfrac{\Delta:0}{\Delta:1\ \Delta:2}$	0.62	0.74	0.12	0.18	1.86	1.26
$\dfrac{\Delta:0}{\Delta:1}$	0.92	1.04	0.27	0.35	4.91	2.50
$\dfrac{\Delta:0}{\Delta:2}$	1.89	2.57	0.20	0.38	3.01	2.55
$\dfrac{18:0}{18:2}$	0.30	0.39	0.03	0.04	1.55	1.24
$\dfrac{18:1}{18:2}$	1.84	2.20	0.65	0.98	0.53	0.91
$\dfrac{18:2}{20:4}$	5.8	1.7	10.8	7.1	1.3	0.8

unsaturated. In Table III the values are presented for the beta fraction. The cholesterolesters have the same ratio as in alpha 1 but this apparent similarity is destroyed when we look at the internal ratios of the several fatty acids. The triglycerides are most modified and the phospholipids which are less saturated in the control than in alpha 1 become still less saturated during the diet. From these two tables we can extract the information for our subsequent discussion.

Table III. The fatty acid ratios of the lipid classes extracted from beta-lipoprotein

Baboon	β TG Control	Ath. diet	β CE Control	Ath. diet	β PL Control	Ath. diet
$\dfrac{\Delta:0}{\Delta:1\ \Delta:2}$	0.60	0.60	0.13	0.18	1.52	1.41
$\dfrac{\Delta:0}{\Delta:1}$	0.90	0.81	0.42	0.44	3.36	2.86
$\dfrac{\Delta:0}{\Delta:2}$	1.79	2.32	0.18	0.30	2.77	2.79
$\dfrac{18:0}{18:2}$	0.19	0.34	0.02	0.06	1.34	1.32
$\dfrac{18:1}{18:2}$	1.74	2.47	0.38	0.60	0.70	0.87
$\dfrac{18:2}{20:4}$	8.7	3.7	5.3	3.2	1.5	1.0

The increase of unsaturation on the atherogenic diet is pronounced for the total alpha fraction as well for oleic as linoleic as arachidonic acid. Going into detail triglycerides are more saturated but their arachidonic content is increased. Cholesterol esters have a higher oleic and arachidonic ratio but a decrease of linoleic. Phospholipids however show better oleic and arachidonic ratios.

For the total beta fraction the increase of saturation is pronounced for linoleic and arachidonic ratios but linoleic is loosing ground. The triglycerides follow the same general pattern as in alpha but with a more pronounced decrease of 18:1 and 18:2. Again the arachidonic acid ratio is favoured. For cholesterolesters the large difference in 18:2 is due to the fact that very little 18:0 is present with a large amount of 18:2. The linoleic to arachidonic ratio is again in favour of arachidonic. The beta phospholipids follow the trend of the alpha phospholipids. We may summarize that diet has increased the unsaturation, especially the 18:1 and 20:4 content of triglycerides and phospholipids.

Acknowledgements

This work was supported by Grant No. 625 of Nationaal Fonds voor Geneeskundig Wetenschappelijk Onderzoek. We wish to thank B. DECLERCQ, D. VANDAMME and M. VANSLAMBROUCK for their technical assistance.

References

1. MAC GILL, H.C.; STRONG, J.P.; HOFMAN, R.L. and WERTHESSEN, J.: Arterial lesions on the Kenya baboon. Circulat. Res. *8:* 670–679 (1960).
2. KAWAMURA, R.: Neue Beiträge zur Morphologie und Physiologie der Cholesterinsteatose, p. 267 (G. Fisher, Jena 1927).
3. HUEPER, W.C.: Experimental studies in cardiovascular pathology. XIV. Experimental atheromathosis in Macacus rhesus monkeys. Amer. J. Path. *22:* 1287–1289 (1946).
4. RHINEHART, J.F. and GREENBERG, L.P.: Atherosclerotic lesions in pyridoxine-deficient monkeys. Amer. J. Path. *25:* pp. 481–485 (1949).
5. HOWARD, A.N.; GRESHAM, G.A.; BOWYER, D.E. (Cambridge) and LINDGREN, F.T. (Berkeley, Cal.): Aortic and coronary atherosclerosis in baboons. Progr. biochem. Pharmacol. *4:* 436–442 (Karger, Basel/New York 1968).
6. BOWYER, D.E.; LEAT, W.M.F.; HOWARD, A.N. and GRESHAM, G.A.: The determination of the fatty acid composition of serum lipids separated by thin layer chromatography and a comparison with column chromatography. Biochim. Biophys. Acta *70:* 423–431 (1963).
7. BLATON, V. and PEETERS, HUB.: Differential fatty acid composition of α_1 and β lipoproteins, in HUB. PEETERS Protides of the Biological Fluids. Proceedings of the 13th Colloquium, Bruges, 1965, pp. 315–319 (Elsevier, Amsterdam 1966).

Authors' address: V. Blaton, H. Peeters, M. D., The Simon Stevin Institute for Scientific Research, *Brugge* (Belgium); G. A. Gresham, and A. N. Howard, M. D., Dept. of Pathology, University of Cambridge, *Cambridge* (England).

From the King Gustaf Vth Research Institute (Head: Prof. G. BIRKE)

Recirculation of H3-Labelled Palmitate in Plasma Lipoproteins

J. BOBERG and E. R. NYE*, Stockholm

Since 1958 [1] it has been known that part of labelled palmitate given intravenously recirculates in the blood plasma in a fatty acid esterified form, mainly as triglyceride fatty acids (TGFA). On the basis of this information many workers have become interested in the study of the isotopic flux in total TGFA and/or TGFA in isolated plasma lipoproteins (LP) after intravenous injection of labelled palmitate in different animals such as the rabbit [2], dog [3, 4], rat [5–7] and man [8–14]. This experimental model has been used to discover in which organ the synthesis of labelled blood plasma triglycerides (TG) occurs and it seems unescapable that this process chiefly occurs in the liver. The model has also been used to estimate the turnover of plasma or perhaps more correctly of plasma-liver TG. When we started experiments of this kind we were faced with the problem of which animal to choose and accordingly we compared the way in which some species behaved in this respect.

Some preliminary results we obtained on recirculation studies of tritium labelled, albumin bound palmitate, given intravenously to horses, rabbits, rats and dogs, are here reported.

Briefly, the general procedure was as follows. All animals were freely fed on their ordinary diet before the experiment. The animals were not anesthetized during the experiment since we had earlier observed that the label incorporated into plasma TGFA in anesthetized rats had a considerably higher peak value and that the incorporation was delayed compared with non-anesthetized rats [15].

Supported by grants from Reservationsfonden, Karolinska Institute, Stockholm, and U.S. Public Health Service (H 7088).

* Visiting research fellow from the Department of Internal Medicine, Dunedin, New Zealand.

Blood samples were withdrawn from indwelling vein catheters in the horses and the dogs and arterial catheters in the rabbits and the rats. In the rats the catheters were inserted into the aorta about a week before the experiment according to the method described by WEEKS et al. [16].

Separation of the LP was carried out in the preparative ultracentrifuge at density 1.006 for 16 h. Radioactivity in the different lipid fractions was determined as described earlier [17]. Further details of the procedure will be described elsewhere.

Results

Almost no radioactivity was incorporated into the cholesterolesters of the blood in the animals studied, which is in agreement with earlier findings.

Fig. 1 shows recirculated radioactivity in the phospholipid fatty acids (PLFA). An appreciable amount of label was found in this fraction but no consistent difference between the animals could be stated.

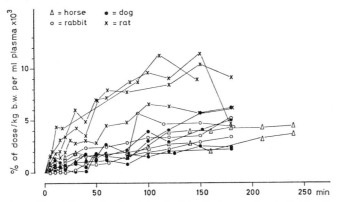

Fig. 1. The radioactivity recirculated in the blood plasma phospholipids after intravenous injection of albumin bound H3-labelled palmitate into two horses; three rabbits, four dogs and four rats.

Radioactivity incorporated into the total TGFA is shown in Fig. 2. As has been found out earlier most label was found within this fraction. In the horses the label started to appear in plasma TGFA 20 min after the injection of labelled palmitate and maximum radioactivity was reached after 100–120 min. The peak value was higher than for the other species studied. TGFA radioactivity in the rabbits entered the plasma pool 10 min after the injection and peak

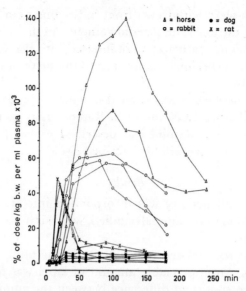

Fig. 2. The radioactivity recirculated in the blood plasma triglycerides after intravenous injection of albumin bound H3-labelled palmitate into two horses, three rabbits, four dogs and three rats.

activity was reached after 50–100 min. The height of the peak came up to about half of the value for the horse. In the rat the lapse of the peak of the curves was more rapid than for any other animal studied. After 10 min label was found in TGFA and the curve had reached its maximum at about 20–30 min. The height of the peaks in two rats was almost the same as for the rabbit. The incorporation of label into plasma TGFA in the dogs differed consistently from the other species. The amount of recirculated radioactivity was very low. The label entered the plasma TGFA about 10 min after the injection of palmitate and after that there was a continuously small increase of label during the whole experiment.

Fig. 3 shows how the TGFA label was distributed between very low density lipoproteins (VLD) and LP of higher density. It has been reported before that the head part of the label found in TGFA is located in the VLD fraction in man, rabbit, and rat. This seems to be true also for the horse but not for the dog, where a considerable amount of the label found in total TGFA was within the LP of higher density than VLD. In this connection it could be mentioned that it has been described [2] that transfer of labelled TGFA from VLD to LP of higher density occurs *in vitro* in the rabbit. These findings we

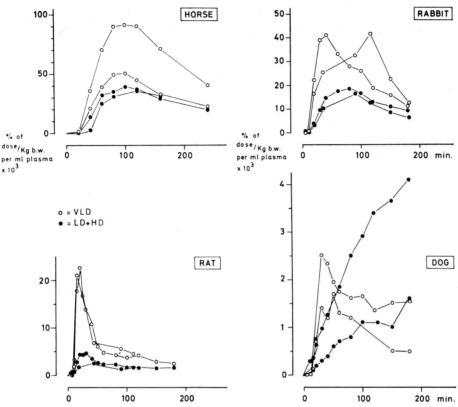

Fig. 3. The radioactivity recirculated in blood plasma triglycerides separated into VLD and LD plus HD. From each species two animals of those presented in Fig. 2 are given.

have been able to confirm but this *in vitro* transfer does not take place in the rat or the dog.

Table I shows the chemical concentration of TG in the different LP fractions of the four species. Even in this respect the dog was

Table I. Triglyceride (TG) concentration of very low density lipoprotein (VLD), low density lipoprotein (LD) and high density lipoprotein (HD) in horse, rabbit, rat and dog.

Animal	moles TG per litre blood plasma		
	VLD	LD	HD
Horse	0.16±0.02	0.08±0.03	0.06±0.08
Rabbit	0.11±0.09	0.08±0.07	0.11±0.05
Rat	0.74±0.17	0.12±0.02	0.01±0.02
Dog	0.09±0.06	0.24±0.05	0.07±0.02

Mean and standard error of the mean for four horses, eight rabbits, eight rats and eight dogs are given.

different from the horse, rabbit and rat. The dog is the only species studied where LD contains a higher TG concentration than VLD.

Comments

These data demonstrate that a species-difference exists in the lapse of the curves of recirculated radioactivity in plasma TGFA after the injection of labelled palmitate. It is a pity we have not yet obtained these data concerning man, but data from the literature show that the recirculated radioactivity reaches its maximum between 2 and 3 h after injection of the labelled palmitate. Furthermore man seems to belong to the group where most of the label is found in the VLD.

To explain these species-differences much more work on this problem is necessary. It is, however, very easy to find factors that might be of importance. The method used in this investigation for separation of LP is that for human blood plasma and there is information in the literature which argues that the LP pattern of other species is quite different from man.

Incorporation of labelled palmitate into TGFA in the blood can change with different diets [12]. Since the animals studied have indeed different diet habits this may very well be one important factor explaining the observed differences.

The general metabolic rate certainly varies in these animals and this factor probably explains the differences in the 'peaking time' of the curves.

References

1. FREDRICKSON, D.S. and GORDON, Jr., R.S.: The metabolism of albumin bound C14 labelled esterified fatty acids in normal human subjects. J. clin. Invest. 37: 1504–1515 (1958).
2. HAVEL, R.J.; FELTS, J.M. and VAN DUYNE, C.M.: Formation and fate of endogenous triglycerides in blood plasma of rabbits. J. lipid. Res. 3: 297–308 (1962).
3. HAVEL, R.J. and GOLDFIEN, A.: The role of the liver and of extrahepatic tissues in the transport and metabolism of fatty acids and triglycerides in the dog. J. lip. Res. 2: 389–395 (1961).
4. GROSS, R.C.; REAVEN, G.M.; EIGENBRODT, E.H. and FARQUHAR, J.W.: Turnover of endogenous plasma and liver triglyceride in man and in the dog. Progr. biochem. Pharmacol. vol. 4 (Karger, Basel/New York 1967).
5. LAURELL, S.: Recycling of intravenously injected palmitic acid-1-C14 as esterified fatty acid in plasma of rats and turnover rate of plasma triglycerides. Acta physiol. scand. 47: 218–232 (1959).

6. Borgström, B. and Olivecrona, T.: The metabolism of palmitic acid-1-C14 in-functionally hepatectomized rats. J. lip. Res. *2:* 263–267 (1961).
7. Baker, K. and Schotz, M. C.: Use of multicompartmental models to measure rates of triglyceride metabolism in rats. J. lip. Res. *5:* 188–197 (1964).
8. Carlson, L.A.: Studies on the incorporation of injected palmitic acid-1-C14 into liver and plasma lipids in man. Acta soc. med. Uppsala *65:* 85–90 (1960).
9. Friedberg, S.J.; Klein, R.F.; Trout, D.L.; Bogdonoff, M.D. and Estes, Jr., E.H.: The incorporation of plasma free fatty acids into plasma triglycerides in man. J. clin. Invest. *40:* 1846–1855 (1961).
10. Havel, R.J.: Conversion of plasma free fatty acids into triglycerides of plasma lipoprotein fractions in man. Metabolism. *10:* 1031–1034 (1961).
11. Nestel, P.J.: Metabolism of linoleate and plamitate in patients with hypertriglyceridemia and heart disease. Metabolism. *14:* 1–9 (1965).
12. Nestel, P.J. and Hirsch, E.Z.: Triglyceride turnover after diets rich in carbohydrate or animal fat. Austr. Ann. Med. *14:* 265–269 (1965).
13. Farquhar, J.W.; Gross, R.C.; Wagner, R.M. and Reaven, G.M.: Validation of an incompletely coupled two-compartment nonrecycling catenary model for turnover of liver and plasma triglyceride in man. J. lip. Res. *6:* 119–134 (1965).
14. Reaven, G.M.; Hill, D.B.; Gross, R.C. and Farquhar, J.W.: Kinetics of triglyceride turnover of very low density lipoproteins of human plasma. J. clin. Invest. *44:* 1826–1833 (1965).
15. Boberg, J. and Carlson, L.A.: Some physiological and pathological factors affecting the plasma lipoproteins. Coll. Protides of the biological fluids, Bruge 1965, p. 303–314 (Elsevier, Amsterdam 1966).
16. Weeks, J.R. and Jones, J.A.: Routine direct measurement of arterial pressure in unanesthetized rats. Proc. Soc. exp. Biol. Med. *104:* 646–648 (1960).
17. Boberg, J.: Separation of labelled plasma and tissue lipids by thin-layer chromatography. A quantitative methodological study. Clin. chim. Acta. *14:* 325–334 (1966).

Authors address: J. Boberg, King Gustaf Vth Research Institute, *Stockholm 60* (Sweden).

Department of Biochemistry, University of Edinburgh Medical School,
Teviot Place, Edinburgh

Studies on Plasma Lipoproteins in the Rat

G.S. Boyd, Funmilayo D. Onajobi and
I.W. Percy-Robb, Edinburgh

Introduction

The circulating plasma cholesterol comprises a heterogeneous metabolic pool of sterol associated with various distinct plasma lipoproteins. This sterol is derived from exogenous (dietary) sources and also from endogenous synthesis (Morris, Chaikoff, Felts, Abraham and Fansah, 1957). The plasma sterol is in a dynamic equilibrium with the sterol in many tissues, particularly the liver, kidney, muscle, etc. (Avigan, Steinberg and Berman, 1962). Thus the concentration of cholesterol in plasma is dictated by the difference between the nett contribution of sterol to this pool and the sum of the sterol withdrawn from the pool. When the organism is in sterol balance the sum of cholesterol absorbed plus cholesterol synthesized will equal the sum of sterols excreted plus cholesterol catabolized to 'non-sterol' products such as bile acids, etc.

When cholesterol is fed to certain mammals there is an increase in hepatic cholesterol concentration and a sharp decrease in the rate of endogenous cholesterol synthesis which is presumably due to feed-back inhibition (Gould, Taylor, Hagerman, Warner and Campbell, 1953; Tomkins, Sheppard and Chaikoff, 1953). If an organism is in positive sterol balance this may or may not be reflected by an increase in the concentration of cholesterol in the circulating plasma. Thus there are factors which influence the rate of cholesterol synthesis and factors which affect the rate of discharge of cholesterol into plasma from tissues such as the liver. From a quantitative standpoint the breakdown of cholesterol to bile acids is of great significance and in this case the end product of sterol catabolism, namely the bile salts, themselves influence cholesterol catabolism, presumably by a feed-back mechanism (BERGSTROM and DANIELSSON, 1958).

The experiments contained in this paper were undertaken to investigate in the rat *in vivo* and in the isolated perfused liver some aspects of the feed-back control exerted by cholesterol and bile salts on the biological survival time of the circulating cholesterol. Consideration was also given to the effect of these control processes on the secretion of plasma lipoproteins under *in vivo* and *in vitro* situations.

Methods

The animals used in these experiments were adult male Wistar strain rats bred in the departmental colony and weighing from 190–250 g. Mevalonic acid-2-^{14}C was purchased from the Radiochemical Centre, Amersham.

The 'stock diet' consisted of 70% wholemeal flour, 25% skimmed milk powder and 5% brewer's yeast. When cholesterol was added to the diet this was dissolved in olive oil so that the resultant stock diet contained 10% by weight of olive oil and 0.5% cholesterol. The appropriate control diet in these experiments was the stock diet supplemented with 10% olive oil. When cholic acid was added to the diet the concentration used in these experiments was 1% by weight in the stock diet. In the experiments in which the bile acid sequestering agent, cholestyramine, was used, this substance was added to the stock diet at a concentration of 3% by weight.

When rats were subjected to biliary drainage, the bile duct cannulation was performed under clean but not sterile conditions. The bile duct was exposed, cannulated with a nylon tube of external diameter 0.025 in, the wound closed in the usual manner, and the animal placed in a restraining cage. Under these conditions the animals were maintained in a healthy and comfortable state for periods up to 48 h. During this period, the animals were given access to food and the drinking water was substituted by 0.95% aqueous sodium chloride solution with added potassium chloride to a final potassium concentration of 5 mEq/l.

The studies involving isolated rat liver perfusions were performed by a method similar to the procedure of MILLER, BLY, WATSON and BALE (1951), but modified for our purpose as described by PERCY-ROBB and BOYD (1966).

Blood samples were obtained from the tail, the blood was allowed to clot and the plasma separated. Cholesterol was determined on plasma samples by the method of ABELL, LEVY, BRODIE and KENDALL (1952). Radioactive assays on cholesterol were performed by liquid scintillation spectrometry employing a Packard liquid scintillation spectrometer type 314EX.

Plasma lipoproteins were separated by zone electrophoresis on Whatman 3 mm filter paper using a veronal buffer, 0.05 Molar, pH 8.6, more or less according to the method of BOYD (1954). The filter paper electrophoretograms were dried and scanned by means of a gas flow proportional counter to identify and quantitate the distribution of radioactive cholesterol throughout the plasma lipoprotein spectrum (ONAJOBI and BOYD, to be published).

Experiments

A. Cholesterol Supplement in the Diet

Male rats were fed the stock diet containing 10% olive oil and similar animals were fed the stock diet containing 10% olive oil and 0.5% cholesterol. This regimen was continued for 28 days and produced a marked increase in the plasma and liver cholesterol concentrations. Selected animals were then injected intraperitoneally with 5 μC mevalonic acid-2-^{14}C. At fixed intervals thereafter the animals were bled from the tail and the specific activity of the plasma total cholesterol and the distribution of radioactivity amongst the plasma lipoproteins were determined. This procedure was repeated at various times after this 'pulse labelling' and the results of these studies are shown in Fig. 1, while the changes in the plasma lipoprotein spectrum are shown in Figs. 2 and 3.

Other control animals and cholesterol fed animals were taken and their livers subjected to the perfusion technique as described previously. When the blood flow and bile flow were observed to be normal, mevalonic acid-2-^{14}C was added to the perfusate. Blood samples were removed at fixed intervals thereafter and the specific activity of the plasma cholesterol determined as described previously. The results are shown in Fig. 4.

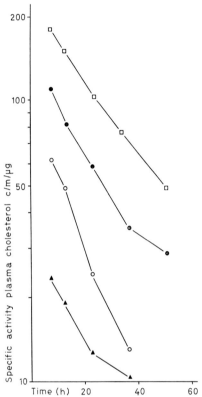

Fig. 1. Specific activity of plasma total cholesterol. Control (●) t ½ 19 hours. Cholate fed (□) t ½ 24 hours. Cholestyramine fed (○) t ½ 13 hours. Cholesterol fed (▲) t ½ 15 hours.

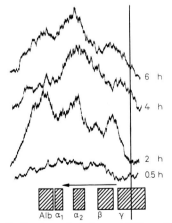

Fig. 2. Time course of distribution of cholesterol-^{14}C on lipoproteins (separated by electrophoresis) in normal rat plasma following intraperitoneal injection of mevalonate-2-^{14}C at zero time.

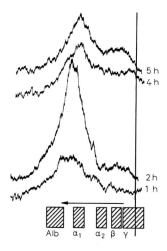

Fig. 3. Time course of distribution of ^{14}C-cholesterol on lipoproteins (separated by electrophoresis) in plasma taken from rats (fed with 0.5% cholesterol in the diet) following intraperitoneal injection of mevalonate-2-^{14}C at zero time.

B. Cholestyramine Supplement in the Diet

Rats were fed the stock diet containing 3% cholestyramine while litter mates were fed the stock diet. This regimen was continued for 10 days and then selected animals were injected with mevalonate-2-^{14}C as before. The plasma cholesterol was studied as described previously and the results are shown in Fig. 1.

C. Rats Subjected to Biliary Drainage

Rats maintained on the stock diet were subjected to bile duct cannulation and biliary drainage allowed to continue for 44 h. The livers were removed from the animals to the perfusion apparatus and when the blood flow and bile flow were properly established mevalonate-2-^{14}C was added to the perfusate. The control animals in this case were litter mates which had had the bile duct cannulated immediately prior to the liver perfusion procedure. At various time intervals after this 'pulse labelling' procedure the perfusate plasma cholesterol specific activity was determined and the results are shown in Fig. 4.

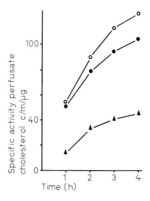

Fig. 4. Time course of specific activity of perfusate plasma cholesterol from control liver perfusions (●), livers from animals previously subjected to biliary drainage (○) and livers from animals previously fed 0.5% cholesterol in the diet (▲). Mevalonate-2-^{14}C added to the perfusate at zero time.

D. Cholic Acid Supplement in the Diet

Rats were fed the stock diet containing 1% cholic acid while litter mates were fed on the stock diet. This regimen was continued for 14 days and then selected animals were injected with mevalonate-2-^{14}C. The plasma cholesterol specific activity was studied as described previously and the results are shown in Fig. 1.

Discussion

The incorporation of relatively small supplements of cholesterol into the diet of the rat results in an increase in the concentration of cholesterol in liver and plasma. There is also an associated inhibition of hepatic cholesterol synthesis and the present studies suggest that the animal responds by increasing the rate of cholesterol turnover in plasma. From *in vivo* studies WILSON (1962) suggests that dietary cholesterol may influence the rate of bile acid formation. From our *in vitro* liver perfusion experiments there is direct evidence that livers from animals previously fed a cholesterol supplement produce more bile acids per unit time than do livers from litter mate controls fed a low cholesterol diet (PERCY-ROBB and BOYD, in preparation).

The incorporation of bile acids or bile salts into the diet of the rat results in an increase in the bile salt pool of the animal. This change is associated with an alteration in the rate of turnover of the plasma total cholesterol which may be due to a diminished rate of conversion of cholesterol to bile acids. Conversely, the administration of a bile salt sequestering agent such as cholestyramine to an intact rat or the use of a biliary drainage technique produces a situation in which the enterohepatic circulation of the bile salts is markedly reduced. Under these circumstances the turnover of the plasma total cholesterol is increased and the rate of production of bile salts from cholesterol is accelerated as measured by the isolated perfused rat liver preparation (PERCY-ROBB and BOYD, 1966). There is therefore a feed-back control of bile salts on the reactions which lead to the conversion of cholesterol into bile salts. Similarly the rate of synthesis of cholesterol in the rat is affected by the feedback of bile salts on the biosynthetic pathway (MYANT and EDER, 1961).

By the use of mevalonate-2-^{14}C pulse labelling it has been possible to study the distribution of newly synthesized cholesterol between different lipoprotein fractions in rat plasma. With this technique the lipoprotein patterns obtained from animals maintained on a normal diet have been shown to be different from the patterns obtained from animals given a small dietary cholesterol supplement. Little radioactivity was found at any time in the β-lipoprotein fraction while most of the activity was found in the a-lipoprotein portion. In normal animals a short time after the administration of mevalonate the a_1- and a_2-lipoproteins were either equally labelled or the activity of the a_1-exceeded the activity of the a_2- lipoprotein. After a few hours the activity on the a_1-lipoprotein declined and the activity on the a_2-lipoprotein increased. In rats previously fed a small dose of cholesterol the lipoprotein pattern was quite different in that the a_1-lipoprotein had the highest initial activity and retained this pattern throughout the experimental period, there being little activity associated with the a_2-lipoprotein at any time.

The endoplasmic reticulum of the liver cell is important in protein and sterol synthesis and hence is a key organelle in lipoprotein production. This same sub-cellular fraction is also important in certain of the early steps in the degradation of cholesterol to bile acids (DANIELSSON and EINARSSON, 1964; MITTON and BOYD, 1965). Thus it is possible that the feedback control of cholesterol and lipoprotein synthesis, together with control of cholesterol catabolism could be exerted on the endoplasmic reticulum by cholesterol itself, bile acids or bile salts.

Summary

The incorporation of small amounts of cholesterol into the diet of the adult rat produced an increased turnover rate of plasma cholesterol and a markedly altered plasma lipoprotein secretion pattern. The administration of a bile salt sequestering agent increased the turnover rate of the plasma cholesterol. Conversely, the addition of bile acids to the diet decreased the turnover rate of the plasma cholesterol.

Acknowledgements

This work was supported by a grant from the Scottish Hospital Endowments Research Trust (HERT 128).

References

ABELL, L.L.; LEVY, B.B.; BRODIE, B.B. and KENDALL, F.E.: A simplified method for the estimation of total cholesterol in serum and demonstration of its specificity. J. biol. Chem. *195:* 357–366 (1952).

AVIGAN, J.; STEINBERG, D. and BERMAN, M.: Distribution of labelled cholesterol in animal tissues. J. lip. Res. *3* (2): 216–221 (1962).

BERGSTROM, S. and DANIELSSON, H.: On the regulation of bile acid formation in the rat liver. Acta physiol. scand. *43:* 1–7 (1958).

BOYD, G.S.: The estimation of serum lipoproteins. A micromethod based on zone electrophoresis and cholesterol estimations. Biochem. J. *58:* 680–685 (1954).

DANIELSSON, H. and EINARSSON, K.: The enzymic formation of 7α-hydroxycholesterol from cholesterol in rat liver homogenates. Acta chem. scand. *18:* 831–832 (1964).

GOULD, R.G.; TAYLOR, C.B.; HAGERMAN, J.S.; WARNER, I. and CAMPBELL, D.J.: Effect of dietary cholesterol on the synthesis of cholesterol in dog tissue *in vitro*. J. biol. Chem. *201:* 519–528 (1953).

MILLER, L.L.; BLY, C.G.; WATSON, M.L. and BALE, W.F.: The dominant role of the liver in plasma protein synthesis. J. exp. Med. *94:* 431–451 (1951).

MITTON, J.R. and BOYD, G.S.: The hydroxylation of cholesterol by liver microsomes. Biochem. J. *96:* 60 P (1965).

MORRIS, M.D.; CHAIKOFF, I.L.; FELTS, J.M.; ABRAHAM, S. and FAMSAH, N.O.: The origin of serum cholesterol in the rat diet versus synthesis. J. biol. Chem. *224:* 1039–1045 (1957).

MYANT, N.B. and EDER, H.A.: The effect of biliary drainage upon the synthesis of cholesterol in the liver. J. lip. Res. *2:* 363–368 (1961).

PERCY-ROBB, I.W. and BOYD, G.S.: Isolated rat liver perfusion as a technique for the study of hepatic cholesterol and bile acid metabolism. Progr. biochem. Pharmacol. *2.* (1966).

TOMKINS, G.M.; SHEPPARD, H. and CHAIKOFF, I.L.: Cholesterol synthesis by liver – its regulation by ingested cholesterol. J. biol. Chem. *201:* 137–141 (1953).

WILSON, J.D.: Relation between dietary cholesterol and bile acid excretion in the rat. Amer. J. Physiol. *203:* 1029–1032 (1962).

Author's address: G. S. Boyd, M. D., Dept. of Biochemistry, University of Edinburgh, Medical School, Teviot Place, *Edinburgh 8* (Scotland).

From the Government Hospital Donolo, Jaffa, Israel

Alpha-Cholesterol Percentage in Coronary Patients and in Healthy Controls

D. BRUNNER and K. LOEBL, Jaffa

It is assumed that alpha-cholesterol percentage is a valuable parameter indicating atherogenic activity on the clinical and subclinical level, especially in subjects without increased serum total cholesterol. It is suggested that the first step in the development of atherogenic lipoprotein pattern is a re-distribution of the cholesterol between the alpha- and the beta-lipoproteins.

Alpha-cholesterol percentage, expressed as a relative value and indicating the percentage of serum total cholesterol in the alpha-lipoprotein fraction, was markedly decreased in middle-aged patients suffering from ischaemic heart disease, irrespective of the level of total cholesterol.

82% of all male patients, and 78% of all female patients had alpha-cholesterol percentage less than 20%. In 89 male patients with total cholesterol values less than 220 mg%, 72% had alpha-cholesterol percentage under 20%. The corresponding figures for healthy controls were 23%. For women the corresponding figures for patients were 63%, and 6% in the control group.

In comparing coronaries with healthy controls matched in sex and serum cholesterol level, we found that alpha-cholesterol percentage for each cholesterol value (from 160 mg% to 280 mg%) was significantly higher in healthy women than in healthy men, and in healthy men than in male coronary patients.

The alpha-cholesterol percentage of female coronaries was in the same low range as in male patients with identical serum-cholesterol values.

Author's address: D. Brunner, M. D., Government Hospital Donolo, *Jaffa* (Israel).

Variations in Human Serum Alpha-1-Lipoproteins*

L. Cohen and Juliana Djordjevich, Chicago, Ill.

Many proteins exist in different molecular forms. These differences are demonstrable in a variety of ways. For example, hemoglobin varieties can be discriminated by electrophoresis, and different forms of human serum beta-lipoproteins can be detected by antigen-antibody reactions. This report describes evidence which suggests the existence of more than one form of human serum alpha-1-lipoprotein.

Three different common patterns of alpha-1-lipoprotein have been recognized by the development of improved methods for the staining and electrophoresis of whole serum lipoproteins in the starch gel. The stain consists of equal volumes of a methanolic oil red 0 solution (0.2 g/l) and an aqueous trichloracetic acid (3 M) solution, mixed just before use. The gel is stained for 16 h at 37° C. The starch gel is moulded in the usual way after adding lauric acid $(4.1 \times 10^{-4} \, M)$.

Two examples of each alpha-1-lipoprotein pattern revealed by these methods are shown in the accompanying figure: one is called alpha-1-F; another, alpha-1-M; and a third, alpha-1-S. F, M and S, respectively, refer to faint, moderate and strong staining with oil red 0. In addition to these staining differences, the F type has four component bands, and the M and S types have five component bands.

The frequency of each alpha-1-lipoprotein type is different in the two sexes. The alpha-1-S type occurs about three times more frequently in females; whereas the alpha-1-F type, common in males, is uncommon in females.

These new methods also reveal two beta-lipoprotein components. Albumin and the $alpha_2$ macroglobulin can also be seen to stain,

---

* Supported by grants from the Chicago, Illinois and American Heart Associations and United States Public Health Service (H-0119).

as do the chylomicrons, which however, are too large to enter the gel and remain at the origin.

The basis and biological significance of this newly demonstrated alpha-1-lipoprotein variation are currently under study.

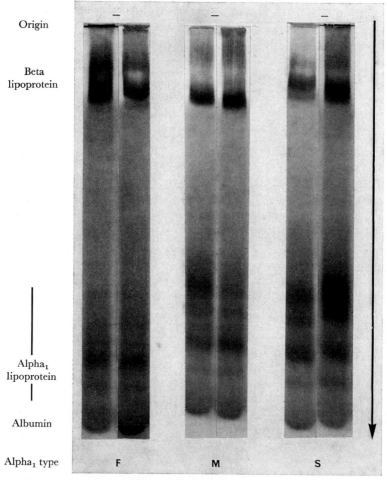

Fig. 1. Human serum alpha-1-lipoprotein patterns. The oil red 0 stained material in six different whole serum samples after electrophoresis is shown. Two examples of each of the three common alpha-1-lipoprotein patterns, F, M and S are depicted. These patterns are defined in the text. Several $alpha_1$ lipoprotein components are evident. Two beta lipoprotein components are also seen. The fastest migrating stained component is albumin. The direction of the electrophoresis is toward the anode and is indicated by the arrow.

Author's address: Louis Cohen, M. D., and Juliana Djordjevich, Dept. of Medicine, University of Chicago, School of Medicine, 950 E. 59th Street, *Chicago*, Ill. (USA).

A Hypothetical Transformation of Chylomicrons and VLD Lipoproteins into Lipoproteins of Higher Density

A. Gustafson, Göteborg

The serum lipoproteins compose a continuous spectrum of macromolecular compounds of increasing density from the chylomicrons to the alpha lipoproteins. The lipoprotein protein distribution among the four major groups, the chylomicrons, the VLD, the beta and the alpha lipoproteins indicates that chylomicrons may transform by the way of VLD lipoproteins into alpha lipoproteins and the more dense VLD lipoproteins into beta lipoproteins.

Hypothetical transformation products of chylomicrons, derived by the elimination of triglycerides and the polymerization of protein, agree well in molar composition and density with that of the experimental VLD lipoproteins D 0.93–0.97 g/cc and the alpha lipoproteins. Similarly, judged by their molar composition, VLD lipoproteins D 0.97–0.98 g/cc may transform into beta lipoproteins simply by loss of triglycerides.

Author's address: A. Gustafson, Dept. of Internal Medicine, Medical School, University of Göteborg, Göteborg, (Sweden).

From the Simon Stevin Institute for Scientific Research, Brugge

Lipid Fatty Acid Relationships in Electrochromatographic Lipoprotein Fractions

H. Peeters and V. Blaton, Brugge

Introduction

Several workers have studied the fatty acid (FA) spectrum of lipoproteins, separated according to density. Philips [1] determined the FA-spectrum of lipoproteins of D <1.109. The phospholipids in human serum lipoproteins of d <1.2, separated on SiO_2 columns, were analysed for their FA composition by Nelson [2]. Differences between the fatty acid pattern of phospholipids from Cohn fractions were observed by Böttcher [3]. Wood [4] studied the fatty acid composition of chylomicron phospholipids.

As was shown by Keler-Bacoka [5] and Peeters [6, 7, 8] electrochromatography is a useful tool for the separation of lipoproteins, and yields alpha 1, alpha 2, beta-lipoprotein and chylomicron fractions. In a previous paper differences between the fatty acid spectrum of the alpha 1 and beta-lipoproteins and a characteristic FA-spectrum for their cholesterolesters, triglycerides was established for the reference subjects by Blaton [9]. In this work the FA-spectrum of serum lipids of reference and aged subjects are compared to the FA-spectrum of alpha 1 and beta-lipoproteins and to the spectra of alpha 1 and beta-lipids. The significance of details in the lipid and fatty acid composition of lipoproteins and their informational potential are being investigated.

Materials and Methods

Five aged male subjects (41, 61, 66, 72 and 78 years of age) were investigated and selected on the base of their lipid values. None of this patients was receiving any medical therapy.

As reference values the serum of three normal male subjects, aged 24, has been examined twice at three months interval. Average values for the two groups are given in Table I.

Table I. Averaged lipid values of reference and aged persons

Lipid classes	Av. 25 years	Av. 65 years
Total lipids	577	859
Phospholipids	169	228
Triglycerides	85	127
Cholesterolesters	229	333
Cholesterol	39	72
TEFA	271	415
% PL	32.4	30.0
% TG	16.3	16.7
% CE	43.9	43.8
% Ch.	7.5	9.5

In this study electrochromatography has been used as a method for the separation of serum lipoproteins in view of the subsequent analysis of their lipid content. The principle of this method is shown in Fig. 1 and the separation obtained is compared with electrophoretic and ultracentrifugal fractionation in Fig. 2.

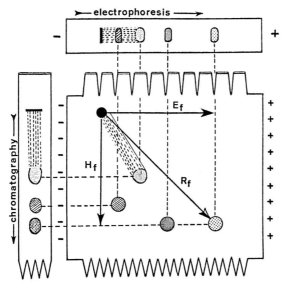

Fig. 1. Principle of the electrochromatography.

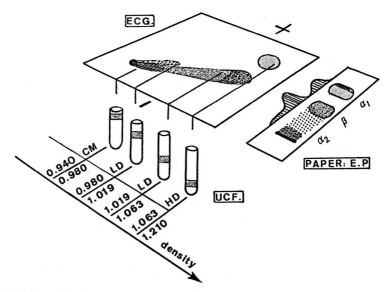

Fig. 2. Comparison between ECG, paper EP and U.C.F. separated lipoproteins.

For electrochromatography (ECG) two ml serum are prestained with Sudan Black B and concentrated under nitrogen on an ultrafilter (L.S.G. 60, Membran Filter Gesellschaft, Göttingen) for 16 h [8]. The concentrated serum is then separated on a curtain of Whatmann paper 3 MM × 30 cm against a Veronal buffer of pH 8.6 and μ 0.033 [9]. After separation the prestained alpha 1 and beta-lipoproteins are cut out and eluted with chloroform-methanol (2/1, V/V). The extract is separated on SiO_2-plates 10% $CaSO_4$ (Kieselgel-G) developed at room temperature for 45 min in petroleum-ether-diethylether-acetic acid 80/20/1 (V/V/V), and the lipids are localized by detecting the edges with Iodine vapours. The lipids are eluted with chloroform-methanol (2/1, V/V) and their fatty acid pattern determined by gas liquid chromatography (GLC) at 200° C on Anakrom ABS 110/120 Mesh (Analabs Inc., Connecticut) coated with 20% butane 1.4-diol succinate.

Way of Procedure

The procedure is aimed at the establishment of the fatty acid (FA) patterns of different lipoproteins and their lipids within the same serum (Fig. 3). The procedure was described earlier by BLATON

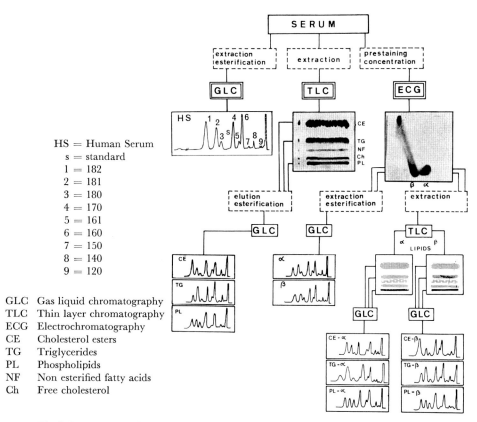

Fig. 3. Integrated lipid and lipoprotein analysis.

[9]. The validity of this procedure inasfar as the fatty acid analysis is concerned depends on the preservation of unsaturated fatty acids during the procedure and is proved by the stability of $\frac{18:0}{18:1}$, $\frac{18:0}{18:2}$ ratio (Table II).

Table II. Validity of the electrochromatographic method

	$\frac{18:0}{18:1}$	$\frac{18:0}{18:2}$
Serum	0.39	0.26
Prestained serum	0.39	0.25
Curtain eluate after electrochromatographic separation under vapour saturated air	0.38	0.25
Same separation under vapour saturated N_2	0.38	0.25

Results

The distribution of polar and apolar lipids in the electrochromatographic alpha 1 and beta-lipoproteins. In Table III the percentual concentrations of polar and apolar lipids in the electrochromatographic alpha and beta-lipoproteins are described. The alpha 1 lipoproteins show a large amounts of phospholipids and nefas while the less polar lipids such as triglycerides and cholesterolesters dominate in the beta-lipoproteins. These differences are a first step in the establishment of the electrochromatographic alpha 1 and beta-lipoprotein entities.

The composition of the lipoproteins obtained through electrochromatographic fractionation is comparable to values of alpha and

Table III. Percentual lipid distribution in alpha 1 and beta-lipoproteins of reference and aged persons

Fraction	Alpha 1 LP		Beta LP	
	ref.	aged	ref.	aged
PL	39	46	17	16
TG	11	10	18	12
CE	44	38	55	61
Ch.	6	6	10	11

beta-lipoproteins as separated by ultracentrifugation and Cohn precipitation.

The differences in the fatty acid content of the total lipids from alpha and beta-lipoproteins. Differences in the fatty acid ratios of the total lipids extracted from alpha and beta lipoproteins are given in Table IV. All values were determined individually and averaged. Ratios are presented because they express a relationship between fatty acids

Table IV. Fatty acid ratios in alpha 1 and beta-lipoproteins

Human male	Alpha 1 LP		Beta LP	
	Av. 25 years	Av. 65 years	Av. 25 years	Av. 65 years
$\dfrac{\Delta:0}{\Delta:1\ \Delta:2}$	0.73	0.77	0.49	0.51
$\dfrac{\Delta:0}{\Delta:1}$	1.24	1.37	1.02	1.11
$\dfrac{\Delta:0}{\Delta:2}$	1.77	1.77	0.93	0.96
$\dfrac{18:0}{18:2}$	0.42	0.49	0.20	0.23
$\dfrac{18:1}{18:2}$	1.00	1.08	0.73	0.75

independent of variations occurring in other sectors of the gas-chromatographic run.

Comparing the results as a whole, we realize immediately that alpha and beta-lipids have lower saturation ratios than the alpha-lipids. If we compare the details of the alpha 1 lipid composition in the young and old groups we observe an increase in saturation in the aged group due to increase of linoleic and of oleic acid. In the beta fraction there is also an increase in saturation more pronounced as a loss of oleic than of linoleic acid. The percentual difference between the indices of alpha and beta-lipids is identical in the two age groups (Table V), and indicates that saturation increase proportionally to the same extent in total lipids extracted from the alpha and beta lipoproteins.

Table V. Percentual difference of the fatty acid ratios of alpha 1 and beta-lipoproteins

Human male	Av. 25 years			Av. 65 years		
	a_1 LP	β LP	%Δ	a_1 LP	β LP	%Δ
$\dfrac{18:0}{18:1}$	0.42	0.27	−36	0.45	0.31	−31
$\dfrac{18:0}{18:2}$	0.42	0.20	−52	0.49	0.23	−53
$\dfrac{18:1}{18:2}$	1.00	0.73	−27	1.08	0.75	−30

The differences in the fatty acid content of the different lipid classes extracted from alpha 1 and beta-lipoprotein fractions. The experimental results are summarized and presented in Table VI and VII. The alpha unsaturation is highest in the cholesterolesters and lowest in the phospholipids. The unsaturation decreases only slightly with age in the cholesterolesters, but to a larger extent in the phospholipids (Table VI).

Table VI. Fatty acid ratios in the lipids of alpha 1 and beta-lipoproteins

Human male	a_1 TG		a_1 CE		a_1 PL		β TG		β CE		β PL	
	Av. 25 years	Av. 65 years	Av. 25 years	Av. 65 years	Av. 25 years	Av. 65 years	Av. 25 years	Av. 65 years	Av. 25 years	Av. 65 years	Av. 25 years	Av. 65 years
$\dfrac{\Delta:0}{\Delta:1\ \Delta:2}$	0.59	0.76	0.44	0.48	0.86	1.28	0.49	0.60	0.22	0.24	0.96	1.11
$\dfrac{\Delta:0}{\Delta:1}$	0.84	0.97	1.13	1.00	1.51	2.44	0.61	0.77	0.55	0.60	2.32	2.29
$\dfrac{\Delta:0}{\Delta:2}$	2.06	3.58	0.74	0.98	2.19	2.70	2.67	2.76	0.36	0.40	1.67	2.17
$\dfrac{18:0}{18:2}$	0.29	0.49	0.09	0.13	0.55	0.92	0.43	0.34	0.04	0.06	0.48	0.77
$\dfrac{18:1}{18:2}$	1.88	2.86	0.41	0.66	1.20	0.95	3.57	3.08	0.44	0.52	0.62	0.83

The beta unsaturation is highest in the cholesterolesters, twice as large as in the alpha-cholesterolesters and does not vary with age. The beta-triglycerides are more unsaturated than alpha-triglycerides while the beta-phospholipids have a high 16:0 content. Saturation increases with age (Table VII).

Table VII. Percentual difference between reference and aged persons in the fatty acid ratios of the different lipids of alpha 1 and beta-lipoproteins

	a_1 LP	a_1 TG	a_1 CE	a_1 PL	β LP	β TG	β CE	β PL
$\dfrac{18:0}{18:1}$	7.1	13.3	— 9.1	110.9	14.8	— 8.4	31.9	20.8
$\dfrac{18:0}{18:2}$	16.6	69.0	44.4	67.3	15.0	—20.9	50.0	60.4
$\dfrac{18:1}{18:2}$	8.0	52.1	61.0	— 20.8	2.7	—13.7	18.2	33.9

From these two tables we can extract the information necessary for our further discussion (Table VIII). The ratios 18:0/18:2 and 18:1/18:2 show important differences in the triglycerides. In the alpha-triglycerides we note a decrease of 18:2 and 18:1 whereas in the beta-triglycerides the linoleic content is improved. It is important to note this opposite behaviour.

Table VIII. Differences between FA ratios of aged and reference subjects

Ratio		TG % difference between the two groups	$a_1 - \beta$	CE % difference between the two groups	$a_1 - \beta$	PL % difference between the two groups	$a_1 - \beta$
$\dfrac{\Delta:0}{\Delta:1\ \Delta:2}$	a_1	29	7	9	0	49	33
	β	22		9		16	
$\dfrac{18:0}{18:1}$	a_1	13	21	— 9	—41	111	90
	β	— 8		32		21	
$\dfrac{18:0}{18:2}$	a_1	69	90	44	— 6	67	7
	β	—21		50		60	
$\dfrac{18:1}{18:2}$	a_1	52	66	61	43	— 21	—55
	β	—14		18		34	

For cholesterolesters the total ratio remains unchanged, however the 18:1/18:2 ratio is modified due to loss of linoleic acid and this occurs simultaneously in the two fractions. For the phospholipids the total ratio differs widely in the alpha fraction. The alpha 18:0/18:1 ratio indicates a relative decrease of oleic and an increase in stearic acid. This does not occur in the beta fraction. The decrease in linoleic acid is identical for the two phospholipids.

Discussion

The proposed analytical scheme contributes to an efficient separation of well defined lipoprotein entities with internal differences in their composition namely alpha 1-lipoproteins, beta-lipoproteins and also alpha 2-lipoproteins and chylomicrons. A large amount of phospholipids and nefas are localized in the alpha 1-lipoproteins while the less polar lipids such as triglycerides and cholesterolesters dominate in the beta-lipoproteins.

From the observed fatty acid ratios we may conclude that the 18:0/18:1 and 18:1/18:2 ratios appear to be important to follow the atherosclerotic syndrome inasfar as aging leads to this condition. The alpha and beta-triglycerides present a contradictory evolution in these ratios. For the alpha 1 and beta-lipoproteins separated by means of electrochromatography an increase of saturation is noted in both fractions during aging and this corresponds to the results presented by Böttcher [3]. In our findings however, this increase is lower for the beta-lipoproteins. The cholesterolesters are fairly unsaturated.

When the phospholipids are considered, the differences between alpha and beta are more pronounced. The normal beta-phospholipids show a high concentration of dienoic fatty acid which does not drop during aging. But in the alpha 1-phospholipids the percentual saturation increases during aging. A further differentiation of the phospholipids into their separate compounds is needed for thorough evaluation of their behaviour.

These findings are to be completed by consideration of the higher unsaturated fatty acids and by further separation of the individual phospholipids but demonstrate at this stage the selectivity of the way of procedure followed in this study.

Acknowledgement

This work was supported by Grant No. 625 of Nationaal Fonds voor Geneeskundig Wetenschappelijk Onderzoek. We wish to thank B. Declercq and D. Vandamme for their technical assistance.

Summary

The significance of details in the lipid composition of lipoproteins and their informational potential are being investigated.

The electrochromatographic separation of serum lipoproteins yields alpha 1, alpha 2 and beta-lipoproteins and chylomicron fractions. The alpha 1-lipoprotein coincides with the presence of large amounts of phospholipids and nefas, while less polar lipids such as triglycerides and cholesterolesters dominate in the beta-lipoproteins. The fatty acid content shows differences between the alpha and beta-lipoprotein triglycerides, phospholipids and cholesterolesters from young reference patients. The differences are more pronounced in elderly subjects. During aging, the largest relative increase in saturation occurs in the alpha fraction stressing a fundamental difference between the lipid behaviour of alpha and beta-lipoprotein.

References

1. PHILIPS, G.: Phospholipids composition of human serum lipoprotein fraction separated by ultracentrifugation. J. clin. Invest. *38:* pp. 489–493 (1969).
2. NELSON, G.: Studies on human serum lipoprotein phospholipids and phospholipid fatty acid composition by silicic acid chromatography. J. Lipid Res. *3:* pp. 71–79 (1962).
3. BÖTTCHER, C. and WOODFORD, F.: Lipid and fatty acid composition of plasma lipoproteins in cases of aortic atherosclerosis. J. Atheroscler. Res. *1:* pp. 434–443 (1961).
4. WOOD, R.; IMAICHI, K.; KNOWLES, J. and MICHAELS, G.: The lipid composition of human plasma chylomicrons. J. Lipid Res. *5:* pp. 225–231 (1964).
5. KELER-BACOKA, M.; PUCAR, Z. and PETEK, M.: Zweidimensionale elektrochromatographische Trennung von Eiweissen, Fetten und Kohlenhydraten der pathologischen humanen Sera. Clin. chim. Acta *3:* pp. 335–342 (1958).
6. PEETERS, H.: Paper Electrophoresis: Principles and Techniques, in SOBOTKA and STEWART's Advances in Clinical Chemistry, vol. 2, pp. 2–132 (Academic Press, New York 1959).
7. PEETERS, H. and VAN LAETHEM, D.: Electrochromatographic pattern of serum lipids, in H. Peeters: Protides of the Biological Fluids. Proc. 8th Colloquium, Bruges 1960, pp. 222–226 (Elsevier, Amsterdam 1961).
8. PEETERS, H.; DE KEERSGIETER, W. and LAMOTE, E.: The detection of lipoprotein after electrochromatography. Chromatographie Symposium II, J. Pharm. Belg., pp. 113–123 (1962).
9. BLATON, V.; PEETERS, H.; DE KEERSGIETER, W.; DECLERCK, B.; DEPICKERE, D. and VANDAMME, D.: Differential fatty acid composition of a_1 and β lipoproteins, in H. Peeters: Protides of the Biological Fluids. Proc. 13th Colloquium, Bruges 1965, pp. 315–319 (Elsevier, Amsterdam 1966).

Authors' address: H. Peeters, M. D., and V. Blaton, Dr. sc., Simon Stevin Institute, *Brugge* (Belgium).

Departments of Medicine and Biochemistry, The University of Chicago, Chicago

Comparative Optical Properties of Human Serum Low- and High-Density Lipoproteins before and after Delipidation

A. Scanu and J. L. Granda, Chicago, Ill.

Measurement of optical rotatory power in the region between 600 and 185 mμ has been proven of value in assessing the conformation of polypeptides and proteins in solution [9]. In the infrared, frequencies of the amide I and II have been shown to depend on conformation and their analysis has provided additional knowledge on the structure of polypeptides in their solid state [1]. Only recently have infrared studies been applied to polypeptides in aqueous solutions and the information derived is comparatively less defined [5].

In the present investigation both techniques of optical rotatory dispersion (ORD) and infrared spectroscopy were applied to the analysis of the conformation of human serum low- and high-density lipoproteins before and after removal of their lipid moiety by organic solvents. Aim of the study was to establish the comparative structure of these two lipoprotein classes in both aqueous and solid phase and the effect of lipid removal on the conformation of their apoprotein.

The data obtained suggest that both lipoproteins are made of a mixture of a helix and random coil and this conformation is to a large extent independent from the bound lipids.

Material and Methods

Lipoproteins of density 1.019–1.063 (LDL_2) and 1.063–1.21 (HDL) were separated from normal human sera and purified of non-lipoprotein contaminants by preparative ultracentrifugation [4, 8]. After extensive dialysis against 0.15 M NaCl containing

0.1% EDTA, they were checked for purity by starch gel and immunoelectrophoresis [8] and aliquots extracted with ethanol-ethyl ether at —10° in the presence (LDL_2) or absence (HDL) of sodium dodecylsulphate (SDS) following previously described procedures [3, 8]. The final delipidated residues, LD_2-P and HD-P, contained about 1% lipid by weight and had characteristic solubility properties in aqueous media [4, 8].

For ORD analysis LDL_2, LD_2-P, HDL and HD-P were dialyzed against phosphate buffer, pH 8.6, ionic strength 0.1. The studies were conducted in a Cary Mod 60 Spectropolarimeter at 27° following a previously described methodology [7]. The observed rotations, a, were corrected for the refractive index of the solvent, the mean residue weight (LDL_2: 115; HDL: 118), protein concentration and cell path length and then analyzed, in the visible and near ultraviolet region, according to the single Drude and Moffitt-Yang equations [2].

Infrared studies in the frequency range of 1800–1450 cm^{-1} were conducted in a Perkin-Elmer Mod 21 Recording Spectrophotometer at 25°. Solutions (0.1 M KCl, pH 8.6) of the various lipoproteins in either H_2O or D_2O were studied in special jacketed silver chloride cells. For solid state analysis, lipoprotein preparations containing 3 mg protein were brought to dryness under N_2 and then ground in a mortar with 300 mg of solid KBr. Pellets were made in a Carver Laboratory Press (Fred S. Carver, Inc., Summit, N.J.) and analyzed immediately.

Results and Discussion

A direct comparison of the dispersion data between LDL_2 and HDL was only possible in the ultraviolet and far-ultraviolet because of the strong absorbancy of LDL_2 in the visible region. The optical rotatory power of LDL_2 was consistently and significantly less than HDL and the difference was also noted in the delipidated specimens, LD_2-P and HD-P (Table I, Fig. 1). In the dispersion curves shown in Fig. 1 this difference is particularly evident at the maximum points of negative (233 mμ) and positive (198 mμ) deflections of the Cotton effect (Fig. 1). Further, a shoulder between 210 and 215 mμ is clearly seen only with HD-P. The type of dispersion curves, the calculated values of λ_c (from simple Drude equation)

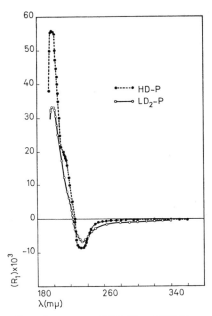

Fig. 1. Optical rotatory dispersion curves of LD$_2$-P and HD-P.

Table I. Parameters of optical rotatory dispersions of LDL$_2$ and HDL before and after delipidation

ORD Parameters	LDL$_2$	LD$_2$-P	HDL	HD-P
λc	*	249	289	280
b_o	*	—185	—514	—460
$[R_1]_{233}$	—8,000	—6,500	—9,000	—8,400
% helix	41.3	30.1	69.6	61.8

* Dispersion not obtained because of high absorbancy.

and b_o (from Moffitt-Yang equation) and the negative sign of the latter, are all compatible with a protein existing as a misture of a-helix and random coil. Estimates of helical content based on b_o parameters ($b_o^H = 630$, $b_o^R = 0$) and on the depth of the trough at 233 mμ (H = 16,500; R = 2,000) gave values fo LDL$_2$ or LD-P$_2$ about 50% less than those with HDL or HD-P.

From the ORD data it could be concluded that either LDL$_2$ or HDL exist as a mixture of a-helix and random coil and that the apoprotein conformation is not largely affected by the presence of lipid. Infrared analyses provided corroboration to this conclusion. Polypeptides in the solid state are known [6] to give characteristic vibration bands of amide I and II which, although they do not differentiate the α-helix from the random configuration provide a distinction between these two forms and the β-structure [6]. The observed frequencies for amide I and II in lipoprotein pellets (Table II) are compatible with those reported for either α-helix (amide I: 1650–1655; amide II: 1530–1560) or random oil (amide I: 1660). They do not support, however, the existence of a β-structure whose frequencies for amide I are between 1628–1635 and those for amide II between 1521–1525 [6]. No data could be obtained with LDL_2 on the solid state owing to the difficulty of obtaining satisfactory pellet preparations.

In the aqueous phase the band frequencies of amide I were significantly lower than those observed in the solid state (Table II).

Table II. Frequencies of amide I and II of LDL_2 and HDL before and after delipidation (frequencies in cm^{-1})

| Material | Solid State | | Liquid State | | | |
| | | | H_2O | | D_2O | |
	Amide I	Amide II	Amide I	Amide II	Amide I	Amide II
LDL_2	*	*	1625	+	n.d.	n.d.
LD_2-P	1645	1530	1620	+	1640	+
HDL	1650	1540	1625	+	n.d.	n.d.
HD-P	1645	1535	1620	+	1640	+

* Material found unsuitable for pellet preparation.
+ Band not clearly detected.
n.d. = not determined.

The dependence of this finding on the strong absorption by water, is indicated by the higher frequency values in D_2O closely approximating those in the solid state. Amide II band was barely detectable when water was replaced by D_2O. In either of the two analyses (liquid and solid) LDL_2 and HDL gave comparable results. They were not affected by removal of lipids.

The results of ORD and infrared analysis, taken together, suggest the following conclusions: (1) LDL_2 and HDL are made of a mixture of a-helix and random coil having no detectable β-structure, (2) LDL$_2$ has a helical content lower than HDL, (3) delipidation does not qualitatively influence the apoprotein conformation of either of the two lipoproteins. These conclusions and the knowledge that LDL$_2$ and HDL are made of subunits [4, 8] support the view that the latter may be held together by lipid bridges. The possibility, however, that in the native products the subunits may have points of contact cannot be ruled out.

Summary

A structural analysis of human serum low- ($1.019 < \rho < 1.063$) and high-density ($1.063 < \rho < 1.21$) lipoproteins before and after removal of lipid, was carried out by the techniques of optical rotatory dispersion and infrared spectroscopy. The data suggest that the two lipoproteins exist as a mixture of a-helix and random coil and that this conformation is not qualitatively affected by delipidation. The helical content of the low-density lipoprotein was significantly lower than that of the high-density class. The possible influence of SDS on the conformation of LD$_2$-P was not established.

Acknowledgements

This work was supported by grants HE-08727 from the USPHS, C65-1 from the Chicago and Illinois Heart Associations, and G-64-42 from the Life Insurance Medical Research Fund.

A. SCANU is the recipient of a Research Career Development Award No. HE-24,867 from the USPHS.

The authors wish to thank Drs. G. SMITH and K. NAKAMOTO, Department of Chemistry, Illinois Institute of Technology for permitting the use of the Cary Mod 60 Spectropolarimeter and the Perkin-Elmer Mod 21 Recording Spectrophotometer.

Note added in proof

Presence of a-helix in both HDL and LDL$_2$ has recently been found in this laboratory by measurements of circular dichroism. In the delipidated state the conformation of HD-P and LD$_2$-P was sensitive to solvent changes.

References

1. ELLIOTT, A.; BRADBURY, E.M.; DOWNIE, A.R. and HANBY, W.E.: Conformation of polypeptide chains in the solid state. In Polyamino acids, polypeptides and proteins. pp. 255–269 (University of Wisconsin Press, Madison 1962).
2. FASMAN, G.: Optical rotatory dispersion. In Methods in enzymology. pp. 928–957 (Academic Press, New York 1963).

3. Granda, J.L. and Scanu, A.: Studies on the protein moiety of human serum low-density lipoproteins. Fed. Proc. (Abst.) *24:* 224 (1965).
4. Granda, J.L. and Scanu, A.: Solubilization and properties of the apoproteins of human serum very low and low-density lipoproteins. Biochem. *5:* 3301–3308 (1966).
5. Jenks, W.P.: Infrared measurements in aqueous media. In Methods in enzymology. pp. 914–928 (Academic Press, New York 1963).
6. Miyazawa, T.: Characteristic amide bands and conformations of polypeptides. In Polyamino acids, polypeptides and proteins. pp. 201–217 (University of Wisconsin Press, Madison 1962).
7. Scanu, A.: Studies on the conformation of human serum high-density lipoproteins HDL_2 and HDL_3. Proc. Natl. Acad. Sci. *54:* 1699–1705 (1965).
8. Scanu, A.: Forms of human serum high density lipoproteins. J. Lip. Res. *7:* 295–306 (1966).
9. Urnes, P. and Doty, P.: Optical rotation and the conformation of polypeptides and proteins. Adv. Prot. Chem. *16:* 401–544 (1961).

Author's address: A. Scanu and J. L. Granda, 950 East 59th Street, *Chicago, Ill. 60637* (USA).

From the Department of Experimental Pathology, University of Birmingham, England

The Role of Low-Density Lipoproteins in the Pathogenesis of Human Atherosclerosis

K.W. WALTON, Birmingham

The following evidence appears to support the contention that low-density lipoproteins (LDL) are implicated in the pathogenesis of human atherosclerosis:

1. The relatively higher incidence of atherosclerosis in males than in females and the increased incidence with age in both sexes have been shown to be paralleled by the variation in serum levels of LDL with age and sex [1].

2. Studies with isotopically labelled isologous LDL in males and females in apparent health have shown that the differences in serum concentrations between males and females are reflected in differences between the fractional and absolute catabolic rates between the sexes [2].

3. Similar studies in a number of diseases associated with a high incidence of atherosclerosis have established that LDL metabolism is abnormal. In some instances, as for example, in hypothyroidism, the fractional catabolic rate (biological half-life) of LDL may be prolonged [2]. In other instances, as for example, essential hypercholesterolaemia, the fractional metabolic rate may show little alteration from normal [3]. But in all cases the total exchangeable pool is greatly enlarged and this is reflected principally in marked increase of the intravascular pool.

4. In one case, evidence was obtained which established clearly that circulating LDL can equilibrate with material in atheromatous lesions in the arterial wall [3]. This patient died of myocardial insufficiency while isotopically labelled lipoprotein was still present in her circulation. At autopsy, when portions of various organs were removed and their radio-activity measured, greater radio-activity

was found in the grumous material from an atheromatous lesion in the aorta than in any other organ examined.

5. In other instances [4], material antigenically identical with LDL has been identified in atheromatous lesions in various arteries, and at varying stages of development of atherosclerosis, by the fluorescent-antibody technique.

References

1. WALTON, K.W. and SCOTT, P.J.: Estimation of the low-density (beta) lipoproteins of serum in health and disease using large molecular weight dextran sulphate. J. clin. Path. *17:* 627–643 (1964).
2. WALTON, K.W.; SCOTT, P.J.; DYKES, P.W. and DAVIES, J.W.L.: The significance of alterations in serum lipoproteins in thyroid dysfunction II. Alterations of the metabolism and turnover of ^{131}I-low-density lipoproteins in hypothyroidism and thyrotoxicosis. Clin. Sci. *29:* 217–238 (1965).
3. WALTON, K.W.; SCOTT, P.J.; VERRIER JONES, J.; FLETCHER, R.F. and WHITEHEAD, T.: Studies on low-density lipoprotein turnover in relation to Atromid therapy. J. Atheroscler. Res. *3:* 396–414 (1963).
4. WALTON, K.W. and WILLIAMSON, N.: Studies by the fluorescent antibody technique in human and experimental atherosclerosis (in preparation).

Author's address: K. W. Walton, M. D., Dept. of Experimental Pathology, University of Birmingham, *Birmingham* (England).

IVth Medical Clinic, Gastroenterological Research Unit and Angiological Laboratory,
Faculty of Medicine, Charles University, Prague

Ultracentrifugal Fractionation of Post Heparin Lipolytic Activity in Human Plasma

J. Skořepa, P. Hrabák, Š. Novak, H. Todorovičová and K. Marčan, Prague

1. One of the factors which influences the lipoprotein spectrum is lipolytic activity in the plasma and tissues. Several enzymes are known today which be taken into consideration:

a) Lipoprotein lipase – splits triglycerides of very low density lipoproteins;

b) transferase of Glomset [1] – this is not a true lipase, but influences the lipoprotein spectrum. It transfers fatty acids from phosphorylcholin to cholesterol;

c) post-heparin phospholipase which was demonstrated recently by Vogel and Zieve [2].

2. In this connection we will draw attention to the post heparin esterase, which appears in the plasma after the intravenous injection of heparin. This esterase splits ethyl butyrate. It can be demonstrated in different ways as we have shown in previous papers [3, 4].

3. One of the facts supporting the existence of this esterase is evident from the figure (Fig. 1).

Enzymes were partially purified by adsorption on calcium phosphate gel [5]; they act on the mixture of substrates: Ediol® (Schenlabs Pharmaceuticals Inc., New York, N.Y.) and ethyl butyrate.

Fatty acids liberated are fractionated by chromatography on a silica gel column [6]. The production of short-chain as well as of long-chain fatty acids can be observed. Using protamine or sodium fluoride as inhibitors the production of long-chain or short-chain fatty acids can be inhibited separately.

4. We attempted to fractionate the post heparin lipolytic activity by separation in the ultracentrifuge. In solutions containing

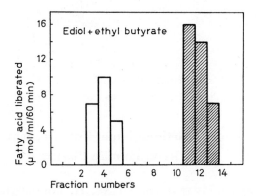

Fig. 1. Production of long-chain (open columns) and short-chain (closed columns) fatty acids from the mixture of substrates, Ediol® and ethyl butyrate by post-heparin enzymes partially purified from human plasma.

post heparin enzymes partially purified by adsorption on calcium phosphate gel we are able to separate an activity which liberates long-chain fatty acids from the one, which splits ethyl butyrate.

In this way we confirm the results of PAYZA and collaborators [7]. These authors also isolated the clearing factor by fractionation in the ultracentrifuge. The clearing factor purified by this method was inactive against the ethyl butyrate.

5. Post-heparin enzyme which splits ethyl butyrate is not a transferase; we were not able to prove the production of cholesteryl butyrate in incubation solutions.

References

1. GLOMSET, J.A.: The mechanism of the plasma cholesterol esterification reaction: plasma fatty acid transferase. Biochem. Biophys. Acta 65: 128 (1962).
2. VOGEL, W.C. and ZIEVE, L.: Post-heparin phospholipase. J. Lipid Res. 5: 177 (1964).
3. SKOŘEPA, J. and TODOROVIČOVÁ, H.: Post-heparin esterase in man. Experientia 12: 228 (1956).
4. SKOREPA, J.; PROCHÁZKA, B. and TODOROVIČOVÁ, H.: Differentiation of post-heparin esterase and heparin clearing factor by means of adsorption to the calcium phosphate gel. Clin. Chim. Acta 5: 256 (1960).
5. NIKKILÄ, E.A.: Partial purification of clearing factor of post-heparin human plasma. Biochim. Biophys. Acta 27: 612 (1958).
6. HARPER, W.J.; SCHWARZ, D.P. and EL-HAGARAWY, I.S.: A rapid silica gel method for measuring total free fatty acids in milk. J. Dairy Sci. 39: 46 (1956).
7. PAYZA, A.N.; EIBER, H.B. and DANISHEFSKY, I.: Studies on plasma clearing factor. I. Fractionation and characteristics. Biochim. Biophys. Acta 111: 159 (1965).

Author's address: Dr. J. Skorepa, Dr. P. Hrabák, Dr. Š. Novák, Dr. H. Todorovičová and Dr. K. Marčan, IVth Medical Clinic, Gastroenterological Research Unit and Angiological Laboratory, Faculty of Medicine, Charles University, *Prague* (Czechoslovakia).

Discussion Session 3

K. W. WALTON (Birmingham): I have two comments to make: (1) This relates to Dr. ALAUPOVIC's paper and to the evidence he has presented to suggest that there are 3 protein moieties (or 'apolipoproteins') related to the soluble lipoproteins. I think most of us would accept that there are indeed two. In relation to the evidence presented for a third (or apolipoprotein C) I wondered whether you had taken into account the possibility of this (a) being either a degradation product or a polymer to one of the other lipoproteins, resulting from delipidation. Such changes would markedly alter both its immunological and physicochemical characteristics. I shall be presenting evidence later. (b) Of a protein not originating from a soluble lipoprotein, but from platelets [1]. A fine particulate material containing protein and lipid is almost invariably present in both plasma and serum. This material which we have called 'platelet-dust' can be distinguished from chylomicra, alpha and beta lipoproteins both immunologically and by physico-chemical methods.

(2) There have been a number of comments in various papers about the way in which the behaviour of the serum lipids in the dog differs from that of other species. Some years ago we compared the lipoproteins of various species [1]. We found that, in case of low density (beta) lipoproteins there were immunological cross-reactions throughout all the mammals examined. Similar cross-reactions, to a lesser degree were encountered among high-density (alpha) lipoproteins. But there were marked differences in the relative proportions of these two lipoproteins, and in the nature of the lipids transported by each, between species, suggestive of variations in their biological role. The most conspicuous exception was the dog in which the quantitatively predominant lipoprotein and the one apparently concerned predominantly in the transport of cholesterol and triglycerides was a lipoprotein cross-reacting with human alpha and *not* beta lipoprotein.

References

1. WALTON, K.W. and WOLF, P.: Progr. biochem. Pharmacol. Vol. 4 (Karger, Basel/New York 1967).
2. WALTON, K.W. and DARKA, S.J.: Proteins of the Biological fluid, *10:* 146 (1963).

P. ALAUPOVIC (Oklahoma City, Okla.): Thank you very much for your comments and questions. Now, in respect to the presence of the apoliproptein C it should be emphasized that the proof for its occurrence was not based only on the immunochemical behavior. Time does not allow for a longer discussion, but be assured that the differentiation between the apolipoproteins A and B and the newly identified apolipoprotein C was based on chemical as well as immunochemical evidence. The chemical methods used were the determination of terminal amino acids, peptide patterns of tryptic and peptic hydrolysates and the binding properties of partially delipidized lipoproteins. I tried to emphasize in my presentation that the phospholipid/protein ratio of partially delipidized apolipoprotein C-containing lipoproteins was 2.4, whereas those of apolipoproteins A and B were between 0.8 and 0.9 (published in Biochemistry *5:* 632, 1966). Your comment concerning the differences in the distribution and absolute amounts of serum apolipoproteins in various species is well taken. May I add that not only dog but also horse is characterized by an extremely high content of apolipoprotein A.

P. ALAUPOVIC (Oklahoma City, Okla.): Although in some cases the presence of small amounts of apolipoproteins in a 'narrow' lipoprotein subfraction could be considered as a 'simple trace contamination', the occurrence of apolipoproteins A, B and C in the 'broad' very low-density lipoproteins ($S_f > 20$), and apolipoproteins A and B in low- and high-density lipoproteins is due to the presence of corresponding apolipoprotein A, B and C-carrying lipoprotein families characterized by the same flotation coefficients or hydrated densities. Studies on the distribution of three phospholipid-protein residues of very low-density lipoproteins isolated from various hyperlipemic sera (Biochemistry 5: 632, 1966) showed in 'mixed type' hyperlipemia 14% of apolipoprotein A, 48% of apolipoprotein B and 48% of apolipoprotein C. In 'carbohydrated-accentuated' hyperlipemia the corresponding values were 33%, 53% and 14%, respectively. These values were obtained by actual isolation and separation of phospholipid-protein residues and were not based on the results of immunochemical methods.

In reply to the second question, it should be pointed out that for example high-density lipoproteins obtained by preparative ultracentrifugation between solvent densities of 1.110–1.210 g/ml usually are free of apolipoprotein B. On the other hand, the VLD-lipoprotein with S_f 20–50 contains only apolipoprotein B. The purification of low-density or β-lipoproteins (d 1.020–1.063 g/ml) could be achieved by precipitating the apolipoprotein A-carrying lipoproteins with their corresponding antibodies, but it represents a very tedious and difficult task.

A.M. SCANU (Chicago, Ill.): You raised an important question and this concerns the structural interrelationship between beta and alpha-lipoproteins. The information at hand does not allow a definite answer. I believe, however, that it remains an interesting possibility open to investigation.

A.M. SCANU (Chicago, Ill.): In lipoprotein work, and this bears on the interesting data by Dr. COHEN, we must be aware of possible genetic variations of each of the lipoprotein classes. This may, perhaps, account for differences in results from laboratory to laboratory on both protein and lipid moieties. For the past three years our chemical analyses have been restricted to serum lipoproteins isolated from a very narrow segment of normal population.

D. FREDRICKSON (Bethesda, Md.): With regard to the last question (Dr. WALTON's) and the excellent papers of Drs. ALAUPOVIC and GUSTAFSON, I should like to comment on the relationship of their work on apoprotein C to the simplified approach to classification of lipoproteins that my colleagues and I have been promoting following its first presentation at Heidelberg and Bruges last year. We have tried to relate the entire lipoprotein 'spectrum' to α and β lipoproteins and triglycerides. This leaves out apoprotein C, but deliberately, since it is a functional scheme and it may yet require the discovery of a mutation associated with 'C' deficiency to indicate the requirement for this apoprotein. We must all look hard to determine the distribution and other signs of the implications of C and for this we need antisera. The available commercial anti a_2 sera actually seem to contain anti β lipoprotein activity only.

P. ALAUPOVIC (Oklahoma City, Okla.): It has been our experience that some batches of the *Behringwerke* anti- a_2 lipoprotein sera contained also antibodies to apolipoprotein C-containing lipoproteins. We have a very limited supply of the antisera at the present time.

E.B. SMITH (London): We have been presented with a large amount of detailed work on the protein part of the lipoproteins, but very little on the lipids. There is some disagreement in the literature on the cholesterol ester fatty acid pattern in different lipoprotein fractions. LINDGREN found that the Sf 0–12 cholesterol ester fatty acid pattern differed from that of the VLD fractions, but not from the HDL fractions; I found differences between all the fractions, and, recently, DE WITT GOODMAN has reported no differences. I also found a marked difference in the proportion of sphingomyelin in the phospholipids. This was highest in the Sf 0–12 LP and lowest in the HDL fractions. I think that more work is required to resolve these discrepancies, and that the differences in lipoprotein lipid must be taken into consideration in evaluating the ideas put forward by Dr. GUSTAFSON.

H.J. PEETERS (Brugge): We find differences in the FA ratios of lipid fractions depending on their lipoprotein origin, not only in normals and in elderly people but also during drug therapy and in experimental animals. If other people did not find these differences it maybe due to the fact that they did not examine detailed fractions.

J.L. BEAUMONT (Paris): I wonder if the occurrence of several apolipoproteins in the same flotation class may be the result of contamination by 'traces' of the material used for immunization. What are the criteria of purity of the material used for immunization?

P. ALAUPOVIC (Oklahoma City, Okla.): The very low density lipoprotein fractions are purified by repeated ultracentrifugations at the solvent density of 1.006 g/ml until free of albumin, gamma globulin, a_1-glycoprotein and fibrinogen as tested by immunochemical methods. Furthermore they are characterized by protein and lipid analyses and by determination of flotation coefficients and quantitation of N-terminal amino acids.

R.W. WISSLER (Chicago, Ill.): I would like to comment on Dr. BOBERG's presentation. Our experience in the rat, monkey and, to a limited extent, in man would indicate that the diet being consumed has an important influence on lipoprotein synthesis and composition. I wonder to what extent his results reflect a difference in dietary background rather than species differences. Furthermore, I would like to ask whether the differences could in part be due to use of an uniform method of separation of proteins from differing species. Our data would indicate that a different set of ultracentrifugal conditions should be used for the rat as compared to man or the monkey.

J. BOBERG (Stockholm): We have not done any systematical studies on how the recirculation of labeled palmitic acid into plasma triglycerides may change with different diets. But as you indicate in your question it is possible that the species differences might be due to the different diets.

The suggestion in your second question is also possible. However, since we do not know exactly how lipoprotein fractions in the various species are metabolised, we have chosen to do the separations in the same way as other authors in rat, rabbit and dog.

J.M. THORP (Macclesfield): Dr. BOBERG showed marked difference in responses of anaesthetized and conscious animals. Since animals were catheterized for 4–6 hours, contribution of 'stress' could be considerable.

J. BOBERG (Stockholm): The catheters were inserted 6–8 days before the experiment. FFA concentration in the conscious rats was not high.

E.H. AHRENS (New York, N.Y.): Recently my colleagues GRUNDY, HOFMANN, DAVIGNON and I reported (J. clin. Invest., June 1966) that in man, cholesterol synthesis is much more impressively controlled by bile acids than by cholesterol itself. Our evidence was of two sorts: sterol balance data and analysis of isotope decay curves, in patients *deprived* of bile acids through administration of cholestyramine or by performance of a surgical by-pass of the terminal ileum, or *supplemented* by peroral administration of bile salts. When bile acids were removed, the excretion of *newly synthesized* fecal steroids increased very greatly. And when bile acids were fed, cholesterol synthesis and conversion of cholesterol to bile acids abruptly ceased or greatly decreased. On the contrary feeding up to 3 g/day of cholesterol (in the form of egg yolk) failed to shut off endogenous cholesterol synthesis, which in 2 patients remained at the normal levels of about 750 mg/day. These new data in man appear to confirm Dr. BOYD's results in rats, but of course they greatly extend the previously held views on cholesterol feedback mechanisms which in rats, rabbits and dogs (but *not* in man) operate via ingested cholesterol.

G. BOYD (Edinburgh): I am pleased to hear of this confirmation in man of our experiments in rats.

A. BORGSTRÖM (Stockholm): In both these situations giving cholestyramine or taurocholate affects absorption of cholesterol.

E. H. AHRENS (New York, N.Y.): We were using a cholesterol free diet.

A. BORGSTRÖM (Stockholm): But there is approximately 1 gram of endogenous cholesterol in the intestine. If you feed cholestyramine it is obvious that the reabsorption of the cholesterol is diminished, and if you feed taurocholate it can be increased.

E. H. AHRENS (New York, N.Y.): This is why we quote the balance method data.

L. W. KINSELL (Oakland, Cal.): Change in the exponential slope as result of cholestyramine or other agents which decrease plasma cholesterol pool are comparable with but not necessarily indicative of change in rate of cholesterol synthesis. If the synthesis rate remains unchanged and the plasma pool decreases, the rate of fall of plasma cholesterol SA will increase.

E. H. AHRENS (New York, N.Y.): Dr. KINSELL states that the reason his results on fecal steroid excretion differ from ours rests in what he calls our 'low' values for fecal bile acids. I have 3 answers to this statement:
1. The much higher values for fecal bile acids found by other workers cannot be used against me. The report by MIETTINEN, AHRENS and GRUNDY (J. Lipid Res., July 1965) discusses the shortcomings of previous methods.
2. That same paper also showed 100% recovery of 24-C^{14}-taurocholate fed to one of our patients. *All* of the labeled conversion products derived from this primary bile acid were recovered by our method. This could not have occurred if we regularly suffered losses of fecal bile acids by our method.
3. The spec. act. of fecal bile acids, in which the mg denominator is obtained by our GLC method, corresponds almost exactly to the spec. act. of plasma cholesterol, in which the mass (in mg) is obtained by the Liebermann-Burchard reaction. Such correspondences were obtained in materials from patients labeled many weeks previously with intravenously administered 4-^{14}C-cholesterol. This fit of the two sets of data could only occur if our mass measurement of fecal bile acids by GLC were highly precise. This final and conclusive confirmation of the completeness of our extraction and precision of our measurement of fecal steroids will be published in September 1966 in J. clin. Invest.

Therefore, I do not agree with Dr. KINSELL's conclusion that we are *under*estimating the fecal bile acids. Whether the conclusions Dr. KINSELL derives from his own sterol balance studies are rendered invalid, because his methods do not allow him to measure the degree of sterol degradation in the gut, remains for future work to tell. At present we do not think there is proper experimental data on which to base any conclusion regarding increased sterol excretion on unsaturated fat diets, and we intend to repeat most of our own work in order to correct for sterol degradation.

G. BOYD (Edinburgh): As a result of cholestyramine administration the enterohepatic circulation of cholesterol is impaired and, thus, the 'plasma-liver' cholesterol pool decreases. Presumably, there would be an increased fecal excretion of sterol, but we have not measured fecal sterols in these rats.

G. GOULD (Palo Alto, Cal.): So much has been said at this meeting about apolipoproteins that it is perhaps appropriate to mention some work which Dr. H. S. SODHI and I have done on the recombination of lipids with apo-HDL. When a solution of apo-HDL is equilibrated with a concentrated petroleum ether solution of mixed plasma lipids, or of pure plasma phospholipids, by gently inverting sixteen times a minute on a rotating plate at 0–4°C for 16 hours, phospholipids and free cholesterol have been found to combine to give almost as high a content as in the original HDL.

Triglycerides also combined in small amounts but only traces of esterified cholesterol combined. When apo-HDL was equilibrated with free cholesterol by itself, only a small amount combined; but when phospholipids were also present, about 6 times as much cholesterol combined. The ultracentrifugal pattern of HDL was changed by delipidization to a low, broad peak with at least 2 components, consistent with its known tendency to aggregate. Equilibration with lipids resulted in a return towards the HDL pattern indicating that when lipids combine with apo-HDL a reversal of aggregation occurs.

A.M. SCANU (Chicago, Ill.): Dr. GOULD, in your experiments on the recombination of HDL apoprotein with lipids, have you noted any effect of the solvent used to dissolve the lipids in the protein? We have evidence that nonpolar solvents alone can affect structurally apo-HDL. For this reason, in our experiments on recombination we use aqueous sols of phospholipids obtained by sonication. Under our experimental conditions a recombination between apo-HDL and HDL phospholipids can be demonstrated.

G. GOULD (Palo Alto, Cal.): The equilibration of apo-HDL with petroleum ether under our conditions does not change the behaviour of the proteins on filter paper electrophoresis, in the analytical ultracentrifuge or in any other test that we have made. The protein solution remains clear and colorless and the protein concentration decreases only slightly. The interfacial zone, which may contain small amounts of precipitated protein, is discarded. Only when *polar* organic solvents are present is the protein damaged to any considerable extent.

We have also combined lipids with apo-HDL in the absence of non-polarsolvents but this is apt to give non-specific combination of lipids with other proteins, like albumin, and lipids may be present in the aqueous solution that are not bound to protein but are difficult to remove. Results obtained in this way are a less reliable indication, we believe, of a specific combination of lipids with apo-HDL than by use of protein ether.

RUTH PICK (Chicago, Ill.): What is the clinical significance of your new findings, Dr. COHEN?

L. COHEN (Chicago, Ill.): It would be premature to answer this question now. The examples shown here are from normal persons, who have normal levels of serum lipids. Such differences are not discernable in other electrophoretic systems. When we understand the fundamental reason for the differences noted here, it will be possible to answer this question more precisely. This means for separating among otherwise similar persons may eventually prove useful in explaining different responses, for example, to the same metabolic load or even perhaps differences in disease susceptibility.

M. MANCINI (Milan): As Ig A type paraprotein tends to interact easily with other plasma components, is it possible that the complex observed by Dr. BEAUMONT may be due to 'nonspecific' physical interaction between this particular paraprotein and the plasma beta-lipoproteins, rather than to 'specific' antigen-antibody reaction?

J. BEAUMONT (Paris): Yes, Dr. MANCINI, it is possible. However, I do not think it is so. The interaction between the Ig A globulin and the β-lipoprotein found in case I looked like a specific antigen antibody reaction:
1. It is still specific for β-lipoprotein even if a large amount of the pure Ig A globulin is used in mixtures with whole serum.

2. The purified Ig A globulin (extracted from the Ig A β-Lp complexes) has a very high specific activity. 10×10^9 g of this protein still gives PHA responses when reacted with 250×10^6 RBC coated with purified β-lipoprotein.

3. This specific activity is completely inhibited by β-lipoprotein and only by them.

4. The active protein is an 'immunoglobulin'.

5. We have not found this specific activity in several other batches of Ig A globulin coming from pooled human serum.

Now, if you move to case II, there is no marked elevation of Ig A globulin in the whole serum. However, a very active immunoglobulin was extracted in this case from the low density lipoproteins.

L.A. CARLSON (Stockholm): I would like to comment on Dr. BRUNNER's finding of low a-cholesterol in coronary heart disease. In studies that we did in Stockholm 6 years ago (Acta med. scand. *167:* 377, 1960) we observed several different plasma lipid abnormalities. In the younger patients we found elevated triglycerides and decreased content of cholesterol and phospholipids in the a_1-lipoproteins while the older ones mainly had elevated cholesterol levels. There was a tendency for high triglyceride values to be associated with low a_1-lipoproteins. I certainly think that Dr. FURMAN and co-workers would agree on this point. Have you seen any correlation between low a_1-lipoproteins and high triglyceride levels?

D. BRUNNER (Jaffa): In our studies we could not find such close correlation between alpha-cholesterol percentage and triglyceride.

In 52 coronary patients 81% had less than 20% a-cholesterol which we consider as a lower limit of normal range. In the same group 58% had triglycerides higher than 120% mg.

In 350 normal people, male members of Kibbutzim, 35% had triglyceride values higher than 120 mg%. In normal middle-aged people about 40% a-cholesterol percentage lower than 20% but they were not concentrated in the group of subjects with high triglycerides.

B. RIFKIND (Glasgow): 1. Are type I cases susceptible to ischaemic heart disease?

2. How do you discriminate between type V and type IV in which there may be a secondary failure to clear dietary fat?

3. We have found atromid-S can produce a marked fall in serum triglycerides in some cases of type I. How is this explained in terms of lipoprotein lipase deficiency?

D. FREDRICKSON (Bethesda, Md.): 1. The available evidence suggests that parents with the Type I defect are not more susceptible to accelerated atherosclerosis. The sample is small; more information is needed.

2. This is not simple and we do not know whether IV and V actually represent different genotypes or simply different expressions of the same diseases. Patients with Type IV can have transient hyperchylomicronemia if given high carbohydrate diets and then a sudden heavy dietary fat load after one week. It is our impression that the response to such experimental manipulation varies considerably, and is worse in patients who are Type V on *regular* diets. There is also some differences between the distribution of Type IV and Type V patterns in kindreds. The data are too small to answer your question as firmly as we would like.

3. I cannot explain the effect of atromid-S in Type I. We have only begun to study the response of all five types. I would wonder if this patient was truly a type I and not really a Type V.

R.H. FURMAN (Oklahoma City, Okla.): Great care must be taken in the isolation of various 'lipoproteins' by means of the preparative ultracentrifuge when subjects with abnormalities of the serum globulins are under study. The use of various salt solutions necessary to increase solvent density to the desired level, may be associated with the formation of flocculates or 'complexes' consisting of β-lipoproteins, immune globulins and various other globular proteins of such serum. These may undergo sedimentation in rather large amounts in a solvent of d 1.063 g/ml, carrying down with them varying amounts of cholesterol and phospholipid. Under these circumstances analysis of the d 1.063 infranatant fraction for cholesterol would lead to the erroneous conclusion that there was an increased concentration of 'a-lipoproteins' in the serum of such subjects.

J. BEAUMONT (Paris): I agree with this remark. However, in 'auto immune hyperlipidemia' (case I) the Ig A, which was complexed with the 'heavy β-lipoprotein' (d 1.063–1.080), once purified, was found to be specifically 'anti β-lipoprotein' in immunological tests, run with physiological saline.

P. ALAUPOVIC (Oklahoma City, Okla.): What are the physical-chemical properties, in particular the flotation coefficients, of the origin and pre-β-lipoprotein bands?

To study the distribution of very low-density lipoproteins (VLDL) in various hyperlipemic syndromes Drs. GUSTAFSON, FURMAN and I developed a standardized method of differential ultracentrifugation at a solvent density of 1.006 g/ml (Biochemistry 4: 596, 1965). Three major fractions with flotation coefficients of $S_f > 400$, S_f 100–400 and S_f 20–100 could be obtained. The lipoprotein subfractions are extracted with organic solvents and the total lipids determined gravimetrically. It has been found that, on a standard diet providing 40% of calories from fat, the 'fat-induced' hyperlipemic subjects contain over 90% of the total lipid content of VLDL in the fraction with $S_f > 400$, the 'carbohydrate-accentuated' hyperlipemics have over 60% of the total lipid content in the fraction with S_f 100–400, and the so-called 'mixed type' hyperlipemics contain 40% in the fraction with $S_f > 400$ and 40% in the fraction with S_f 100–400. The normal postabsorptive subjects contain over 60% of the total lipid content of VLDL in the fraction with S_f 20–100. The advantage of this method consists in a more precise characterization of VLDL subfractions with respect to the particle size and density distribution. Since even a more elaborate subfractionation of VLDL is possible, this method represents a potential means for determination or of additional as yet unrecognized hyperlipemic types.

D. FREDRICKSON (Bethesda, Md.): We have made a number of comparisons of the bands obtained by the paper electrophoretic system used in the approach to the recognition of fat transport aspects by lipoprotein patterns. These appear in several publications, mainly in J. clin. Invest., December 1965, January and April 1966. Lack of time prevents my giving a furthermore detailed answer here to Dr. ALAUPOVIC's question.

Plasma Lipids

Department of Internal Medicine, Karolinska Sjukhuset, and King Gustaf V Research Institute, Stockholm, Sweden

Recent Advances in the Metabolism of Plasma Lipids

L. A. CARLSON, Stockholm

Introduction

A vast amount of new knowledge has been accumulating during the last decade on the metabolism of plasma lipids in animals and in man. To select one or more of these recent discoveries to represent the major recent advances is certainly difficult. I rather think that the most important recent advances should be related to the fact that we are now beginning to get a unifying functional and biochemical concept of the plasma lipids as related to transport of fatty acids.

Fig. 1 schematically summarizes what I would like to call the recent advances in the metabolism of plasma lipids. It illustrates that from the lipid transport point of view we can distinguish three different forms of fatty acid transport in blood plasma:

1. *Chylomicrons* transport the newly absorbed dietary lipids.
2. *Lipoproteins*, mainly synthesized in the liver and the intestines, transport endogenous fatty acids to the periphery.
3. *Free fatty acids (FFA)*, continuously mobilized from adipose tissue, transport fatty acids to the liver and peripheral tissues.

Table I describes that these functionally different plasma lipid transport classes also are quite different physico-chemically. While e.g. the chylomicrons contain about 1% of proteins the FFA-

Abbreviations used: FFA = free fatty acids, TG = triglycerides.

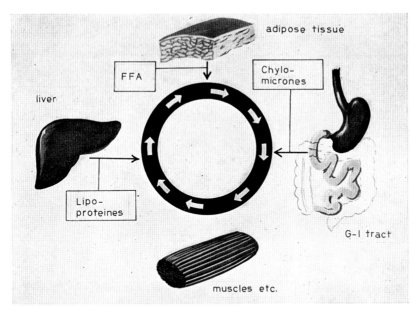

Fig. 1. The three major functional classes of blood lipid transport.

Table I. The three major lipid transport classes and their chemical form

Lipid transport class	Chemical form	Source of fatty acids
Chylomicrons	emulsion	exogenous
Lipoproteins	macromolecules	endogenous
FFA	albumin-fatty acid complex	endogenous

albumin complex consists of about 99% protein. The lipoproteins, which are made up of a spectrum of different classes, have an intermediate position with regard to protein content.

Coronary Heart Disease and Plasma Lipids

What do we know now about the plasma lipids in atherosclerotic diseases? In Table II I have summarized our knowledge about the three major lipid transport classes in one of the clinical manifestations of atherosclerosis, coronary heart disease (CHD). In interpreting these data several points should be kept in mind. CHD may well have *additional pathogenetic* elements besides atherosclerosis

Table II. The three major lipid transport classes and their behaviour in coronary heart disease

Lipid transport class	Findings in CHD	Remarks
Chylomicrons	post prandial hyperlipemia	proportional to endogenous, fasting TG-level in plasma
Lipoproteins	elevated	a) LD-LP hypercholesterolemia b) VLD-LP hypertriglyceridemia c) a+b
FFA	elevated	fasting, venous levels

CHD = coronary heart disease
LD-LP = low density lipoproteins
TG = triglycerides
VLD-LP = Very low density lipoproteins

such as thrombosis. Most of the findings in Table II are based on studies made on patients *after development* of such a severe disease as coronary heart disease. A third important point finally is that the findings in this figure apply to *group means* and not to an individual patient.

It is obvious that we may encounter abnormalities of all lipid transport classes in CHD. To what extent may such abnormalities be interrelated? It is already indicated in the figure that the abnormalities of chylomicron metabolism, as reflected in exaggerated post prandial hypertriglyceridemia, is directly related to the amount of triglyceride-rich endogenous lipoproteins present in plasma in the fasting state [1, 2]. The two major types of lipoprotein abnormalities – essential hypercholesterolemia and hypertriglyceridemia – are on the other side not related to each other. Hypertriglyceridemia may however be related to the rate of mobilization of FFA from adipose tissue. The rest of my paper will be confined to this problem and discuss possible connections between plasma FFA and plasma triglycerides (TG) which problem has interested us for the last few years.

Mobilization and Metabolism of Plasma FFA

The mobilization of FFA from adipose tissue and their uptake and subsequent metabolism in various tissues is depicted in Fig. 2.

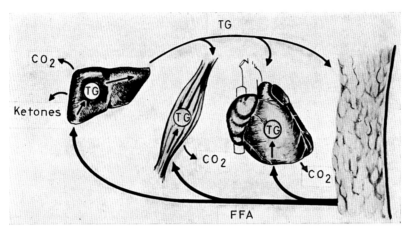

Fig. 2. The mobilization of free fatty acids (FFA) from adipose tissue and their subsequent uptake and metabolism in heart, skeletal muscle and liver. FFA are immediately esterified and can then either be stored or oxidized. In the liver they may also be resecreted into plasma as lipoprotein triglycerides (TG).

One important aspect of this metabolism is that the uptake of FFA in various tissues is proportional to the amount of FFA perfusing the organ. This implies that, in general, when the rate of mobilization of FFA from adipose tissue increases, the amount of FFA taken up in various tissues will also increase. Increased uptake of FFA may in most tissues cause either an increased oxidation of fatty acids or an increased storage.

It is however of special interest in connection with plasma TG that FFA in the liver may be incorporated into TG which in turn may be secreted into plasma as TG of the very low density lipoproteins as shown schematically in Fig. 2. It is conceivable that enhanced mobilization of FFA from adipose tissue can cause hypertriglyceridemia. Factors regulating the mobilization of FFA as well as the metabolism of FFA has been discussed in detail elsewhere [3].

Excessive Mobilization of FFA

Several studies have shown that hyperlipoproteinemia develops when the mobilization of FFA is stimulated in various ways so as to become excessive, i.e. the amount of FFA mobilized exceeds the amount of fatty acids oxidized in the body [4, 5, 6, 7, 8, 9].

We have used dogs infused with noradrenaline for the purpose of studying excessive mobilization of FFA [9, 10]. Table III summarizes some of the findings obtained with these dogs. The time course for the development of fatty liver and hypertriglyceridemia in these dogs [9] fits well with the hypothesis in Fig. 2 that excessive FFA mobilization may cause hypertriglyceridemia. It has to be realized of course that some of the results listed in Table III may be

Table III. Excessive FFA mobilization. Observations in dogs infused with noradrenaline 8–24 hours. FFA mobilization increased about five times [9, 10]

Whole animal	Tissues	
Increased O_2 consumption	*Blood:*	Hypertriglyceridemia, ketonemia,
Increased body temperature	*Liver:*	Increased triglyceride content
Dyspnea	*Lungs:*	Atelectasis, fat infiltration
∞ 50% mortality 24 hours	*Myocardium, skeletal (red) muscle, kidneys, gastrointestinal tract:*	Fatty infiltration

Summarizing data on the effects of continuous intravenous infusion of noradrenaline (0.5 μg/kg/min) to dogs.

ascribed to the cardio-vascular or other actions of noradrenaline rather than to the stimulation of FFA mobilization as discussed in more detail elsewhere [9, 10].

Inhibition of Mobilization of FFA

The mobilization of FFA may be inhibited by various means [11]. Nicotinic acid is a potent inhibitor of this mobilization under various conditions [12, 13] and we have used nicotinic acid under some of these conditions to study the effect of inhibition of FFA mobilization.

In the rat inhibition of FFA mobilization with nicotinic acid decreases the TG pools in plasma, liver, skeletal muscle and heart as summarized in Table IV [14, 15, 16]. These results are also in good agreement with the scheme in Fig. 2. *It would certainly be of greatest interest in connection with atherosclerosis to know if TG pools in arterial walls also would decrease after nicotinic acid.*

Table IV. Inhibition of FFA mobilization. Observations in fasted rats within 2–4 hours after nicotinic acid [14, 15, 16]

Blood	Red skeletal muscle
Glucose slight decrease	TG marked decrease
TG marked decrease	Ch no change
Ch slight decrease	Ph no change
Ph slight decrease	

Liver	Myocardium
TG marked decrease	TG marked decrease
Ch no change	Ch no change
Ph slight decrease	Ph no major change

TG = triglycerides	Ch = cholesterol	Ph = phospholipids

Summarizing data on the effects of subcutaneous administration of nicotinic acid (250 mg/kg) to rats.

Table V summarizes acute effects of nicotinic acid in the basal state in man [17, 18, 19]. The most marked effect besides the lowering of plasma FFA levels is a rapid decrease in ketone bodies (Fig. 2).

It is of general interest in the context of the relationship between plasma FFA and TG, and also in relation to clinical atherosclerosis, that in emotional stress accompanied by raised levels of FFA also the plsma TG increase [20]. Nicotinic acid inhibits this rise as seen in Table VI [20].

Table V. Inhibition of FFA mobilization. Observations in the basal, fasting state in man within 1–6 h after nicotinic acid

Blood	Oxygen consumption	Respiratory quotient
Glucose no major change	No major change	Slight increase
TG slight decrease		
Ch no effect		
Ph no effect		
β-HBA marked decrease		

TG = triglycerides	Ch = cholesterol	Ph = phospholipids
	HBA = hydroxybutyric acid	

Summarizing data on the effects of administration of nicotinic acid to fasting, human subjects.

Table VI. Inhibition of FFA mobilization. Observations during various conditions in man within 1–6 h after nicotinic acid

Emotional stress	*Noradrenaline infusion*
Increase in plasma TG inhibited	Increase in oxygen consumption inhibited

Summarizing data on the effects of nicotinic acid when FFA mobilization has been increased in man either by emotional stress or by infusion of noradrenaline.

Plasma FFA and Coronary Heart Disease

There are several ways in which excessive mobilization of FFA may play a role in the development of coronary heart disease. Fig. 3 illustrates three different mechanisms by means of which excessive mobilization can accelerate and/or precipitate CHD starting possibly with the early fatty streak and possibly ending the process by initiating thrombus formation. Elevated levels of triglyceride-rich lipoproteins may also play a role in the pathogenesis of the disease in several ways. However, high concentrations of TG in plasma may only be a sign of excessive FFA mobilization, the TG level being a result of an integration in time of the rate of FFA mobilization. A more detailed discussion on the possible connections between excessive FFA mobilization and CHD has been given elsewhere [3].

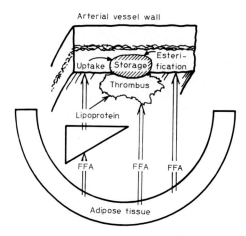

Fig. 3. Possible hypothetical mechanisms where an excessive FFA mobilization may play a role for the development of coronary heart disease. A challenge for future research [3].

Summary

A concept of three different plasma lipid transport classes is briefly discussed from the functional and biochemical point of view.

The relationship between plasma lipids and coronary heart disease is shortly reviewed against the background of three major plasma lipid transport classes.

The mobilization and metabolism of plasma free fatty acids is outlined with main emphasis on consequences in the intact organism of excessive and inhibited rate of mobilization from adipose tissue.

Finally the possible role of FFA mobilization in coronary heart disease is discussed.

Acknowledgement

Supported by grants from Svenska Nationalföreningen mot Hjärt – Kärlsjukdomar and Syskonen Wesséns Stiftelse.

References

1. ANGERVALL, G.: On the fat tolerance test. Acta med.scand. Suppl. 424 (1964).
2. PELKONEN, R. and NIKKILÄ, E.A.: Plasma transport of alimentary triglycerides in coronary heart disease. Acta med.scand. *178:* 511–575 (1965).
3. CARLSON, L.A.; BOBERG, J. and HÖGSTEDT, B.: Some physiological and clinical implications of lipid mobilization from adipose tissue. Handbook of Physiology V. Adipose Tissue, pp. 625–644. Amer.Physiol.Soc. (Washington 1965).
4. DURY, A.: Lipide distribution and phospholipide turnover on tissues of rabbit shortly after growth hormone. Proc.Soc.exp.Biol.Med. *90:* 623–628 (1955).
5. DURY, A.: Effect of epinephrine on lipid partition and metabolism of the rabbit. Circulat.Res. *5:* 47–53 (1957).
6. DURY, A. and TREADWELL, C.R.: Effect of epinephrine on plasma lipid components and interrelationships in normal and epileptic humans. J.clin.Endocrin. *15:* 818–825 (1955).
7. GILDEA, E.F. and MAN, E.B.: The effect of epinephrin on the blood lipoids of normal man. J.clin.Invest. *15:* 295–300 (1936).
8. SHAFRIR, E.; SUSSMAN, K.E. and STEINBERG, D.: The nature of epinephrine induced hyperlipidemia in dogs and its modification by glucose. J.Lipid Res. *1:* 109–117 (1959).
9. CARLSON, L.A.; LILJEDAHL, S.-O. and WIRSÉN, C.: Blood and tissue changes in the dog during and after excessive free fatty acid mobilization. A biochemical and morphological study. Acta med.scand. *178:* 81–102 (1965).
10. BIRKE, G.; CARLSON, L.A.; LILJEDAHL, S.-O. and ZETTERQUIST, E.: Unpublished results.
11. CARLSON, L.A. and BALLY, P.R.: Inhibition of lipid mobilization. Handbook of Physiology V. Adipose Tissue, pp. 557–574. Amer.Physiol.Soc. (Washington 1965).

12. Carlson, L.A. and Orö, L.: The effect of nicotinic acid on the plasma free fatty acids. Demonstration of a metabolic type of sympathicolysis. Acta med. scand. *172:* 641–645 (1962).
13. Carlson, L.A.: Studies on the effect of nicotinic acid on catecholamine stimulated lipolysis in adipose tissue in vitro. Acta med. scand. *173:* 719–722 (1963).
14. Carlson, L.A. and Nye, E.R.: Acute effects of nicotinic acid in the rat. I. Plasma and liver lipids and blood glucose. Acta med. scand. *179:* 453–461 (1966).
15. Carlson, L.A.; Fröberg, S.O. and Nye, E.R.: Acute effects of nicotinic acid on plasma, liver, heart and muscle lipids. Nicotinic acid in the rat. II. Acta med. scand. *180:* 571–579 (1966).
16. Carlson, L.A.: Consequences of inhibition of normal and excessive lipid mobilization. Studies with nicotinic acid. Progr. biochem. Pharmacol., vol. 3, pp. 151–166 (Karger, Basel/New York 1967).
17. Havel, R.J.; Carlson, L.A.; Ekelund, L.-G. and Holmgren, A.: Studies on the relation between mobilization of free fatty acids and energy metabolism in man: Effects of norepinephrine and nicotinic acid. Metabolism *13:* 1402–1412 (1964).
18. Carlson, L.A. and Östman, J.: Plasma β-hydroxybutyric acid response to nicotinic acid-induced plasma free fatty acid decrease in man. Diabetologia *2:* 127–129 (1966).
19. Carlson, L.A. and Östman, J.: Unpublished results.
20. Carlson, L.A.; Levi, L. and Orö, L.: Plasma lipids and urinary excretion of catecholamines in man during experimentally induced emotional stress with and without administration of nicotinic acid (in preparation).

Author's address: Lars A. Carlson, M.D., King Gustaf V Research Institute, *Stockholm 60* (Sweden).

From the Department of Medicine (Ludolf-Krehl-Clinic), University of Heidelberg, Germany (Director: Prof. Dr. G. Schettler)

Conjugated Bile Acids in Serum of Patients with Essential Hyperlipemia and Hypercholesterolemia

B. Frosch and H. Wagener, Heidelberg

In 1952 Friedman and co-workers reported increased plasma cholate concentrations in patients with essential hypercholesterolemia and proposed that elevated plasma bile acid levels were related to the pathogenesis of this metabolic disease. According to their theory diminished transfer of plasma cholesterol into liver cells would result from alteration of cholesterol-protein binding caused by the increase in bile acid concentration. However, in 1964 Osborn and Wootton showed that methodological difficulties in bile acid quantitation may have influenced the theory of Friedman and co-workers. Sandberg and co-workers (1965) did not detect increased plasma bile acid concentrations in two patients with essential hyperlipemia.

We developed a method for quantitative determination of conjugated serum bile acids which consists of extraction of serum with hot ethanol, elimination of interfering neutral lipids by distribution with ether-ethanol-heptane-water (1:1:1:1, v/v), diminution of phospholipids, monoglycerides, and bilirubin and separation of conjugated bile acids by thin-layer chromatography (Frosch, 1965).

Fig. 1 shows a thin-layer chromatogram of conjugated serum bile acids with the solvent system butanol-acetic acid-water (10:1:1, v/v). The individual spots correspond to taurocholic acid, taurochenodeoxycholic and taurodeoxycholic acids, glycocholic acid, glycochenodeoxycholic and glycodeoxycholic acids respectively. Bilirubin remains at the starting line, phospholipids and monoglycerides migrate with the solvent front. After elution of the bile acids from the corresponding silica gel areas they are individually

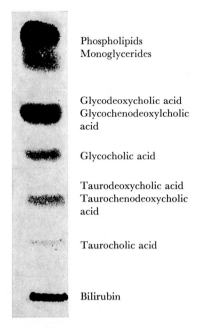

Fig. 1. Thin-layer chromatogram of conjugated serum bile acids (solvent system: butanol-acetic acid-water, 10:1:1, v/v, spray: phosphomolybdic acid).

estimated by spectrophotometry. Pure bile acids when added to plasma samples are quantitatively recovered. The sensitivity of this method permits the determination of 0.5 µg of each conjugated bile acid per milliliter serum. In normals no conjugated bile acids are present in serum.

Using this method conjugated bile acids were determined in serum of 12 patients with essential hyperlipemia and essential hypercholesterolemia (Table I). In all cases increased concentrations of conjugated bile acids were found ranging from 25 to 139 µg/ml. The main fractions in most cases were conjugates of chenodeoxycholic acid. In 8 of these patients liver biopsies were performed which revealed fatty livers of different degrees in 7 cases. In only one hyperlipemic patient (No. 95) liver steatosis was absent.

These observations led us to determine conjugated serum bile acids in 6 normolipemic patients who had been found by biopsy to have fatty livers (Table II). In all of these patients conjugated serum bile acids were also elevated with values ranging from 13.5 to 72 µg/ml.

Essential Hyperlipemia and Hypercholesterolemia

Table I. Concentrations of lipids and conjugated bile acids in serum of patients with essential hyperlipemia and essential hypercholesterolemia

No. Diagnosis*		90 e.Hl.	91 e.Hl.	92 e.Hl.	93 e.Hl.	94 e.Hl.	95 e.Hl.		96 e.Hl.	97 e.Hl.	98 e.Hl.		99 e.Hch.	100 e.Hch.		101 e.Hch.
							a)	b)			a)	b)		a)	b)	
Taurocholic acid	(μg/ml)	2.5	4.5	4.5	6	2.5	6	4	4.5	5	6	10	6	5	3	4
Taurochenodeoxycholic acid	(μg/ml)	13.5	0	25.5	84	14	61	76	37.5	23	29	23	52.5	35	36	32
Taurodeoxycholic acid	(μg/ml)	5.5	2	1	6	4	0	3	0	0	5	1.5	9.5	4	1.5	8
Taurine conjugates	(μg/ml)	21.5	6.5	31	96	20.5	67	83	42	28	40	34.5	63	44	40.5	44
Glycocholic acid	(μg/ml)	2	3.5	4	6	5	3	2	4	3	4.5	6	2	1	1	4
Glycochenodeoxycholic acid	(μg/ml)	0	10	9	32	7	25.5	16	27.5	4	15.5	9.5	16	30	12	30
Glycodeoxycholic acid	(μg/ml)	1.5	11.5	0.5	5	4	0	2	0	0	1	1.5	7.5	1.5	1	10
Glycine conjugates	(μg/ml)	3.5	25	13.5	43	16	30.5	20	31.5	7	21	17	25.5	32.5	14	40
Total conjugated bile acids	(μg/ml)	25	31.5	44.5	139	36.5	97.5	103	73.5	35	61	51.5	93.5	76.5	54.5	88
Total lipids	(mg%)	890	1315	1060	1620	1095	1125		1060	905	2615		900	1950	1090	1015
Total cholesterol	(mg%)	300	355	199	500	342	260		240	196	385		364	550	277	339
Fatty liver					+	+			+		+				+	+

* e.Hl. = essential Hyperlipemia, e.Hch. = essential Hypercholesterolemia.

Table II. Concentrations of conjugated bile acids in serum of normolipemic patients with fatty livers

No.		84	85	86	87	88	89
Taurocholic acid	(μg/ml)	4	4	8	5	7	4.5
Taurochenodeoxycholic acid	(μg/ml)	12	0	0	24	34	37.5
Taurodeoxycholic acid	(μg/ml)	3	5.5	0	0.5	2	0
Taurine conjugates	(μg/ml)	19	9.5	8	29.5	43	42
Glycocholic acid	(μg/ml)	2	1.5	6	3	4	2.5
Glycochenodeoxycholic acid	(μg/ml)	5	0	0	4.5	0	27.5
Glycodeoxycholic acid	(μg/ml)	0.5	2.5	3	1	6	0
Glycine conjugates	(μg/ml)	7.5	4	9	8.5	10	30
Total conjugated bile acid	(μg/ml)	26.5	13.5	17	38	53	72

The finding of elevated concentrations of conjugated bile acids in serum of normolipemic and hyperlipemic patients with fatty livers requires further investigation with particular emphasis on liver cell function. So far our results confirm and extend the above mentioned findings of FRIEDMAN and co-workers (1952).

References

FRIEDMAN, M.; BYERS, S.O. and ROSENMAN, R.H.: The accumulation of serum cholate and its relationship to hypercholesteremia. Science *115:* 313–315 (1952).
FROSCH, B.: Die quantitative Bestimmung der konjugierten Gallensäuren des Serums nach dünnschichtchromatographischer Trennung. Arzneimittelforsch. *15:* 178–184 (1965).
OSBORN, E.C. and WOOTTON, J.D.P.: Serum bile acid levels in hypercholesterolaemia. J.clin. Path. *17:* 156–159 (1964).
SANDBERG, D.H.; SJÖVALL, J.; SJÖVALL, K. and TURNER, D.A.: Measurement of human serum bile acids by gas-liquid chromatography. J.Lipid Res. *6:* 182–192 (1965).

Authors' address: B. Frosch, M. D. and H. Wagener, M. D., Bergheimerstr. 58, *69 Heidelberg* (Germany).

From the IIIrd Department of Medicine, University of Budapest
Director: Prof. S. Gerö

Gastric Mucosubstances and Serum Lipid Pattern

S. Gerö, M. Bihari-Varga and J. Székely, Budapest

It is known that a lipid metabolism disorder occupies an important role in the process of atherogenesis. According to the findings of Bronte-Stewart and others [1, 2, 3], after a fat meal, alimentary lipaemia is greater and more prolonged in coronary patients than in matched controls. However these differences on feeding fat orally were abolished when the same fat was given intravenously [2, 4]. These findings pointed to the possibility that the differences shown, probably arose from absorptive mechanisms [5]. Furthermore, it is known that serum lipid levels are extremely low in patients with Addisonian anaemia, gastric atrophy, or after gastrectomy. Again, Cantone and his coworkers [6, 7, 8] and others [9, 10, 11] found, that the oral administration of an aminopolysaccharide mixture extracted from hog's gastric mucosa reduced the plasma lipid levels in patients with atherosclerosis. Similar effect of heparinoid substances of duodenal origin has been established too [12, 13, 14].

These previous findings drew our attention on the possibility that some mucoid substances of gastrointestinal origin may play a role in the *physiological* regulation of plasma lipid pattern.

In our experiments the correlation between the concentration of some gastric mucopolysaccharide (MPS) components and that of some serum lipid parameters was investigated in the same individuals. On the other hand the relationships between the MPS values in the gastric juice and the MPS values in the serum were studied. The concentration of hexosamine, sialic acid, fucose, hexuronic acid, protein-bound hexose and that of Winzler's mucoid substance according to the methods of Elson-Morgan [15], Ayala and coworkers [16], Jacubeit and coworkers [17], Dische [18], Stary and coworkers [19], Weimer and Redlich-Moshin [20] - was determined. From among the serum lipid parameters the total lipid, cholesterol, and beta-lipoprotein values were estimated, according to the methods of KUNKEL and coworkers [21], RAPPAPORT and EICHHORN [22], BOYLE and MOORE [23]. Gastric juice, obtained from 150 control subjects, without coronary, gastrointestinal or metabolic disease, was aspirated without any stimulation, after an overnight fast. The same parameters were determined in the serum and in the gastric juice of 75 hypercholesterolaemic patients with serum cholesterol values above 250 mg%, suffering of different diseases, except gastrointestinal disturbances.

The hypercholesterolaemic patients were subdivided according to the clinical diagnosis: patients with ischaemic heart disease (after myocardial infarction, – 24 cases), patients with diabetes mellitus (15 cases) and subjects with hyperlipaemia of unknown origin (36 cases).

Fig. 1a shows the relation between the concentration of gastric hexosamine (mg/g protein) and serum cholesterol (mg%) level in

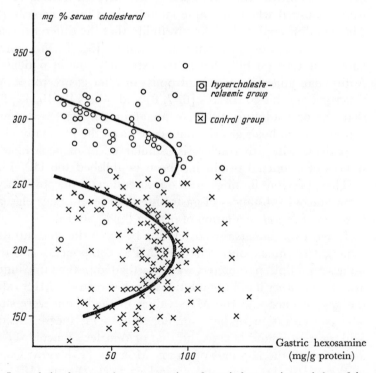

Fig. 1a. Interrelation between the concentration of gastric hexosamine and that of the serum cholesterol.

hypercholesterolaemic and control subjects. Upper curve is a scatter diagram for hypercholesterolaemic, lower one for control subjects. Curves are drawn by eye. The bulk of the MPS values cluster around 90 mg/g protein values. In the case of the hypercholesterolaemic patients only the upper branch of the curve could be detected.

Similar results were obtained when investigating the relation between sialic acid or mucoid content of gastric juice and serum cholesterol values (Fig. 1b and 1c).

No connection could be detected between the other MPS components (hexose, fucose, hexuronic acid) investigated by us and the serum cholesterol values. There appeared no interdependence either when plotting the serum total lipid or beta-LP values against the concentration of gastric MPS components.

Table I represents the average values of gastric MPS components in control and hypercholesterolaemic subjects. The difference between the average values of gastric hexose, hexosamine, sialic acid

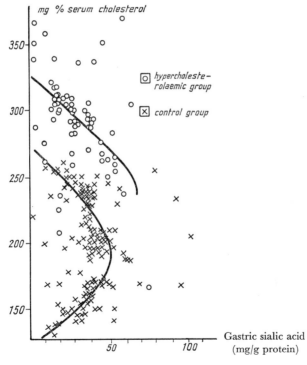

Fig. 1b. Interrelation between the concentration of gastric sialic acid and that of the serum cholesterol.

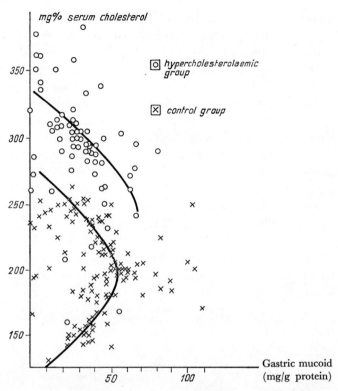

Fig. 1c. Interrelation between the concentration of gastric mucoid and that of the serum cholesterol.

Table I. Average values of MPS components in gastric juice (mg/g protein)

	Hexose	Hexos-amine	Sialic acid	Winzler mucoid	Fucose	Hexuronic acid
Control group	143.2	76.4	37.7	42.0	2.23	3.35
Hypercholesterol-aemic group	123.6	62.0	28.8	30.1	2.56	2.82

Statistical significance of the deviation between the values obtained in the hypercholesterolaemic and control group

	Hexose	Hexos-amine	Sialic acid	Winzler mucoid	Fucose	Hexuronic acid
t	1.99	3.3	3.6	3.99	not significant	not significant
P %	5*	1**	0.1***	0.1***		

* = significant   ** = highly significant   *** = very highly significant.

and Winzler-mucoid components in the hypercholesterolaemic and control group, proved to be statistically significant.

Again, some interesting results were obtained when subdividing the hypercholesterolaemic patients according to the clinical diagnosis.

Table II shows the average values of gastric MPS components obtained in the three groups of hyperlipaemic patients and in that

Table II. Average values of MPS components in gastric juice (mg/g protein)

Diagnosis	Hexose	Hexosamine	Sialic acid	Winzler mucoid	Fucose	Hexuronic acid
Control	143.2	77.7	37.7	42.0	2.23	3.35
Coronary heart disease	111.9	47.0	23.2	23.1	2.34	2.86
Diabetes mellitus	157.1	71.3	39.0	30.6	2.51	3.57
Lipaemia of unknown origin	114.9	62.9	27.6	36.0	2.50	2.62

Statistical significance of the deviation between the values obtained in the hypercholesterolaemic groups and the control group

		Hexose	Hexosamine	Sialic acid	Winzler mucoid	Fucose	Hexuronic acid
Coronary heart disease	t	2.0	4.8	3.9	4.0	not significant	not significant
	P %	5.0*	0.1***	0.1***	0.1***		
Diabetes mellitus	t	0.6	0.7	0.2	1.7	not significant	not significant
	P %	50.0	40.0	80.0	5.0		
Lipaemia of unknown origin	t	2.1	2.6	3.1	1.4	not significant	not significant
	P %	5.0*	5.0*	1.0**	10.0		

* = significant   ** = highly significant   *** = very highly significant.

of the control subjects. It can be seen that the difference between the MPS concentration in the gastric juice of patients with coronary heart disease and of control ones is highly significant. Similarly in the group with hyperlipaemia of unknown origin the concentration of the same MPS components was significantly lower, than in the control group. No significant differences could be detected when examining the average values of MPS components obtained in the case of diabetes mellitus.

There were some relations recognized also between the concentration of some MPS components of the gastric juice and that of the serum in the same subjects.

Fig. 2 shows a scatter diagram of gastric hexosamine (mg/g protein) and serum hexosamine (mg %) levels. Upper line for hypercholesterolaemic, lower one for control subjects. Lines are drawn by eye. The regression of the serum hexosamine values on the gastric ones is linear and the statistical analysis proved the b value significantly different from zero. The same holds for Winzler's mucoid materials.

No interrelation could be established for the other MPS components examined by us.

The important role of gastric mucus is indicated by the circumstance that intrinsic factor, blood group activity are linked to some if it's carbohydrate components.

Our findings seem to support the concept, that there may be a *functional* correlation between the MPS content of serum and gastric juice and that some gastric MPS components may play a role in the physiological regulation of blood lipid pattern.

Any explanation for the findings reported above would be highly speculative in our present state of knowledge.

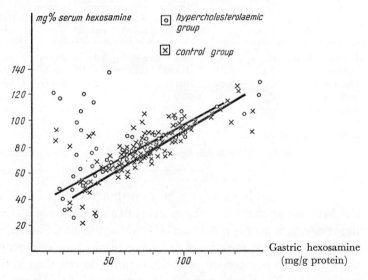

Fig. 2. Interrelation between the concentration of hexosamine in gastric juice and in serum.

Acknowledgement

Authors are greatly indepted to Dr. I. JUVANCZ and Dr. P. CSÁKY (Biometric Department of the Mathematical Institute of the Hungarian Academy of Sciences) for performing the statistical analysis.

Summary

Correlations between the concentration of some gastric mucopolysaccharide (MPS) components and that of some serum lipid parameters were investigated in the same individuals. The relation between the hexosamine, Winzler's mucoid and sialic acid content of the gastric juice and the serum cholesterol level was found to be significant in both groups. On the other hand the relationship between the MPS values in the gastric juice and the MPS values in the serum was studied: there was significant interrelation between gastric and serum hexosamine content. – Some further significant differences were established between the average values of gastric MPS components of the control subjects and the hypercholesterolemic patients. The findings seem to support the concept, that some gastric MPS components may play a role in the physiologic regulation of blood lipid pattern.

References

1. BRONTE-STEWART, B. and BLACKBURN, H.J.: Essential fatty acids. pp. 180–181 (Livingstone, London 1958).
2. BOUCHIER, J.A.D. and BRONTE-STEWART, B.: Alimentary lipaemia and ischaemic heart disease. Lancet i: 363–366 (1961).
3. BARRITT, D.W.: Alimentary lipaemia in men with coronary artery disease and in controls. Brit.med.J. ii: 640–644 (1956).
4. NESTEL, P.J. and DENBOROUGH, M.A.: Alimentary lipaemia and ischaemic heart disease. Lancet ii: 828–829 (1961).
5. MITCHELL, J.R.A. and BRONTE-STEWART, B.: Alimentary lipaemia and heparin clearing in ischaemic heart disease. Lancet i: 167–169 (1959).
6. CANTONE, A.; RULLI, V. and ROSSI, B.: Antilipaemic activity of an extract of hog gastric mucosa. Circulat.Res. 7: 291–294 (1959).
7. CAPRARO, V. et CANTONE, H.: The possible physiological rate and mechanism of action of an aminopolysaccharide isolated from hog gastric mucosa. Arch.ital. Biol. 96: 187–196 (1958).
8. CAPRARO, V.; CRESSERI, A. und CANTONE, A.: Über die Anwesenheit einer resorbierbaren mit 'clearing' Vermögen versehenen Substanz in normalem Magensaft. Naturwissenschaften 43: 36–37 (1956).
9. ROSSI, B. and RULLI, V.: Plasma clearing effect of gastric mucin in healthy and atherosclerotic subjects under basal conditions. Circulation 18: 397–399 (1958).
10. RULLI, V. and ROSSI, B.: Plasma turbidity changes and electrocardiographic alterations induced by alimentary hyperlipemia in anginal patients before and after the administration of gastric mucin. Circulation 18: 400–404 (1958).
11. GOLDECK, H.: Klinische Studien zur lipotropen Muzin-Aktivität. Ärztl. Forsch. 12: 53–58 (1958).
12. BIANCHINI, P. et OSIMA, B.: Studi su alcune sostanze eparinoidi. Atti Soc.Lomb.Sci. Med.Biol. 13: 71–75 (1958).

13. GERÖ, S.; RÉTSÁGI, GY. und GERGELY, J.: Erfahrungen mit anhaltender Ateroidbehandlung. Z. ges. inn. Med. *17:* 603–604 (1962).
14. RÉTSÁGI, GY.; KELLER, L. and GERÖ, S.: Observations on the possibility of influencing hyperlipaemia by different drugs. Geront. clin. *6:* 22–27 (1964).
15. ELSON, L.A. and MORGAN, W.T.J.: A colorimetric method for the determination of glucosamine and chondrosamine. Biochem. J. *27:* 1842–1844 (1933).
16. AYALA, W.; MOOR, L.V. and HESS, B.L.: Purple color reaction given by diphenylamine reagent. J. clin. Invest. *30:* 781–785 (1951).
17. JACUBEIT, M.; BRUNGER, P. und KNEDEL, M.: Untersuchungen über den Gehalt an Fucose im Serum und in Proteinfraktionen. Klin. Wschr. *37:* 460–465 (1959).
18. DISCHE, Z.: A new specific color reaction of hexuronic acids. J. biol. Chem. *167:* 189–193 (1947).
19. STARY, Z.; BURSA, F.; KALEOGLÜ, O. et BILEN, M.: Méthode pour la détermination des carbohydrates. Bull. Fac. Med. Istanbul *13:* 243–247 (1950).
20. WEIMER, H.E. and MOSHIN, J.R.: Serum glycoprotein concentrations in experimental tuberculosis of guinea pigs. Amer. Rev. Tuberc. *68:* 594–602 (1953).
21. KUNKEL, A.G. and AHRENS, E.H.: Relationship between serum lipids and electrophoretic pattern. J. clin. Invest. *28:* 1575–1579 (1949).
22. RAPPAPORT, F. and EICHHORN, F.: Sulfosalicylic acid as a substitute for paratoluene sulfonic acid. A. In the estimation of cholesterol. B. In the diagnostic test for systemic lupus erythematosus. Clin. chim. Acta *5:* 161–163 (1960).
23. BOYLE, E. and MOORE, R.V.: A new precipitation method for estimating serum beta lipoproteins. J. lab. clin. Med. *53:* 272–281 (1959).

Authors' address: S. Gerö, Prof., M. Bihari-Varga, J. Székely, IIIrd Department of Medicine, University of Budapest, VIII. Mezö Imre ut 17., *Budapest* (Hungary).

From the Department of Medicine, Stanford School of Medicine, Palo Alto, Cal.

Metabolism of Plasma Lipids

R.G. Gould and E.A. Swyryd, Palo Alto, Cal.

The metabolism of plasma lipids is inseparably connected with the metabolism of plasma lipoproteins since all the lipids in plasma are bound to specific proteins. Any change in level of a plasma lipid is generally a result of a change in level of one or more lipoprotein classes, rather than of a change in composition of a lipoprotein, although both may occur. A decrease in the plasma level of a lipoprotein must be due either to: 1) a decreased rate of secretion from the liver, incompletely compensated by a decrease in the rate of disappearance or 2) to an increased rate of disappearance from plasma incompletely compensated by an increase in the rate of secretion from liver. Changes in the rate of secretion of lipoproteins from liver may be due to a variety of causes; for example, large increases in the supply of free fatty acids to liver stimulate the synthesis of triglycerides in liver and may result in an increase in the release of triglyceride-rich lipoproteins into plasma. Dietary cholesterol results in storage of cholesterol esters in liver and may result in an increase in plasma level of cholesterol (and of phospholipids) due primarily to an increase in low density lipoproteins. Conversely, inhibition of hepatic cholesterol synthesis often results in decreased plasma levels of cholesterol and of phospholipids as well.

Investigation of the effects of lipid lowering drugs may help to reveal not only the mechanism of action of the drug but is also a useful tool in elucidation of the physiological mechanisms of control of plasma lipoprotein levels.

When the plasma lipid levels are altered by means of drugs, diet or other methods, the cholesterol and phospholipid levels tend to change in the same direction and often to more or less the same degree whereas triglyceride levels may behave quite differently than cholesterol levels. For example, BEST and DUNCAN have recently reported results on the responses of groups of 8–12 patients to four hypocholesterolemic drugs [1]. CPIB (Atromid S, Clofibrate) decreased cholesterol levels by about 16% and triglycerides by 35%. Dextrothyroxine decreased cholesterol by 18% and triglycerides by only 9% and neomycin by 35% and 25% respectively. β-sitosterol decreased cholesterol by 12% but did not cause consistent decrease in triglycerides. This evidence suggests that CPIB decreases primarily the very low density and β-sitosterol primarily the low density lipoproteins.

Although a great many studies have been done on the effects of these and other drugs on plasma lipid levels, relatively few studies have been concerned with their mechanism of action. During the past few years, we have been studying the effects of CPIB in rats, particularly the effects of this drug on the synthesis of liver lipids and on their rate of appearance in serum in lipoprotein form. In 1965 we reported that when CPIB was fed to rats at a level of 0.3% of the diet for several days, a decrease of about 50–75% was observed in the hepatic synthesis of cholesterol per g of liver [2]. However, this drug increases the liver size in rats by about 25–30% [3, 4, 2]; the rate of synthesis of cholesterol calculated for the whole liver, when estimated by incubation of liver slices with acetate-^{14}C, decreased by about 60%. A similar, although somewhat smaller effect was observed by *in vivo* methods, using various precursors of cholesterol, including acetate-^{14}C and tritium water.

The next question we investigated was the location of the decreased rate of synthesis in the long chain of reactions involved in cholesterol synthesis from acetate. When mevalonate was used as a precursor instead of acetate, we found no inhibition [2]. Results of a more recent experiment are shown in Table I and indicate that CPIB decreases the formation of cholesterol from acetate, but not from mevalonate. These results are expressed in terms of equal weights of liver and if corrected for the change in liver weight induced by CPIB would show an increase in cholesterol synthesis from mevalonate.

AZARNOFF *et al.* [4], on the other hand, have found a decreased conversion of mevalonate to cholesterol but the conditions of their experiment were different from ours since they used fasted rather than normally fed rats and longer periods of drug treatment. They did not investigate the effect of CPIB on cholesterol synthesis from

Table I. Effects of CPIB administration on cholesterol synthesis from acetate and from mevalonate by liver homogenates

Synthesis in DPM per gram of liver:

	Acetate	Mevalonate
	Cholesterol	
Control	8.36±2.68	53.35±4.41
CPIB	1.51±0.42	62.05±4.35
% Change	—82%	+16%
P	<.05	<.05
	Non-saponifiable fraction	
Control	11.18±3.35	114.5±3.08
CPIB	3.19±1.14	112.5±2.82
% Change	—71.5%	—1.7%
P	<.05	

Nine male Sprague-Dawley rats in the weight range of 172–207 were used. Four were fed stock diet *ad lib*. and five were fed a stock diet containing 0.3% CPIB for 6 days *ad lib*. Aliquots of liver homogenates from each rat were incubated for 2 hours with acetate-1-^{14}C at a concentration of 20 μm = 2 μc per g of liver or with mevalonate-2-^{14}C at a concentration of 4 μm = 0.4 μc per g of liver. After precipitation with perchloric acid, saponification of the precipitate with alcoholic alkali and extraction with petroleum ether, one aliquot of the non-saponifiable extract was counted. This would contain all the intermediates from farnesol on, in addition to cholesterol. From another aliquot, cholesterol was isolated as the digitonide and counted.

acetate so it is not clear whether the effect on mevalonate incorporation may be secondary to a more prompt effect on the conversion of acetate to mevalonate. Dietary cholesterol has a definite inhibitory effect on the conversion of mevalonate to cholesterol but it is secondary to the more pronounced effect on the conversion of acetate to mevalonate and develops more slowly [5].

It is of interest to note that a number of other drugs that inhibit cholesterol synthesis act very late in the synthetic chain and give rise to the accumulation of abnormal intermediates, particularly desmosterol. CPIB inhibits very early in the synthetic chain at a site before mevalonate and in the same general area as such physiological regulators of cholesterol synthesis as dietary cholesterol or the cholesterol concentration in liver. Since CPIB does not inhibit fatty acid synthesis or ketone body formation, it may be postulated that CPIB blocks cholesterol synthesis, either directly or indirectly, at a point very close to mevalonate, probably at the reduction of HMGCoA to mevalonate.

The question of whether the effect of CPIB in decreasing the plasma level of cholesterol is due to its depressing effect on cholesterol synthesis in liver cannot be answered at present. However, it is worth noting that a number of drugs which do inhibit cholesterol synthesis in liver also result in a decrease in the plasma cholesterol level and generally also in a decrease in the phospholipid level. Triparanol for example, is known to be an inhibitor of the Δ^{24} sterol reductase necessary for the conversion of desmosterol to cholesterol and its administration is associated with a definite decrease in the cholesterol level in both plasma and liver.

Since CPIB decreases the plasma level of triglycerides even more than that of cholesterol, it was of interest to see if CPIB produces a decrease in triglyceride synthesis in liver. In the experiments already cited in which cholesterol synthesis was decreased, triglyceride synthesis was found to be not decreased and, in fact, was significantly increased. This was measured both *in vitro* and *in vivo* and it was also estimated by means of several different precursors including glycerol-^{14}C, tritium water and acetate-^{14}C [6].

The rates of appearance in plasma of newly synthesized triglycerides and cholesterol were nevertheless decreased in these experiments. When the plasma and liver lipids were isolated four hours after the injection of labeled acetate it was found that the amounts of labeled triglycerides and cholesterol present per ml of serum were considerably less in the CPIB-treated rats.

Recently we have been studying the time curves of newly synthesized plasma triglycerides to obtain information on whether the disappearance rate from plasma is changed. Figure 1 shows the time curves expressed in terms of disintegrations per minute per ml of serum for triglycerides. The rate of disappearance from plasma was not detectably different in the drug treated animals than in the controls but the peak values reached were consistently lower in the drug treated animals. Since the specific activity of liver triglycerides was higher in the drug treated animals than in the controls, a considerably smaller amount of newly synthesized triglycerides were present in plasma at the time the measurements were made. Corresponding curves for phospholipids gave very similar results. In other experiments in which leucine-^{3}H was used as a precursor of the protein components of the lipoproteins as well as of the lipids, it was found that both low density and high density lipoproteins showed a decrease in all the constituents of the lipo-

DPM of ^{3}H in serum triglycerides per ml of serum following glycerol -^{3}H injection.

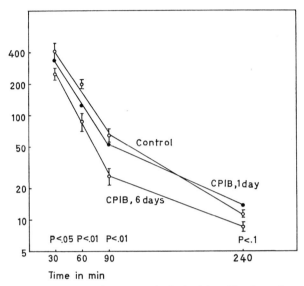

Fig. 1. Rates of disappearance of newly synthesized triglycerides from plasma in control and crib-treated rats. Eighteen male Sprague-Dawley rats in the weight range of 220–250 grams were randomized into three groups:
  I. Control – maintained on stock diet.
  II. CPIB – 1 day – on stock diet containing 0.3% CPIB for one day, fed ad lib.
  III. CPIB – 6 days – on stock diet containing 0.3% CPIB for six days, fed ad lib.

Glycerol-2-^{3}H was injected intraperitoneally in a dose of 100 μc per 100 g of body weight. Blood samples (0.3 ml) were obtained from the tail tip at 30, 60 and 90 minutes and the animals were killed at 240 minutes.

Triglycerides and phospholipids were isolated from aliquots of serum by the method of Mendelsohn and Antonis [9] and counted in a liquid scintillation counter.

The ordinate gives disintegrations per minute (in thousands) per ml of serum on a logarithmic scale.

Vertical lines at each point indicate ±1 standard error of the mean and the P values indicate the confidence limits for the difference between the control and CPIB-6 day values.

proteins [7]. These results suggest that the drug may interfere with the release of lipoprotein molecules from liver.

Nestel et al. [8] have recently measured the effect of CPIB on the rate of disappearance of labeled cholesterol from plasma in several patients and found a small but definite decrease in the rate of disappearance. This agrees with the data on rats discussed above since a decrease in synthesis would be expected to cause a slower rate of dilution of labeled plasma cholesterol with unlabeled and in consequence a slower rate of disappearance.

DPM of ^{3}H in serum phospholipids per ml of serum following glycerol -^{3}H injection.

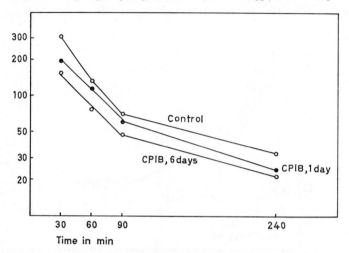

Fig. 2. Rates of disappearance of newly synthesized phospholipids from plasma in control and CPIB-treated rats. Phospholipid data from the same experiment described in Fig. 1.

The CPIB-treated animals had only 64% as high a value at 240 minutes as the control and the P value was <0.02.

References

1. BEST, M.M. and DUNCAN, C.H.: Effects of cholesterol-lowering drugs on serum triglycerides. J. Amer. Med. Assoc. *187:* 137–40 (1964).
2. AVOY, D.R.; SWYRYD, E.A. and GOULD, R.G.: Effects of α-p-chlorophenoxyisobutyryl ethyl ester (CPIB) with and without androsterone on cholesterol biosynthesis in rat liver. J. Lipid Res. *6:* 369–76 (1965).
3. BEST, M.M. and DUNCAN, C.H.: Hypolipemia and hepatomegaly from ethyl chlorophenoxyisobutrate (CPIB) in the rat. J. lab. clin. Med. *64:* 634–42 (1964).
4. AZARNOFF, D.L.; TUCKER, D.R. and BARR, G.A.: Studies with ethyl chlorophenoxyisobutyrate (Clofibrate). Metabolism. *14:* 959–65 (1965).
5. GOULD, R.G. and SWYRYD, E.A.: Sites of control of hepatic cholesterol biosynthesis. J. Lipid Res. *7:* 698–707 (1966).
6. GOULD, R.G.; SWYRYD, E.A.; COAN, B. and AVOY, D.R.: Effects of chlorophenoxyisobutyrate (CPIB) on liver composition and triglyceride synthesis in rats. J. Atheroscler. Res. (in press).
7. GOULD, R.G.; SWYRYD, E.A.; AVOY, D.R. and COAN, B.: The effects of CPIB on the synthesis and release into plasma of lipoproteins in rats. Progr. biochem. Pharmacol., vol. 2, pp. 345–357 (Karger, Basel/New York 1967).
8. NESTEL, P.J.; HIRSCH, E.Z. and COUZENS, E.A.: The effects of chlorophenoxyisobutyric acid and ethinyl estradiol on cholesterol turnover. J. clin. Invest. *44:* 6 (1965).
9. MENDELSOHN, D. and ANTONIS, A.: A fluorimetric micro glycerol method and its application to the determination of serum triglycerides. J. Lipid Res. *2:* 45–50 (1961).

Author's address: R. G. Gould, Ph. D. Stanford University School of Medicine, *Palo Alto, Cal.* (USA).

From the Department of Medicine, Stanford University School of Medicine, Palo Alto, California

Turnover of Endogenous Plasma and Liver Triglyceride in Man and the Dog

R. C. Gross, G. M. Reaven, and J. W. Farquhar,
Palo Alto, Cal.

In recent years a number of investigators have reported an association between elevated concentrations of plasma triglyceride and premature atherosclerosis. It has also been shown that the liver is the chief source of plasma triglyceride of non-dietary origin. In addition, Havel and his co-workers [4] have found that hepatic triglyceride is related in a precursor fashion to the triglyceride contained in the $S_f > 20$ (very low density) fraction of plasma lipoproteins, and that this fraction comprises the majority of triglyceride newly secreted into plasma.

In order to define causes of hypertriglyceridemia in man, particularly the variety termed 'carbohydrate-induced' by Ahrens and his associates [1], we studied triglyceride turnover using radioactive glycerol to label triglycerides in liver and plasma. From these studies a model for the kinetics of endogenous triglyceride metabolism in man was defined and validated [2] and used to calculate turnover rates of hepatic and plasma triglyceride under several conditions [2, 3, 5]. We have shown that carbohydrate-induced hypertriglyceridemia is associated with increased turnover rates of this lipid in plasma, and is therefore due to increased secretion of $S_f > 20$ triglyceride by the liver.

To further elucidate factors influencing synthesis of hepatic triglyceride and its subsequent secretion into plasma, we turned to the dog as an experimental animal. In doing this we were able to study more closely the dynamics of triglyceride metabolism in the liver. Radioactive glycerol was again used to label hepatic and plasma triglycerides. In this report our findings in the dog are discussed and compared with those previously reported in man.

All studies were performed on anesthetized dogs receiving intravenous glucose. Dietary fat had been restricted for three days. Liver biopsies and blood samples for determination of specific activity of plasma $S_f > 20$ and liver triglycerides were taken up to four hours after injection of the isotope.

Figure 1 shows typical specific activity-time curves of these triglycerides. In this study radiopalmitate was injected simultaneously with radioactive glycerol in order to compare the turnover of triglycerides labeled with each material. Maximal specific activity occurred at the same time with both, but slopes of disappearance of glycerol-labeled triglycerides were significantly steeper. This is presumably due to the greater recycling of labeled fatty acids as discussed by FARQUHAR et al. [2]. For this reason radioactive glycerol was used in subsequent studies.

In all studies specific activity of liver triglyceride was maximal by the time the first biopsy was taken, and decline of radioactivity was

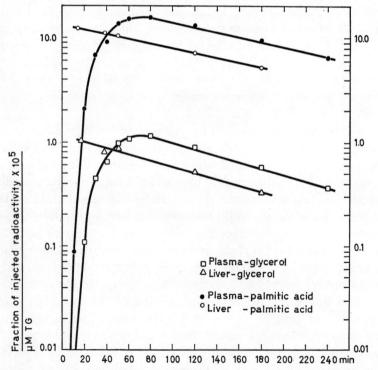

Fig.1. Curves of specific activity vs. time of plasma $S_f > 20$ and liver triglycerides after simultaneous injection of glycerol-2-^{3}H (lower curves) and palmitic acid-1-^{14}C (upper curves) in a single dog.

first-order. Radioactivity appeared in plasma $S_f > 20$ triglycerides about ten minutes after injection of the isotope, reached a maximum between 60 and 100 minutes, and then declined in a first-order manner with a slope parallel to that of the corresponding liver triglyceride specific activity in all studies. This parallelism suggests that the fractional turnover of the precursor compartment, liver triglyceride, is reflected in the disappearance of label from the product compartment, plasma $S_f > 20$ triglyceride; i.e., hepatic triglyceride is rate-determining in this system.

The liver and plasma specific activity-time curves did not intersect at the point of maximal specific activity of plasma triglyceride as they would have, according to the criteria of ZILVERSMIT et al. [7], if *all* hepatic triglyceride were in a precursor relationship to plasma $S_f > 20$ triglyceride. We interpret this as evidence for the presence of triglyceride in the liver which does not become labeled during the study.

To further define the characteristics of the plasma $S_f > 20$ triglyceride compartment, we injected radioactive glycerol-labeled $S_f > 20$ lipoproteins or labeled whole blood in two dogs. Figure 2 shows the specific activity-time curves of plasma $S_f > 20$ triglyceride in these studies. The decline of specific activity again approximates first-order, but the slopes are two to three times greater than those of plasma $S_f > 20$ triglycerides observed after labeled glycerol in the same experiments. This provides further evidence for the rate-determining role of liver triglyceride in the dog, for when *only* plasma triglycerides are labeled, radioactivity disappears from them much more rapidly than when *both* liver and plasma triglycerides are labeled with radioactive glycerol. In addition, fractional turnover rates obtained from these curves agreed quite well with those calculated from the liver and plasma specific activity curves after labeled glycerol, thus validating these calculations.

In similar studies in man, the disappearance of labeled triglyceride from plasma after the injection of prelabeled $S_f > 20$ lipoproteins was much slower than in the dog, and was parallel to the slope of decline of specific activity of plasma $S_f > 20$ triglyceride after injection of labeled glycerol. This suggests that in man the fractional turnover of the product compartment, plasma $S_f > 20$ triglyceride, is rate-determining [2].

Figure 3 shows a simple compartmental model for endogenous triglyceride turnover in liver and plasma of the dog. It does not

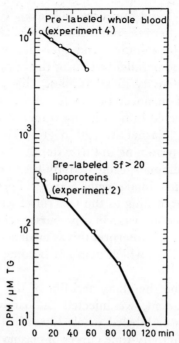

Fig. 2. Curves of specific activity vs. time of plasma $S_f > 20$ triglycerides after injection of radioactive glycerol-labeled $S_f > 20$ lipoproteins (lower curve) and radioactive glycerol-labeled whole blood (upper curve) in two dogs.

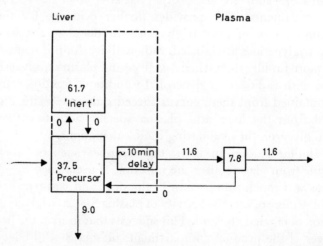

Fig. 3. A model for endogenous triglyceride turnover in liver and plasma of the dog. Numbers in boxes refer to compartment size in μmole triglyceride/kg body wt. Arrows represent transfer between and numbers with them are transfer rates in μmole triglyceride/h per kg body wt. Values are means of data from seven dogs.

depict all possible pools or pathways, but only those sufficient to account for the foregoing data. In liver, the 'precursor' compartment comprised about 40% of total triglyceride. The analogous compartment in man accounted for approximately 75% of total hepatic triglyceride [2]. The remainder of hepatic triglyceride (60% in the dog, 25% in man) was considered to be in an 'inert' compartment, inactive with respect to plasma triglyceride synthesis. In the present studies there was no evidence for communication between these compartments, but this may occur under other conditions.

Approximately 60% of the turnover of hepatic precursor triglyceride is secreted into plasma in the dog. In man less than 3% of this turnover is directed to plasma. In both species the remainder of liver precursor triglyceride turnover apparently is metabolized within the liver.

Because of the first-order nature of the specific activity-time curves of each compartment, we show no recycling of plasma $S_f > 20$ triglyceride back to the liver. This was also true in man.

In Figure 4, the relationship of turnover rate (or rate of influx from the liver) to concentration of plasma $S_f > 20$ triglyceride in the dog is compared to that in man. In man the relationship is not linear and

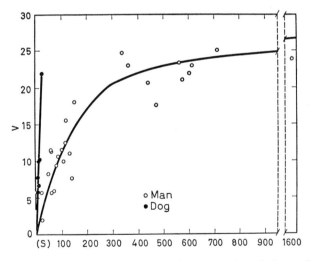

Fig. 4. Relationship between turnover rate and concentration of plasma $S_f > 20$ triglyceride compared in dogs and men. 'V' on the ordinate refers to turnover rate (or rate of influx) in mg triglyceride/h per kg body wt. 'S' on the abscissa represents plasma concentration in mg triglyceride/100 ml. Each point represents a separate turnover study.

concentration rises with increasing rapidity as turnover rate increases, finally approaching a zero-order reaction. We have interpreted this relationship as following the kinetics of a saturable, enzyme-dependent system [5]. In the dog the relationship is clearly linear in the range of turnover rates and concentrations observed in the present studies. This could result from a process of simple diffusion, but more likely represents a saturable system in a region of the curve where turnover rate is very small compared with maximum attainable rates [6]. Whether it is possible to increase influx by experimental means sufficiently to saturate removal mechanisms and produce lipemia in the dog remains to be studied.

Though the range of turnover rates of plasma $S_f>20$ triglyceride is similar in the two species, plasma concentration is always lower in the dog than in man (Fig. 4). Thus a given variable could increase triglyceride secretion equally but would result in only an insignificant increase in plasma concentration in the dog while producing lipemia in man.

The exact connection between differences in the models for triglyceride turnover and the difference in the functions relating turnover rate to concentration in plasma is yet unknown. It is possible that the two species differ in number or 'efficiency' of removal sites or in some characteristic of the substrate lipoprotein. However, *within* each species, increased plasma concentration is invariably associated with an increase in turnover rate, and is therefore due to increased influx of triglyceride, not secondary to changes in efficiency of removal of this lipid from plasma; if this were the case, concentration would rise with a concomitant decrease in turnover rate.

These findings may help to explain the difference in susceptibility of man and the dog to lipemia. In addition they emphasize the need to study turnover rates as well as concentration in order to understand more fully the factors influencing steady-state concentration.

Summary

In order to determine rates of turnover of endogenous triglyceride in plasma and liver, a model for this turnover is proposed and validated for the dog. Radioactive glycerol was used to label triglycerides.

This model is compared with that previously reported for man. An important difference in the relationship between hepatic secretion rate and plasma concentration of $S_f > 20$ triglyceride is noted between the two species, and the possible implications of this are discussed.

Acknowledgements

This work was supported in part by U.S. Public Health Service Grants AM 05972 and HE 08506 from the National Institutes of Health, and by U.S.P.H.S. Post-Doctoral Fellowship Grant HE-23,446 from the National Heart Institute.

References

1. AHRENS, E.H.; HIRSCH, J.; OETTE, K.; FARQUHAR, J.W. and STEIN, Y.: Carbohydrate-induced and fat-induced lipemia. Trans. Ass. amer. Physicians 74: 134–146 (1961).
2. FARQUHAR, J.W.; GROSS, R.C.; WAGNER, R.M. and REAVEN, G.M.: Validation of an incompletely coupled two-compartment nonrecycling catenary model for turnover of liver and plasma triglyceride in man. J. Lipid Res. 6: 119–134 (1965).
3. FARQUHAR, J.W.; FRANK, A.; GROSS, R.C. and REAVEN, G.M.: Glucose, insulin, and triglyceride responses to high and low carbohydrate diets in man. J. clin. Invest. 45: 1648–1656 (1966).
4. HAVEL, R.J.; FELTS, J.M. and VAN DUYNE, C.M.: Formation and fate of endogenous triglycerides in blood plasma of rabbits. J. Lipid Res. 3: 297–308 (1962).
5. REAVEN, G.M.; HILL, D.B.; GROSS, R.C. and FARQUHAR, J.W.: Kinetics of triglyceride turnover of very low density lipoproteins in man. J. clin. Invest. 44: 1826–1833 (1965).
6. WHITE, A.; HANDLER, P. and SMITH, E.L.: Principles of biochemistry. pp. 225–226 (McGraw-Hill, New York, 1959).
7. ZILVERSMIT, D.B.; ENTENMAN, C. and FISHLER, M.C.: On the calculation of 'turnover time' and 'turnover rate' from experiments involving the use of labeling agents. J. gen. Physiol. 26: 325–331 (1943).

Authors' addresses: R. C. Gross, M. D., King Gustaf V Research Institute, *Stockholm 60* (Sweden); G. M. Reaven, M. D., and J. W. Farquhar, M. D., Department of Medicine, Stanford University School of Medicine, 300 Pasteur Drive, *Palo Alto*, Cal. (USA).

Changes in Serum Free Fatty Acid Levels after Acute Vascular Occlusion

V.A. Kurien and M.F. Oliver, Edinburgh

Studies of changes in serum free fatty acids (FFA) during the 72 h following acute vascular occlusions have not been previously reported.

Serum FFA levels were estimated in 20 patients with acute myocardial infarction without shock, 12 patients with acute cerebral vascular occlusion and 12 patients with acute coronary insufficiency. Significant elevation of serum FFA was observed throughout the 72 h in patients with acute myocardial infarction, and in those with acute cerebral vascular occlusion. A significant but transient rise, lasting less than 24 h was seen in patients with acute coronary insufficiency.

In an attempt to elucidate the significance of these findings, studies were made to assess the effects on serum FFA of a preceding meal, of short term starvation, and of the stress of admission to hospital. The changes noted could not explain the magnitude and duration of the elevation of serum FFA in episodes of acute vascular occlusion. Changes similar to those found in patients with acute coronary insufficiency were also seen in patients suffering from severe pain of nonmyocardial origin confirming that pain can contribute to the elevation of serum FFA levels. Changes observed in association with pain are insufficient to account for the sustained elevation of serum FFA levels in episodes of acute vascular occlusion, but the high levels in patients with acute myocardial infarction during the first 12–24 h might be due to the association of the occlusive event with pain. Furthermore, patients with cerebral vascular occlusion did not experience pain.

Since the prolonged elevation of serum FFA observed in acute vascular occlusion cannot be explained by features clinically associated with the event, two alternative hypotheses are put forward to account for the changes. One is that the intravascular occlusion is precipitated by a sudden rise of serum FFA. The other and more plausible hypothesis is that the changes are due to the release of noradrenaline from anoxic myocardial and cerebral tissues, and as a result of a general sympathetic response.

Authors' address: Dr. V. A. Kurien and Dr. M. F. Oliver, Department of Cardiology, The Royal Infirmary Edinburgh 3 (Scotland).

Royal Infirmary and Medical Research Council Atheroma Unit, Western Infirmary, Glasgow

Blood Lipid Levels as Related to Adiposity and Obesity

B. M. RIFKIND, T. BEGG, I. JACKSON and M. GALE, Glasgow

In 401 apparently healthy men and in 97 males with occlusive vascular disease of the lower limbs, serum triglyceride levels measured by a direct method bore a moderate to low-grade positive correlation with adiposity, the latter expressed as relative body weight. Similar, though rather lower, correlations were observed for serum cholesterol and beta-lipoprotein levels. The degree of correlation tended to fall with age. Serum phospholipid and plasma free fatty acid levels were not significantly related to relative body weight. In a small group of subjects, adipose tissue mass, derived from measurements of total body water using tritiated water, also directly correlated with serum triglyceride levels.

However, the relationship between blood lipid levels and adiposity was not a simple linear one. Inspection of the blood lipid levels in relation to relative body weight revealed:

(1) The majority of subjects with hypertriglyceridaemia (>150 mg/100 ml), hypercholesterolaemia (>275 mg/100 ml) or hyperbetalipoproteinaemia (>3.2 'immunocrit' units) to exceed 110% of their standard weight;

(2) many overweight subjects, however, to have normal or low lipid levels;

(3) the majority of 'underweight' subjects ($<100\%$ standard weight) to have low lipid levels.

Raised triglyceride levels were also observed in a group of 39 markedly obese females ($>130\%$ standard weight) compared to a non-obese control group. Differences between the groups were most pronounced for the younger subjects. Serum cholesterol and phospholipid levels were normal. Considerable variation in the triglyceride levels of equally obese subjects was noted. When the obese subjects were divided into two groups according to whether they had high or low triglyceride levels they were found to be very similar with respect to their pattern of body fat distribution as assessed by measurement of skinfold thickness at various sites with skinfold calipers.

It is suggested that the frequent association of high relative body weight with hypertriglyceridaemia and hypercholesterolaemia may by relevant to the production of hyperlipidaemia; it is likely that the factor(s) leading to obesity are also responsible, directly or indirectly, for the high lipid levels.

Authors' address: Dr. B. M. Rifkind, Dr. T. Begg, Dr. I. Jackson and Dr. M. Gale, Royal Infirmary and Medical Research Council, Atheroma Unit, Western Infirmary, *Glasgow* (Scotland).

From the IIIrd Department of Medicine, University of Helsinki, Helsinki, Finland

Plasma Insulin Response to Glucose in Endogenous and Alimentary Hyperglyceridemia

E.A. Nikkilä, T.A. Miettinen, R. Pelkonen and
M.-R. Taskinen, Helsinki

Plasma insulin was determined by immunoassay during oral glucose tolerance test in 30 non-diabetic patients with different degrees of hyperglyceridemia (150 to 1000 mg/100 ml). The maximal postglucose level was abnormally high in 70% of cases. However, the correlation between plasma insulin response and fasting triglyceride level was poor.

To show the effect of alimentary hyperglyceridemia on glucose-stimulated insulin secretion the following experiments were carried out in normal persons. Plasma insulin was assayed serially during intravenous glucose tolerance test in fasting state. The test was repeated on the next day 3 h after a 60 g oral fat load. It was observed that at 5 min after the injection of glucose the plasma insulin level was higher and at 90 min lower during the alimentary hyperglyceridemia than in fasting state. The fractional disappearence rate of glucose (K) was correspondingly increased by the fat load in every subject. It is thus apparent that the insulin secretion stimulated by glucose is augmented by alimentary hyperglyceridemia.

Authors' address: E. A. Nikkilä, M. D., IIIrd Dept. of Medicine, University of Helsinki, *Helsinki* (Finland).

Discussion Session 4

A.N. HOWARD (Cambridge): Is Dr. GOULD suggesting that CPIB directly inhibits cholesterol synthesis? CPIB is known to affect plasma FFA levels and this may depress synthesis of β-lipoprotein cholesterol and hence cholesterol.

G. GOULD (Palo Alto, Cal.): Proving that any drug effect is direct rather than indirect is very difficult and I don't believe we have enough information on the effects of CPIB to say which are direct effects. If the depressing effect of CPIB on the mobilization of FFA from adipose tissue can be shown to be responsible for the decrease in plasma lipoprotein levels – and this has not yet been done, it would be compatible with the increased hepatic synthesis of triglycerides from acetate-C^{14} and from H^3OH.

S. GARATTINI (Milan): I have several questions with regard to the paper of Dr. CARLSON. We have obtained similar data as he mentioned about the decrease of blood ketone bodies after administration of drugs which are lowering FFA including nicotinic acid, salicylic acid, 3.5 dimethylpyrazole, particularly in animals which have been submitted to prolonged fasting. I would also like to have Dr. CARLSON's opinion to a couple of other questions. He mentioned that after nicotinic acid administration, there is a decrease in heart TG. I think this was in dogs. We could not find much change in heart TG in rats, particularly after administration of drugs or conditions which increased plasma FFA. We found that the heart TG levels are quite stable in rats. You mentioned the effect of nicotinic acid not only on FFA and TG but also on cholesterol, and apparently you were trying to relate – if I understood it correctly – the decrease of plasma FFA with the decrease of cholesterol. This seems to be unique for nicotinic acid and not for other drugs which are lowering plasma FFA. They lower plasma TG, as for instance 3.5 dimethylpyrazole, but there is no change in the plasma cholesterol so that we could see that two factors are really related. And, finally, I should like to draw your attention to our studies on the decrease of plasma FFA by various drugs. It is quite interesting that frequently one encounters a rebound effect. Particularly with rather small doses of these drugs, you get a marked increase in FFA after a time and also an increase in plasma TG and liver TG. I was wondering whether you have encountered the same rebound effect or if you have any explanation.

L.A. CARLSON (Stockholm): 1. I am pleased to hear that you find similar effects in rats as we do in man with regard to ketone bodies when FFA mobilization is inhibited.
2. The reduction of heart triglycerides when FFA mobilization is inhibited was in rats (Acta med.scand. 1966, in press). The increase in heart lipids during excessive FFA mobilization was in dogs (Acta med.scand. *178:* 81, 1965), as reported also by other authors after adrenaline and noradrenaline (e.g. MALING, H.M. and HIGHMAN, B., Amer.J.Physiol. *194:* 590, 1958).
3. I did not try to relate the small, acute decrease of plasma cholesterol to the inhibition of FFA mobilization although part of it of course is due to the reduction of the triglyceride rich very low density lipoproteins which contains cholesterol. This question is discussed in more detail elsewhere (Acta med.scand. *179:* 453, 1966).
4. With regard to the puzzling rebound phenomenon we see it regularly with nicotinic acid in man (Acta med.scand. *172:* 641, 1962), but have no explanation, to offer.

T. ZEMPLÉNYI (Prague): Dr. CARLSON, you mentioned, if I understood you well, that in your 100 patients you did not observe any relationship between lipid levels and carbohydrate metabolism. My question is whether you followed in your glucose tolerance tests also the FFA levels; in view of the 'fatty acid glucose cycle', one could expect that, at least, along this line, one should find some relationship.

L.A. CARLSON (Stockholm): No, we did not follow FFA levels during the glucose tolerance tests. I am not sure that one could expect to find any relationship here but one could always hope.

P. ALAUPOVIC (Oklahoma City, Okla.): Dr. SKOREPA, how were the esterase and lipase activities differentiated? Has physical-chemical state of the substrate been taken into consideration?

J. SKOREPA (Prague): The substrate for the lipase was a triglyceride emulsion (Ediol); esterase activity was estimated by splitting ethyl butyrate which was present in a solution.

J.J. GROEN (Jerusalem): Dr. RIFKIND has shown very convincingly that there is only a relatively small correlation between serum lipid levels and obesity. Epidemiological studies have shown the same. May I suggest as an explanation that, just as in normal individuals, the *type* of nutrition (perhaps together with physical activity) is the main factor which determines the serum lipid level. In other words, when a person becomes fat because of a too high consumption of saturated fat and sugar, his serum lipids are high but when in another person the obesity is the result of overeating bread or peas or beans, his serum cholesterol is low. We found this most strikingly illustrated in the Trappist monks; many of them are fat but their serum cholesterol is low because they eat a high bread, low saturated fat, low sugar and low cholesterol diet.

From the Departments of Pathology and Biochemistry of The Bowman Gray School of Medicine, Winston-Salem, N.C.

Recent Advances in Arterial Metabolism: The Whole Artery

H.B. LOFLAND, Winston-Salem, N.C.

The World Health Organization's definition of atherosclerosis includes the phrase '...a variable combination of changes in the intima of arteries...'. Within the limits of this definition, one must consider a variety of factors which might influence arterial metabolism, leading to the accumulation of lipids, complex carbohydrates, calcium, fibrous tissue, etc. Such factors must include the nature of the substance accumulated, vascular dynamics, the focal distribution of lesions, species differences and genetic differences within species, and various injuries or insults to the arterial wall. Some of these factors will be discussed today.

From the work of many investigators, using tissues from man and from several animal species, we now have a reasonably good understanding of the chemical nature of the substances which accumulate in atherosclerotic lesions. The work of BÖTTCHER (1963), some of which will be presented during this meeting, and that of SWELL and TREADWELL (1963), and many others, have provided us with an understanding of the biochemical changes which occur in normal and diseased human arteries. Such studies are an absolute requisite to any further understanding of the mechanisms by which phospholipids, and cholesterol and its esters accumulate during atherogenesis. These studies have also shown the existence of both similarities and dissimilarities between serum lipids and those of the atherosclerotic plaque. It seems apparent that some alteration in serum lipid levels or types is the *sine qua non* for atherogenesis in many animal species. Likewise, in diseased arteries, certain metabolic pathways can be shown to differ from those of normal arteries. However, the relative importance of these two factors to the disease process has yet to be established with certainty. Few investigators today would accept the concept that either factor, operating alone, can explain the observed facts for all animal species; this, in turn, suggests the existence of some sort of interaction between the vessel wall and the medium which is perfusing it. In this regard, however, a word of caution appears to be desirable, namely that major species differences exist and if, indeed, there is a single most important factor – e.g., elevated cholesterol levels, altered arterial metabolism, filtration, or synthesis, such a factor may not be the same for all species. Likewise, it seems important in our further discussion to distinguish between mechanisms which might result in 'fatty streak' formation and those which are operative in, or characteristic of, a complicated plaque.

There is little doubt that serum lipids, including lipoproteins, are capable of entering the intact arterial wall. GERÖ et al. (1961) have shown, in atherosclerotic lesions of man, the presence of certain plasma components (albumin, globulins and other proteins). Likewise, WATTS (1963), using fluorescent antibody techniques, has shown the presence of low density lipoproteins in atheromatous lesions. More recently KAO and WISSLER (1965), using a similar technique with higly purified lipoprotein fractions, showed that relatively little high density lipoprotein occurs in such lesions, whereas low density fractions were readily demonstrable. The work of DAVIS, ADAMS and BAYLISS (1963) and of CHRISTENSEN (1964), some of which will be described later in this meeting, confirms the viewpoint that plasma lipids readily penetrate the arterial wall.

Similarly, cholesterol and other plasma lipids (or lipoproteins) can be shown to leave the artery. NEWMAN et al. (1962) have described the flux of cholesterol in the arteries of cholesterol-fed rabbits. Their results showed that the rate of transfer in such arteries exceeds the accumulation, suggesting the removal of cholesterol even during the process of lesion formation. It is of some interest, however, that these same authors report that such flux occurs even in arteries which have been boiled, or poisoned with cyanide (1966). GOULD, JONES and WISSLER (1959) administered tritium-labeled cholesterol over a several day period to humans during terminal illnesses, and showed the appearance of labeled cholesterol in atherosclerotic plaques, and concluded that cholesterol must also be leaving the lesions. Similar conclusions were reached by FIELD et al. (1960), who administered to human beings cholesterol-4-C^{14} in a single dose. It seems clear from these studies, and others, that the arterial wall permits the passage, in both directions, of plasma lipids. Still unexplained, however, is their accumulation in the artery.

Several aspects of the metabolism of the intact aorta have been studied using partially eviscerated animals and arterial perfusion *in situ*. Thus it has been possible to establish that arterial tissue is capable of synthesizing, in significant amounts, certain lipid classes, such as phospholipids, and that such synthesis may, in fact, be the predominant mechanism by which they accumulate in arteries (SHORE, ZILVERSMIT and ACKERMAN, 1955). Later in the session we will hear the results of similar studies by Drs. BOWYER, GRESHAM and HOWARD of Cambridge.

In our laboratories we have used atherosclerosis-susceptible White Carneau pigeons, and their resistant counterpart, the Show Racer pigeon (LOFLAND and CLARKSON, 1959; LOFLAND, MOURY, HOFFMAN and CLARKSON, 1965a), for studies on arterial metabolism. Several facets of the disease have suggested to us the importance of the role of the arterial wall in this species. First, the White Carneau pigeon develops, in 100% of the birds, naturally-occurring atherosclerosis of the aorta even when kept on cholesterol-free diets. The Show Racer, on the same diet, develops almost no aortic disease. Secondly, in both breeds of pigeons, the administration of cholesterol in the diet results in marked hypercholesterolemia (often in excess of 3000 mg%). All parameters of plasma lipids studied appear to be identical in the two breeds, yet only the aortas of White Carneaux respond with rapid lesion formation (CLARKSON and LOFLAND, 1961). On the other hand, we find the two breeds to be equally susceptible to coronary artery atherosclerosis (LOFLAND, 1965b). We interpret these observations to mean that factors other than elevated serum lipids must play an important role in these animals – such factors are probably under genetic control, probably operating at the tissue level.

We have previously described a technique for the simultaneous perfusion of 12 isolated, intact, pigeon aortas (LOFLAND, MOURY, HOFFMAN and CLARKSON, 1965a; LOFLAND and CLARKSON, 1965c). A schematic diagram of the apparatus is shown in Figure 1. In

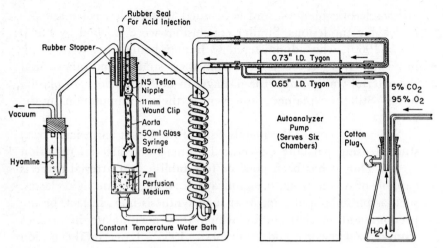

Fig. 1. Schematic diagram of perfusion system (LOFLAND, MOURY, HOFFMAN and CLARKSON, 1965a).

these studies, the thoracic aorta of the pigeon is removed from the bird, a process requiring about 4 min, and attached to a teflon nipple in the chamber. The entire chamber is immersed in a constant temperature water bath. The perfusion medium, generally blood serum containing the isotopically-labeled subtrate to be studied, is delivered to the artery by a peristaltic pump. Before entering the artery under slight positive pressure, the stream of perfusion medium is segmented with a 95% oxygen, 5% carbon dioxide gas mixture. Only the intimal surface of the artery is exposed to the medium, and each chamber is completely independent of all others. After several hours of perfusion, the arteries are removed, cleaned, weighed, and graded for the extent of atherosclerosis. Lipids are extracted from the aortas, separated into various classes by thin layer chromatography and the fractions obtained are used for analytical and radioactivity measurements.

When pigeon aortas were perfused with blood serum containing radioactive acetate, it was apparent that there was an active synthesis of lipid which was postively correlated with the severity of aortic atherosclerosis. In these studies, the severity of the disease was determined by both visual measurement of plaque size and by aorta cholesterol determinations. Table I shows results obtained in a typical experiment. The relationship between lipid synthesis and severity of atherosclerosis is obvious.

Table I*. Relationship between lipid synthesis and severity of aortic atherosclerosis

Severity of disease	Radioactivity of newly synthesized lipid**
	cpm/g of aorta
Slight	35,800±3,400
Moderate	95,300±6,700
Severe	536,800±250,000

* Table 4 in Lofland, Moury, Hoffman and Clarkson (1965a).
** Mean±SEM (n = 4 in each group).

The focal distribution of lesions in pigeon aortas allows one to excise the plaques in such a way that little normal tissue is included in the sample. When lesions were excised from aortas which had been perfused with serum plus C^{14}-acetate (Table II), it appeared that the plaque itself is the site of the enhanced lipid synthesis. Since obtaining these earlier results, we have perfused several hundred aortas, in various stages of disease. The results have invariably confirmed the positive correlation between severity of atherosclerosis and the rate of lipid synthesis. It should be emphasized, however, that we do not obtain markedly enhanced lipid synthesis until the plaques are relatively severe. In our opinion, we are probably observing events associated with lesion progression rather than lesion initiation.

Table II*. Lipid synthesis in different areas of the same aorta

Source of material	Radioactivity of newly synthesized lipid
	cpm/g of tissue
Undiseased area of aorta No. 1	24,100
Plaque from aorta No. 1	160,000
Undiseased area of aorta No. 2	46,900
Plaque from aorta No. 2	212,700

* Table 5 in Lofland, Moury, Hoffman and Clarkson (1965a).

Another facet of these studies of interest to us has been the distribution of newly synthesized lipid among the various classes. Table III shows an example of the typical distribution among

Table III*. Distribution of radioactivity among classes of lipid synthesized during perfusion

Source of material	expt. No.	% of total radioactivity			
		phospho-lipid	free sterol	tri-glyceride	sterol ester
Undiseased artery	1	35	6	51	8
	2	38	8	50	4
Severely diseased artery	3	37	5	27	31
	4	24**	4**	37**	35**
Excised plaque	5	12	2	42	44
	6	19	6	39	36

* From Table 6 in LOFLAND, MOURY, HOFFMAN and CLARKSON (1965a).
** Arteries perfused with tissue culture medium, plus acetate-1-C^{14}, with no blood serum.

phospholipid, free sterols, triglycerides and sterol esters. A striking feature is the increased percentage of newly synthesized lipid which is found in the sterol ester fraction in severely diseased aortas, or in plaques therefrom. When this fraction is saponified and the sterol and fatty acids counted separately, most of the radioactivity is found to be in the fatty acid fraction. In this table, those values labeled with an asterisk were obtained by perfusing the arteries with C^{14}-acetate in a cholesterol-free tissue culture medium, instead of blood serum. The distribution is almost identical with that obtained using blood serum. From these studies we conclude that in pigeons, the increasing severity of atherosclerosis is characterized by a markedly enhanced fatty acid synthesis. The fate of a large proportion of this newly synthesized fatty acid is esterification with cholesterol pre-existing in the lesion. This finding appears to be in contrast to those reported by NEWMAN and ZILVERSMIT (1966), who used rabbits, and concluded that aortic sterol esters originate in the plasma. It also seems apparent that the aorta of pigeons contains the enzyme systems necessary to esterify fatty acids to cholesterol, as well as to synthesize the latter, at least to a limited extent.

References

1. BÖTTCHER, C.J.F.: Phospholipids of atherosclerotic lesions in the human aorta, in JONES Evolution of the atherosclerotic plaque, pp. 109–116 (University of Chicago Press, Chicago 1963).
2. CHRISTENSEN, S.: Transfer of labeled cholesterol across the aortic intimal surface of normal and cholesterol-fed cockerels. J. Atheroscl. Res. 4: 151–160 (1964).

3. Clarkson, T.B. and Lofland, H.B.: Effects of cholesterol-fat diets on pigeons susceptible and resistant to atherosclerosis. Circulat. Res. 9: 106–109 (1961).
4. Davis, J.N.; Adams, C.W.M. and Bayliss, O.B.: Gradient in cholesterol concentration across human aortic wall. Lancet 7320: 1254–1255 (1963).
5. Field, H.; Swell, L.; Schools, P.E. and Treadwell, C.R.: Dynamic aspects of cholesterol metabolism in different areas of the aorta and other tissues of man and their relationship to atherosclerosis. Circulation 22: 547–558 (1960).
6. Gerö, S.; Gergely, J.; Jakab, L.; Székely, J. and Virág, S.: Comparative immunoelectrophoretic studies on homogenates of aorta, pulmonary arteries and inferior vena cava of atherosclerotic individuals. J. Atheroscl. Res. 1: 88–91 (1961).
7. Gould, R.G.; Jones, R.J. and Wissler, R.W.: Lability of labeled cholesterol in human atherosclerotic plaques. Circulation 20: 967 (1959).
8. Kao, V.C.Y. and Wissler, R.W.: A study of the immunohistochemical localization of serum lipoproteins and other plasma proteins in human atherosclerotic lesions. Exp. Molec. Path. 4: 465–479 (1965).
9. Lofland, H.B., Jr.; Moury, D.M.; Hoffman, C.W. and Clarkson, T.B.: Lipid metabolism in pigeon aorta during atherogenesis. J. Lipid Res. 6: 112–118 (1965a).
10. Lofland, H.B., Jr.: Experimental atherosclerosis in pigeons, in Roberts and Straus Comparative atherosclerosis, pp. 50–57 (Hoeber, New York 1965b).
11. Lofland, H.B., Jr. and Clarkson, T.B.: A biochemical study of spontaneous atherosclerosis in pigeons. Circulation Res. 7: 234–237 (1959). – Certain metabolic patterns of atheromatous pigeon aortas. Arch. Path. 80: 291–296 (1965c).
12. Newman, H.A.I. and Zilversmit, D.B.: Quantitative aspects of cholesterol flux in rabbit atheromatous lesions. J. Biol. Chem. 237: 2078–2084 (1962). – Uptake and release of cholesterol by rabbit atheromatous lesions. Circulat. Res. 18: 293–302 (1966).
13. Shore, M.L.; Zilversmit, D.B. and Ackerman, R.F.: Plasma phospholipide deposition and aortic phospholipide synthesis in atherosclerosis. Amer. J. Physiol. 181: 527–531 (1955).
14. Swell, L. and Treadwell, C.R.: Interrelationships of lipids in blood and tissues, in Sandler and Bourne Atherosclerosis and its origin, pp. 201–247 (Academic Press, New York 1963).
15. Watts, H.F.: The mechanism of arterial lipid accumulation in human coronary artery atherosclerosis, in Likoff and Moyer Coronary heart disease, pp. 98–113 (Grune and Stratton, New York/London 1963).

Author's address: Hugh B. Lofland, Jr., Ph.D., Department of Pathology, The Bowman Gray School of Medicine, *Winston-Salem, North Carolina 27103* (USA).

From the Departments of Pathology and Medicine, Guy's Hospital Medical School,
University of London, England

Quantitative Histochemical Observations on Certain Oxidative and Lipolytic Enzymes in Human Aortic Wall*

C.W.M. Adams, Y.H. Abdulla, O.B. Bayliss, R.F. Mahler and M.A. Root, London

Qualitative histochemistry and biochemistry reveals decreased activity of certain dehydrogenases, tetrazolium reductases and ATPase in the ageing and atherosclerotic human aorta (see reviews 1, 12, 15, 21, 26). Histochemical studies have shown that dehydrogenase and reductase activity usually increases in the macrophages of the intima, but we observed that such enzymic activity decreases in the middle zone – and subsequently in the inner zone – of the human aortic media [1, 2, 5, 6].

This zonal change may be evident even by the third decade of life. Sandler and Bourne [20–22] observed an *early* patchy loss of ATPase from subintimal medial muscle fibres in the human aorta, but their observations do not relate to this zonal loss of enzyme from the mid-media of the senescent vessel. We suggested that this mid-medial enzymic impairment is due to ischaemic damage at the critical watershed between the vasal circulation and direct permeation of nutriments from the lumen [1, 2, 5, 6]. We also suggested that this mid-medial enzymic failure impairs the synthesis of lipotrophic agents required for the outward flow of cholesterol through the aortic wall [4].

Recently we have adapted Linderström-Lang's [14] quantitative histochemical technique, so that fixed or unfixed serial sections of flattened human aortic wall can be cut in consecutive layers from the inner intima down to the beginning of the adventitia. Such 20–50 μ thick sections can be cut on the constant temperature thermoelectric microtome with only 4.8% coefficient of variation in their weight [3, 8, 23]. Using this technique, enzyme activity can be estimated in multiple consecutive layers from the inside to the outside of the aorta. Lactic dehydrogenase (LDH), with and with-

* Supported by the British Heart Foundation and the Tobacco Research Council.

out phenazine methosulphate (PMS) as intermediate electron acceptor, was estimated with INT tetrazolium by a modification of Laatsch's method. $NADH_2$-tetrazolium reductase ($NADH_2$-TR) activity was estimated by a basically similar technique [23].

Esterase activity was estimated by the β-naphthyl laurate method [24]. Lipase was estimated by glycerol release from tributyrin [16].

Dehydrogenase Activity. Slide histochemical methods show that LDH, $NADH_2$-TR and ATPase activities decline in the mid-medial zone of the ageing human aorta. However, biochemical estimation of 'dummy-incubated' sections shows that 63% of LDH in the aortic wall is water-soluble and must, accordingly, be leached out during histochemical incubation [23]. On the other hand, only 4% of $NADH_2$-TR activity is dissolved during such 'dummy incubation' and it is, therefore, validly demonstrated by histochemical tetrazolium methods [23]. This artefactual localization of LDH in the arterial wall is due to the formation of reduced NAD in the incubating medium, which acts as substrate for the $NADH_2$-TR that is retained in the tissue-section [9, 19].

Biochemical estimation of consecutive sections of human aortic wall confirms that $NADH_2$-TR activity declines in the mid-media of the senescent human aorta, as does that of LDH when mediated through the *endogenous* $NADH_2$-TR of the vessel wall (Fig. 1). Conversely, LDH activity, when mediated through *exogenous* PMS, shows an actual increase in the mid-medial zone of the ageing aorta (Fig. 1).

This increased activity of specific LDH can be regarded as an adaptation to hypoxic conditions in the ischaemic aorta [13]. Nevertheless, it can serve no useful metabolic purpose, as the $NADH_2$-TR required for electron transport to the cytochrome system is deficient in this region of the degenerating aorta (Fig. 1). Previously, from our slide histochemical results, we had regarded the generally reduced enzymic activity in the mid-media as an indication of ischaemic *necrosis* of the muscle cells in this layer. However, we have modified this view following our quantitative cytochemical demonstration of *increased* specific LDH activity in this zone. It can now be inferred that the muscle fibres in this zone suffer ischaemic metabolic damage that falls short of complete necrosis [7].

Lipolytic activity. Histochemical methods reveal esterase activity in the muscle fibres of the rat and human aortic media – as well as

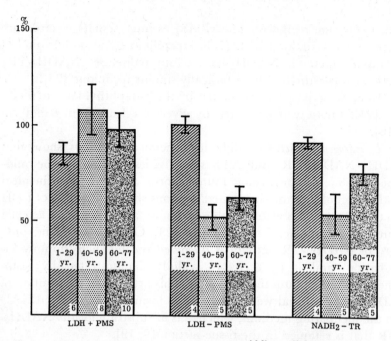

Fig. 1. Human aorta. Oxidative enzyme activity ratio: $\dfrac{\text{middle medial zone}}{\text{outer medial zone}}$

in the macrophages or modified smooth muscle cells of the atherosclerotic plaque – but esterase activity in rat aorta is much higher than in other mammalian species [25]. The reaction in human aorta is particularly weak with either α-naphthyl acetate or 5-bromo-o-indoxyl acetate as substrates. Our quantitative histochemical studies showed that 48% of esterase activity is extracted from unfixed sections of human aorta during 'dummy' histochemical incubation, so the weak reaction of the human vessel could be attributed to leaching of enzyme during incubation. An attempt to preserve this esterase in cryostat sections by brief preliminary fixation with Holt's formalin-sucrose was unsuccessful.

Lipase activity cannot be demonstrated histochemically in human and rat aorta by Gomori's method [10, 11]. Lipase cannot be localized in arterial tissue because 98% of the enzyme is soluble and is lost from unfixed sections of human aorta during 'dummy' histochemical incubation. The enzyme is either inhibited or extracted by treatment with formaldehyde (77% loss), acetone (98%) and ethanol (99% loss). It was, therefore, concluded that it is not possible to retain sufficient active lipase in the sections – after 'fixation' with these agents – to demonstrate the enzyme histochemically.

Biochemical estimation of consecutive serial sections of human aorta, cut on the thermo-electric microtome, indicates that the ratio of *esterase* activity in the *middle* media to that in the *outer* media significantly declines in older women and shows a similar *trend* in older men (Fig. 2). Lipase activity in the *middle* aortic media, however, shows no relative decline with age when compared with that in the *outer* layer of the media (Fig. 3). It is apparent, therefore,

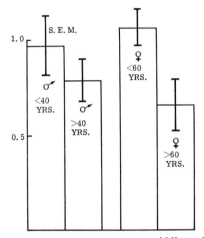

Fig. 2. Human aorta. β-naphthyl laurate esterase ratio: $\frac{\text{middle medial zone}}{\text{outer medial zone}}$

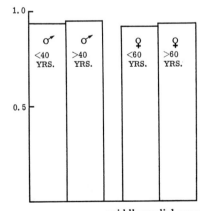

Fig. 3. Human aorta. Lipolytic activity ratio: $\frac{\text{middle medial zone}}{\text{outer medial zone}}$

that esterase activity declines in somewhat the same way as does that of $NADH_2$-TR in the ischaemic middle layers of the ageing human aorta. Conversely, lipase activity is not affected by the postulated hypoxic conditions in this middle zone.

Esterase activity (β-naphthyl laurate) declines with advancing age in homogenates of *whole* human aorta; this enzymic decline appears to occur somewhat earlier in men than in women (Fig. 4).

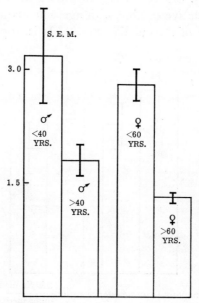

Fig. 4. Human aorta. β-naphthyl laurate esterase activity in whole homogenate (mg/β-naphthol/100 mg wet wt/h).

Lipase activity (tributyrin) also decreases in the ageing male aorta, but in the female this fall is not statistically significant with the sample studied (Fig. 5).

Although it is said that both lipase and esterase hydrolyse β-naphthyl laurate and similar substrates [17, 18], our aortic 'esterase' appears to be a different enzyme to the 'lipase'. Thus, the esterase is substantially inhibited by M NaCl, 0.1 M Zn^{++} and 10^{-3} M E 600, while the lipase is only slightly inhibited by these agents and 1% protamine sulphate. Both enzymes are unaffected by 0.1% sodium taurocholate and 10^{-4} M eserine. These differential inhibitor studies are essentially consistent with those recorded by ZEMPLÉNYI and his colleagues [25, 26].

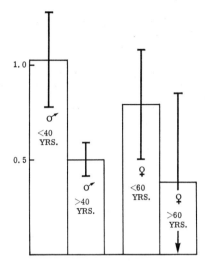

Fig.5. Human aorta. Lipolytic activity (μM glycerol/100 mg wet wt/h).

The quantitative histochemical observations reported in this communication show that, in the senescent human aorta, $NADH_2$-TR and esterase activities decline in the *middle* zone of the media when compared with the *outer* zone. Conversely, lipase and specific lactic dehydrogenase do not show this mid-zonal impairment. Lipolytic (lipase+esterase) activity declines in the ageing male aorta, as does that of esterase in older women. These quantitative results confirm our previously expressed view that respiratory function decreases in the middle layer of the senescent aortic media [1, 2, 5, 6] and also show that enzyme systems concerned in triglyceride degradation are impaired in the ageing human aorta.

References

1. ADAMS, C.W.M.: Arteriosclerosis in man, other mammals and birds. Biol. Rev. *39:* 372–423 (1964a).
2. ADAMS, C.W.M.: Histochemical studies on the distribution of lipids and enzymes in the normal and atherosclerotic artery; in CHALMERS and GRESHAM's Symposium on Biological Aspects of Occlusive Vascular Disease, pp. 41–45 (Cambridge University Press, Cambridge 1964b).
3. ADAMS, C.W.M.: Passage of cholesterol through the arterial wall. Proc. Roy. Soc. Med. *57:* 31–34 (1964c).
4. ADAMS, C.W.M.; BAYLISS, O.B.; DAVISON, A.N. and IBRAHIM, M.Z.M.: Autoradiographic evidence for the outward transport of ^{3}H-cholesterol through rat and rabbit aortic wall. J. Path. Bact. *87:* 297–304 (1964).
5. ADAMS, C.W.M.; BAYLISS, O.B. and IBRAHIM, M.Z.M.: A hypothesis to explain the accumulation of cholesterol in atherosclerosis. Lancet *i:* 890–892 (1962).

6. ADAMS, C.W.M.; BAYLISS, O.B. and IBRAHIM, M.Z.M.: The distribution of lipids and enzymes in the aortic wall in dietary rabbit atheroma and human atherosclerosis. J. Path. Bact. *86:* 421–430 (1963).
7. ADAMS, C.W.M.: Vascular Histochemistry (Lloyd-Luke, London, in press).
8. DAVIS, J.N.; ADAMS, C.W.M. and BAYLISS, O.B.: Gradient in cholesterol concentration across human aortic wall. Lancet *ii:* 1254–1255 (1963).
9. FARBER, E.: Control studies on the histochemical localization of specific DPN-linked dehydrogenases. J. Histochem. Cytochem. *10:* 657–658 (1962).
10. GEORGE, J.C. and IYPE, P.T.: Improved histochemical demonstration of lipase activity. Stain Tech. *35:* 151–152 (1960).
11. GOMORI, G.: Microscopic Histochemistry (Chicago University Press, Chicago 1952).
12. KIRK, J.E.: Intermediary metabolism of human arterial tissue and its changes with age and atherosclerosis; in SANDLER and BOURNE's Atherosclerosis and Its Origin, pp. 67–117 (Academic Press, New York 1963).
13. LEHNINGER, A.L.: The metabolism of the arterial wall; in LANSING's Arterial Wall, pp. 136–160 (Baillière, Tindall and Cox, London 1959).
14. LINDERSTRÖM-LANG, K.: Distribution of enzymes in tissue and cells. Harvey Lect. *34:* 214–245 (1939).
15. LOJDA, Z.: The enzyme topochemistry of the arterial wall. Morfologie (Cesk.) *10:* 46–61 (1962).
16. MAHLER, R.F.: The demonstration of arterial lipolytic activity with tributyrin substrate. J. biol. Chem. (in press).
17. MAIER, N. and HAIMOVICI, H.: Metabolism of arterial tissue with special reference to esterase and lipase. Proc. Soc. exp. Biol., N.Y. *118:* 258–261 (1965).
18. MUNEAKI, A.; KRAMER, S.P. and SELIGMAN, A.M.: The histochemical demonstration of pancreatic-like lipase and comparison with the distribution of esterase. J. Histochem. Cytochem. *12:* 364–383 (1964).
19. NOVIKOFF, A.B.: Electron transport enzymes: biochemical and tetrazolium staining methods; in WEGMANN's Histochemistry and Cytochemistry, pp. 465–481 (Pergamon Press, Oxford 1963).
20. SANDLER, M. and BOURNE, G.H.: Some histochemical observations on the human aortic wall in atherosclerosis. Circulat. Res. *8:* 1274–1277 (1960).
21. SANDLER, M. and BOURNE, G.H.: Histochemistry of atherosclerosis; in SANDLER and BOURNE's Atherosclerosis and Its Origin, pp. 515–532 (Academic Press, New York 1963).
22. SANDLER, M. and BOURNE, G.H.: Dietary effects on the arterial wall. Angiology *16:* 375–378 (1965).
23. SAUDEK, C.C.; ADAMS, C.W.M. and BAYLISS, O.B.: The quantitative histochemistry and cytochemistry of lactic dehydrogenase and $NADH_2$-tetrazolium reductase in human aortic wall. J. Path. Bact. (in press).
24. SELIGMAN, A.M. and NACHLAS, M.M.: The colorimetric determination of lipase and esterase in human serum. J. clin. Invest. *29:* 31–36 (1950).
25. ZEMPLÉNYI, T.; GRAFNETTER, R. and LOJDA, Z.: Some problems connected with the lipolytic and esterolytic activity of tissues; in DESNUELLE's Enzymes of Lipid Metabolism, pp. 203–212 (Pergamon Press, Oxford, 1961).
26. ZEMPLÉNYI, T.; LOJDA, Z. and MRHOVÁ, O.: Enzymes of the vascular wall in experimental atherosclerosis in the rabbit; in SANDLER and BOURNE's Atherosclerosis and Its Origin, pp. 459–513 (Academic Press, New York 1963).

Authors' address: C.W.M. Adams, M.D.; Y.H. Abdulla, M.B.; O.B. Bayliss, R.F. Mahler, M.D. and M.A. Root, Ph.D, Depts. of Pathology and Medicine, Guy's Hospital Medical School, University of *London* (England).

Westminster Medical School, London

Factors Affecting the Synthesis of Individual Phospholipids in the Rat Aorta

J.D. BILLIMORIA and T.J. ROTHWELL, London

In rabbits injected with radioactive ^{32}P-phosphate, ZILVERSMIT, SHORE and ACKERMAN (1954) showed that larger quantities of ^{32}P labelled phospholipids were found in the aortae of animals fed with cholesterol than in normal animals. NEWMAN, McCANDLESS and ZILVERSMIT (1960) further showed that following ^{32}P-phosphate injection, the specific activity of the phospholipids was higher in the aortic intima than in plasma and that this difference was accentuated in eviscerated animals. More recently the metabolic activity of the isolated aorta in synthesising cholesterol esters has been shown in the White Carneau pigeon by LOFLAND, MOURY, HOFFMAN and CLARKSON (1965). On the other hand, the rat is known to be resistant to atherosclerosis and apart from the early experiments of CHERNICK, SRERE and CHAIKOFF (1948), little is known of the synthesis of lipids by rat aorta.

We have recently investigated the metabolism of phospholipids in rat tissues (MACLAGAN, BILLIMORIA and HOWELL, 1966) and the present paper is directed to the synthesis of individual phospholipids by the isolated rat aorta. Since it is now possible to induce hypercholesterolaemia into rats by certain diets (GRESHAM and HOWARD, 1960) the effect on aortic phospholipid synthesis after such a diet is also reported.

Experimental and Results

The separation of all phospholipid fractions was carried out by linear gradient elution chromatography from silicic acid columns with methanol and chloroform and the fractions were further identified and estimated by thin layer chromatography and chemical estimations. Practical details are beyond the scope of the present communication and will be reported elsewhere.

The composition of aortic phospholipids is compared with that of plasma, intestine and liver in Table I. A striking difference appears in the low lecithin content of the aorta (30%) which is com-

Table I. Rat tissue phospholipid

	aorta	% of total phospholipid plasma	intestine	liver
PA	21	1.5	4	6
PE	34	7	29	35
LPE	10	1.5	8	3
(CEPH)	(44)	(9)	(37)	(38)
LEC	29	81	51	53
SPHING	5	9	8	4
CEPH/LEC	1.5	0.11	0.73	0.72

PA = Phosphatidic acid; PE = Phosphatidyl ethanolamine; LPE = Lysophosphatidyl ethanolamine; CEPH = Cephalin; LEC = Lecithin; SPHING = Sphingomyelin.

pensated by a total cephalin (44%) and an elevated phosphatidic acid fraction (21%). The cephalin to lecithin ratio is much higher in this than in other tissues or plasma. The cephalin is made up of phosphatidyl ethanolamine (PE), lysophosphatidyl ethanolamine (lyso PE) and trace amounts of phosphatidyl serine. Phosphatidyl inositol was not separated on the column but appeared with the lyso PE fraction and constituted some 20% of this fraction.

The isolated rat aortae were slit longitudinally and incubated in a suitable tissue culture medium or saline with added $Na_2H^{32}PO_4$. Alternatively, the whole thoracic aorta was perfused with this medium in an apparatus similar to that described by LOFLAND *et al.* (loc. cit.). The effect of glucose added to the medium was also studied. In both types of experiments (perfusion and incubation) the rate of *de novo* synthesis of phospholipids, as observed from their increasing specific activity (SA) or incorporation of total radioactivity, was not affected by concentration of glucose from 0 to 100 mg%. However, when the glucose concentration was raised from 100 to 200 mg%, an optimum rate of synthesis was obtained and the SA of the phospholipids was almost doubled. Concentrations of glucose above 200 mg% did not further influence the rate of synthesis.

The percentage distribution of the radioactivity and the SA of phospholipid fractions found in the incubation and perfusion experiments at the optimum glucose concentration are shown in Table II. With strip incubation the SA and incorporation of label were much higher than in the perfusion experiments. Although the

Table II. Aortic phospholipid radioactivity after incubation and perfusion

	% total radioactivity and SA strip incubation		(μC/mg P) perfusion	
Total phospholipid	–	(42)	–	(25)
PA	9	(4)	29	(39)
PE	76	(80)	54	(43)
LPE	4	(16)	6	(8)
CEPH	80	(69)	61	(35)
LEC	11	(18)	10	(9)
SPHING	1	(4)	1	(2)
CEPH/LEC	7.5	(3.8)	6.2	(4.0)

amount of radioactivity incorporated into the fractions depended on the technique the cephalin to lecithin ratio was similar in the two experiments. A notable difference was observed in the phosphatidic acid fraction where the SA and radioactivity were found to be higher with perfusion than with incubation.

Preliminary experiments indicated that a freshly dissected aorta was metabolically active for up to 6 h; the synthesis of phospholipids by incubation was therefore studied over this period and the results are shown in Table III. The SA of the PE and lecithin fractions

Table III. Specific activity of aortic phospholipids synthesised after various times of incubation
(Aortae incubated in glucose/saline containing 500 μC ^{32}Pi for various time intervals)

Hours incubation	Specific activity of phospholipid fractions μC/mg P						Total PL
	PA	PE	LPE	Total cephalin	L	S	
1	0.86	4.30	0.79	3.80	4.20	0.52	2.93
2	1.70	9.71	1.83	8.45	7.81	2.04	6.66
3	3.20	25.0	8.45	22.1	12.0	4.42	14.2
6	8.13	45.3	7.65	32.5	9.43	2.83	17.5

were similar in the first hour but the activity of PE increased rapidly between two to six hours. The choline group of lipids, lecithin and sphingomyelin reached their maximum SA at 3 h whereas PE and PA values increased throughout the experiment. The maximum values of the choline group were much lower than that of the cephalin or PE.

The effects of diets on the phospholipid synthesis by the isolated aorta are shown in Table 4. The cholesterol diet of HOWARD ET AL. (loc. cit.) as modified by CONSTANTINIDES (1965) consisted of 1.5%

Table IV. Effect of diet on radioactivity of aortic phospholipids. Incubated 16 h with 200 μC $Na_2H^{32}PO_4$ and 200 mg% glucose

	Specific activity μC/mg P and (counts/aorta/min)			
	normal	cholesterol	olive oil	cholesterol and olive oil
TPL	0.4	3.2	1.5	3.8
	(13,600)	(44,000)	(27,400)	(88,200)
PA	0.19	1.7	0.6	2.1
PE	0.87	7.3	3.6	8.1
LPE	0.26	3.8	0.4	2.9
CEPH	0.81	7.5	3.0	7.7
LEC	0.17	1.5	0.5	2.3
SPHING	0.03	0.1	0.2	0.4
CEPH/LEC	4.8	5.0	6.0	3.4

cholesterol, 0.5% cholic acid and 0.2% propylthiouracil with normal rat diet ad 100%. This was fed for 8–10 weeks. A second group of rats was also force-fed with a single dose of olive oil (450 mg), killed 4 h after feeding and the isolated aortae incubated as described previously. A third group of rats, kept on the cholesterol diet were fed with a single dose of olive oil as above. In the cholesterol-fed group total phospholipid (TPL) radioactivity was increased threefold and the SA increased eightfold over that of the control group. The olive-fed rats also showed a striking increase in TPL radioactivity and with the cholesterol and olive oil-fed group the increases due to each diet were additive.

Discussion

The diet of GRESHAM and HOWARD as modified by CONSTANTINIDES, in our hands, resulted in a loss of body weight in the rat. Although a few aortae were found to contain yellow plaque-like material, on the average, there was no increase in aortic lipids after the diet. A general thickening of arterial wall and gross discolouration of the aorta were observed and blood cholesterol was increased by 30%. The results of the present experiments show that after induction of hypercholesterolaemia (if not atheroma?) the rate of phospholipid synthesis was accelerated as compared with synthesis by the normal rat aorta. The accelerated phospholipid synthesis observed by increasing the glucose concentration from 100–200 mg% further suggests that phospholipid synthesis may be increased in conditions giving rise to hyperglycaemia. Normal rat blood glucose levels are 80–100 mg% and recently BARANOV and SOKOLOVEROVA (1966) have shown that in alloxan diabetes the rat glucose levels are 150–200 mg%. Experiments along these lines are now in progress.

The higher specific activities obtained in incubation experiments than in perfusion experiments indicate that both intima and adventitia were responsible for the synthesis using the former technique; whereas synthesis was restricted only to the intima with perfusion. In both processes the rate of cephalin synthesis was much higher than that of lecithin.

The present experiments support the suggestion of ZILVERSMIT and co-workers that aortic synthesis may play an important role in the formation of aortic plaques. It would also account for the location of the plaques below the intimal surface without recourse to an 'infiltration' hypothesis.

However, in view of the implication of the cephalin group of lipids in blood coagulation, the higher cephalin content of the aorta and the increased rate of synthesis of this group of lipids in aortic tissue, it is tempting to speculate that the Duguid hypothesis of atheroma may still be a valid explanation for surface plaques in the aorta. Whichever hypothesis prevails, dietary factors play an important role in the aetiology of atheroma, either by accelerating blood coagulation or by disturbing the metabolic activity of the arterial tree.

Summary

1. The metabolic activity of the isolated thoracic aorta in synthesising phospholipids has been studied by incubation and perfusion with $Na_2{}^{32}PO_4$ in a suitable medium.

2. Concentration of glucose below 100 mg% does not affect the rate of phospholipid synthesis but this rate is almost doubled when the glucose concentration reaches 200 mg%.

3. The individual ^{32}P phospholipids synthesised from 1–6 hours have been characterised; the amount of incorporation of radioactivity and the specific activity of the fractions have been assessed. The specific activity of phosphatidyl ethanolamine synthesised was much higher than that of lecithin. More phosphatidic acids were synthesised in perfusion than in incubation experiments. The significance of these findings is discussed.

4. With rats rendered hypercholesterolaemic, aortic synthesis was highly activated. A single dose of olive oil had a similar but smaller effect. The latter effects were additive when olive oil was fed to hypercholesterolaemic rats. These findings are discussed in relation to the aetiology of atheroma.

Acknowledgements

We are indebted to the British Heart Foundation and the Central Research Fund of the University of London for gerous financial assistance.

It is a pleasure to acknowledge the encouragement and advice of Professor N.F. MACLAGAN throughout this work.

References

BARANOV, V.G. and SOKOLOVEROVA, I.M.: Experimental model of latent diabetes mellitus in rats and factors promoting transition to overt diabetes. Fed. Proc. 25: T 55–58 (1966).

CHERNICK, S.; SRERE, P.A. and CHAIKOFF, I.L.: The metabolism of arterial tissue, II. J. biol. Chem. 179: 113–118 (1949).

CONSTANTINIDES, P.: Experimental atherosclerosis (Elsevier, Amsterdam 1965).

GRESHAM, G.A. and HOWARD, A.N.: Atherosclerosis produced by semi-synthetic diet with no added cholesterol. Arch. Path. 74: 1–5 (1962).

LOFLAND, H.B.; MOURY, D.M.; HOFFMAN, C.W. and CLARKSON, T.B.: Lipid metabolism in pigeon aorta during atherogensis. J. Lipid Res. 6: 112–118 (1965).

MACLAGAN, N.F.; BILLIMORIA, J.D. and HOWELL, C.: Phospholipids of rat tissues after feeding pure phosphatidyl ethanolamine and lecithin. J. Lipid Res. 7: 242–247 (1966).

NEWMAN, H.A.I.; MCCANDLESS, E.L. and ZILVERSMIT, D.B.: The synthesis of C^{14} lipids in rabbit atheromatous lesions. J. biol. Chem. 236: 1264–1268 (1961).

ZILVERSMIT, D.B.; SHORE, M.L. and ACKERMAN, R.F.: The origin of aortic phospholipid in rabbit atheromatosis. Circulation 9: 581–585 (1954).

Authors' address: Dr. J.D. Billimoria and Dr. T.J. Rothwell, Westminster School of Medicine, Reserach Laboratories, Udall Street, Vincent Square, *London S.W. 1* (England).

Gaubius Institute, University of Leiden

Origin and Development of Atherosclerotic Lesions

C.J.F. BÖTTCHER, Leiden

Since the present symposium is mainly concerned with the advanced stages of atherosclerosis – an emphasis understandable from a clinical point of view – I would like to restrict myself almost entirely to the origin and earliest developmental stages of human atherosclerosis. Since as yet, only some vague contours are discernible in the hazy dawn of atherosclerosis, this restriction has no disadvantages because the available time of ten minutes is almost sufficient to survey what we know of the subject.

Most of us will agree that the earliest *manifestations* of human atherosclerosis are found in the subendothelial layer of the intima. This is one of the many striking differences from experimental atherosclerosis in animals, in which the conditions are so unphysiological that the process begins with a simple physicochemical phenomenon: the adhesion to the arterial wall of lipidloaded phagocytes from the blood stream. Endothelium grows over the adsorbed cells, and thus the earliest stage of atherosclerosis in experimental animals resembles a much later stage of *human* atherosclerosis when, at sites at which there has been damage to the endothelium, adsorption of platelets, formation of micro-thrombi, and even adsorption of lipids, take place, followed by the events so well described by DUGUID.

Only under extreme conditions, for instance with extreme hyperlipaemia, could adsorption of particles to the endothelium play a rôle in the initial phase of human atherogenesis. In most of the cases it starts in a much more sophisticated fashion by fundamental changes in the subendothelial layer under the intact endothelium.

As a physical chemist I visualize the subendothelial layer of a medium size human artery as a system of elastin fibres, collagen fibres, fibroblasts, and 'smooth muscle cells', all embedded in an aqueous gel of mucopolysaccharides and proteins. In other words: I prefer to approach the structure of the subendothelial layer by starting from the three-dimensional network of two of its mucopolysaccharides, hyaluronic acid and chondroitin sulphate C, with protein molecules attached to the latter.

This network of mucopolysaccharides and attached protein molecules forms, together with water, a continuous gel, reinforced by elastin and collagen fibres, which considerably improves the mechanical properties of the system, in particular its elasticity.

Electron microscopy has revealed in recent years that not only fibres but also cells are embedded in the aqueous mucopolysaccharide gel. Morphologically, a great majority of these cells (more than 90%) have all the characteristics of a smooth muscle cell, particularly the shape and the presence of many myofilaments (contractile elements) per cell. A minority of the cells have the appearance of fibroblasts, and it is quite likely that these cells produce the mucopolysaccharides, the elastin, and the collagen of the intima.

Less obvious is the task of the cells which morphologically have exactly the appearance of smooth muscle cells and are therefore classified as such by practically all the investigators who have examined them.

We should not be surprised if it is discovered that these cells do not have the ordinary task of smooth muscle cells – to contract and relax – but instead prove to be the active metabolic centres so badly needed by the subendothelial layer, particularly for its energy supply, i.e. for its ATP production. It is difficult to see how a small number of isolated smooth muscle cells could make any contribution to the elasto-mechanical properties of the mucopolysaccharide gel in the presence of a wealth of elastin and collagen fibres, whereas the subendothelial layer could hardly do without metabolic centres.

I pay so much attention to the so-called smooth muscle cells of the subendothelial layer because – as I shall show in a minute – we find typical metabolic disturbances in them as soon as something goes wrong. But let me first finish the picture of the healthy subendothelial layer.

Dissolved in the mucopolysaccharide gel are ions and numerous types of molecules, varying greatly in size. This gel is the major transport regulator of the arterial wall and it acts to a certain extent as an ultrafilter as well. These facts must be emphasized since the flexibility of transportation in the subendothelial layer, a flexibility which decreases with age, is in my opinion one of the most important factors in connection with the origin of atherosclerotic lesions.

The picture I have drawn so far is much too static. Actually, everything is in rapid biochemical transition. All the time, macromolecular structures, such as mucopolysaccharides and proteins, are being broken down and newly formed, and transport takes place in all directions.

Of the two acid mucopolysaccharides I have mentioned, chondroitin sulphate C is the strongest transport regulator. It is a polymer with a molecular weight of about 50,000. Each of its units has two ionic groups, $COOH$ and OSO_3H, which have a strong electric interaction with many types of molecules and ions. The other transport regulator, hyaluronic acid, has only one ionic group and thus relatively less electric interaction with other particles. It has a molecular weight of a few millions.

The direct evidence for the function of these two acid mucopolysaccharides as transport regulators is that the application of hyaluronidase, an enzyme which attacks both hyaluronic acid and chondroitin sulphate C, leads to a marked increase of the permeability. Because of this, hyaluronidase is even known as an additive for the improvement of the diffusion of subcutaneously injected drugs.

Other evidence for the transport-regulating function of the two mucopolysaccharides is that depolymerization of hyaluronic acid and chondroitin sulphate C by means of X-ray irradiation leads to a drastic increase of the permeability of the arterial wall, as shown by LAMBERTS and DE BOER [1] for the carotid arteries of rabbits.

In an advanced stage of atherosclerosis, damage to the mucopolysaccharides-network might promote the so-called infiltration of undesirable components from the lumen into the arterial wall, particularly if the endothelium has also been damaged.

In the earliest stage of atherosclerosis, however, it is more likely that *lack* of transport flexibility contributes to the initiation and development of lesions since, although the total percentage of acid mucopolysaccharides in the intima does not increase with atherosclerosis, there is a considerable shift in its composition in that chondroitin sulphate C and hyaluronic acid increase at the cost of the other two components: heparitin sulphate and chondroitin sulphate B (BÖTTCHER and KLYNSTRA [2]).

The electron-microscopical investigations of Dr. Sachs of our institute have demonstrated the presence of 'clumps' of accumulated lipids in the so-called smooth muscle cells of the intima in the earliest stage of atherosclerosis. Chemical analysis of such lipid accumulations has shown that they are of a primitive composition, particularly as regards the fatty acids of the cholesterol esters. Hence, we have come to the conclusion that these lipids are synthesized *in situ* (BÖTTCHER [3]).

As regards the sequence of events, I would like to mention two of the many possibilities.

The undesirable sequence of events might start with the so-called smooth muscle cells doing the wrong thing, which could be caused, for instance, by an unbalanced ratio of nutrients offered to their mitochondria. This possibility could please those of you who are looking for a connection with nutrition, even in an early stage of the disease.

A second possibility is that changes in enzyme activity due, for instance, to hormonal factors, ultimately cause changes in the mucopolysaccharides leading to a decrease in the transport flexibility followed by a lack of nutrients in certain smooth muscle cells.

There could also be a coincidence of various disturbing factors.

It is not difficult to understand how the lipid accumulation in smooth muscle cells can lead to lesions. It seems a reasonable assumption that such smooth muscle cells would gradually be disrupted, perhaps via an intermediate stage in which they become some kind of foam cells. After disrupture, extracellular fat droplets occur, and this would lead to the formation of the earliest lesions, the fatty streaks.

I hope that some of the following lectures will contribute to the clarification of the many uncertainties still existing with respect to the earliest stages in the development of atherosclerotic lesions.

References

1. LAMBERTS, H.B. and BOER, W.C.R.M. DE: Int.J.Radiat.Biol. *6:* 343 (1963).
2. BÖTTCHER, C.J.F. and KLYNSTRA, F.B.: Lancet *ii:* 439 (1963).
3. BÖTTCHER, C.J.F.: Proc.roy.Soc.Med. *57:* 792 (1964).

Author's address: Dr. C.J.F. Böttcher, Gaubius Institute, University of Leiden, *Leiden* (The Netherlands).

Aortic Perfusion in Experimental Animals
A System for the Study of Lipid Synthesis and Accumulation

D. E. Bowyer, A. N. Howard, G. A. Gresham, D. Bates
and B. V. Palmer, Cambridge

Introduction

The accumulation of lipids in atherosclerotic lesions in man was first demonstrated histologically by Mettenheimer in 1857. Studies on the composition of the arterial lipids were hampered for many years by the lack of suitable techniques for handling and separating the small amounts of lipid obtainable. Recently, however, with the advent of sensitive techniques such as thin – layer chromatography and gas – liquid chromatography reliable analyses have been made. Conflicting results have been obtained (particularly with respect to the fatty acid composition of the cholesteryl esters), but these discrepancies have arisen on two accounts: (1) Some authors have studied separated intima and media, whilst others have taken intima plus media preparations. (2) Comparisons between diseased and undiseased tissue have been made on material from different aortae, rather than between adjacent tissues from the same aorta. In consequence, normal tissue has been derived from young aortae, whilst diseased tissue has been obtained from older aortae, and changes of composition due to age may have been superimposed upon changes due mainly to the atherosclerotic processes. However, where similar tissues have been analysed, there is agreement upon changes of intimal lipid composition accompanying the development of the atherosclerotic plaque.
In summary the changes are as follows:
In the intimal fatty streak there is an increase in the total cholesterol, and the cholesteryl esters increase more than the free cholesterol. There is little change in the amount of triglyceride. Sphingomyelin increase markedly and to a lesser extent lecithin. There is little increase in phosphatidyl ethanolamine. There are also changes

in the fatty acid composition of the accumulating lipids in the plaque compared with the adjacent normal intima, the most notable of which is the increase in the percentage of cholesteryl oleate in the early fatty streak.

Similar changes in lipid composition are seen in the lesions in arteries of experimental animals fed atherogenic diets. Although the lesions produced in some systems are histologically different from those seen in man (for example in the cholesterol fed rabbit) in many instances the lesions are similar to the human ones [8] and the chemical changes parallel those seen in the human fatty streak (for example the rabbit fed a semi-synthetic diet without added cholesterol).

A summary of typical values for arterial lipids and for the composition of the cholesteryl esters are shown in Tables I and II.

Table I. Lipids of normal aortic intima and atherosclerotic lesions

Ref.	Species	Tissue	Free chol.	Esterified chol.	Sph.	Lec	P.E.
[1, 2]	man	undiseased	3.0	4.8	1.5	2.5	0.9
		diseased	8.8	25.3	2.6	4.5	1.2
[4]	man	undiseased	2.3	1.3	2.1	1.5	2.4
		diseased	14.8	14.8	5.9	2.8	2.4
[3]	rabbit	normal	0.9	0.4	1.1	2.1	0.9
[5, 6]	rabbit	normal	1.4	0.3	1.2	2.8	2.0
[6]	rabbit (cholesterol diet)	diseased	4.7	10.5	5.1	6.0	3.1
[3]	rabbit (semi-synthetic diet)	undiseased	2.8	3.6	1.5	2.6	1.2
		diseased	4.9	6.7	5.9	9.4	1.4

Values in mg/g of wet tissue. chol = cholesterol; Sph. = spyingomyelin; Lec = lecithin; P.E. = Phosphatidyl ethanolamine.

The accumulation of lipids in plaques could occur by two mechanisms:
(1) Deposition of lipids, lipoprotein, or larger blood components such as platelets without their subsequent return to the blood;
(2) local synthesis again without removal to the blood.

The accumulation of cholesteryl oleate could be the result of either preferential *in situ* synthesis, or preferential deposition of cholesteryl oleate or preferential hydrolysis of the other cholesteryl esters (whether arriving by synthesis or deposition).

Table II. Percentage of major fatty acids of cholesteryl esters of aortic intima

Ref.	Species	Tissue	% Fatty acid			
			16:0	18:0	18:1	18:2
[3]	man	undiseased intima	13	2	29	26
		fatty streaks	9	2	43	17
		atheroma	11	1	33	35
[3]	rabbit	normal	22	8	12	11
[3]	rabbit (semi-synthetic diet)	diseased	14	8	60	5
[7]	rabbit (cholesterol diet)	diseased	13	3	54	15

In the experiments reported here, we have used a perfusion technique to study further the synthesis of arterial lipids and to try to see why the fatty acid composition of the cholesteryl esters becomes bizarre.

Materials and Methods

A. Perfusion. We have used the technique of arterial perfusion for two reasons: (1) Arterial lipid synthesis may unambiguously be distinguished from deposition. This is in contrast to the situation in the whole animal where the extra-arterial metabolism followed by the deposition of the altered precursor may lead to misleading results. (2) The anatomical structure of the arterial wall is unaltered, and this is of great importance if the endothelial membrane acts as a selective barrier between the inner layers of the arterial wall and the blood in the lumen.

Two methods for perfusion have been studied; either the aorta is left intact and therefore largely undisturbed in the dead animal, which is kept at 37°C by a heating box, or it is excised and perfused in a heating chamber. The latter method has the advantage that a smaller perfusate volume is required and that the control of the perfusate and aorta temperature is simplified. In outline the method is similar to the LOFLAND [10] and is as follows:

The animal is killed by the intravenous injection of an overdose of nembutal and opened ventrally. All the branches of the abdominal aorta are ligated and a catheter is inserted just above the bifurcation. The aorta is removed together with the aortic arch without ligation of the intercostal arteries. Any blood is washed from the outside of the aorta with 0.9% saline at 37°C and the aorta is suspended inside the perfusion chamber. 5 ml of pregassed buffer at 37°C without labelled buffer is then pumped through the artery in order to remove blood from the lumen. 5 ml of the perfusion buffer with the labelled substrate is then put into the chamber and the perfusion started. The time required to set up the technique is about 10 min. A diagram of the apparatus is shown in Figure 1 and the composition of the bicarbonate buffer in Table III.

B. Preparation of tissue; lipid extraction and separation; estimation of radioactivity. After perfusion the aorta is removed from the chamber, opened longitudinally and washed twice with ice cold saline. Intima, with approximately one half of the media is then stripped with fine forceps, weighed and disintegrated by freeze presseing. The lipids are

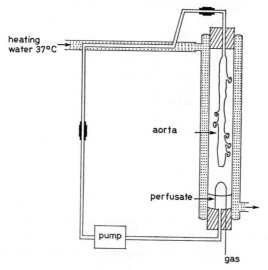

Fig. 1. Perfusion apparatus.

Table III. Perfusion buffer

0.90% NaCl	100.4 ml
1.15% KCl	4.0
1.22% CaCl$_2$	1.5
2.11% KH$_2$PO$_4$	1.0
3.80% MgSO$_4 \cdot$ 7H$_2$O	1.0
1.30% NaH CO$_3$	21.0
5.40% glucose	1.1
Defatted bovine serum albumin (Armour-Cohn fraction V)	4 g/100 ml

After addition of serum albumin the buffer is dialysed at 4°C for 16 h against buffer without albumin.

The buffer is gassed with O$_2$/5% CO$_2$ at 37°C 10 min before use and throughout perfusion.

pH. 7.4 at 37°C.

extracted by the method of Folch and separated into individual phospholipids and neutral lipids by thin layer chromatography on silica gel H. Phospholipids are separated in chloroform: methanol: acetic acid: water: 60:20:1:3 v/v, and neutral lipids in petroleum ether: diethyl ether: acetic acid: methanol: 90:5:1:2 v/v. The separated lipids are located by spraying the plate with a mixture of 2' 7' dichlorofluorescein and Rhodamine B. For scintillation counting, appropriate areas are placed directly into an aqueous scintillator. Complete elution of all of the lipids into solution with greater than 95% recovery of the applied counts has been demonstrated. The counting efficiency is measured by an internal standard and is neither affected by the presence of the dyes nor the silicic acid. With the solvents used, less than 0.5% tailing of the free fatty acids into other lipids occurs. Absolute amounts of compounds incorporated into the lipid fractions can be calculated from a knowledge of the specific activity of the perfused material assuming negligible dilution by the arterial pools.

C. Results. (1) The perfusion of serum albumin bound 1-C^{14} palmitic, stearic, oleic and linoleic acids in the normal rabbit (New Zealand White).

All of the fatty acids were perfused at a concentration of 0.1 µeq./ml and the fatty acid to serum albumin ratio was 1 to 6. Perfusion time was 1 h. The results shown in Table IV are the means of four perfusions for each fatty acid. Similar patterns of labelling of the lipid fractions with each fatty acid was observed, with greatest incorporation into glycerides and lecithin. Activity was observed in the sphingomyelin fraction as removed from the plate (these figures are shown in parentheses in the table) but on alkaline hydrolysis of this material and subsequent separation on a second plate the activity in the sphingomyelin disappeared and was most likely due to contaminating lecithin. The concept of arterial synthesis of sphingomyelin from exogenous fatty acid precursors should be carefully reconsidered. Triglyceride and diglyceride synthesis occurred, but was markedly lower using stearic acid as precursor. This effect is not so marked in the diglyceride fraction, indicating that the incorporation of stearic acid into diglyceride is limiting. This might be expected in view of the low percentage of stearic acid in the triglyceride fraction. There was extremely small synthesis of cholesteryl esters and no preferential acylation of cholesterol by oleic acid could be demonstrated.

Table IV. The uptake of serum albumin bound 1-C^{14} fatty acids by normal rabbit aorta

Fatty acid	Sph	Lec	PE	DG	FFA	TG	CE
	*						
16:0	0 (293)	7742	281	607	1326	214	17
18:0	0 (243)	3233	160	238	1371	35	0
18:1	0 (97)	2065	253	327	537	228	0
18:2	0 (119)	2771	239	513	528	317	0

Figures are µµmoles of fatty acid/g wet tissue and are means of four perfusions for each fatty acid. Fatty acid concentration 0.1 µE/ml. Molar ratio of serum albumin to fatty acid 6 to 1. Perfusions times = 1 h.

* see text.

Sph = sphingomyelin; Lec = lecithin; PE = phosphatidyl ethanolamine; DG = diglycerides; FFA = free fatty acids; TG = triglycerides; CE = cholesteryl esters.

(2) The perfusion of normal rabbit aortae with sodium 1-C^{14} acetate.

Sodium 1-C^{14} acetate was added to the perfusate to give a concentration of approximately 1 mg per 100 ml. No cold fatty acids were present. The results are shown in Table V. As with perfusion of fatty acids, the greatest incorporation was into the glycerides and lecithin. The sphingomyelin fraction was not further purified by differential hydrolysis and may contain contaminating lecithin. There was a small amount of activity in the cholesterol and cholesteryl ester fractions, but the sterol was not purified by digitonide precipitation or by dibromination and this activity cannot be taken as proof of the synthesis of cholesterol by this system. We wondered whether the activity in the cholesteryl esters was oleic acid, but on separation of the methyl esters of the total

Table V. Perfusion of normal rabbit aorta with 1 C¹⁴ sodium acetate

Sph	Lec	PE	DG	C	FFA	TG	CE
*(520)	2320	250	170	20	120	840	70

Figures are uptake in μμmoles acetate/g wet tissue and are means for four perfusions. The distribution of C^{14} in the synthesised fatty acids was: Myristic (14:0) 5%; Palmitic (16:0) 66%; Stearic (18:0) 29%.
* see text.
Sph = sphingomyelin; Lec = lecithin; PE = phosphatidyl ethanolamine; DG = diglycerides; FFA = free fatty acids; TG = triglycerides; CE = cholesteryl esters.

Table VI. Uptake and percentage hydrolysis of cholesteryl – 1-C¹⁴ palmitate, stearate, oleate and linoleate by perfused normal rabbit aortae

	C 16:0	C 18:0	C 18:1	C 18:2
Uptake μμmoles/g wet tissue/h	274 (181–425)	632 (110–1487)	483 (109–1235)	414 (288–498)
% Hydrolysis in 1 h	12% (7–18)	9% (1–16)	4% (2–6)	33% (20–52)

The figures are means for four perfusions of each ester. Ranges are shown in parentheses.

lipid fatty acids by gas chromatography followed by determination of their C^{14} activity by flow counting, only the synthesis of saturated fatty acids could be demonstrated.

(3) The perfusion of cholesteryl – 1-C¹⁴ palmitate, stearate, oleate, and linoleate in the normal rabbit.

The cholesteryl esters were dissolved in a small amount of ethanol which was then injected rapidly into the perfusate at 37°C. The esters dispersed to give a slightly cloudy solution which was stable on standing. The concentration of esters was approximately 6 mg/100 ml. During one hour perfusion, there was a similar uptake of the four esters, but the percentage hydrolysis (measured as the percentage of the incorporated activity found in the free fatty acids) was markedly different. There was a very small hydrolysis of cholesteryl oleate. The figures are shown in Table VI. If removal of cholesteryl esters from the arterial wall requires prior hydrolysis to free cholesterol, then lack of cholesteryl oleate hydrolase activity would account for the accumulation of this ester in the rabbit, which is susceptible to atherosclerosis. A similar study in the rat which is resistant to atherosclerosis shows that the normal rat aorta does have considerable cholesteryl oleate hydrolase activity, but on feeding an atherogenic diet this is depressed. The results are shown in Table VII.

Table VII. Perfusion of normal (N) and atherosclerotic (A) rat aortae with 4-C^{14} cholesteryl stearate, oleate and linoleate

	C 18:0	C 18:1	C 18:2
Uptake, μμmoles/g wet tissue/2 h			
N	4400	1270	375
	(470–8270)	(820–2180)	(145–734)
A	5010	820	330
	(1160–2580)	(400–1530)	(189–464)
Percentage hydrolysis in 2 h			
N	0	26%	51%
		(22–34)	(48–56)
A	0	11%	51%
		(8–12)	(37–64)

The figures are means for three perfusions for each ester. Ranges are shown in parentheses.

Discussion

Analyses of the lipids of atherosclerotic lesions in man and in experimental animals reveal the accumulation of free and esterified cholesterol and of lecithin and sphingomyelin. It is clearly of great importance to determine the source of these lipids and see whether local synthesis is the most important factor causing accumulation.

It has been demonstrated for a number of species that free cholesterol exchanges between the arterial wall and the blood, and FIELD and coworkers [11] estimate that more than 60% of the free arterial cholesterol in man is derived from the blood.

Further, *in situ*, synthesis, as demonstrated by homogenates, slices and perfusions is very small. Similarly, there is little synthesis of cholesteryl esters and we confirm and extend here, the results of STEIN [12] that there is very little acylation of free cholesterol. We have been able to demonstrate the uptake of cholesteryl esters in the perfused normal rabbit aorta and suggest that these may also be derived from the plasma. Further in the rabbit, there is a very small hydrolysis of deposited cholesteryl oleate compared with the other cholesteryl esters. This data supports the hypothesis that the increase in the percentage of cholesteryl oleate in the rabbit lesions is the result of selective hydrolysis. A similar study in the rat, which is more resistant to atherosclerosis shows that there is considerable hydrolysis of cholesteryl oleate, and that this is reduced, but not completely inhibited on feeding an atherogenic diet. It is interesting to note that every atherogenic diet raises the amount of cholesteryl oleate in the plasma. It is possible that in those species susceptible to atherosclerosis, a low activity of the arterial cholesteryl oleate hydrolase leads to a rapid deposition of cholesteryl oleate.

Phospholipid synthesis has been studied in a number of laboratories both by *in vivo* [9] techniques, and in slices and homogenates [12]. We confirm the observations that there is incorporation of fatty acid into lecithin and to a lesser extent into phosphatidyl ethanolamine. However, we find only a very small incorporation into sphingomyelin, purified by alkaline hydrolysis.

It has been suggested by DIXON [13] that, in liver, neutral lipids are kept in the form of soluble micelles by phospholipid and that in situations where insufficient phospholipid is synthesised the neutral lipids appear as globules in the cells. If a similar situation exists in arterial tissue, then synthesis of phospholipid might occur in response to deposited cholesterol and cholesteryl esters. Since sphingomyelin and phosphatidyl ethanolamine are poorly synthesised by the normal artery compared with lecithin, only lecithin would seem to fill this role.

Acknowledgements

This work was supported in part by grants from the Tobacco Research Council, the National Institutes of Health, Bethesda P.H.S. Grant No. H6300 and Eli Lilly, Indianapolis, U.S.A.

References

1. SMITH, E.B.: Intimal and medial lipids in human aortas. Lancet *i:* 799–803 (1960).
2. SMITH, E.B.: The influence of age and atherosclerosis on the chemistry of aortic intima. Part I, The lipids. J. Atheroscler. Res. *5:* 224–240 (1965).
3. BOWYER, D.E.: Unpublished data.
4. BUCK, R.C. and ROSSITER, R.J.: Lipids of normal and atherosclerotic aortas. Arch. Path. *51:* 224–237 (1951).
5. MCCANDLESS, E.L. and ZILVERSMIT, D.B.: The effect of cholesterol on the turnover of lecithin, cephalin and sphingomyelin in the rabbit. Arch. Biochem. Biophys. *62:* 402–410 (1956).
6. DURY, A.: Effect of cortisone on lipid metabolism of plasma, liver and aorta and on retrogression of atherosclerosis in the rabbit. Amer. J. Physiol. *187:* 66–74 (1956).
7. ZILVERSMIT, D.B.; SWEELEY, C.C. and NEWMAN, H.A.I.: Fatty acid composition of serum and aortic intimal lipids in rabbits fed low- and high-cholesterol diets. Circulat. Res. *9:* 235–241 (1961).
8. GRESHAM, G.A. and HOWARD, A.N.: Comparative histopathology of the atherosclerotic lesion. J. Atheroscler. Res. *3:* 161–177 (1963).

9. ZILVERSMIT, D.B.; SHORE, M.L. and ACKERMAN, R.F.: The origin of aortic phospholipid in rabbit atheromatosis. Circulation 9: 581–585 (1954).
10. LOFLAND, H.B.; MOURY, D.M.; HOFFMAN, C.W. and CLARKSON, T.B.: Lipid metabolism in pigeon aorta during atherogenesis. J.Lipid Res. 6: 112–118 (1965).
11. FIELD, H.; SWELL, L.; SCHOOLS, P.E. and TREADWELL, C.R.: Dynamic aspects of cholesterol metabolism in different areas of the aorta and other tissues in man and their relationship to atherosclerosis. Circulation 22: 547–558 (1960).
12. STEIN, Y. and STEIN, O.: Incorporation of fatty acids into lipids of aortic slices of rabbits, dogs, rats and baboons. J.Atheroscler.Res. 2: 400–412 (1962).
13. DIXON, K.C.: Fatty deposition: A disorder of the cell. Quart.J.exp.Physiol. 43: 139–159 (1958).

Authors' address: Dr. D. E. Bowyer, Dr. A. N. Howard, Dr. G. A. Gresham, Dr. D. Bates and Dr. B. V. Palmer, Department of Pathology, University of Cambridge, Tennis Court Road, *Cambridge* (England).

Department of Physiology, University of Aarhus

Intimal Uptake of Plasma Lipoprotein and Atherosclerosis*

S. CHRISTENSEN, Aarhus

The significance of a plasma lipoprotein uptake by intimal tissue for the development of atheromatosis was investigated in the thoracic aorta of cockerels. Experimental conditions were established for the measurement of the rates of transfer across the surface of intimal endothelium for three plasma substance groups such as (1) phospholipids, (2) phosphoproteins and (3) cholesterol plus cholesterol esters. The rates obtained are discussed relative to the rates of accumulation of phospholipids and total cholesterol in intima-media during atherogenesis, and also relative to the rate of phospholipid synthesis *in situ*. Lipoprotein movement through the canine aortic wall was studied by DUNCAN, BUCK and LYNCH (1963). Various other references to pertinent work are given in ref. 1–5.

Methods

Technical details were presented previously [1–5]. Shortly, the experimental procedure was as follows. Plasma phospholipids and phosphoproteins were labelled biosynthetically with ^{32}P and then introduced by intravenous injection of 3–6 ml of the labelled plasma per kg of the experimental animals. These animals were killed at intervals up to 5 h. In cholesterol studies whole blood from the experimental animal was incubated with 4-^{14}C-cholesterol for 2 h and then reinjected. These animals were killed at intervals up to 24 h.

The uptake of label at the intimal surface was measured as the amount of label present in the inner 'half layer' of the thoracic aortic wall. The inner 'half layer' consists of intima plus part of the media and is easily separated by hand. From general statements as to vascularization of arteries and from differences in uptake-time curves for the inner and outer 'half layers' [1, 2] it is justified to assume that the inner one (exclusively used for the following) is very poorly vascularized, and that the uptake by this

* Supported in part by 'Statens alm. videnskabsfond' and grant HE 08263, N.I.H., U.S.A.

layer is primarily due to the intimal surface. For a calculation of rate constants for the intimal transfer of the substance groups studied it is necessary to know the average concentration of the appropriate label in the plasma during the uptake process. In 132 of about 170 experiments the plasma concentrations were therefore followed. In the remaining experiments use was made of mean values from comparable experiments.

Hyperlipaemia was provoked in cockerels by cholesterol feeding or by stilboestrol treatment. During stilboestrol treatment great amounts of phosphoproteins appear in the plasma together with phospholipids, triglycerides and cholesterol. The phosphoprotein is probably all phosvitin (HEALD, 1962, 1963). At the degree of hyperlipaemia used here, about two thirds of the phosphoprotein were probably present in lipoprotein and one third in various complexes (conf. expt. 2, Table I, ref. 8). Electrophoresis studies supported that statement [2]. The labelling of both phospholipid and phosphoprotein with ^{32}P represent a double labelling of the macromolecular system as a whole, and this double labelling was useful for a first approximation of a study of intimal lipoprotein transfer. The complicating metabolic incorporation of plasma TCA-soluble ^{32}P-labelled compounds into intima-media phospholipids and proteins was roughly estimated from orthophosphate incorporation experiments and is believed to be of no importance for the results.

Table I. Rate constants for the transfer of three plasma substance groups from plasma to aortic intima-media

Exptl. animal, and (in brackets) the type of donor animal	Phospholipid $\mu l/cm^2/h$	Phosphoprotein $\mu l/cm^2/h$	Cholesterol $\mu l/cm^2/h$
Normal birds (own blood)			0.09±0.02 (5 expts.)
Normal birds (Stilboestrol-treated birds)	0.10±0.01 (33 expts.)	0.11±0.01 (25 expts.)	
Stilboestrol-treated birds (similar birds)	0.08±0.01 (27 expts.)	0.10±0.01 (27 expts.)	
Cholesterol-fed birds (similar birds)	0.11±0.02 (15 expts.)		
Cholesterol-fed birds (own blood)			0.05±0.01 (25 expts.)
Normal rabbits (own blood)			0.04±0.01 (7 expts.)

Expression of results. The results are to a great extent presented as rate constants for the process of transfer. For phospholipids the individual rate constants were calculated as

$$\frac{\text{lipid }^{32}P \text{ found per } cm^2 \text{ intima-media layer}}{(\text{average lipid-}^{32}P \text{ per } \mu l \text{ of plasma}) \times \text{hours of expt.}}$$

and similarly for phosphoproteins and cholesterol. The designation becomes $\mu l/cm^2/h$. For all three labelled substance groups a relatively high uptake was found in short-time experiments resulting in high-rate constants. However, provided that the uptake proceeded for more than 1 h, the rate constants calculated were rather independent upon time, for which reason the term rate constant was adopted. In the case that a plasma droplet uptake is the only mechanism of endothelial macromolecular transfer it appears that the rate constant has a substantial meaning as the plasma volume taken up per cm² per h. This is mentioned mostly for the clarification of concepts. The mechanism of the uptake is unknown..

It is easy to imagine ideal systems in which identical transfer rate constants could be obtained for several labelled lipoprotein components. The simplest consist of 'plasma' with only one type of lipoprotein, which is transferred intact to another compartment ('intima-media') and trapped there for at least the time of experimentation. Identical rate constants come out also if the labelled lipoprotein components exchange, molecule for molecule, with other pools of these components, provided that the resulting fall in specific activity of the individual components of the plasma lipoprotein is accounted for in the calculation of rate constants (vide formula).

Rates of transfer are derived as the product of the appropriate rate constant and plasma concentration.

Results and Discussion

Rate constants. The average values obtained in various series of experiments are presented in Table I, including one value from experiments with rabbits.

The cockerels were partly normal and partly hyperlipaemic ones. Two hyperlipaemic conditions were included. Further, a spectrum of lipoproteins are always present in the plasma. On this background it seems that the differences between the averages presented in Table I are not evidence against the concept that the labelled substances appeared in the intima-media as a consequence of a transfer of intact lipoprotein molecules from the plasma. On the whole, the results rather seem to support that concept. A further support was found in a very close correlation between the rate constants for phospholipid- and phosphoprotein-^{32}P in the double labelling experiments [1].

Rates of transfer and atheromatosis. In a series of 15 birds treated for 16 days with stilboestrol a rate of phospholipid transfer (calculated per g intima-media) of 22.6 µg/g/h was found. The rate of accumulation of phospholipid in intima-media during the 16 days was about 3.0 µg/g/h, i.e. one seventh of the rate at which plasma phospholipid penetrated into this tissue. Similarly, in cholesterol-fed birds the rates obtained were 6.8 µg/g/h and 0.6 µg/g/h, respectively. The rates obtained for cholesterol in cholesterol-fed birds were 5.3 and 0.9 µg/g/h, respectively. In all series transfer rates thus exceeded the accumulation rates several times.

Transfer rates, intima-media lipid concentrations and atheromatosis. Significant positive correlation was present between the intima-media lipid concentrations (phospholipid or cholesterol) and the rates of transfer of phospholipid [3]; the P values for no regression of intima-media concentration upon transfer rate were <0.01 and 0.001, respectively. In cholesterol-fed birds the P value was <0.001 for intima-media concentration of cholesterol versus the rate of transfer of cholesterol from the plasma; in the latter series the intima-media concentrations also showed significant regression upon the plasma concentrations. This was not the case in the first mentioned series. It is in accordance with these findings, and appears from the tables of previous work [2, 3], that intima-media lipid concentrations could be high in birds showing a moderate degree of hyperlipaemia. However, in such birds high rate constants were measured, and accordingly a relatively high rate of transfer of phospholipid was calculated. Similar tendencies were obtained in the cholesterol transfer studies. Reversely, low intima-media lipid concentrations in animals with high plasma lipid concentrations were combined with low transfer rate constants. Thus, the rate constants link together, in a physiologically reasonable way, the plasma lipid concentrations and the intima-media lipid concentrations in the present experiments. The intima-media lipid concentrations were highly correlated with visible atheromatosis.

Phospholipid synthesis in situ versus plasma phospholipid transfer. The rate values and thus also the ratio between the rates of (1) intima-media accumulation of phospholipid, (2) phospholipid transfer from the plasma and (3) phospholipid synthesis *in situ*, are in fair agreement with the values obtained in the cholesterol-fed rabbit by SHORE, ZILVERSMIT and ACKERMAN (1955) and supported by NEWMAN, MCCANDLESS and ZILVERSMIT (1961).

In the present material the ratio between the three rates is roughly about 1:10:100. Thus, during the lipid accumulation associated with atherogenesis, the amounts of phospholipid appearing in the intima-media due to the processes of synthesis *in situ* and transfer from the plasma must somehow be eliminated to almost 100%, say 99%. As to the 1% accumulated, nothing can be stated about the relative importance of the two sources for phospholipid, *i.e.* synthesis *in situ* and transfer from the plasma. However, some support for the concept that the plasma phospholipid play the greater role during phospholipid accumulation is found in the correlation between intima-media phospholipid concentration and the rate of transfer of phospholipid from the plasma [3]. Theoretically, all phospholipid accumulated might be derived from the plasma (conf.

the ratio between the rates for accumulation and transfer from the plasma). Within the frames of this concept the very considerable increase in rate of phospholipid synthesis *in situ* becomes only one aspect of an accelerated rate of phospholipid synthesis-degradation (metabolic turnover) without net production.

If, as suggested in the present work, plasma phospholipid is transferred as an integral component of intact plasma lipoproteins it appears to be an alternative concept – likewise explaining the correlation between intima-media concentration and the rate of transfer – that plasma lipoprotein molecules within the intima-media tissue provoke a *net* increase in the rate of phospholipid synthesis proportional to the rate of penetration of the cholesterol containing lipoprotein. Some support for such a concept was presented by DAY *et al.* [6] studying metabolic responses of macrophages to serum lipoprotein.

References

1. CHRISTENSEN, S.: Plasma phospholipid and phosphoprotein transfer across the intimal surface of the normal and slightly atherosclerotic aorta of the cockerel. J. Atheroscl. Res. *1:* 140–147 (1961).
2. CHRISTENSEN, S.: Plasma lipoprotein transfer across the intimal surface of arteries. Thesis (Universitetsforlaget, Aarhus 1961).
3. CHRISTENSEN, S.: Transfer of plasma phospholipid across the aortic intimal surface of cholesterol-fed cockerels. J. Atheroscl. Res. *2:* 131–138 (1962).
4. CHRISTENSEN, S.: Transfer of labelled cholesterol across the aortic intimal surface of normal and cholesterol-fed cockerels. J. Atheroscl. Res. *4:* 151–160 (1964).
5. CHRISTENSEN, S. and JENSEN, J.: Uptake of labelled cholesterol from plasma by aortic intima-media in control and insulin-injected rabbits. J. Atheroscl. Res. *5:* 258–259 (1965).
6. DAY, A.J.; GOULD-HURST, P.R.S.; STEINBORNER, R. and WAHLQUIST, M.L.: Removal of double-labelled lipid mixtures and double-labelled lipoprotein preparations by reticulo-endothelial cells. J. Atheroscl. Res. *5:* 466–473 (1965).
7. DUNCAN, L.E.; BUCK, K. and LYNCH, A.: Lipoprotein movement through canine aortic wall. Science *142:* 972–973 (1963).
8. HEALD, P.J.: Isolation of phosphorylserine form the plasma phosphoproteins of the laying hen. Biochem. J. *83:* 212–216 (1962).
9. HEALD, P.J.: Isolation of phosvitin form the plasma of the laying hen. Biochem. J. *87:* 571–576 (1963).
10. NEWMAN, H.A.I.; MCCANDLESS, E.L. and ZILVERSMIT, D.B.: The synthesis of C-14-lipids in rabbit atheromatous lesions. J. biol. Chem. *236:* 1264–1268 (1961).
11. SHORE, M.L.; ZILVERSMIT, D.B. and ACKERMAN, R.F.: Plasma phospholipid deposition and aortic phospholipid synthesis in experimental atherosclerosis. Amer. J. Physiol. *181:* 527–531 (1955).

Author's address: Dr. S. Christensen, Department of Physiology, University of Aarhus, *Aarhus* (Denmark).

Department of Medicine and Department of Physiology, University of Cape Town, Observatory, Cape, South Africa

The Filtration of Plasma Constituents into the Wall of the Aorta

L. H. Krut and J. A. Wilkens, Cape

Atherosclerosis is a patchy process with lesions forming at characteristic sites throughout the arterial tree. In the aorta, for example, plaques typically develop in the region of the arch, in relation to sites of branching and, in general, are more pronounced in the abdominal aorta than in the descending thoracic aorta. Other areas show little or no involvement by this disease process. These features emphasize that one cannot consider the whole vessel wall to be exposed and/or predisposed homogeneously to those factors relevant to the genesis of this disease. It is apparent that an understanding of the processes determining the localization of plaques is a prerequisite for establishing the mechanism of their formation.

Many hold the view that the filtration of certain plasma constituents into or across the arterial wall, in some way, gives rise to the development of the atheromatous plaque. To date, there are no data which could account for the localization of plaques in terms of this filtration concept. If the development of atheroma is determined by this process, one may propose that there is a greater filtration of plasma constituents at those sites where plaques are prone to form. We have tested this hypothesis in the following manner.

Healthy rabbits which had been on a standard laboratory diet were each given 1.0 mC of ^{32}P in the form of sodium ortho-phosphate in isotonic buffer (Radiochemical Centre, Amersham) via a peripheral ear vein. The animals were killed 3.5 h later, their aortae removed, washed several times in N saline, split longitudinally, again washed and then fixed overnight in 10% formal-saline containing 1% calcium acetate. The adventitia was then stripped off and any arterial branches protruding beyond the surface of the outer media were trimmed flush with this surface. The specimen was again washed and excess water removed with blotting paper. It was then laid flat on a glass plate with the intimal surface uppermost, covered with a polythene sheet and lightly pressed to lie flat and without folds or distortion by a second glass plate plus a small weight and frozen in this position at —15°C. A standard non-screen X-ray film was placed between the polythene sheet and the upper glass plate and allowed to expose for empirically determined times. A total of 3 experiments were done at different times and all aortae showed the same autoradiographic features. The figure is a photograph of one of the autoradiographs obtained. This shows one continuous length of aorta which could be made to lie flat without folds of distortion. To achieve this it was necessary to exclude the ascending aorta, some parts of the arch had to be trimmed and the bifurcation excluded. No lesions suggestive of atheroma were visible in any of the aortae.

It can be seen from the figure that there is a greater exposure in the region of the aortic arch than in the rest of the thoracic aorta in general. There is also a dense band of exposure running along the arch. In other preparations of the arch region alone a very dense exposure around the mouths of the vessels arising in this region can also be seen. There is also an increased exposure in the region of the origins of the intercostal branches. In the abdominal aorta the exposure is dense in general and areas corresponding to the origins of arteries are outlined.

A feature of interest is the well defined spot just distal to the inner arc of the arch. This area corresponds to the attachment of the fibrosed ductus arteriosus. This too was a constant feature and cannot be ascribed to an artefact produced by a collection of blood in the small scar which is usually visible on the intimal surface at this site.

While it is almost certain that some of the ^{32}P had exchanged with endogenous phosphorus in the time period of this study, it was not our purpose to measure such processes. Although the whole of the aorta was evenly exposed to this label, in the various forms in which it was circulating, there was a greater penetration into the wall of the aorta at some sites than at others.

We consider these studies to have demonstrated that the aortic intima allows an increased filtration of at least some plasma constituents at those sites which are prone to develop atheromatous lesions.

Those who believe that atheroma forms as the result of the accretion of blood elements on the intimal surface have presented arguments in terms of flow mechanics to account for the localization of atheromatous plaques. The experiments of HAIMOVICI and MAIER [1] raise other possibilities. They exchanged segments of abdominal and thoracic aortic homografts in dogs put onto an atherogenic regimen and found more atheroma in the abdominal segment than in the adjacent thoracic aorta and less atheroma in the thoracic segment than in the adjacent abdominal aorta. Their findings cannot be explained in terms of flow mechanics but do demonstrate that there are regional differences in the wall of the aorta which determine atheroma formation.

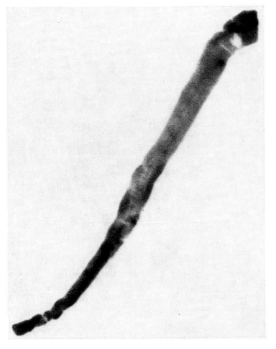

Fig.1. Autoradiograph of aorta from a rabbit which had been given 1.0 mC of ^{32}P 3.5 h prior to death. Note the increased exposure in the region of the arch compared with the descending thoracic aorta and the dense band of exposure along the arch; the demarcation of sites relating to the intercostal branches; the marked exposure over the abdominal aorta and the demarcation of the origins of vessels arising from the abdominal aorta.
The dense spot at the distal end of the inner arc of the arch corresponds to the attachment of the fibrosed ductus arteriosus.
The unexposed gaps in the region of the arch reflect tears in the vessel wall.

We suggest that the increased filtration of plasma constituents in certain regions of the arterial wall is the process which, in the first instance, determines the localization of plaques and possibly influences their development.

One circumstance which favours increased filtration is a defect in the arterial wall such as that left by the fibrosed ductus arteriosus. The increased filtration into the wall of the aorta at sites not directly related to existing arterial branches may be determined by such lesions in the arterial wall. DALITH [2] has pointed out that the early embryonic aorta contains numerous vessels in that part of it which becomes abdominal and which are not evident later in foetal development. Embryonic vessels present in the region of the arch also disappear during intra-uterine life. The degeneration of such vascular structures could conceivably leave defects in the vessel wall similar to that left by the fibrosed ductus ateriosus. Such sites of embryonic degeneration, although not readily demonstrable, may yet have altered the filtration characteristics at those sites. Another possible consequence of these developmental processes is the presence of cells at those sites with inadequate capacity to metabolise the increased load presented to them.

References

1. HAIMOVICI, H. and MAIER, N.: J. Atheroscl. Res. *6:* 62 (1966).
2. DALITH, F.: J. Atheroscl. Res. *4:* 239 (1964).

Acknowledgements

This work was carried out in the Clinical Nutrition Research Unit, which is supported in the Department of Medicine by the South African Council for Scientific and Industrial Research. It was also supported in part by a research grant (HE-03316) from the National Heart Institute, Public Health Service, USA.

Authors' address: Dr. L. H. Krut, Department of Medicine and Dr. J. A. Wilkens, Department of Physiology, University of Cape Town, Observatory, *Cape* (South Africa).

The Department of Medicine B, Hadassah University Hospital,
Departments of Biochemistry and Experimental Medicine and
Cancer Research, Hebrew University-Hadassah Medical School, Jerusalem

Metabolism of Lysolecithin by Human Umbilical and Dog Carotid Arteries[*]

Y. STEIN, S. EISENBERG and O. STEIN, Jerusalem

In a previous study it was found that arterial segments of various mammals utilize labelled fatty acids for complex lipid synthesis [1]. The enzymic pathways involved in these processes were investigated in homogenates prepared from rabbit and dog aortic tissue [2]. As shown in Fig. 1, one of the pathways leading to lecithin synthesis involves acylation of lysolecithin, the origin of which has not been elucidated so far in arterial tissue. Since no known pathway of lysolecithin synthesis has been described, this lysophospholipid could have been derived either (a) through degradation of arterial lecithin or (b) through uptake from the circulating blood. The latter possibility was investigated first, since it was recently shown that labelled lysolecithin is taken up from the circulation, very rapidly, by various mammalian organs and is acylated to lecithin [3]. To that end arterial segments of human umbilical artery, prepared as described before [4], or of dog carotid artery were incubated *in vitro* with labelled lysolecithin. The incubation medium which consisted of Krebs Ringer phosphate buffer (calcium^{++} omitted), pH 7.4, 5 mM glucose and labelled lysolecithin, bound to serum albumin was introduced into the lumen of the artery, in order to simulate *in vivo* conditions. The incubation procedure and post-incubation washing is depicted in Fig. 2. Segments of human umbilical artery took up lysolecithin from the incubation medium and converted it to lecithin (Table I). These processes were stable for at

---

[*] Supported in part by a research grant H-5705 NIH. USP HS.

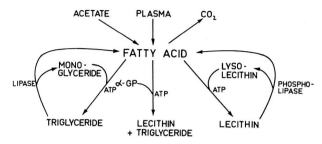

Fig. 1. Tentative scheme of triglyceride and lecithin metabolism in the aortic wall (from STEIN et al., Biochim. biophys. Acta *70:* 33 1963).

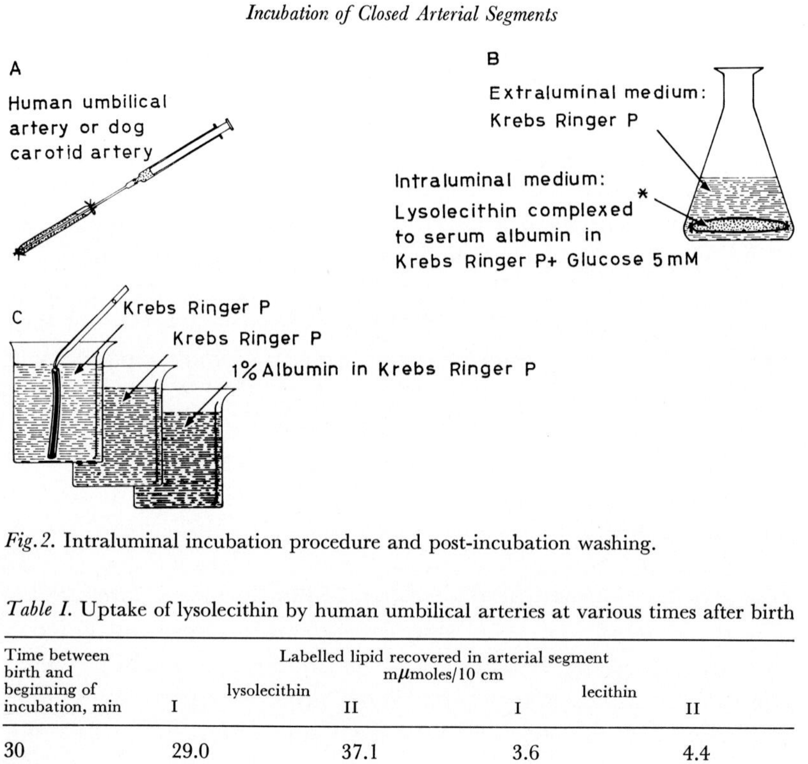

Incubation of Closed Arterial Segments

Fig. 2. Intraluminal incubation procedure and post-incubation washing.

Table I. Uptake of lysolecithin by human umbilical arteries at various times after birth

Time between birth and beginning of incubation, min	Labelled lipid recovered in arterial segment mμmoles/10 cm			
	lysolecithin		lecithin	
	I	II	I	II
30	29.0	37.1	3.6	4.4
90	25.1	44.2	3.2	4.9
180	24.2	48.1	3.0	6.1
360	16.8	40.0	3.9	5.5

The incubation medium consisted of in exp. I ^{32}P-lysolecithin (0.65 mM), in exp. II ^{14}C-palmitoyl lysolecithin (1.0 mM); bound to serum albumin (3:1 M/M), Krebs Ringer phosphate buffer, pH 7.4, (calcium^{++} omitted); 5 mM glucose (K-R-P-B). Incubated for 30 min at 37°.

least 3 h after birth, if the tissue was kept at 0°. Since after incubation with $Na^{32}PO_4$, under the experimental conditions used (Krebs Ringer phosphate buffer) no labelled lipids were found in the artery, it seems safe to conclude that the ^{32}P-phospholipids isolated from the artery following incubation with lysolecithin were not derived from inorganic phosphate (Table II).

Table II. Incorporation of $Na^{32}PO_4$ into lipids by segments of human umbilical artery

Segment	Labelled substrate	Counts/min incubation medium	Counts/min arterial lipids	^{32}P-lipids in artery, percent lysolecithin	^{32}P-lipids in artery, percent lecithin
1	$Na^{32}PO_4$	50,000	10	–	–
2	$Na^{32}PO_4$	50,000	10	–	–
3	^{32}P-lysolecithin	17,000	1,425	75	25
4	^{32}P-lysolecithin	17,000	1,560	80	20

The incubation medium consisted of $Na^{32}PO_4$ or ^{32}P-lysolecithin (0.45 mM) bound to albumin (3:1 M/M) in K-R-P-B.
Incubated for 30 min at 37°.

In order to be able to compare the metabolic events occurring in different arterial segments, a correlation was sought between the extent of uptake, segment length or intimal surface area. Table III shows that the uptake of labelled lysolecithin was proportional to the length of the artery if the ratio of the intraluminal volume (incubation medium) to segment length was kept constant. If the

Table III. Effect of length of arterial segment and intraluminal volume on the uptake of ^{32}P-lysolecithin by human umbilical arteries

Experiment No.	Temperature °C	Arterial segment length cm	Results calculated per 10 cm arterial length intraluminal medium ml	Results calculated per 10 cm arterial length intimal surface, cm²	^{32}P-lysolecithin uptake by arterial segment mμmole per 10 cm	^{32}P-lysolecithin uptake by arterial segment mμmole per cm²
1	0	4.5	0.22	1.67	19.1	11.4
	0	5.5	0.36	2.14	28.6	13.5
	0	5.5	0.54	2.62	33.8	12.8
2	0	4.5	0.55	2.65	27.3	10.6
	0	7.0	0.57	2.68	31.2	11.7
3	37	5.5	0.54	2.62	46.5	17.7
	37	5.0	0.50	2.45	53.3	21.7

The incubation medium consisted of ^{32}P-lysolecithin (1.08–1.37 mM) bound to serum albumin (3:1, M/M), in K-R-P-B. Incubation time 30 min.

intraluminal volume was varied while the segments were of comparable length the uptake became proportional to the surface area. Hence, in all subsequent experiments a constant volume to length ratio was kept and results are expressed per 10 cm length of arterial segment. Lysolecithin was taken up by the arterial segment also at 0°C (Table III) but its conversion to lecithin showed a clear temperature dependence and was progressive with time (Table IV).

Similar results were obtained also when dog carotid arteries were incubated with ^{32}P-lysolecithin (Table V). The mode of conversion of lysolecithin to lecithin was investigated using lysolecithin labelled both in ^{32}P and with ^{14}C in the fatty acid moiety. A conservation of the ^{14}C/^{32}P ratio in the resulting lecithin would indicate that acylation of lysolecithin occurred [5], while a rise in the ratio would point to a condensation reaction of two lysolecithins [6]. The results

Table IV. Conversion of ^{32}P-lysolecithin to lecithin as affected by temperature and time of incubation

Incubation medium temperature °C	lysolecithin mM	Umbilical artery ^{32}P-lysolecithin converted to lecithin mμmoles/10 cm arterial segment		
		min 5	15	30
0	0.45	–	–	0.8
22	0.45	1.2	2.0	2.4
37	0.45	2.1	3.8	4.4
37	1.37	2.4	3.3	6.2

Conditions of incubation as in Table I.

Table V. Uptake of ^{32}P-lysolecithin and its conversion to lecithin by dog carotid arteries

Experiment No.	Temperature °C	Labelled lipid recovered in arterial segment mμmoles/10 cm	
		lysolecithin	lecithin
1	0	11.3	1.3
	37	41.8	5.3
2	0	12.5	1.4
	37	27.1	5.2
3	0	13.5	1.5
	37	37.0	6.0

Incubation medium consisted of ^{32}P-lysolecithin 0.62 mM bound to serum albumin (3:1 M/M), in K-R-P-B. Incubation time 30 min.

of representative experiments shown in Table VI indicate that the addition of unlabelled fatty acid to the incubation medium resulted in conservation of the $^{14}C/^{32}P$ ratio in the newly formed lecithin. Hence it seems that lysolecithin is converted to lecithin, via the acylation reaction, but lecithin is formed also from labelled fatty acid derived from hydrolyzed lysolecithin. In the dog carotid artery lysolecithin was also converted to lecithin via the acylation reaction. In order to learn about the uptake of lysolecithin from a natural complex of phospholipids, as encountered in the serum, rat serum phospholipids were labelled biosynthetically with ^{32}P [3]. Segments of human umbilical arteries were incubated with such serum at

Table VI. Conversion of lysolecithin to lecithin in arterial segments incubated with 1-^{14}C palmitoyl-^{32}P-lysolecithin

Exp.	Incubation medium lysolecithin mM	carrier fatty acid mM	$^{14}C/^{32}P$ in lysolecithin	Arterial segment labelled lipid recovered mμmoles/10 cm lysolecithin	lecithin	$^{14}C/^{32}P$ ratio lysolecithin	lecithin
				umbilical artery			
1	0.65	0	1.0	12.0	4.2	1.0	1.7
	0.65	1.5	1.0	45.0	6.0	1.0	0.9
2	0.60	0	1.0	23.4	7.0	1.0	1.6
	0.60	1.5	1.0	39.0	6.0	1.0	1.0
3	0.60	0	1.0	32.1	6.4	1.0	2.2
	0.60	1.5	1.0	58.2	6.7	1.0	1.0
				dog carotid artery			
4	0.93	0	1.0	51.0	5.3	1.0	0.9
5	0.65	0	1.0	24.8	6.2	1.0	1.0

Conditions of incubation as in Table I.

both 0° and 37°. Two batches of dialyzed serum were used, which varied in their lysolecithin content. As seen in Fig. 3 the arterial segments extracted labelled lysolecithin preferentially, especially at 0°. This affinity of vascular tissue for lysolecithin at low environmental temperatures, when its acylation to lecithin is reduced, might have deleterious effects on arterial tissue during surgery under hypothermia. The finding of selective binding of lysolecithin by arterial tissue indicates that, in addition to *in situ* synthesis [7], arterial lecithin might be derived in part from serum lysolecithin.

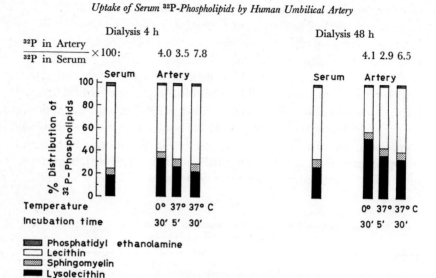

Fig. 3. Conditions of incubation: the incubation medium consisted of rat serum in which the phospholipids had been labelled bio-synthetically. Prior to incubation the serum was dialyzed for either 4 or 48 h resulting in an enrichment in the lysolecithin content.

Acknowledgement

The excellent technical assistance of Mr. G. Hollander is gratefully acknowledged.

Summary

The present study deals with the utilization of lysolecithin by arterial tissue. Labelled lysolecithin, introduced into the lumen of human umbilical and dog carotid arteries, *in vitro*, was taken up by the arterial wall. The uptake was found to be proportional to the segment length and intimal surface area, and occurred also at 0°. A part of the label was recovered in arterial lecithin and the conversion of lysolecithin to lecithin was temperature dependent. Using ^{14}C palmitoyl-^{32}P-lysolecithin it was shown that the newly synthesized lecithin had a $^{14}C/^{32}P$ ratio similar to the introduced lysolecithin, indicating that it was formed by acylation. The umbilical artery extracted lysolecithin preferentially when incubated with biosynthetically labelled serum ^{32}P-phospholipids. These findings might have a bearing on the origin of arterial phospholipids under normal and pathological conditions.

References

1. STEIN, Y. and STEIN, O.: Incorporation of fatty acids into lipids of aortic slices of rabbits, dogs, rats and baboons. J. Atheroscler. Res. *2:* 400–412 (1962).
2. STEIN, Y.; STEIN, O. and SHAPIRO, B.: Enzymic pathways of glyceride and phospholipid synthesis in aortic homogenates. Biochim. Biophys. Acta *70:* 33–42 (1963).
3. STEIN, Y. and STEIN, O.: Metabolism of labeled lysolecithin, lysophosphatidyl ethanolamine and lecithin in the rat. Biochim. Biophys. Acta *116:* 95–107 (1966).
4. STEIN, O.; SELINGER, Z. and STEIN, Y.: Incorporation of (1-^{14}C)-inoleic acid into lipids of human umbilical arteries. J. Atheroscler. Res. *3:* 189–198 (1963).
5. LANDS, W. E. M.: Metabolism of glycerolipids. II. The enzymatic acylation of lysolecithin. J. Biol. Chem. *235:* 2233–2237 (1960).
6. ERBLAND, J. F. and MARINETTI, J. B.: The enzymatic acylation and hydrolysis of lysolecithin. Biochim. Biophys. Acta *106:* 128–138 (1965).
7. ZILVERSMIT, D. B. and MCCANDLESS, E. L.: Independence of arterial phospholipid synthesis from alterations in blood lipids. J. Lipid Res. *1:* 118–124 (1959).

Author's address: Dr. Y. Stein, Department of Medicine B, Hadassah University Hospital, *Jerusalem* (Israel).

Discussion Session 5

F. BÖTTCHER (Leiden): Dr. STAMLER used in his remark the expression 'huge mass of accumulation of cholesterol and its esters', when he was referring to the deposits in the human arterial wall. In this connection, I would like to draw your attention to the fact that the amount of cholesterol in a human aorta with severe atherosclerosis is never more than a few hundred milligrams. If we then take into account that the accumulation requires at least 20–30 years, we come to quantities of 10–20 mg per year for the whole aorta. It would be extremely difficult to prove that at least part of the accumulation could not be due to synthesis *in vitro*. Hence, this point cannot be considered as definitely settled.

E.B. SMITH (London): I should like to ask Dr. LOFLAND what type of lesions show this marked increase in synthesis of sterol ester fatty acid? Are they fatty streaks consisting of fat filled cells, or are they fibrous type plaques?

H.B. LOFLAND (Winston-Salem, N.C.): The lesions in these birds ranged in severity from fatty streaks to extremely complicated plaques, containing intra- and extra-cellular lipid, calcium and fibrous tissue, and were frequently ulcerated. I would like to emphasize the fact that we see the markedly enhanced synthesis of fatty acid only in severly diseased aortas.

E.B. SMITH (London): I would like to ask Dr. BOWYER if he has measured the specific activities of the lipid fractions. In current studies on duckling aortae grown under conditions of tissue culture, we find a high specific activity in the cholesterol ester fatty acids. The amount of sterol ester is very small, thus the total counts are low, but the specific activity is very high, indicating a rapid turnover.

D.E. BOWYER (Cambridge): No, we have not measured the specific activity of the sterol ester fractions which have C^{14} activity.

C.W.M. ADAMS (London): I would like to add a comment about Glomsett's transacylating enzyme, which my colleague Dr. Y.H. ABDULLA has recently identified in normal rabbit and rat aortic wall. This enzyme in arterial tissue is active at pH 5.0 but not at neutral pH. Its presence implies that the aorta can esterify cholesterol by transfer of fatty acid from lecithin.

P. ALAUPOVIC (Oklahoma City, Okla.): Dr. LOFLAND, 1. Did you determine the type and distribution of lipoproteins and compare the patterns between the atherosclerosis-susceptible and non-susceptible pigeons? We have found significant differences between normal and hypercholesterolemic human subjects with respect to the content of their low-density lipoprotein subfractions. Hypercholesterolemic subjects have more than 75% of low-density lipoproteins between solvent density of 1.020–1.050 g/ml, whereas normal subjects have approximately 70–80% of low-density lipoproteins between solvent density of 1.050–1.063 g/ml. It would be quite interesting to know whether such differences could also be demonstrated in birds or other animal species.

2. How could the clearly demonstrated occurrence of low-density or β-lipoproteins in atherosclerotic lesions be explained?

H.B. LOFLAND (Winston-Salem, N.C.): We have not made detailed comparisons of lipoprotein fractions in our susceptible and resistant strains of pigeons. This should be done. However, several years ago we published some observations on the behavior of lipoproteins which showed that female pigeons ovulate about every 30 days. At this time, the transport of cholesterol is shifted from the alpha to the beta fraction. This does not, of course, happen in male pigeons. Since we can find no difference between males and females in extent of atherosclerosis, we felt that, probably, differences in lipoproteins are not of primary importance in this species.

A.L. ROBERTSON (Cleveland, Ohio): I would like to ask Dr. ADAMS if the enzymatic changes he described in the midportion of the tunica media, particularly reduction in $NADH_2$-tetrazolium reductase and lipolytic 'non specific esterase'[2] activity were also present in the tunica intima in the presence of severe atheroma. In our own studies, with isolated human intimal cells from arteries with grade 2 and 3 lesions, we found a significant reduction of these enzymes following development of lipidladen cells and increased extracellular lipids, suggesting a reduction of oxidative pathways in contrast to the increased respiratory rates found in 'early' or incipient fatty streaks.

C.W.M. ADAMS (London): We find – as have many other workers (ZEMPLÉNYI et al. in SANDLER and BOURNE's 'Origin of Atherosclerosis', 1963) – that oxidative and esterolytic activities are at first very high in intimal cells in atheromatous plaques in both man and the rabbit. Subsequently, over-all activity in histochemical slide preparations appears to decline as a result of lipid accumulation and 'dilution' of the tissue with such lipids. Quantitative histochemical estimation of the atherosclerotic tunica intima reveals that oxidative, esterolytic and lipolytic activities are usually somewhat higher in this layer than in the tunica media.

P. ALAUPOVIC (Oklahoma City, Okla.): What do you mean by lipolytic activity? Are you referring to the lipoprotein lipase or to the triglyceride lipase? It would be useful if you could differentiate between these various activities. What was the percentage of tributyrin? Was it in soluble form or in emulsified state?

C.W.M. ADAMS (London): *Lipolytic* activity was estimated against an 0.4% aqueous emulsion of tributyrin, while *esterolytic* activity was measured against an emulsion of 3-naphthyl laurate. (If Desnuelle's classification is followed, the latter could be regarded as a lipase!) Our two aortic enzymes differ from each other in their responses to activators and inhibitors (see Table I). The *lipolytic* enzyme has similar characteristics to the aortic lipolytic enzyme described be ZEMPLÉNYI et al. (in SANDLER and BOURNE's 'Atherosclerosis and Its Origins', Academic Press, 1963) and it is not lipoprotein lipase. Possibly it may be less confusing (but very cumbersome) to refer to *Lypolytic* enzyme as tributyrin hydrolase and to the esterolytic enzyme as β-naphthyl laurate hydrolase.

Table I. Inhibition characteristics of aortic wall 'lipase' and 'esterase' (% inhibition)

	Lipase	Esterase
Bile salt (1%)	0	0
Protamine sulphate (1%)	18.6	†
NaCl (M)	18.6	66
Cu (0.01 M)	35	†
Zn (0.01 M)	38	91.5
Eserine ($10^{-4} M$)	0	0
E600 ($10^{-3} M$)	36.5	57
E600 ($10^{-5} M$)	33.3	5.5

† Interfered with colour reaction with diazo coupler.

O.J. POLLAK (Dover, Del.): Accumulation of cholesterol is an 'offensive' process; it is followed by the 'defensive' phenomenon of phospholipid synthesis in the arterial wall. Some of the cholesterol increase in complicated plaques is due to intra-plaque hemorrhage, whether of visible or of microscopic magnitude.

P. CONSTANTINIDES (Vancouver, B.C.): I wonder whether Dr. ADAMS has had the opportunity to study the ultrastructural localisation of his enzymes with the electron microscope. Dr. BARNETT and his pupil MARCHESI located ATPase in the pinocytotic vesicles of cells. If Dr. ADAM's esterases could also be shown to reside in pinocytotic vesicles, we might find out whether they are concerned with uptake of extraneous material into the cell.

C.W.M. ADAMS (London): Thank you, Dr. CONSTANTINIDES, for this comment. My colleague, Dr. R.O. WELLER, has been studying acid phosphatase, ATPase and other phosphatases in atheromatous aorta by electron-histochemical means. He has not, as yet, tried the electron-histochemical thiolesterase and tetranitroblue dehydrogenase methods on aortic tissue.

K.W. WALTON (Birmingham): I wish to make 3 comments:
1. In relation to experiments with ^{32}P, I have confirmed the findings of others [1, 2] that ^{32}P incorporated into α- and β-lipoproteins shows evidence of exchange between these lipoproteins, and also exchange with the phosphorus in the phospholipid of red cell membranes. This makes the interprebation of the significance of localisation of this label very difficult.
2. In relation to the accumulation of 'non-specific esterase' activity [3] when tested against the same substrates as those used by ADAMS for his histochemical tests. The finding of such activity in the intima of atheromatous lesions might thus be due to lipoproteinous infiltration of the intima rather than changes intrinsic to this tissue.
3. In reply to the general question posed by ALAUPOVIC, it may be relevant to remark that when serum diffuses into agar, a medium containing sulphated polysaccharide, the lipoprotein, is bound firmly and cannot be eluted whereas when serum diffuses into agarose (free from sulphated polysaccharide) or gelatine no binding occurs [4]. It is also known that lipoproteins form insoluble complexes with sulphated polysaccharides *in vitro* [5, 6]. It seems possible that as serum diffuses through the arterial intima, the lipoproteins may be selectively bound or complexed by the acid micropolysaccharides which are known to be abundant in the intima.

References

1. KUNKEL, H.G. and BEARN, A.G.: Proc. Soc. exp. Biol., N.Y. *86:* 887 (1954).
2. FLORSHEIM, W.H. and MORTON, M.E.: J. appl. Physics *10:* 301 (1957).
3. URIEL, J.: In 'Immunoelectrophoretic Analysis', p. 30 (Elsevier, Amsterdam 1964).
4. WALTON, K.W.: Immunochemistry *1:* 279 (1964).
5. ONCLEY, J.L.; WALTON, K.W. and CORNWELL, D.G.: J. amer. chem. Soc. *79:* 4666 (1957).
6. CORNWELL, D.G. and KRÜGER, F.A.: J. lipid Res. *2:* 110 (1961).

R. WISSLER (Chicago, Ill.): I too would like to make an appeal, as Dr. CONSTANTINIDES, for carrying out studies of phospholipid synthesis by the artery wall to the cell organelle or ultrastructural level. Much of phospholipid synthesis is related to membrane formation in an injured cell and not to the accumulation of lipid associated with atherogenesis.

T. ZEMPLÉNYI (Prague): In connection with the finding of the higher activity of lactate dehydrogenase, as demonstrated by Dr. ADAMS, I would like to mention that, according to quite recent findings of LOJDA (J. Atheroscler. Res.) as well as our findings, the arterial wall displays – similarly as in some other tissues-aerobic (H) as well as anaerobic (M) LDH isoenzymes together with 'hyeride' fractions. The increased total activity of LDH in ADAM's findings probable reflects an adaptation phenomenon: an increased activity of the anaerobic 'M' isoenzyme under hypotonic conditions.

C.W.M. ADAMS (London): Drs. ZEMPLÉNYI and LOJDA's results are very consistent with our findings and may well explain them.

J. SKOREPA (Prague): Two different lipolytic activities exist in the tissues and they seem to have quite different physiological function. One is heparin sensitive and may be responsible for the uptake of fatty acids by tissues. The other lipase is cyclic 3', 5'-AMP dependent, is regulated by the hormone system and mobilizes fatty acids from the tissues. Both types of lipases differ in pH optima and inhibition pattern. Have you some suggestion as to which of these activities is detected in your studies?

C.W.M. ADAMS (London): Neither our *lipolytic* or *esterolytic* enzymes required heparin for activation. We did not add cyclic 3', 5'-adenosine monophosphate to the incubating medium.

D. E. BOWYER (Cambridge): May I make a plea for the use of the correct terminology in the description of enzymes (as recommended by the enzyme commission); in particular, correct description of the substrate attacked and the type of bond cleaved.

J. STAMLER (Chicago, Ill.): Dr. BILLIMORIA has implicitly – or perhaps explicitly – arrived at the conclusion in his paper that the infiltration theory of atherogenesis is invalid, and that the excess lipids in atherosclerotic plaques are produced by endogenous synthesis in the arterial wall. As I heard him, he tended to present this conclusion in a rather definitive fashion, and not as an hypothesis. Since this is one of the most important questions in atherosclerosis research, it must be given due attention. To be frank, I was unable to grasp the relationship between Dr. BILLIMORIA's data and his sweeping conclusion on this basic issue. As far as I could judge, his findings really have no bearing on this question. It would be valuable if he could clarify this matter.

Dr. LOFLAND's fine contribution also left me unclear on this point. I would appreciate it if he would clarify further for the symposium his interpretation of his data on this crucial question of the major and primary source of the excess cholesterol in the atherosclerotic plaque. Does he mean to infer that in the main it is derived by arterial wall synthesis?

This whole issue was discussed in a very similar fashion, after a series of similar papers (including a presentation by Dr. BÖTTCHER, among others) at the Symposium on the Evolution of the Atherosclerotic Plaque held in Chicago, in March, 1963. To me at least, it is somewhat discouraging to see questions – apparently long since settled – re-opened again and again, without significant new data to justify the effort. A vast collection of research findings supports the thesis that the excess lipid in atherosclerotic plaques – particularly the critical lipid, cholesterol – is in a major way derived from the plasma. Correspondingly, careful work in recent years – including that of ZILVERSMIT and GOULD – indicates that little cholesterol synthesis occurs in normal or atherosclerotic arteries, and that it is virtually impossible to account for the large accumulations of free and esterified cholesterol in atherosclerotic plaques based on endogenous synthesis by the arterial wall. The evidence further indicates that the increased arterial wall phospholipid synthesis accompanying atherogenesis is a reaction to injury, i.e. an aspect of the cellular reparative effort, in response to the injurious effects of excess cholesterol. It is most difficult for me to grasp how any of the facts currently available on the metabolism of the arterial wall can be interpreted as refuting the infiltration theory. I strongly believe this is an erroneous deduction from the facts, i.e. a misinterpretation.

Everyone here should appreciate the significance of this faulty deduction. If indeed the excess lipids (including cholesterol) in atherosclerotic plaques are all derived by endogenous synthesis, then all the animal-experimental and human data on the association among dietary cholesterol-lipid, hyperlipidemia and atherogenesis become irrelevant and meaningless. In fact most of the current knowledge in this field loses significance. We are left with a mysterious enigmatic problem: Why do the arteries of certain people and certain populations take it upon themselves to synthesize excess lipid and become atherosclerotic? This is a matter concerning which next to nothing is known – properly so, since this is not what actually happens in the primary pathogenesis!

Please forgive the pointed nature of my remarks. The issue is so important that it needs to be dealt with directly. One of our tasks as a community of scientists is to go forward in our research endeavor, in order to make progress toward a solution – theoretical and practical – of the problem. This is especially urgent in atherosclerosis research, in view of the great toll exacted by the disease. One of the prerequisites for progress is to arrive at firm conclusions, as a foundation for next steps forward, and to reject tendencies to re-open questions – unless these tendencies are founded on solid information. It is appropriate to demand the solid evidence – and to be insistent – when a call is being made to discard a basic conclusion, lest the field go around in circles, become entrapped by 'exquisite' minutiae and retrogress.

J. D. BILLIMORIA (London): We have not done any work on cholesterol. I was quoting ZILVERSMIT as saying that plaque phospholipids originate from aortic synthesis and not from plasma. Although our results support this aortic synthesis they do not contradict the Duguid hypothesis which could also account for the accumulation of lipids. I should probably have specified more clearly that as far as phospholipids go, the infiltration theory was not required to account for the accumulation of these lipids.

Dover Medical Research Center, Dover, Delaware, USA

Recent Advances in the Metabolism of Arteries
Tissue Cultures, Homogenates and Slices

O. J. POLLAK, Dover, Del.

My review of recent advances in arterial metabolism is based both on the abstracts of papers to be read in this session and on personal observations. I shall point out some gaps in our knowledge, and pose several new questions.

The papers represent a few samples from many efforts toward the same goal: To elucidate some of the intricate facets of atherogenesis.

There are many differences between species, between members of the same species, between parts of the circulatory system, and between segments of a blood vessel. These differences concern anatomic structure, chemical composition, and metabolism.

According to ZEMPLÉNYI *et al.*, some differences in *arterial enzymes* are sex-linked: Aortae of females contain less LDH, MDH and 5'NU than aortae of males. This situation can be reversed by gonadectomy or hormone treatment. PATELSKI and coll. isolate lipolytic enzymes from pig aorta. They evaluate the effect of various diets on lipolytic enzymes in aortae of rats and rabbits. It takes courage to extrapolate between species, especially if one discusses hormonal pathways, enzymes, and metabolism.

In trying to link the localization of atherosclerotic lesions to the presence or absence of certain enzymes, we must keep in mind that in some species lesions appear first in the ascending aorta, while in other species they appear first in the terminal aorta. One must define each atherosclerotic lesion that is subjected to study of enzymes or of other chemical constituents, since there is a definite relationship between the anatomic character and the chemical composition of lesions. I believe that, at first, chemical changes precede anatomic alterations; later, they follow the structural alterations.

The quantity of *beta-glucuronidase* and of the enzymes of the Krebs cycle increases with aging and with the severity of atherosclerosis, according to ZEMPLÉNYI *et al.* There is a relationship between beta-glucuronidase and mast cells: Both are scanty in atherosclerosis-prone species but abundant in resistant species. I reported that the number of intimal mast cells decreases with aging of man and with the progression of atherosclerotic lesions.

Analytic results differ, whether based on study of the inner layers or of the whole arterial wall. Results of chemical assays, whether calculated on the basis of dry or wet weight, vary with wall thickness. They are not the same in extracts, dialyzates, and homogenates. Histochemical methods allow us to localize enzymatic activity. Since two methods for the same enzyme do not yield the same results, those who plan to study enzymes should agree on a common methodology. Enzyme studies of arteries of various species performed to date should be extended to arterial cell cultures. Lipid-laden macrophages are metabolically more active than other cells. Mitotic activity, absent in the intact intima of rabbits, is present in the growing experimental plaque, according to McMILLAN and STARY.

Most errors of metabolism are due to a genetic defect or absence of an enzyme. Some enzymes, which have been linked to atherogenesis by one or another author, have not been sufficiently studied, or have not been studied at all: *Phospholipodiesterase* has a bearing on phospholipid metabolism. It is related to *thromboplastinase* which, in the presence of calcium ions, converts prothrombin to thrombin. There seems to be an antagonism between *acetylphosphatase* and *adenosinetriphosphatase*. The availability of AMP, ADP, and ATP is significant for nucleoside synthesis. ADP plays a role in some phases of platelet adhesiveness. *Hyaluronidase* reduces the viscous mucopolysaccharides of mesenchymal interfibrillar matter. *Elastase* has been recovered in extracts of arterial mucopolysaccharides. *Deoxyribonuclease* should be studied on various levels, including lysosomes. *Ribosomal ribonucleic acid polymerase* should be investigated. There is still more to be learned about *cytochrome oxidase*.

The *chemical composition of the vessel wall* has not been sufficiently studied except for the lipoid constituents. We recovered fewer free amino acids in rabbits' aortae with plaques than in normal aorta.

We observed a shift of methionine from the altered part to the intact, distal aortic sectors. A similar shift has been reported for succinic and cytochrome oxidases. I found that in induced aortic plaques cystathionine replaces glutathione. We were able to identify regularly cystathionine in the nutrient medium after growth of cells from atherosclerotic plaques. *Cystathionine synthetase* may play an important role in atherogenesis. Enzymatic control of amino acid metabolism deserves more attention.

THOMAS and associates report that some *atherogenic diets depress protein synthesis* in rat aortae and cause a reduction in the incorporation of labeled amino acids into microsomal protein, and that depressed RNA acts as a limiting factor in protein synthesis. Protein synthesis, DNA, and RNA are increased in monkeys' atherosclerotic aortae. Objections to experiments with 'drastic' diets are invalid as long as we are aware that they represent extremes: It is important to establish maximum as well as minimum deviations from the norm.

In experimental lesions of rabbits, total protein increases, together with total lipoids and with cholesterol. However, the composition of aortic proteins is not well known. This is one of many gaps in our knowledge. Increased protein synthesis and protein content in certain lesions could be ascribed to a numeric increase of cells. It is heartening to learn that more sophisticated methods are being introduced, instead of or in addition to chemical analysis, which depends on regard or disregard for intercellular fluid. I was pleased to read in the abstract of THOMAS *et al.* that 'cells appear in response to injury', and I would welcome clarification as to whether this includes chemical injury and, especially, enzymatic and pH derangement.

The relative role of lipoid incorporation versus arterial synthesis of lipoids has not yet been clarified. Those working with MEAD and with ROTHBLAT extended *metabolic studies of lipoids,* sterols, and fatty acids to *cell cultures.* Differences in capability to desaturate fatty acids, and to adsorb, absorb, bind, degrade or synthetize sterols can be largely explained by the origin and age of cell lines. It is well known that fibroblasts and also macrophages from multiple sites of an animal differ. It is hoped that these interesting studies will be extended to various arterial cells. We have demonstrated that corneal endothelial cells and corneal fibroblasts differ, and that aortic endothelial cells and fibroblasts from the same arterial segment differ *in vitro.*

I have been interested for several years in *tissue cultures* and have published many reports in this area. I have stated repeatedly that 'the study of arterial cell cultures is fascinating and holds great promise with regard to elucidation of atherogenesis.' May I now caution against over-enthusiasm? All the features of extracorporeal cell growth can be modified by changing the culture technique, the nutrient, the temperature, the humidity, or the oxygen supply. Whether we use cell cultures to study the transport of sugars or lipoids across cell membranes, to study amino acid metabolism, enzymes, or respiration, or to study reaction to antibodies or toxins, we must realize that cells *in vitro* multiply at a rapid rate and that their metabolic requirements differ dramatically from *in vivo* conditions.

LAZZARINI-ROBERTSON studies *oxygen uptake by human aortic intimal cells grown in vitro*. The results of oxygen consumption by tissues and cells depend on the character of atherosclerotic lesions used for study. Unfortunately, no two investigators use the same technique or express the results of respiration studies in the same manner. We found that the higher the blood cholesterol of a rabbit, the lower the respiration quotient of the animal's aorta. Oxygen uptake decreases as tissue lipidosis increases, regardless of the type of tissue. In my opinion, tissue and cell anoxia are the results, not the causes, of lipidosis.

For comparison with studies of tissue homogenates a mixed cell population grown from such tissue may be suitable. For most studies, proper characterization of isolated cell strains is most important. POLLAK and ADACHI identified four cell types grown from rabbit and from human arteries. They delineated the extent to which extrapolation of such tissue cultures is feasible.

A further topic of interest is the production of *mucopolysaccharides* by aortic cells. Fibroblasts are most active in this respect. HOLLANDER and co-workers found that MPS increase in the earliest stages of atherosclerosis, before the appearance of lipoids in the arterial wall and independent of the increase in low-density lipoproteins. Mucopolysaccharides also accumulate with aging, and, together with hydroxyprolin and calcium, under lowered oxygen tension. Sodium, potassium, magnesium, and manganese must not be overlooked in studying the cellular environment, and in studying the metabolism of the arterial wall.

In my efforts to find a common denominator for the papers to be delivered here today, I became aware that they are all related to the

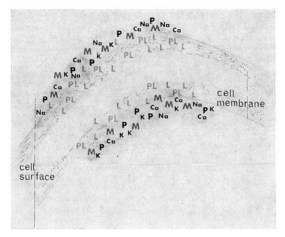

Fig. 1. L = lipids, PL = phospholipids, P = proteins, M = mucopolysaccharides, Ca = calcium, Na = sodium, K = potassium.

cell surface (Fig. 1). The interplay between organic and inorganic components of the extracellular space and the cell membrane influences homeostatic regulation of trans-membrane movements. The study of these phenomena will create not only new questions, but will also lead to answers pertinent to atherogenesis.

Author's address: O. J. Pollak, M. D. Ph. D., 9 Kings Highway, *Dover*, Del. 19901 (USA).

The Metabolism of Cholesterol, Lipoproteins, and Acid Mucopolysaccharides in Normal and Atherosclerotic Vessels[1]

W. Hollander[2], D.M. Kramsch[3] and G. Inoue[4], Boston, Mass.
and with the technical assistance of
M.F. Sullivan and M.A. Colombo

Introduction

One of the major lipid changes that occurs in atherosclerotic vessels is an accumulation of cholesterol in the arterial intima. The manner in which cholesterol is deposited in the arteries has not been clearly established although a number of studies have demonstrated that a major portion of arterial cholesterol is derived from the plasma [1-3]. The arterial intima also may be a source of cholesterol in some animal species but there is no evidence, as yet, that the human intima can synthesize cholesterol [4]. However, *in vitro* studies indicate that the human intima is capable of synthesizing lipoprotein, a form in which cholesterol circulates in the plasma [5].

Studies of experimental atherosclerosis in rabbit [2] suggest that the deposition of cholesterol in arterial plaques may be due to an increased influx of plasma cholesterol into the lesion. Recent studies in our laboratory indicate that the accumulation of cholesterol in the human fatty plaque also is associated with an increased influx of plasma cholesterol as well as with an increase in the low density lipoprotein content of the involved intima [6].

The present studies in experimental atherosclerosis of dog and

---

[1] Supported by the U.S. Public Health Service Grants HE-07299-05 and HE-01536-13 and by Mr. U.A. Whitaker.
[2] Associate Professor of Medicine, Boston University School of Medicine, Boston, Mass. USA. Head of the Hypertension and Atherosclerosis Research Section, University Hospital, and Director of the U.A. Whitaker Laboratories For Blood Vessel Research, Boston, Mass.
[3] Instructor of Medicine, Boston University School of Medicine.
[4] Research Fellow, Boston University School of Medicine.

in normal and diseased human arteries were undertaken to clarify some of the mechanisms which might be involved in the deposition of cholesterol and lipoproteins in the arterial intima.

Methods

Atherosclerosis was induced in 5 mongrel dogs by methods similar to those employed by Steiner and Kendall [7] by feeding thiouracil and a diet high in cholesterol and fat for 12–30 weeks. At 5 days before sacrifice with intravenous pentobarbital, the experimental dogs and the paired normal control dogs were injected intravenously with C^{14}-labeled cholesterol, 2 $\mu c/kg$. At 24 h before sacrifice the dogs also were injected intravenously with the sodium salt of S^{35}-labeled sulfate, 14 $\mu c/kg$. The aorta was removed immediately after sacrifice and cleaned of surrounding tissue. The abdominal aorta of the experimental dogs contained atherosclerotic plaques which were excised and analyzed separately from the adjacent normal appearing arterial tissue. Corresponding aortic segments were excised from the control dogs which showed no evidence of atherosclerosis.

Human aortae were removed at surgery or at autopsy within 4 h of the demise of the patient. In a number of surgical or moribund patients C^{14}-labeled cholesterol was injected intravenously in trace doses from 6–60 days before the removal of the aortae. The inner layer of the aorta was stripped from the intact aorta, cleaned of contaminating blood with a moist saline sponge and separated into atherosclerotic and a normal appearing tissue.

Lipoproteins were extracted from the aortic intimal layer according to the method of Hanig [8] and were separated into lipoprotein fractions by differential ultracentrifugation in a Spinco preparative ultracentrifuge by methods similar to those employed by Havel, Eder and Bragdon [9] for plasma lipoproteins.

Subcellular fractionation of the arterial intima was performed by differential ultracentrifugation as previously described [10] by a modification of the method of Schneider and Hogeboom [11].

The acid mucopolysaccharides in the arterial tissue were extracted and isolated with cetavlon as previously described [12, 13] by a modification of the method of Schiller and Dorfman [14].

Lipids were extracted from the tissues and subcellular and lipoprotein fractions with chloroform and methanol (2/1 v/v) and analyzed radiochemically for lipid content as previously described [3, 12, 13].

Results

Table I compares the aortic cholesterol content and the rate of influx of plasma cholesterol into the aorta of 5 normal control dogs and 5 dogs with experimentally induced atherosclerosis. The influx of cholesterol was calculated by dividing the mean specific activity of the plasma into the C^{14}-cholesterol radioactivity of the aorta. As has been reported by others in the rabbit [2], the accumulation of cholesterol in the diseased aorta of dog was associated with a 5–15

Table I. Influx of plasma cholesterol into the dog aorta

	C^{14}-cholesterol of aorta % dose x 10^{-3}/g.d.t.	Total cholesterol of aorta mg/g.d.t.	Sp. activity of aortic cholesterol	Total plasma cholesterol mg %	Sp. activity of plasma cholesterol	Ratio: sp. act. aortic chol. sp. act. plasm. chol.	Cholesterol influx mg/day
Control dog (5)	1.74 ±0.55	3.42 ±0.60	0.54 ±0.16	169 ±24	2.24 ±0.12	0.26 ±1.1	0.16 ±0.02
Experimental dog (5) 'normal'	4.16 ±1.50	8.38 ±1.60	0.50 ±0.16	1510 ±602	0.97 ±0.24	0.51 ±0.08	0.87 ±0.21
Plaque	13.36 ±3.34	24.64 ±7.85	0.53 ±0.05	1510 ±602	0.97 ±0.24	0.61 ±1.9	2.46 ±1.9

fold increase in the influx of plasma cholesterol into the involved artery. It is noteworthy that the cholesterol content and the influx of plasma cholesterol also was strikingly increased in the normal appearing portions of the diseased aorta although to a lesser extent than in the visible lesions. These observations indicate that the disease process in experimental atherosclerosis is not localized to the plaques but is a diffuse process involving the entire aorta including those areas which appear normal.

In Table II the acid mucopolysaccharide (AMP) changes in the normal and experimental aorta are compared. The AMP content, as indicated by the uronic acid and hexosamine composition of the AMP, was markedly reduced in the plaques and in the adjacent normal appearing portions of the diseased aorta. Preliminary studies in our laboratory suggest that the hypothyroid state induced by feeding thiouracil to the experimental dogs may be responsible for some of these changes including the decrease in the arterial content of sulfated acid mucopolysaccharides. The incorporation of

Table II. Acid mucopolysaccharides in experimental atherosclerosis

	Uronic acid mg/g.d.t.	Hexosamine mg/g.d.t.	$S^{35}O_4$ % dose x 10^{-3}/g.d.t.	Ratio: $S^{35}O_4$ uronic acid	Ratio: $S^{35}O_4$ hexosamine
Control dog (5)	2.55 ±0.43	2.26 ±0.40	2.44 ±0.69	0.96 ±0.22	1.07 ±0.24
Experimental dog (5) 'normal'	1.46 ±0.31	1.04 ±0.20	0.86 ±0.45	0.58 ±0.21	0.81 ±0.31
Plaque	1.60 ±0.39	1.25 ±0.36	1.83 ±0.77	1.13 ±0.60	1.45 ±0.32

intravenously administered $S^{35}O_4$ into AMP at 24 h after the radiosulfate injection was reduced in the diseased aorta but was higher in the plaque than in the adjacent normal appearing aortic tissue.

These studies indicate that an increase in the influx and accumulation of cholesterol in the arteries may occur even when the content and turnover rate of the AMP is reduced or unaltered in an involved artery. The findings also suggest that the absolute and fractional turnover rates of AMP in the plaques are higher than those in the adjacent normal appearing arterial segments. The latter interpretation is consistent with the findings of Haus and coworkers [15] who have observed a significant correlation between the incorporation of injected $S^{35}O_4$ into aortic AMP of rat at 24 h and the biological half-life times of aortic AMP.

Table III compares the influx of plasma cholesterol into fatty plaques and adjacent normal appearing arterial intima of human patients who were injected intravenously with radioactive cho-

Table III. Transfer of plasma cholesterol into the human aortic intima

	C^{14}-cholesterol of intima % dose x 10^{-3}/g.d.t.	Total cholesterol of intima mg/g.d.t.	Sp. activity of intimal cholesterol	Sp. activity of plasma cholesterol	Ratio: sp. act. aortic chol. sp. act. plasma chol.
'Normal' (9)	3.9	14.0	0.28	0.79	0.35
Plaque (9)	6.0	56.0	0.11	0.79	0.14

lesterol between 6 and 60 days before removal of the arterial tissue. The results are similar to those found in the dog with experimentally induced atherosclerosis. In all 9 human studies the cholesterol radioactivity was higher in the fatty plaque than in the adjacent normal arterial intima by about 58%. However, the specific activity of the intimal cholesterol including the free and esterified cholesterol, was significantly lower in the plaque than in the normal tissue suggesting a dilution of radioactive cholesterol by a relatively stagnant pool of cholesterol in the plaque.

Recent radioautographic studies of the same arterial tissue support this interpretation. They indicate that there are two pools of intimal cholesterol in the fatty plaque, an inner pool which is relatively immobile and an outer pool which equilibrates rapidly with the plasma cholesterol. An increase in newly synthesized cholesterol by the intima also may operate to reduce the cholesterol specificity in the plaque but this explanation is not supported by the data on sterol synthesis in Table IV.

Table IV compares the incorporation of C^{14}-acetate into sterol and the incorporation of $S^{35}O_4$ into AMP by incubated segments of normal arterial intima and adjacent atherosclerotic plaques in

Table IV. Sterol and AMP synthesis (*in vitro*) in human aorta

	C^{14}-sterol % dose x 10^{-3}/g.d.t.	Sterol sp. activity	$S^{35}O_4$ % dose x 10^{-3}/g.d.t.	Ratio: $S^{35}O_4$ uronic acid	Ratio: $S^{35}O_4$ hexosamine
'Normal' (5)	4.2 ±2.1	0.16 ±0.06	11.4 ±3.2	2.2 ±0.5	2.7 ±0.4
Plaque (5)	4.1 ±2.0	0.07 ±0.03	13.1 ±2.8	2.7 ±0.3	3.1 ±0.3

5 other patients. The studies indicate no significant differences in the synthesis of sterol although the sterol specific activity was significantly lower in the plaque than in the 'normal' intima. Total lipid and fatty acid synthesis also was not significantly different in the normal intima and plaque. The changes in AMP synthesis appeared similar to the findings in experimental canine atherosclerosis with the mean incorporation of $S^{35}O_4$ into AMP being somewhat greater in the plaque than in the adjacent normal appearing arterial intima. However, the AMP content was comparable in the normal and diseased intima (Table V).

Table V. Relationship of cholesterol and AMP content of human aortic intima

	Cholesterol mg/g.d.t.	Acid mucopolysaccharides uronic acid mg/g.d.t.	hexosamine mg/g.d.t.
'Normal' (9)	14.0 ±4.5	5.2 ±0.6	4.1 ±0.3
Plaque (9)	56.0 ±20.5	5.0 ±0.5	4.0 ±0.3

Before discussing the lipoprotein distribution of the intravenously administered C^{14}-cholesterol, it might be well to describe the lipoprotein content and composition of the human aorta. About 50%

of the total cholesterol or lipid content of the aorta was extractable as lipoprotein lipid. In Table VI the extractable lipoprotein content of the uninvolved intima of 3 patients and the atherosclerotic intima of 5 other patients are compared. Similar results were obtained in

Table VI. Extractable lipoprotein content of human aortic intima

	Low density fractions mg/g.d.t.			High density mg/g.d.t.
	d<1.063	d<1.019	d 1.019-1.063	d 1.063–1.210
Uninvolved intima	20.1	7.9	9.7	7.8
Atherosclerotic intima	86.0	40.0	42.5	6.3

the diseased intima of 5 additional patients. The extractable low density lipoprotein content (d <1.063) was about 4 times higher in atherosclerotic lesions than in the normal intima. However the extractable high density lipoprotein content (d 1.063–1.210) was comparable in the diseased and normal intima. The ratio of low density to high density lipoproteins in the uninvolved intima was about 2.5 to 1 whereas in the diseased intima it was about 14 to 1.

In Table VII the composition of the lipoprotein fractions in the uninvolved and the atherosclerotic intima are compared. The major differences in the composition of the lipoproteins in these tissues

Table VII. % composition of aortic intimal lipoproteins

	Uninvolved intima low density fractions			high density	Atherosclerotic intima low density fractions			high density
	d<1.063	d<1.019	d 1.019–1.063	d 1.063–1.210	d<1.063	d<1.019	d 1.019–1.063	d 1.063–1.210
Cholesterol	30	17	15	6	56	57	48	10
Phospholipids	33	20	26	30	26	19	29	41
Triglycerides	22	47	28	0	8	15	7	0
Protein	15	16	31	64	10	9	16	49

were in the low density lipoprotein fractions. The low density fractions of the diseased intima contained a greater percentage of cholesterol but a smaller percentage of triglycerides than did the uninvolved intima. Over 90% of the extractable lipoprotein cholesterol, phospholipid and triglycerides in the diseased intima were contained in the low density fractions. The distributions of lipids in the uninvolved intima were similar except for the phospholipids, 64% of which were contained in the low density fractions and 36% of which were in the high density fractions.

The studies shown in Table VIII and other similar studies on the incorporation of C^{14}-labeled leucine and C^{14}-acetate into lipo-

Table VIII. Average incorporation of C^{14}-acetate into lipoproteins by aortic tissue

	Total radioactivity (cpm/g.d.t.)	Low density fractions		High density fractions
		d<1.019	d 1.019–1.063	d 1.063–1.210
		(% of Radioactivity)		
C^{14}-lipid	62,000	29%	60%	11%
C^{14}-fatty acid	54,000	29%	61%	10%
C^{14}-sterol	852	37%	62%	1%
C^{14}-protein	16,000	26%	50%	24%

proteins by incubated human aortic tissue indicate that the normal and diseased arterial intima is capable of synthesizing the protein and lipid components of the lipoprotein fractions with most of the newly synthesized lipoproteins being of the low density type.

Table IX shows the distribution of intravenously administered C^{14}-cholesterol in the lipoprotein fractions of a relatively normal aorta of a patient who died following corrective surgery 14 days after the injection of the labeled cholesterol.

Table IX. Distribution of I.V. C^{14}-cholesterol in aortic lipoprotein fractions in an uninvolved aorta

	Low density fraction (d < 1.063)	High density fraction (d < 1.063)
Cholesterol mg/g.d.t.	7.2	0.5
C^{14}-cholesterol cpm/g.d.t.	810	112
Cholesterol sp. activity	112	224

J.S. ♂, 48.

The incorporation of the injected C^{14}-cholesterol into the low density lipoprotein fractions was about 8 times higher than into the high density fractions and was roughly proportional to the cholesterol content of the fractions.

Table X shows the distribution of intravenously administered C^{14}-cholesterol in the lipoprotein fractions of an atherosclerotic aorta of a patient who died 13 days after the injection of the labeled cholesterol.

Table X. Distribution of I.V. C^{14}-cholesterol in lipoprotein fractions of an atherosclerotic aorta

	Low density fractions			High density fractions	
	$d < 1.063$	$d < 1.019$	$d\ 1.019–1.063$	$d\ 1.063–1.210$	$d > 1.063$
Cholesterol mg/g.d.t.	46.7	12.7	34.6	2.0	2.0
C^{14}-Cholesterol cpm/g.d.t.	2647	651	2156	218	298
Cholesterol sp. activity	57	51	62	109	149

M.S. ♀, 52.

Similar to the findings in normal aorta, the total incorporation of the injected C^{14}-cholesterol into the low density fractions was about 13 times higher than into the high density fractions and was roughly proportional to the cholesterol content of the fractions.

It is not clear from these data whether the distribution of the injected C^{14}-cholesterol represents the relative rates of influx of low and high density plasma lipoproteins into the arteries since the observed distribution of the labeled cholesterol may have resulted from an exchange of cholesterol between the lipoprotein fractions in the arterial intima. However, it is noteworthy that in 1 patient in whom I^{131}-labeled low density lipoprotein was intravenously administered 5 days before removal of an upper segment of a diseased thoracic aorta, the injected lipoprotein was isolated by ultracentrifugation from the aorta.

Table XI compares the cholesterol content and the distribution of intravenously administered H^3-cholesterol in the subcellular fractions of the uninvolved and adjacent atherosclerotic intima of a patient who died 21 days following the injection of the labeled cholesterol.

Most of the cholesterol in the normal intima was contained in the nuclear debris fraction which consisted not only of fragmented nuclei debris and cell membranes but also of elastic and collagen fibers. In the diseased intima the cholesterol content of all the fractions increased with the greatest increases being in the nuclear

Table XI. Subcellular distribution of i.v. administered H^3-cholesterol in human aortic intima

Fraction	Uninvolved intima			Atherosclerotic plaque		
	cholesterol mg/g.d.t.	H^3-cholesterol cpm/g.d.t.	cholesterol sp. activity	cholesterol mg/g.d.t.	H^3-cholesterol cpm/g.d.t.	cholesterol sp. activity
Nuclear debris	15.6	8946	573	71.3	7653	107
Mitochondria	2.4	574	239	7.6	1436	189
Microsome	1.5	416	277	13.8	1894	137
Supernatant	3.7	1686	455	50.4	4328	86

A.C. ♀, 34.

debris and supernatant fractions. Similar results were obtained when the cholesterol data were expressed per mg of protein or DNA.

In general the uptake of the intravenously administered C^{14}-cholesterol was roughly proportional to the cholesterol content of the individual fractions. These and other studies indicate that the intimal cells of normal arteries as well as diseased arteries incorporate plasma cholesterol with the intracellular incorporation of plasma cholesterol being higher in the diseased intima than in the normal intima. Recent radioautographic studies in our laboratory also indicate that intravenously administered radiactive cholesterol is deposited extracellularly particularly over the elastic membranes of the aorta [16].

Summary

In experimental atherosclerosis of the dog, accumulation of cholesterol in the aorta was associated with an increase in the influx of plasma cholesterol but a decrease in the mucopolysaccharide content of the aorta. The incorporation of intravenously administered $S^{35}O_4$ into AMP was increased in the plaques as compared to the adjacent normal appearing aortic tissue.

In man, the accumulation of cholesterol and low density lipoproteins in atherosclerotic plaques was also associated with an increase in the influx of plasma cholesterol which occurred without a change in the aortic AMP content. The extractable low density lipoprotein content but not the high density lipoprotein content was strikingly increased in the plaque. In both normal and diseased arteries intravenously administered cholesterol was deposited intracellularly as well as extracellularly and was incorporated mainly into low density lipoproteins. The reduced cholesterol specific activity in the plaque together with recent radioautographic findings suggest that an impaired transport of arterial cholesterol in addition to an increase in the influx of plasma cholesterol and low density lipoproteins may play a role in the accumulation of cholesterol in the arterial intima.

References

1. BIGGS, M.W. and KRITCHEVSKY, D.: Observations with radioactive hydrogen (H^3) in experimental atherosclerosis. Circulation *4:* 34 (1951).
2. NEWMAN, H.A.I. and ZILVERSMIT, D.B.: Quantitative aspects of cholesterol flux in rabbit atheromatous lesions. J.biol.Chem. *237:* 2078 (1962).
3. CHOBANIAN, A.V. and HOLLANDER, W.: Body cholesterol metabolism in man. I. The equilibration of serum and tissue cholesterol. J.clin.Invest. *41:* 1732 (1962).
4. AZARNOFF, D.L.: Species differences in cholesterol biosynthesis by arterial tissue. Proc.Soc.exp.Biol., N.Y. *98:* 680 (1958).
5. HOLLANDER, W. and KAPLAN, R.N.: Studies of lipoprotein metabolism in atherosclerotic tissue. J.clin.Invest. *42:* 943 (1963).
6. HOLLANDER, W.; KRAMSCH, D. and INOUE, G.: The metabolism of acid mucopolysaccharides and lipoproteins in normal and diseased human arteries. J.clin. Invest. *45:* 1025 (1966).
7. STEINER, A. and KENDALL, F.E.: Atherosclerosis and arteriosclerosis in dogs following ingestion of cholesterol and thiouracil. Arch.Path. *42:* 433 (1946).
8. HANIG, M.; SHAINOFF, J.R. and LOWY, A.D.: Flotational lipoproteins extracted from human atherosclerotic aortas. Science *124:* 176 (1956).
9. HAVEL, R.J.; EDER, H.A. and BRAGDON, J.H.: The distribution and chemical composition of ultracentrifugally separated lipoproteins in human serum. J.clin. Invest. *34:* 1345 (1955).
10. HOLLANDER, W.; KRAMSCH, D.M.; CHOBANIAN, A.V. and MELBY, J.C.: Metabolism and distribution of intravenously administered d-aldosterone-1,2-H^3 in the arteries, kidneys, and heart of dog. Circulat.Res. *18:* Suppl. 1-35 (1966).
11. SCHNEIDER, W.C. and HOGEBOOM, G.H.: Intracellular distribution of enzymes: V. Further studies on the distribution of cytochrome in rat liver homogenates. J. biol.Chem. *183:* 123 (1950).
12. HOLLANDER, W.; YAGI, S. and KRAMSCH, D.M.: *In vitro* effects of vasopressor agents on the metabolism of the vascular wall. Circulation (Suppl.2) 30, II-1 (1964).
13. HOLLANDER, W.; MADOFF, I.M.; KRAMSCH, D.M. and YAGI, S.: Arterial wall metabolism in experimental hypertension of coarctation of the aorta. Hypertension *13:* 191 (1965).
14. SCHILLER, S.; SLOVER, A. and DORFMAN, A.: Method for the separation of acid mucopolysaccharides: Its application to the isolation of heparin from the skin of rats. J.biol.Chem. *236:* 983 (1961).
15. HAUSS, W.H.; JUNGE-HULSING, G. and HOLLANDER, H.J.: Changes in metabolism of connective tissue associated with ageing and arterio- or atherosclerosis. J.Atheroscl.Res. *2:* 50 (1962).
16. KRAMSCH, D.M.; CHOBANIAN, A.V.; GORE, I. and HOLLANDER, W.: Distribution of intravenously administered H^3-cholesterol in the arteries and other tissues: II. Radioautographic findings. Fed.Proc. *24:* 325 (1965).

Authors' address: Dr. William Hollander, Dr. Dieter M. Kramsch and Dr. Gosuke Inoue. Boston University Medical Center, 750 Harrison Avenue, *Boston, Mass.* (USA).

From the Department of Pathology, School of Medicine, McGill University, Montreal, Canada

Radioautographic Observations on DNA Synthesis in the Cells of Arteriosclerotic Lesions of Cholesterol-Fed Rabbits*

G. C. McMillan and H. C. Stary, Montreal, Que.

It has been known for some time that there is mitotic activity in the aortic atherosclerotic lesions of cholesterol-fed rabbits (McMillan and Duff, Arch. Path. *46*: 179, 1948). This paper presents preliminary observations on the DNA synthetic phase of such mitosis studied by the use of tritiated thymidine and radioautography in 14 New Zealand white rabbits fed cholesterol with transisomerized olive oil for 42 days, and in 2 control rabbits.

The rabbits weighed from 2 to 2.5 kg at this time. The dose of ^{3}H-thymidine was 0.5 μc/g body weight administered intravenously. The injection was given between 8:30 and 9:00 a.m. and the animals were killed 1 h later. The dissection time before tissue fixation was about 20 min. The development times for the radioautographs was either 7 or 14 days. The sections for radioautography were made after the aorta was rolled up like a snail-shell so that the entire length was available for study. Heavy labelling of some nuclei was obtained.

We are aware of two other studies in the literature in which experimental atherosclerotic lesions were labelled with ^{3}H-thymidine (Sparagen *et al.*, Circulation Res. *11*: 329, 1962 and Richie *et al.*, Surg. Forum *14*: 306, 1963).

We have observed the DNA-synthetic phase. Some cells in an early post-synthetic or G_2 phase may also have been observed.

In these experiments, preliminary – and I would stress the word preliminary – observations indicate that about one out of every

---

* Supported by Grant-in-Aid from the Medical Research Council of Canada.

25 nuclei related to the atherosclerotic plaques are labelled. These include the nuclei of stellate and foam cells that make up the bulk of the plaque, the endothelial cells and the adjacent subintimal smooth muscle cells. If only the cells constituting the bulk of the plaque are considered, about 1 nucleus of every 30 has been labelled as it synthesized DNA.

The experiment disclosed labelling at a frequency of from 1 of 10 to 1 of 50 cells in various plaques and animals. To date, we have detected no correlation between the labelling index of cells and the size of plaques, their location in the aorta, their severity as judged by macroscopic grading, or the serum cholesterol of the rabbit prior to death.

Endothelium and smooth muscle cells located away from plaques are seldom labelled. Near plaques, these cells are often labelled and smooth muscle cells my be labelled deep in the media beneath a plaque. On the other hand, the cells of the adventitia do not appear to show an increased frequency of labelling beneath a plaque.

Where the intimal lesion consists only of a slight swelling or thickening of the intima in the absence of foam cells, increased labelling of endothelial, stellate and smooth muscle cells is seen. Whether these slight lesions represent early or incipient plaques is not certain, but this seems probable.

It is not possible to use the data presented here to predict the growth rate of the plaques by cellular proliferation, since it would be necessary to make several important assumptions for the calculations. Nevertheless, it is apparent that cells in the plaques are proliferating and that the frequency of cellular labelling is high enough to allow its study under a variety of experimental conditions.

Authors' address: G. C. McMillan, M. D., Dept. of Pathology, School of Medicine, University of Toronto, *Toronto, Ont.* (Canada).

Department of Biophysics and Nuclear Medicine
and Department of Biological Chemistry,
School of Medicine, Center for the Health Sciences,
University of California, Los Angeles

Recent Advances in Polyunsaturated Fatty Acid Metabolism

J.F. MEAD, D.F. HAGGERTY, L.E. GERSCHENSON
and I. HARARY, Los Angeles, Cal.

So many functions have been ascribed to the polyunsaturated fatty acids that it is difficult, at present, to decide which are of most importance in atherosclerotic disease. Certainly, the transport of cholesterol has some relationship and the activities of many lipoprotein enzymes are at least indirectly related. The aspect I should like to stress at this time is their function as structural elements in the many membranous portions of the cell. These functions are of such importance that it can be appreciated that alterations in the composition of the fatty acid portions of the phospholipids of membrane structures could easily have a profound influence on the cells that contain them.

For this brief presentation I will have to limit my discussion to some of the recent advances from our laboratory and neglect the many important findings from other organizations.

It seemed obvious to us that the whole animal is too complex a system for a study of the function of the polyunsaturated acids, but, on the other hand, the cell homogenate may be missing the very structures of interest. We therefore chose, for experimental subject, the mammalian cell in culture – in this case, largely the HeLa cell. Admittedly, this is an abnormal cell from several points of view, but the information obtained with it has been very revealing.

In agreement with the report of HAM (1963), it was found that HeLa cells could not grow on an otherwise adequate medium if the fetal bovine serum were replaced by albumin. However, when 2.5×10^{-8} M linoleate as the albumin complex was included in the medium, growth was resumed and with 1.4×10^{-8} M arachidonate, it approached normality.

As a measure of membrane integrity, oxidative phosphorylation and respiratory control were checked in cells on the different media. As shown in Table I, replacement of serum by 10^{-8} M albumin resulted in a fall in ADP:O ratios and loss of respiratory control.

Table I. ADP:O and respiratory control as measured with β-hydroxybutyrate

	Fetal calf serum (20%)		Albumin (10^{-8} M)		Linoleic acid (2×10^{-8} M)		Arachidonic acid (10^{-8} M)	
	ADP:O	R.C.	ADP:O	R.C.	ADP:O	R.C.	ADP:O	R.C.
Exp. 1	2.8	2.2	1.2	1.1	2.4	1.3	2.8	2.0
Exp. 2	2.7	1.9	0.8	1.3	2.6	1.8	2.5	1.9

Both were brought nearly to normal with 2×10^{-8} M linoleate and to normal with 10^{-8} M arachidonate. These results, with β-hydroxybutyrate were qualitatively similar to those obtained using α-ketoglutarate and succinate as substrates (GERSCHENSON et al., 1966). Thus it appears that a lack of polyunsaturated fatty acid may affect mitochondrial integrity in these cells.

Analysis of the fatty acids of the deficient and supplemented cells revealed a fatty acid spectrum reminiscent of the fat deficiency state in whole animals in that palmitoleate and oleate were high and linoleate and arachidonate lower than in the serum- or fatty acid-supplemented cells (Table II). However, linoleate supplementation did not appear to increase the arachidonate concentra-

Table II. % of total methyl ester extract

	Serum	Albumin	Linoleic acid	Serum	Albumin	Arachidonic acid
18/1	19	29	21	16	19	13
18/2	6.4	3.8	9.8	6.8	0.7	1.2
20/3	2.9	1.5	1.9	6.5	4.9	2.8
20/4	15	4.4	4.9	14	1.1	9.5

tion and, in separate experiments, it was found that increased linoleate actually resulted in decreased arachidonate (Table III).

In an attempt to explain this curoius finding, HeLa cells were incubated in the presence of 10^{-7} M linoleate-1-C^{14} and the label was followed in the fatty acids of lipids extracted from the cells after

Table III. % of total methyl ester extract

Incubation concentration of linoleic acid	2 days		7 days	
	% linoleic acid	% arachidonic acid	% linoleic acid	% arachidonic acid
2.5×10^{-8} M	8.2	16	9.8	4.9
2.5×10^{-7} M	22	10	36	0.49

2 days (HAGGERTY et al., 1965). For a comparison, newborn rat heart cells cultured for 2 days according to HARARY and FARLEY (1963) were treated in an identical manner.

In Fig. 1 can be seen the results of a radio-gas-chromatographic analysis of the heart cells. First, it can be seen that the linoleate was well assimilated by the cells and appears in the cellular lipids. Second, the linoleate was evidently subjected to β-oxidation since

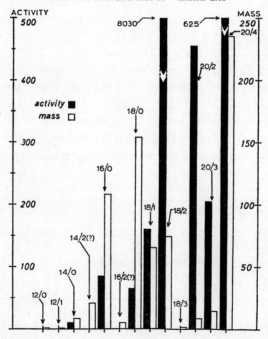

Fig. 1. Uptake of radioactivity from linoleic-1-^{14}C acid by fatty acids of heart cells. Amount (from thermal conductivity detector) and radioactivity (from ionization detector) in arbitrary units on the ordinates, plotted against chain-length and unsaturation of fatty acids on the abscissa. Heights of bars reflect actual peak heights multiplied by the relative retention time in order to represent actual area of the GLC peaks.

several fatty acids such as palmitate and stearate that are synthesized from acetate contain the label. Third, it is evident that arachidonate was synthesized from the linoleate and there is some indication that the route followed is 18:2 → 20:2 → 20:3 → 20:4 rather than 18:2 → 18:3 → 20:3 → 20:4 since there is high activity in the 20:2 and none in the 18:3. It is, of course, possible that the 18:3 → 20:3 conversion is so rapid that it would not appear in the isolated fatty acids (MEAD, 1957) but the relative Michaelis constants for 18:3 and 18:2 (possibly similar to 20:2) conversion (BRENNER, 1966) are not so different as to favor the 18:3 to this extent.

In Fig. 2 can be seen the results for the HeLa cells. In this case, the linoleate was also well absorbed and oxidized but arachidonate formation was absent or negligible as was that of 20:3 and the principle product was 20:2. It appears, therefore, that the HeLa cells can carry out elongation of fatty acids but cannot desaturate them

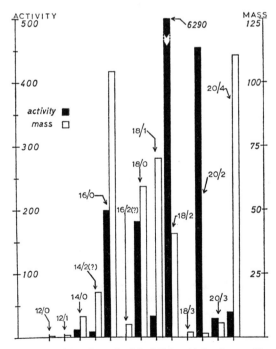

Fig. 2. Uptake of radioactivity from linoleic-1-^{14}C acid by fatty acids of HeLa cells. Coordinates are the same as for Fig. 1.

to form polyunsaturated fatty acids. As a matter of fact, the stearate-oleate conversion also seems to be depressed.

Several conclusions can be reached from these results. First, it appears that although the heart cells can carry on the usual conversions of the polyunsaturated acids, HeLa cells are either incapable of desaturation or carry out this important function at a greatly reduced rate. Second, since the cells grew on linoleate without the ability to convert it to arachidonate, it appears that in this case, linoleate (or 20:2) can function to some extent as an essential fatty acid in its own right. This would also explain the decrease in arachidonate with increasing linoleate as the result of growth stimulation without arachidonate synthesis. Finally, it will be of interest to determine whether this inability is due to the nature of the HeLa cells as cancer cells or as cells in culture. It should be noted that the heart cells, which retained the ability, had been in culture for only a few days.

In conclusion, these findings indicate that cells may become deficient in certain polyunsaturated fatty acids either through lack of an external supply or through loss of the ability to carry out the necessary transformations of available precursors. In either case, the deficiency is disastrous to the cell.

References

BRENNER, R.R. and PELUFFO, R.O.: Effect of saturated and unsaturated fatty acids on the desaturation *in vitro* of palmitic, oleic, linoleic and linolenic acids. Biochim. biophys. Acta (in press).

GERSCHENSON, L.E.; MEAD, J.F.; HARARY, I. and HAGGERTY, D.F., Jr.: Studies on the effects of essential fatty acids on growth rate, fatty acid composition, oxidative phosphorylation and respiratory control of HeLa cells in culture. Biochim. biophys. Acta *131:* 42–49 (1967).

HAGGERTY, D.F., Jr.; GERSCHENSON, L.E.; HARARY, I. and MEAD, J.F.: The metabolism of linoleic acid in mammalian cells in culture. Biochem. Biophys. Res. Comm. *21:* 568–574 (1965).

HAM, R.G.: Albumin replacement by fatty acids in clonal growth of mammalian cells. Science *140:* 802–803 (1963).

HARARY, I. and FARLEY, B.: *In vitro* studies on single beating rat heart cells. Exp. Cell Res. *29:* 451–474 (1963).

MEAD, J.F.: The metabolism of the essential fatty acids. VI. Distribution of unsaturated fatty acids in rats on fat-free and supplemented diets. J. biol. Chem. *227:* 1025–1034 (1957).

Authors' address: Dr. James F. Mead, Dr. Donald F. Haggerty, Dr. Lazaro E. Gerschenson and Dr. Isaac Harary, Laboratory of Nuclear Medicine and Radiation Biology, Department of Biophysics and Nuclear Medicine and Department of Biological Chemistry, UCLA School of Medicine, *Los Angeles, Cal. 90024* (USA).

Department of Physiological Chemistry, Medical Academy, Poznań
Department of Pathology, University of Cambridge, Cambridge

Lipolytic Enzymes of the Aortic Wall

J. Patelski[*], Z. Waligóra and S. Szulc, Poznań; D. E. Bowyer, A. N. Howard and G. A. Gresham, Cambridge

Introduction

Enzymes catalyzing hydrolysis of fatty acid esters in the aortic wall seem to be important in the prevention of their accumulation and of the development of atherosclerosis (Zemplényi, 1964). However, little is known about aortic esterases. As regards specific substrates, the lipase activity in rat aorta (Korn, 1955) and the cholesterol esterase in pig aorta (Patelski, 1964) have been demonstrated.

This paper provides evidence for the existence of the phospholipase A (Phosphatide acyl-hydrolase, E.C.3.1.1.4), lipase (Glycerol-ester hydrolase, E.C.3.1.1.3) and cholesterol esterase (Sterol-ester hydrolase, E.C.3.1.1.13) in the aortic wall, and describes alterations in their activities in experimental atherosclerosis. Some properties of the enzymes in crude extracts from pig aortas (I) and the effect of different diets and of polyenphosphatidyl choline on the enzyme activities in rats and rabbits (II)[**] have been investigated.

Material and Methods

Twenty-four hooded rats and twenty-two albino rabbits were used. The experimental rats were fed on diets containing 5% cholesterol, 2% cholic acid and: Group A: 40%

---

[*] Research fellow of the Polish Ministry of Health and Social Welfare at the Department of Biochemistry, University of Cambridge, Cambridge, England.
[**] Investigations were carried out: Part I at the Department of Physiological Chemistry, Medical Academy, Poznań, Poland, and Part II at the Departments of Biochemistry and Pathology, University of Cambridge, Cambridge, England.

butter for 10 weeks; B: 40% butter and 0.3% thiouracil for 8–18 weeks; C: 40% peanut oil for 14 weeks and D: 40% peanut oil and 0.3% thiouracil for 18 weeks, *ad libitum*. Rabbits were given an atherogenic semi-synthetic diet with 20% beef tallow (HOWARD et al., 1965) for 10 weeks (groups I and II), 20 weeks (group III) and 32 weeks (group IV). Simultaneously Lipostabil* was administered to rabbits, either in 0.1 ml intravenous and intraperitoneal injections, alternately, for 6 days a week (group II), or in the diet supplemented with 1% of the drug (group III). Control animals were given a standard diet of commercial cubes.

Glycerol-water extracts and water extracts from acetone-butanol powder from pig aortas (for details see PATELSKI, 1964) and from wet rat and rabbit aortas were used for enzyme activity assays. Lecithin from egg yolk (Merck, Germany), purified after HANAHAN (1954), polyenphosphatidyl choline (Nattermann, Germany), glycerol trioleate and cholesteryl oleate (British Drug Houses) have been used as hydrosol substrates (cholesterol and glycerol esters in the form of ethanol-water dispersions, PATELSKI, 1964). Enzyme activities were assayed as described for the cholesterol esterase estimations (PATELSKI, 1964), by continuous titration of fatty acid released into the reaction mixture at constant pH (pH-Stat, Radiometer).

Results and Discussion

Highest phospholipase and cholesterol esterase activities were found in glycerol-water extracts. Highest lipase activity appeared in water extracts. These extracts contained more protein and showed only approx. 70% of phospholipase activity of glycerol-water extracts and no activity of cholesterol esterase. Extracts from acetone-butanol powder were approximately twice as active as extracts from wet tissue slices which may be explained by the action of butanol in disrupting the lipoprotein complexes (MORTON, 1955).

pH optima of phospholipase, lipase and cholesterol esterase, examined under substrate saturation, are 8.0, 8.3 and 8.6, respectively. Storage of the enzyme extracts at 4°C results in a decrease in the cholesterol esterase and lipase activities of 7% and 11% per hour, respectively. Both enzymes are inactivated by heating for 10 min. at 60°C. There is a fall in phospholipase activity of 20–30% of the initial value during the first hour of storage at 4°C or after 10 min heating of the extract at 60°C. Further storage or heating brings about only a small further decrease of 0.15%/h. The conversion of lecithin into lysolecithin in the reaction mixture as well as other properties of this thermoresistant enzyme point to its similarity with a phospholipase A of other tissues.

* The drug containing polyenphosphatidyl choline both, for injections and in dragees was kindly supplied by Nattermann & Co., Cologne, Germany.

Several agents affecting lipolytic enzymes form pig aortas are presented in Table I and Fig. 1. The depressant effect of electrolytes seems to be connected with the coagulating action of ions on the substrate hydrosols (ABRAMSON et al., 1964). The activating effect of sodium taurocholate and glycocholate on cholesterol esterase seems to result from the coenzyme-like action, presumably due to the presence of OH-group at 7th C-atom of the steran ring, as was suggested by MURTHY and GANGULY (1962), rather, than from their substrate emulsyfying abilities. Sodium deoxycholate as well as higher concentrations of sodium taurocholate and glycocholate reveal an inhibiting effect on the cholesterol esterase. Both, the lipase and phospholipase are inhibited by these bile acid salts. The observation that heparin may either inhibit or activate (see

Table I. Compounds affecting lipolytic enzymes from pig aorta

Compound		(a) Phosphatide acylhydrolase	(b) Glycerolester hydrolase	(c) Sterolester hydrolase
None	activity (mU)*	250	50	10
Potassium chloride	pI_{50}**	1.7	(0.5–20.0)***	2.2
Sodium chloride		1.9	2.2	2.7
Calcium chloride		4.0	2.8	3.3
Sodium tauro- and glycocholate		2.2	2.3	activation 200–300% (3–6 × 10^{-3} M)
Sodium deoxycholate		3.1	3.2	1.7
Diisopropylfluorophosphate		3.6		
Tetraisopropylpyrophosphoramide			9.3	6.4
p-Chlormercuribenzoic acid (5×10^{-6}, 5×10^{-5} M)	activity (%)	100	85	60
Reduced glutathione (8×10^{-6} M)		91		
(5×10^{-5} M)			221	231

* mU = milliunits of specific activity = 10^{-3} μmoles/mg of protein/min.
** pI_{50} = –lg M concentration at which 50% of the enzyme activity is abolished.
*** 9–13% inhibition.

Reaction mixtures (20 ml of volume): (a) lecithin, (b) glycerol trioleate, (c) cholesteryl oleate, in mM concentration each, and 2 ml of glycerol-water extracts (a and c) or water extract (b) from acetone-butanol powder (approx. 6.0 and 7.0 mg of protein, respectively). Electrolytes and bile acid salts were added to the reaction mixtures. Solutions of compounds affecting OH- and SH-enzyme groups have been preincubated with enzyme extracts at 4°C for 30 min. Reaction temperature 25°C. Titration reagent 0.02 N NaOH.

Fig. 1) the tissue lipase *in vitro* is in agreement with a similar one concerning extract from rat heart acetone powder (KORN, 1954). Factors affecting the aortic lipase are also known to inhibit the 'lipoprotein lipase', whether it is from tissues or postheparin and non-heparin plasma (KORN, 1959; ENGELBERG, 1956).

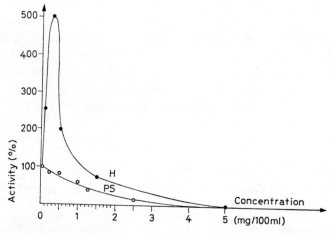

Fig. 1. Effect of heparin (H) and protamine sulphate (PS) on the aortic lipase.

Results of activity estimations of the lipolytic enzymes of rat and rabbit aortas are listed in Table II. All the enzyme activities are much higher in rats than in rabbits.

There is an increase in aortic lipase activity in both the atherosclerotic rabbits and rats and is much higher in rats. This observation is in agreement with similar ones concerning histochemical reactions for non-specific esterases and lipolytic activity of the aorta, and with their small adaptive increase in rabbits fed on an atherogenic diet (ZEMPLÉNYI et al., 1959a, b). Moreover, the degree of increase is altered by polyunsaturated fatty acids as indicated by comparison of results obtained in rats fed on an experimental butter diet and the diet with peanut oil (groups A and C, respectively). This is in agreement with observations in which dietary polyunsaturated fatty acids only elevated the lipolytic activity of rat aorta (PATELSKI et al., 1962). Thiouracil depresses the elevation of lipase activity.

Phospholipase activity is only increased in rats fed on a diet with peanut oil. Thiouracil again depresses the elevation of phos-

Table II. Activities of the lipolytic enzymes of rat and rabbit aortas

Group		Number	(a) Phosphatide acylhydrolase	(b) Glycerolester hydrolase	(c) Sterolester hydrolase
			10^{-3} µmoles/mg/min		
Rats, control diet		5	1058±415*	627±143	276±31
Experimental diets: 5% cholesterol, 2% cholic acid and:	A. 40% butter	5	842±185	1249±504 +	236±25
	B. 40% butter, 0.3% thiouracil	5	682±295	1227±422 ++	102±43 ++++
	C. 40% peanut oil	5	2098±898 +	2884±452 ++++	69±19 ++++
	D. 40% peanut oil, 0.3% thiouracil	4	969±293	1299±266 +++	25±11 ++++
Rabbits, control diet		5	180±44	148±31	116±10
Atherogenic semi-synthetic diet (HOWARD ET AL., 1965)	I. 10 weeks	5	89±16 +++	211±32 +++	52±17 +++
	II. 10 weeks + Lipostabil inj.	3	125	208	116
	III. 20 weeks + Lipostabil in diet	4	104±11 ++	65±31 +++	65±8 ++++
	IV. 32 weeks	5	106±38 +	158±83	23±16 ++++

* Mean ± Standard Deviation. Means statistically different from normal:
+P≤0.05; ++P≤0.02; +++P≤0.01; +++P≤0.001.

Reaction mixtures (5 ml of volume): (a) lecithin 1.25 mM, (b) glycerol trioleate 1.0 mM, (c) cholesteryl oleate 1.0 mM, and water extracts (a and b) from rat and rabbit wet aortas (0.02 mg of protein/0.25 ml and 0.14 mg/0.50 ml, respectively) or glycerol-water extracts (c) from rat and rabbit aortas (0.05 mg of protein/0.40 ml and 0.12 mg/0.80 ml, respectively) which were previously extracted with water. Sodium taurocholate (3.2×10^{-3} M) was used for determinations of Sterol-ester hydrolase activity and reduced glutathione (5×10^{-5} M) for estimations of both, the Sterol-ester hydrolase and Glycerol-ester hydrolase. Reaction temperature 30°C. Titration reagent 0.01 N KOH.

pholipase activity. The enzyme activity values were not altered by differences of several weeks in duration of the diet.

A considerable decrease in the cholesterol esterase activity of the aorta occurs in animals fed on an experimental diet, except for rats fed on a diet with butter and rabbits injected with Lipostabil where there was no change. Unlike the lipase activity, there was never an increase in the cholesterol esterase activity. Thiouracil in this case decreases the enzyme activity.

It can be seen that Lipostabil delays the depression of cholesterol esterase activity. It is interesting that polyenphosphatidyl choline, the major compound of Lipostabil, is not hydrolyzed in the pH optimum for aortic phospholipase which is 8.0. There is a small hydrolysis of polyenphosphatidyl choline in pH 9.2 in normal rat and rabbit aortas (approx. 100×10^{-3} μmoles/mg/min) and is not changed in rabbits treated with Lipostabil. It most probably maintains the proper concentration ratios of phospholipids and cholesterol esters in blood, thus preventing precipitation of these cholesterol esters, and the cholesterol esterase activity of the aorta at an adequate level. Cholesterol esterase activity seems to be negatively correlated with lipase activity.

Results obtained exclusively with lipid substrates not only confirm the presence of lipase in crude extracts from the aortic wall, but also provide evidence for the existence of enzymes which catalyze the hydrolysis of phospholipids and cholesterol esters. The activities of these enzymes cannot be neglected when chylomicrons or other lipoproteins are used in the estimations of 'the lipoprotein lipase' in crude extracts. Moreover, results presented indicate a relationship between the known different intensities in accummulation of main lipid esters on the one hand, and the different activity levels of the lipolytic enzymes in the aortic wall and their different behaviour in the course of atherosclerosis, on the other hand.

Summary

Phospholipase A, lipase and cholesterol esterase and the effect of different diets and of polyenphosphatidyl choline on these enzyme activities in the aortic wall have been demonstrated and discussed regarding atherosclerosis.

References

ABRAMSON, M.B.; KATZMAN, R. and GREGOR, H.P.: Aqueous dispersions of phosphatidylserine. J. biol. Chem. *239:* 70–76 (1964).
ENGELBERG, H.: Human endogenous lipemia clearing activity. Studies of lipolysis and effects of inhibitors. J. biol. Chem. *222:* 601–612 (1956).
HANAHAN, D.J.; RODBELL, M. and FURNER, L.D.: Enzymatic formation of monopalmitoleyl- and monopalmitoyllecithin (lysolecithins). J. biol. Chem. *206:* 431–441 (1954).

Howard, A.N.; Gresham, G.A.; Jones, D. and Jennings, J.W.: The prevention of rabbit atherosclerosis by Soya bean meal. J. Atheroscl. Res. *5:* 330–337 (1965).
Morton, R.K.: Methods of extraction of enzymes from animal tissues, in Colowick's and Kaplan's Methods in enzymology, Vol. 1, pp. 25–51 (Academic Press, New York 1955).
Murthy, S.K. and Ganguly, J.: Studies on cholesterol esterases of the small intestine and pancreas of rats. Biochem. J. *83:* 460–469 (1962).
Patelski, J.: Esteraza cholesterolowa tetnicy głównej – Cholesterol esterase of the aorta (Panstwowy Zaklad Wydawnictw Lekarskich, Warszawa 1964).
Patelski, J.; Rozynkowa, D. and Paluszak, J.: Lipolytic activity of rat aorta in relation to lipid concentrations and degree of saturation of the fatty acids in blood. Acta med. pol. *3:* 417–427 (1962).
Korn, E.D.: Properties of clearing factor obtained from rat heart acetone powder. Science *120:* 399–400 (1954).
Korn, E.D.: Clearing factor, a heparin-activated lipoprotein lipase I. Isolation and characterization of the enzyme from normal rat heart. J. biol. Chem. *215:* 1–14 (1955).
Korn, E.D.: The assay of lipoprotein lipase *in vivo* and *in vitro;* in Glick's Methods of biochemical analysis, Vol. 7, pp. 145–192 (Interscience Publishers, New York/London 1959).
Zemplényi, T.: The lipolytic and esterolytic activity of blood and tissue and problems of atherosclerosis. Adv. Lip. Res., Vol. 2, pp. 238–293 (Academic Press, New York/London 1964).
Zemplényi, T. and Grafnetter, D.: The lipolytic activity of heart and aorta in experimental atherosclerosis in rabbits. Brit. J. exp. Path. *40:* 312–317 (1959a).
Zemplényi, T.; Lojda, Z. and Grafnetter, D.: Relationship of lipolytic and esterolytic activity of the aorta to susceptibility to experimental atherosclerosis. Circulat. Res. *7:* 286–295 (1959b).

Authors' addresses: Dr. J. Patelski, Dr. Z. Waligóra and Dr. S. Szulc, Department of Physiological Chemistry, Medical Academy, *Poznań* (Poland); Dr. D. E. Bowyer, Dr. A. N. Howard and Dr. G. A. Gresham, Department of Pathology, University of Cambridge, *Cambridge* (England).

Dover Medical Research Center, Inc., Dover, Delaware

Human and Rabbit Arterial Cells Compared in Tissue Cultures

O.J. POLLAK and MINORU ADACHI, Dover, Del.

Introduction

For over 50 years the rabbit has been used for research in atherosclerosis. The classical experiment of feeding cholesterol to rabbits in order to produce atherosclerosis has undergone countless modifications. The use of the rabbit as an experimental model for atherosclerosis has been both violently criticized and passionately defended.

In recent years, the study of rabbits' aortic cells *in vitro* has been added as a method of studying atherosclerosis. POLLAK and KASAI (1964) reviewed the bulk of cardiovascular tissue culture work with rabbit cells. Several new questions are being asked: What are the similarities, and what are the differences between arterial cells of rabbits and arterial cells of man? To what extent can the study of rabbit aortic cells *in vitro* contribute to the knowledge about human aortic cells *in vitro*? To answer these questions, parallel *in vitro* studies with human and rabbit tissues were initiated.

Materials

We used 218 human aortic explants, 237 explants of other human arteries, and 252 explants of rabbit aortae. The human tissues were obtained within two hours after death, or were removed in the operating room. The youngest subjects were newborn, the oldest was 86 years old. Human explants from grossly normal (and, in adjacent sectors, also microscopically normal) arteries were used for explants, along with portions of atherosclerotic plaques of varying appearance. Rabbit aortae were either normal, or were altered by induced cholesterol atherosclerosis. The rabbits used were between 2 and 4 months old.

Methods

The tissue culture methods used in this study were those described by KOKUBU and POLLAK (1961). There was only one difference between the technique used for human explants and that used for rabbit explants: Human serum was added to the nutrient media inoculated with human tissues, while calf serum was used for rabbit tissues. Portions of human aortae and of large pulmonary arteries were stripped of the innermost 1–2 mm. The strippings were cut into 2 mm squares, and were positioned with the intima facing the supporting cover slip. Human arteries which had a similar caliber to that of rabbit aortae were cut into 1 mm high ring shaped segments. These were inverted, so that the intima faced outward. This technique, used for rabbit aortae, has been demonstrated by POLLAK and BURNS (1964).

Cells were observed under the light microscope and under the phase contrast microscope, and were studied by time lapse cinephotomicrography under light and phase contrast illumination. Slides were colored with the May-Grünwald-Giemsa stain, with Pollak's trichrome stain, and with the Periodic acid-Schiff (PAS) stain. Six enzymes were studied histochemically: alkaline phosphatase, acid phosphatase, adenosine triphosphatase, succinic dehydrogenase, lactic dehydrogenase, and malic dehydrogenase.

Results

We shall first compare human and rabbit aortic cells *in vitro*, and shall comment on other human arterial cultures later. The nearly identical appearance of four types of cells from human and rabbit aortae is documented by eight photomicrographs (Fig. 1–8). Without the legends, one could not tell which cells grew from human aorta and which grew from rabbit aorta. Human and rabbit aortic cells have many similarities and only few dissimilarities as shown in Table I.

Space does not permit reproduction of the 40 photomicrographs required to document the six enzyme reactions we studied. In each series there were marked differences between the four types of cells. However, there were no major differences in the enzyme activities of paired cells, i.e. human and rabbit endothelial cells, human and rabbit fibroblasts, etc.

The morphology and the cytochemical reactions varied only slightly whether we compared (1) cells from normal arterial tissues of several sites (aorta, pulmonary artery, femoral artery, popliteal artery, posterior tibial artery), or (2) aortic cells from subjects of different age (newborn, adolescent, adult, old), or (3) cells from anatomically normal segments and from atherosclerotic lesions.

The structural similarities between human and rabbit cells of the same type, and the morphologic similarities between human arterial

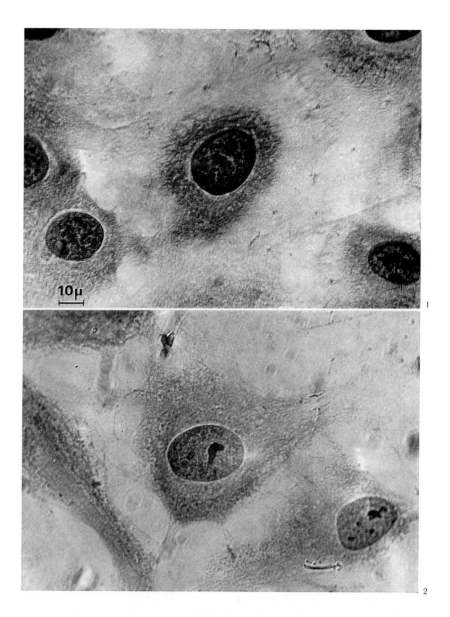

Fig. 1. Endothelial cells; rabbit aorta. Culture, 5 days old; Giemsa stain (×1000).

Fig. 2. Endothelial cells; human aorta. Culture, 10 days old; Giemsa stain (×1000).

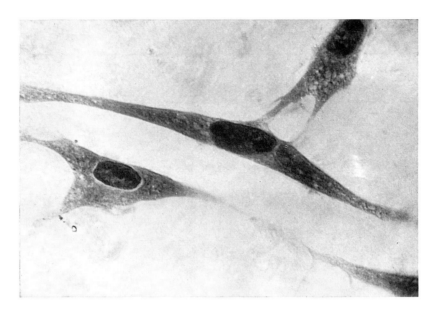

Fig. 3. Fibroblasts; rabbit aorta. Culture, 5 days old; Giemsa stain (×1000).

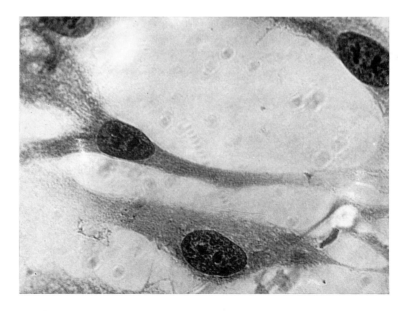

Fig. 4. Fibroblasts; human aorta. Culture, 21 days old; Giemsa stain (×1000).

Fig. 5. Smooth muscle cells; rabbit aorta. Culture, 17 days old; Pollak's Trichrome stain (×1000).

Fig. 6. Smooth muscle cells; human posterior tibial artery. Culture, 25 days old; Pollak's Trichrome stain (×1000).

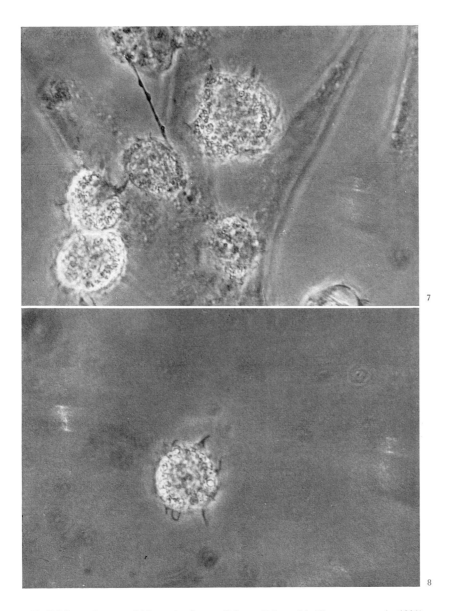

Fig. 7. Macrophages; rabbit aortic plaque. Culture, 5 days old; Phase contrast (×1000).

Fig. 8. Macrophage; human aortic plaque. Culture, 9 days old; Phase contrast (×1000).

Table I

Aortic cells	Rabbit	Human
Endothelial		
Shape	polygonal to round	polygonal to round
Size	variable, 30×50 μ to 50×100 μ	variable, 30×50 μ to 50×100 μ
Nucleus	1, central, round, 20 μ	1, central, round, 20 μ
Nucleoli	2 or more	2 or more
Cytoplasm	lucid perinuclear rim, dense inner zone, transparent outer zone, homogenous, agranular	lucid perinuclear rim, dense inner zone, transparent outer zone, homogenous, agranular
Projections	none	none
Profile	flying saucer – like	flying saucer – like
Luminescence	during mitosis, only	during mitosis, only
Motion	undulating	undulating
Fibroblasts		
Shape	at first, spindle; later, polygonal	at first, spindle; later, polygonal
*Size	80–120 μ × 10–15 μ	80–120 μ × 20–30 μ
*Nucleus	1, oval, with thick membrane, 15–20 μ × 10 × 15 μ	1, oval, with less thick membrane, 15–20 μ × 10–15 μ
Nucleoli	2 to 4, regular	2 to 4, regular
Cytoplasm	at first, homogenous; later, granular	at first, homogenous; later, granular
Projections	long, slender	long, slender
Profile	spindle to rhomboid	spindle to rhomboid
Luminescence	none	none
Motion	gliding	gliding
Smooth muscle		
Shape	ribbon – like	ribbon – like
Size	? 160×20 μ, average	? 160×20 μ, average
Nucleus	1, oval, slender, poorly staining	1, oval, slender, poorly staining
Nucleoli	1 to 3, regular	1 to 3, regular
*Cytoplasm	myofilaments distinct	myofilaments indistinct
Projections	syncytial formations	syncytial formations
Profile	tubular	tubular
Motion	? none	? none
Luminescence	none	none
Macrophage		
Shape	round	round
*Size	20–40 μ	15–30 μ
Nucleus	1, central, large; outline, hazy	1, central, large; outline, hazy
Nucleoli	1 ?	1 ?
*Cytoplasm	coarse granular	coarse granular, less
*Projections	many	many, but fewer
Profile	burr – like	burr – like
Motion	rapid, jerking	rapid, jerking
Luminescence	permanent	permanent

* Dissimilarities.

cells from different subjects and from different parts of the vascular system, were not matched by other cultural characteristics.

The number of positive cultures depended greatly on the source of the explant. The results for rabbit aortae and normal adult human aortae have been compared in Table II which shows the percentage of positive cultures for each of three types of cells.

Table II

Aortic cells	Rabbits	Human*
Endothelial cells	90%	60%
Fibroblasts	100%	90%
Smooth muscle cells	10%	4%**

* Normal, adult.
** By placing explants on the cover slip with the intima facing up, the number of smooth muscle cells will increase, that of endothelial cells will decrease.

Growth from arteries of small caliber was much better than growth from large human arteries. Explants of the aortae of newborn humans yielded results which closely approached those of rabbit aortae. Human atherosclerotic lesions rendered very poor results: The more complex a lesion, the fewer positive cultures of either type of cells. Whereas explants of rabbit aortae with plaques yielded lipid-laden macrophages in substantial numbers in 100% of cultures, these cells rarely grew from explants of lipid lesions of man, and never appeared from explants of fibrous or complicated atherosclerotic plaques.

The lag phase of the four types of cells from rabbit aortae and from adult human aortae differed considerably, as seen in Table III.

Table III

Aortic cells	Rabbits	Human
Endothelial cells	24–48 h	8 days*
Fibroblasts	72 h	9 days*
Smooth muscle cells	10–15 days	21 days*
Macrophages	24–48 h	9 days

* Normal, adult.

We observed considerable differences in time of first outgrowth of cells, and differences in height and slope of growth curves of cell populations from anatomically normal rabbit aortae, newborn human aortae, smaller human arteries, and normal adult human aortae (Fig. 9).

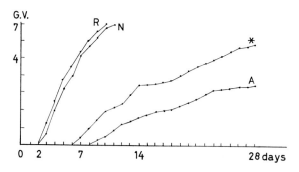

Fig. 9. Growth curves of cell populations from arterial explants: R: rabbit aorta; N: newborn human aorta; A: adult human aorta; *: composite of growth from femoral, popliteal, and posterior tibial human arterial explants. – G.V.: growth values: 0, no new cells; 1, 1–50 new cells; 2, 51–100 new cells; 3, 101–1000 new cells; 4, 0.1×10^4 new cells; 5, $1 \times 10^4 – 6 \times 10^4$ new cells; 6, $6 \times 10^4 – 1 \times 10^5$ new cells; 7, over 1×10^5 new cells.

Discussion

The four types of cells grown *in vitro* from human and rabbit arterial explants are distinct. They can be differentiated by morphologic, tinctorial, histochemical, and growth characteristics, and by time lapse cinephotomicrographic study of locomotion and of light reflection under phase contrast illumination. Some of these techniques can be applied to 'dead' tissues, i.e. to histologic preparations, and to 'live' tissues in culture, as well. Other methods can be used for tissue cultures, only. When one refers to methodology, the connotation of the terms '*in vivo*' and '*in vitro*' could be reversed. There is much speculation about transformation of cells in the living body. There is no extracorporeal transition between endothelial cells, fibroblasts, and smooth muscle cells of either human or rabbit aortae. Although we anticipated seeing *in vitro* transformation of one or more of these cells into macrophages, we have yet to observe such. This is not the only question which is lacking an answer. We have successfully cloned endothelial cells and fibroblasts, but not smooth muscle cells and macrophages. We are well acquainted with the mode of reproduction of endothelial cells and fibroblasts, but not with that of smooth muscle cells or macrophages. Cultures of vascular tissues from healthy and ill subjects of varying age, sex, body build, of different ethnic background, with different blood pressure, dietary habits, and other inborn or acquired characteristics may lead to interesting observations, as could the study of multiple explants from normal and from variously altered sectors of the blood vessels of selected subjects.

The success or failure of human vascular cultures – of the lag phase, of maximum growth values, of the growth curves, and of the crop of various types of cells – depends considerably on (1) the donor's age, (2) the caliber of the blood vessel, and (3) the structural quality of the explant. In contrast, the uniformly vigorous outgrowth, the short lag phase, the good yield of all four types of cells, and the speed of proliferation greatly facilitate *in vitro* study of rabbit aortae. The interpretation of results of rabbit aortic cultures is comparatively easy. When one interprets the results with human material, one has to consider numerous variables which influence the results.

The caged rabbit is a more uniform donor of explants than is the human. The caliber of rabbit's blood vessels changes less with the animal's growth and aging than does the caliber of human blood vessels. The degree and character of induced atherosclerosis in the rabbit can be fairly well regulated, whereas human atherosclerosis shows great variability.

Rabbit and human arterial cells have very similar morphologic characteristics and very similar enzymatic activity. Comparison of oxygen consumption of cell lines from various sources will be difficult because of differences in the rate of propagation. However, results of cytopharmacologic experiments with rabbit aortic cells should have reasonable validity for human aortic cells.

Summary

Human and rabbit arterial cells were compared *in vitro*.

The results of human arterial tissue cultures depend greatly upon the source of the explant: Small caliber arteries yield better results than large arteries; cells from arteries of young humans grow better than cells from the arteries of adults; intact arterial segments render better results than do altered segments. Cultures of rabbit aortae give uniformly good results.

Human and rabbit arterial tissue cultures differ greatly with respect to quantitive growth. Morphologically and enzymatically, the four types of cells grown from human and rabbit arterial tissues are very much alike.

Extrapolation of results of vascular tissue cultures of either species is feasible. The choice between human and rabbit explants for *in vitro* studies will depend upon the questions to which answers are sought.

Acknowledgement

This study was made possible by an institutional grant from the Lilly Research Laboratories, Eli Lilly and Co., *Indianapolis, Ind.*, USA.

References

Kokubu, T. and Pollak, O.J.: *In vitro* cultures of aortic cells of untreated and of cholesterol-fed rabbits. J. Atheroscl. Res. *1:* 229 (1961).

Pollak, O.J. and Burns, D.A.: Tissue cultures in cardiovascular research. Motion picture, sound, color, 16 mm; 22 min (1964).

Pollak, O.J. and Kasai, T.: Appearance and behavior of aortic cells *in vitro*. Amer. J. med. Sci. *248:* 71 (1964).

Authors' address: Dr. O. J. Pollak and Dr. Minoru Adachi, Dover Medical Research Center, 9 Kings Highway, *Dover, Delaware 19901* (USA).

Cleveland Clinic Foundation, Cleveland, Ohio

Oxygen Requirements of the Human Arterial Intima in Atherogenesis*

A. Lazzarini Robertson, Jr., Cleveland, Ohio

Introduction

A better understanding of the metabolism of the arterial wall as an organ is necessary to evaluate the role of perfusion [11, 12] across the avascular zone or intima of the arterial wall in naturally occurring atheroma. Unique to this layer, measuring 0.8–1.3 mm in thickness in normal adult aorta, is its metabolic dependence on the transport of serum components from the arterial lumen due to lack of capillaries for direct blood supply. Local variations on perfusion pressures as well as thickness of intima have profound effects in the concentration gradients of serum components, including oxygen. It has been calculated that the layer of intima supplied with oxygen by diffusion alone is less than 1 mm, suggesting that hypoxic conditions could easily occur if the permeability or thickness of the intima is significantly altered.

The metabolism of the arterial intima has only recently been studied following isolation of the intima lining by mechanical means [21] or by growth in organ culture [19]. It is no exception to the double pattern of catabolic reactions for energy release: aerobic or oxidative pathway and anaerobic or glycolytic cycle. Although the oxygen consumption of the aorta is only one tenth that of the liver, the presence of oxygen causes only a relatively small reduction on the conversion of glucose to pyruvate or lactate, indicating a deficiency in regulatory mechanisms that prevent excessive accumulation of lactic acid under aerobic conditions [8]. This 'aerobic glycolysis'

* These studies were supported by a grant from the American Heart Association and the United States Public Health Service # H-2471.

or lack of Pasteur effect results in uneconomic utilization of glucose by the arterial wall, since only 1/20 of adenosinetriphosphate (ATP) is produced as compared with that provided by complete oxidation of glucose via the tricarboxylic acid cycle [22]. The direct effect of this high rate of lactic acid production is that energy reserves of the arterial intima are lower than most other tissues and easily affected by changes on the supply of metabolites.

In the present studies, evaluation of oxygen requirements by isolated human vascular intimal cells and the effects of hypoxia in their lipid metabolism have been carried out. These studies suggest that local hypoxia may play an important role in the localization and severity of atherosclerotic lesions and that differences in oxygen requirements by different arteries may explain the well known variations on frequency of atherosclerosis in various segments of the arterial bed.

Materials and Methods

Organ cultures from full thickness segments of the vascular wall of large or medium-size arteries or veins taken sterile at autopsy or during reconstructive vascular surgery were used for isolation by growth of intimal cells (intimacytes) from other elements of the blood vessel wall [15]. Isolated intimacytes were harvested and transferred to monolayer cultures preceding preparation of suspension cultures in chemically defined nutrients with 10% homologous serum [18] for respiration and lipid metabolism studies. Similar preparations were made with pseudo-intimal cells lining homografts and man-made vascular prostheses.

For determination of oxygen consumption rates, stock suspension cultures were used. Cells were scraped from the culture flask wall with a rubber 'policeman' and suspensions of atherophils (see below) were counted in a hemocytemeter after staining dead cells with erythrosin. Oxygen consumption was determined by the cartesian diver technique [9] which measures changes in the buoyancy of an open gas-filled diver by observation of pressure changes required to maintain the diver floating at a fixed level [23]. In order to maintain constant carbon dioxide (CO_2) concentrations while measuring respiration, the flotation medium used consisted of a 30% solution of diethylamine (v/v) with potassium iodide to obtain a specific gravity of 1.414 g/ml [2, 3]. The flotation flask was modified by use of a closed constant pressure vessel with mercury seal to prevent CO_2 leakage. Further reduction of CO_2 losses was achieved by use of dibutyl-phthalate seals in the neck of the diver [13]. Each diver was filled with 1.0 ml of cell suspension after adding succinate to a final concentration of 1×10^{-3} M. The gas phase consisted of 5% CO_2 in air. The flotation vessel and diver were kept at $37.10 \pm 0.01°$C. Progressive changes of pressure were plotted against elapsed time [14]. Six simultaneous determinations including control divers with substrate only were made for each experiment.

For lipid incorporation studies, Na acetate-1-^{14}C, cholesterol-4-^{14}C, DL-mevalonic-2-^{14}C and ^{3}H labeled serum beta lipoproteins were used. The labeled compounds were added to cell suspensions in culture flasks with final specific activities of 6–20 μc/mM. The gas phase of the cultures during incubation consisted of 5% CO_2 in air or decreasing concentrations of oxygen (18–2%) in nitrogen with constant 5% CO_2. Culture flasks were maintained in rotary shakers at 40 strokes a minute and 37.0°C. Cell fractions were prepared by sonication in hypotonic salt solution at 4°C followed by differential centrifugation in sucrose gradients. Microsomal-cell sap fractions were extracted by ethanol-ether, the sterol-digitonide fraction was further purified with dibromide and counted by liquid scintillation spectrometry. Squalene concentrations were determined by the sulfuric acid-formaldehyde method [20] and read at 400 μm in a spectrophotometer. Estradiol benzoate was prepared as an aqueous suspension in balance salt solution [15] and added to cultures to obtain a final concentration of 0.4 μg/ml; globin insulin with zinc was added to a final concentration of 0.04 units per ml; Puramycin* and Actinomycin D** were added to a final concentration of 30–60 μg/ml. Adjustments to pH 7.2 for all solutions were made with 0.4% sodium bicarbonate in isotonic saline when necessary.

Results and Discussion

As it has been previously reported, from human intima two distinct cell populations may be isolated that maintain morphological and biochemical characteristics *in vitro* for several generations. The terms 'atherophils' and 'fibrophils' have been proposed to describe them [18].

Atherophils are flat polygonal cells with considerable amounts of acid phosphatase activity and respiratory enzymes of the Krebs cycle. Under the electron microscope they show abundant cell organelles, particularly mitochondria, endoplasmic reticulum with abundant ribosomes and isolated myofilament-like elements. The most important metabolic characteristic of atherophils is that of rapid incorporation of extracellular lipids under unfavorable environmental or metabolic conditions resulting in lipid-laden 'foam cells' or 'atherocytes'.

Fibrophils, in contrast, are characterized by their reluctance to incorporate extracellular lipids and their tendency to synthesize acid mucopolysaccharides and collagen *in vitro*.

Average oxygen consumption rates of human vascular cells shown in Table I and expressed in microliters per hour per 10^7 cells, indicate that atherocytes have higher oxygen requirements than intimacytes from normal artery or vein. These findings are in agreement with similar observations on respiratory rates of aortas with experimental atherosclerosis in rats [10] and rabbits [21].

Even higher oxygen consumptions were found for atherocytes

* Nutritional Biochemical Co., Cleveland, Ohio.
** Cosmegen (Merck, Sharp and Dohme, Div. of Merck & Co., West Point, Pennsylvania).

Table I. Oxygen consumption rates of human vascular intimal cells

Intimacytes from normal aorta (atherophils)	0.094±0.021 µl*
Intimacytes from aorta with 2+ —3+ lesions (atherocytes)	0.148±0.045 µl*
Intimacytes from vena cava	0.016±0.008 µl*

* /h/10^7 cells.
Results of 24 determinations for each cell type – Suspension cultures.

from human coronary arteries [17], suggesting that metabolic differences, as expressed by oxygen requirements, may exist between arteries of different sizes. In both man and hamster an inverse relationship between diameter of the vessel and oxygen consumption has been recently reported [7].

Hormonal effects at cell level may also be expressed in changes in oxygen consumption as indicated in Table II for estradiol benzoate and insulin. It should be noted that the response of human coronary atherophils to estradiol was higher than that of similar cells of the

Table II. Oxygen consumption – human cyto I cells

I. 24 experiments intima of aorta and large arteries	0.112±0.018 µM*
II. Same as I. and estradiol benzoate 0.4 µg/ml	0.098±0.025 µM*
III. Same as I. and insulin Zn 0.040 g/ml	0.196±0.041 µM*
IV. 11 experiments intima coronary arteries	0.226±0.078 µM*
V. Same as IV. and estradiol benzoate 0.4 µg/ml	0.075±0.012 µM*

* /h/10^7 cells.

aorta, since experimental evidence has shown inhibition of atherogenesis in coronary arteries by estrogen therapy but failed to induce regression of aortic lesions [6, 17, 22]. Cell response to estrogens may also be dependent upon species differences since estradiol has been shown to reduce the oxygen consumption of normal rat aorta [6].

The effects of decreasing oxygen concentrations on the intracellular incorporation of labeled serum cholesterol by human atherophils is shown on Table III. It should be noted that only oxygen concentrations below 5% induced significant increases in incorporation of cholesterol by these cells.

Table III. Effects of O_2 on cholesterol-4-^{14}C serum uptake microsome fraction atherocytes c p m**

% O_2 in Gas phase*	h	1	4	8	12	16
15		28	36	42	67	136
10		31	41	53	81	149
5		39	57	69	98	194
3		64	84	107	214	296
1		108	202	345	567	982

* +5% CO_2 and N_2 to balance.
** Average total net counts of 48 experiments.

Lower oxygen concentrations also induced higher incorporation of labeled serum lipoproteins. As shown in Figure 1, increased intracellular concentrations of labeled lipids occurred when oxygen concentrations were below 8%.

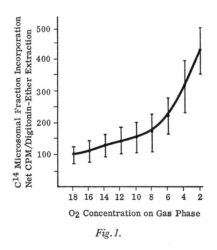

Fig. 1.

Because considerable experimental evidence suggested that the arterial wall is capable of synthesizing sterol from precursors such as acetate or mavelonate, labeled acetate was added to cultures of isolated atherophils at different oxygen concentrations. As shown in Table IV, oxygen levels below 10% produced reduction in acetate utilization [16, 17].

Table IV. Effects of O_2 on sodium acetate 1-^{14}C utilization human atherocytes – severe atheroma
Total lipid fraction c p m per mg cell protein*

% O_2 in Gas phase** h	4	8	12	16	20
15	45	125	248	526	847
10	38	109	198	326	510
5	24	86	107	205	314
3	18	46	59	73	87

* Average total net counts of 27 experiments.
** +5% CO_2 and N_2 to balance.

A simultaneous evaluation of acetate utilization and ^3H labeled serum beta lipoprotein incorporation expressed as radioactivity of ethanol-ether-digitonide extracts of the microsomal-cell sap fractions showed that low oxygen concentrations reduced acetate utilization for cholesterol synthesis but increased uptake of the serum lipoproteins (Table V).

Table V. Effect of oxygen tensions on sodium acetate 1-^{14}C utilization and ^3H lipoprotein incorporation microsome fractions human atherocytes (aorta)*

% O_2 in Gas phase ** h	2		4		6		8		12		16		20	
	^3H	^{14}C	^3H	^{14}C	^3H	^{14}C	^3H	^{14}C	^3H	^{14}C	^3H	^{14}C	^3H	^{14}C
15	32	12	39	21	69	54	128	78	176	91	194	87	278	96
	48	36	54	38	94	71	205	97	296	121	292	134	384	171
10	46	10	86	16	126	26	132	56	356	64	428	73	605	82
	71	31	101	23	310	31	387	64	402	78	516	84	721	98
5	69	8	158	12	288	18	227	32	604	41	691	54	891	67
	94	26	203	21	421	38	526	44	727	58	827	78	986	86
3	121	4	298	8	359	11	320	23	671	31	926	44	1724	51
	189	16	406	17	487	26	598	35	829	46	1046	65	1978	73

* Total lipid fraction net counts × mg/wet cells Spinner chemically defined HACHD # 6.
** +5% CO_2 and N_2 to balance.

Since labeled acetate was still incorporated by atherophils at concentrations below 5% although the amount of labeled cholesterol extracted from the cells was significantly reduced, a quantitative colorimetric micromethod for squalene developed by ROTHBLAT ET AL. [20] was used to determine if under hypoxic conditions the labeled acetate or mevalonate was utilized for squalene synthesis (Figure 2). The increase in squalene content at low oxygen tensions is in agreement with the concept that cyclization of squalene to lanosterol is oxygen dependent.

The effects of decreasing oxygen concentrations on lipid incorporation and *de novo* sterol synthesis by human coronary intimal cells is summarized in Figure 3. It shows that oxygen levels below

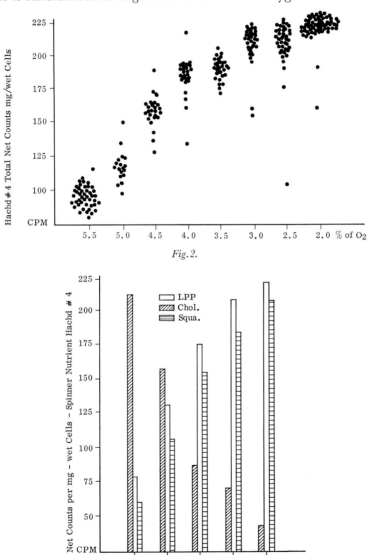

Fig. 2.

Fig. 3.

5% induce increased permeability of the cell membrane for serum lipoproteins (LPP), decreased cholesterol synthesis (CHOL) and increased squalene (SQUA) accumulation.

Since mammalian cells are not able to utilize squalene in large amounts, the increased concentrations of squalene and permeability of the cell membrane for extracellular lipids under hypoxia may thus accelerate formation of 'foam cells' or atherocytes.

The role of local hypoxia in atherogenesis is therefore quite complex and may be represented by a vicious circle (Figure 4).

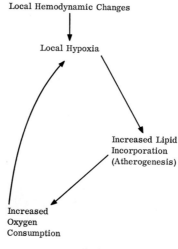

Fig. 4.

Local hemodynamic changes such as variations in perfusion pressures or permeability of the arterial intimal lining due to platelet accumulation, mural thrombus or endothelial changes [5], result in reduction of rate of transport of metabolites and local hypoxia. This in turn increases permeability of the cell membrane of atherophils for extracellular lipids and inhibits the ability of these cells to emulsify and disperse cytoplasmic lipids, usually invisible as 'micellar fat' [4] held inside laminar miscelles formed by phospholipid and protein membranes. By interfering with oxidative phosphorilation, hypoxia inhibits further protein and phospholipid synthesis and excess intracellular cholesterol reduces the emulsifying capacity of preexisting phospholipids. If the increased oxygen demands of these cells are not satisfied, the lipid micelles are transformed into 'globular fat' that cannot be easily mobilized by the cell, inducing formation of lipid-laden cells or atherocytes and accelerating local hypoxia and further lipid incorporation by cells not yet affected. That hypoxia is responsible for induction of globular or unmasked intracellular lipid by inhibition of protein synthesis was demonstrated in studies in which 'foam cells' formation was induced by treatment of the suspension cultures by Puramycin or Actinomycin D [17].

The effects of local hypoxia in the various components of the arterial intima may thus be summarized as follows (Figure 5):

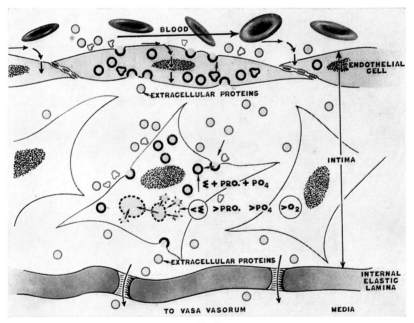

Fig. 5.

Plasma lipoproteins are transported across vascular endothelium by splitting of the protein and lipid moieties at cell membrane level [19]. The lipids are incorporated at a faster rate into the cell cytoplasm, particularly the rough endoplasmic reticulum and cell sap fractions. This incorporation requires and is dependent upon *de novo* protein and phospholipid synthesis by the cell (heavy lines). Similar processes occur at subendothelial cell level where the lipids of extracellular lipoproteins need synthesis of new protein and phospholip 'coats' to be incorporated as micelles in the cytoplasm of intimacytes (atherophils).

Metabolic changes that affect oxidative phosphorilation (PO4) by intimacytes such as enzymatic alterations (Σ) or *reduced oxygen* ($> O_2$) concentrations such as those following *local hypoxia* may thus inhibit protein and phospholip synthesis by the cell, producing abnormal micellar 'coats' and inducing cohesion of these micelles into intracellular vacuoles with formation of lipid-laden foam cells or atherocytes.

A summary of the role of local metabolic changes in human atherogenesis in view of this experimental data in the presence of cell hypoxia is shown in Figure 6.

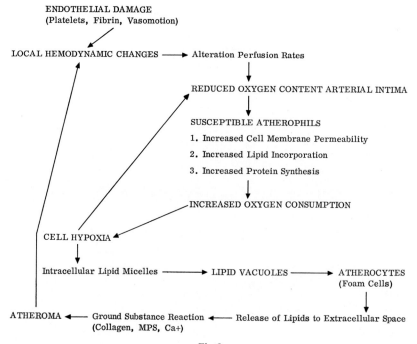

Fig. 6.

In Summary

Utilizing short term cultures of isolated human intimacytes, obtained by cloning from organ cultures of normal and atheromatous human arteries, it was found that lipid transport and metabolism in these cells is oxygen dependent. Significant increase in oxygen consumption rates occurred in *early* stages of lipid incorporation by intimacytes *in vitro*.

Oxygen concentrations below 5% increased permeability of the cell membrane for extracellular lipids and simultaneously interfered with intracellular micelle formation due to inhibition of protein and phospholipid synthesis. Although cholesterol precursors were still incorporated under hypoxic conditions, the pathway of intracellular cholesterol synthesis did not continue beyond squalene.

It is concluded that relative local hypoxia at cell level may play an important role in accelerating human atherogenesis by initiation as a chain reaction of a series of self-sustaining metabolic abnormalities. Many of the present theories on the pathogenesis of naturally occurring atheroscleroris may thus be explained as having as a common denominator reduction of oxygen transport and its local availability to the arterial intima.

References

1. Cornforth, J.W.; Cornforth, R.H.; Pelter, A.; Horning, M.G. and Popjak, G.: Rearrangement of methyl groups in the enzymic cyclization of squalene to lanosterol. Proc.chem.Soc.Lond., p.112 (1958).
2. Danes, B. Shannon and Paul, J.: Environmental factors influencing respiration of strain L cells. Exp.Cell Res. *24:* 344 (1961).
3. Danes, B. Shannon; Broadfoot, M.M. and Paul, J.: A comparative study of respiratory metabolism in cultured mammalian cell strains. Exp.Cell Res. *30:* 369 (1963).
4. Dixon, K.C.: Deposition of globular lipid in arterial cells in relation to anoxia. Amer.J.Path. *39:* 65 (1961).
5. Gutstein, W.H.; Robertson, A. Lazzarini and La Taillade, T.N.: The role of local arterial irritability in the development of arterio-atherosclerosis. Amer.J.Path. *32:* 417 (1956).
6. Hilz, H. and Utermann, D.: Der Sulfatstoffwechsel der Gefäßwand in Beziehung zur Arteriosklerose. Biochem.Z. *332:* 376 (1960).
7. Howard, R.O.; Richardson, D.W.; Smith, M.H. and Patterson, J.L.: Oxygen consumption of arterioles and venules as studied in the cartesian diver. Circulat. Res. *16:* 187 (1965).
8. Kirk, J.E.; Effersoe, P.G. and Chiang, S.P.: Rate of respiration and glycolysis by human and dog aortic tissue. J.Geront. *9:* 10 (1954).
9. Linderstrøm-Lang, K.: Principle of the cartesian diver applied to gasometric technique. Nature, Lond. *140:* 108 (1937).
10. Loomeijer, F.J. and Ostendorf, J.P.: Oxygen consumption of the thoracic aorta of normal and hypercholesterolemic rats. Circulat. Res. *7:* 466 (1959).
11. Page, I.H.: Atherosclerosis, an introduction. Circulation *10:* 1 (1954).
12. Page, I.H.: Introduction to Symposium on atherosclerosis, National Academy of Sciences, National Research Council, Puplication *338:* 3 (1955).
13. Paul, J. and Danes, B.: A modified cartesian diver method permitting measurement of oxygen uptake in the presence of carbon dioxide. Analyt. Biochem. *2/5:* 470 (1961).

14. PHILLIPS, H.J.; ANDREWS, R.V. and SKANK, V.: Expressing tissue culture respiration. Exp. Cell Res. *20:* 607 (1960).
15. ROBERTSON, A. LAZZARINI: Studies on the effects of lipid emulsions on arterial intimal cells in tissue culture in relation to atherosclerosis. Doctoral thesis, Cornell Grad. Med. School, Cornell University, New York City (1959).
16. ROBERTSON, A. LAZZARINI: Effects of O_2 tension on the uptake of labelled lipoproteins by human atheromatous plaques. Circulation Abs. *24/4:* 1096 (1961).
17. ROBERTSON, A. LAZZARINI: Hypoxia and sterol synthesis in human atheromatous plaques. Fed. Proc. *22/2:* 161 (1963).
18. ROBERTSON, A. LAZZARINI: Metabolism and ultrastructure of the arterial wall in atherosclerosis. Cleveland Clin. Quart. *32/3:* 99 (1965).
19. ROBERTSON, A. LAZZARINI: Intracellular incorporation of plasma lipoproteins by arterial intima in relation to early stages of intravascular thrombosis, p. 267, in Biophysical mechanisms in vascular homeostasis and intravascular thrombosis., ed. SAWYER, P.N., Appleton-Century-Crofts, New York 1965).
20. ROTHBLAT, G.H.; MARTAK, D.S. and KRITCHEVSKY, D.: A quantitative colorimetric assay for squalene. Analyt. Biochem. *4:* 52 (1962).
21. WHEREAT, A.F.: Oxygen consumption of normal and atherosclerotic intima. Circulat. Res. *9:* 571 (1961).
22. ZEMPLENYI, T.: Enzymes of the arterial wall. J. Atheroscl. Res. *2:* 2 (1962).
23. ZEUTHEN, E.: Microgasometric methods, cartesian divers. Second International Congress Histo- and Cytochemistry Frankfurt/Main, p. 70 (1964).

Author's address: Dr. Abel Lazzarini Robertson, Cleveland Clinic, 2020 East 93rd Street, *Cleveland, Ohio 44106* (USA).

The Wistar Institute of Anatomy and Biology, Philadelphia, Pennsylvania

Cholesterol Ester Metabolism in Tissue Culture Cells*

By G. H. ROTHBLAT**, R. HARTZELL, H. MIALHE and
D. KRITCHEVSKY***, Philadelphia, Pa.

Introduction

From previous studies on sterol metabolism in tissue culture cells (ROTHBLAT et al., 1966), we obtained data which indicate that free sterol is taken up by these cells through a physical adsorptive process which was temperature- and concentration-dependent and which was reduced by only $30\% \pm 5\%$ when cellular enzyme systems were inactivated by heat. These findings also suggested that the side chain of the steroid molecule is essential for uptake while modifications in the ring structure do not effect adsorption. Cellular-bound sterol could not be removed from the cells by washing with solutions of Tween or serum, nor did the bound sterol exhibit a high degree of molecular exchange when exposed to additional exogenous sterol.

Studies on free sterol metabolism in tissue culture cells have shown that one particular cell line, designated L-5178Y, demonstrated a high cholesteryl ester content and an unusually high cholesteryl ester/cholesterol (CE/C) ratio. Because of this unusual finding we have conducted a comparative study on steryl ester metabolism in tissue culture cells with particular emphasis on the L-5178Y cell line.

---

* This investigation was supported in part by Public Health Service Research Grant HE 09-103-02 from the National Heart Institute.
** Recipient of Public Health Service Special Research Fellowship No. 1F3AM-34, 974-01 from the National Institute of Arthritis and Metabolic Dieseases.
*** Recipient of Career Award USPHS K6-HE-734 from the National Heart Institute.

Materials and Methods

The L-5178Y cell line is a stable line of lymphoblasts derived from a murine leukemia. These cells were routinely grown as an agitated culture in medium containing 2.5% horse serum as previously described (ROTHBLAT et al., 1966).

The L-cells used in this study were grown in double Eagle's medium with Earle's modified spinner salt solution supplemented with 7.5% fetal bovine serum. Cells were harvested by centrifugation in the logarithmic phase of growth.

Horse serum was purchased from Baltimore Biological Laboratory. Calf serum and fetal bovine serum were obtained from Microbiological Associates. Methods for assay of isotopically labeled compounds, gas-liquid chromatography and extraction and quantitation of sterols have been presented in previous publications (ROTHBLAT et al., 1966, 1967).

Results

Table I is a compilation of data demonstrating the general sterol pattern of tissue culture cells. Much of the data is derived from a study published by MACKENZIE et al. Their data, originally expressed as µg sterol/µg cell protein, have been converted to µg sterol/mg dry weight by assuming 50% of the cellular dry weight to be protein.

The CE/C ratio of the representative cell lines cultivated on various sera has a maximum of 0.4 (Table I). In addition, the actual amount of cellular sterol in these cells falls into a narrow range, particularly if one considers that different laboratories use various techniques for the determination of sterol and for the quantitation of cells.

Table II demonstrated similar data obtained from studies on the L-5178Y cell line. The total sterol content of this line, when grown in calf or horse serum, is similar to other cell lines; however, the CE/C ratio is high, being 1.7 when the cells are cultivated in horse serum and 2.7 when cells are grown in calf serum. When fetal calf serum is used, the cholesteryl ester content of the cells drops, although the free sterol content remains constant.

Because of the high cellular ester content, we attempted to determine the origin of steryl esters in L-5178Y cells. Growing the cells

Table I. Sterol content of tissue culture cells

Cell	Serum	Free	μg/mg dry weight Ester	Total	CE/C	Ref.
HeLa	horse	7.5	2.5	10.0	0.33	1
HeLa	rabbit	7.0	3.0	10.0	0.43	1
L	horse	10.5	3.0	13.5	0.29	1
L	rabbit	13.0	4.0	17.0	0.31	1
L	fetal calf	11.0	1.6	12.6	0.15	2
Chang liver	horse	9.5	2.0	11.5	0.21	1
Chang liver	rabbit	8.5	3.0	11.5	0.35	1
Rat liver	horse	9.0	2.0	11.0	0.22	1
Rat liver	rabbit	8.5	2.5	11.0	0.29	1
Rabbit liver	horse	10.0	3.5	13.5	0.35	1
Rabbit liver	rabbit	11.0	4.5	15.5	0.41	1
Monkey kidney	calf	14.2	1.5	15.7	0.11	2
MB III	human	37.4	0.0	37.4	0.00	3
WI-38	calf	21.3	6.4	27.7	0.30	4
EB 2	fetal calf	15.3	3.4	18.7	0.22	2

References
1. Mackenzie, C. G. et al., Exp. Cell Res., *36*, 533 (1964).
2. Rothblat, G., unpublished data.
3. Bailey, J. M., PSEBM, *107*, 30 (1961).
4. Kritchevsky, D. and Howard, B., Ann. Med. exp. Fenn., *44*: 343 (1966).

Table II. Sterol content of L-5178Y cells

Sera	free	μg/mg dry weight ester	total	CE/C
Horse	6.1	10.0	16.3	1.7
Fetal bovine	6.4	5.1	11.5	0.8
Calf	5.7	15.5	21.2	2.7

in the presence of 1-^{14}C-acetate showed that only 7.5% of the total radioactive sterol synthesized could be recovered in the steryl ester fraction of the cellular lipids. In additional experiments, 4-^{14}C-cholesterol was supplied to the growth medium. After 18 hours the amount of cholesterol-^{14}C present in cellular free and ester sterol was determined, and it was found that only 3.4% of the exogenous radioactive cholesterol had been incorporated into the steryl ester fraction. Obviously, then, high cellular esters are not a result of *de novo* synthesis or of esterification of exogenous free sterol. The one other possibility, that cellular esters are a result of the uptake of exogenous serum steryl esters, is confirmed by the data presented in Table III, which shows the fatty acid pattern of L-5178Y cholesteryl esters and cholesteryl esters of the horse serum in which these cells were cultivated. The fatty acid spectra of both cells and serum

Table III. Cholesteryl ester fatty acids of L-5178Y cells and horse serum

Fatty acid	L-5178Y cells %	Horse serum %
14:0	trace	trace
14:1	trace	trace
16:0	12.1	10.9
16:1	2.0	0.7
18:0	2.4	1.2
18:1	14.4	8.2
18:2	67.5	74.7
18:3	trace	trace
20:4	1.6	4.3

are similar, indicating that the cells derive their esters unchanged from the medium. If this be the case, two possibilities exist for the high ester content of L-5178Y cells grown in horse and calf sera: 1) increased uptake of exogenous esters by L-5178Y cells when compared to other cell lines or 2) an accumulation of cellular esters in L-5178Y cells due to an inability to hydrolyze or excrete steryl esters.

In an effort to determine if L-5178Y cells removed more steryl ester from the medium than other cell lines, a series of experiments were conducted in which concentrated cell suspensions in buffer were exposed to serum taken from a rabbit which had been fed 4-^{14}C-cholesterol. Following a one hour exposure period at 37°C, the cells were removed by centrifugation and the loss from the medium of labeled free and esterified sterol was assayed. It was evident from these experiments that L-5178Y cells do not take up appreciably more steryl ester than do L-cells when exposed to rabbit serum under identical conditions (L-cells = 0.48 µg/mg dry wt/1 h; L-5178Y = 0.50 µg/mg dry wt/1 h); however, it should be kept in mind that these experiments were conducted on concentrated cell systems, and thus may not truly reflect the events which take place in a growing culture. Also, more free sterol than ester is taken by the cells under these test conditions. It seems then that our first hypothesis – that the increased content is due to an enhanced uptake – may not be the entire answer to the increased ester content of L-5178Y cells.

In order to test the second possibility – that cholesteryl esters are not removed from the L-5178Y cell at rates similar to other cells – the following experiments were conducted. Either growing cultures of L-cells or L-5178Y cells in heat-inactivated serum were exposed to 4-^{14}C-cholesteryl esters which had been added in isotopic amounts in acetone to the culture medium. Heat-inactivated serum was used in these experiments to eliminate the possibility of sterol interconversions by serum esterases. Following incubation of from 18 to 24 hours, the washed cells were extracted and the amount of both labeled free and ester sterol in the cells was determined. By this method we could calculate the amount of free sterol which was a product of cellular ester hydrolysis. The data for both cholesteryl oleate and cholesteryl acetate are presented in Table IV. L-cells hydrolyzed much more of the cholesteryl ester, both oleate and

Table IV. Hydrolysis of 4-^{14}C-cholesteryl esters by growing tissue culture cells

Serum* Cholesteryl ester	Percent recovered as free cholesterol					
	L-cells			L-5178Y cells		
	Fetal calf (7.5%)		Horse (7.5%)	Fetal calf (5.0%)		Horse (2.5%)
	Acetate	Oleate	Oleate	Oleate	Oleate	Acetate
Culture supernatant	1	2	1	2	1	1
Uninoculated medium	1	2	1	1	1	1
Cells	56	66	48	11	12	10

* Heat inactivated, 60° C/30 min

acetate, than did L-5178Y cells. The type of serum in which the cells were grown did not effect the percent hydrolysis in either cell line. It is possible, then, that the cellular ester differences we noted between L-5178Y cells and other cell lines could be due to this pronounced difference in the amount of ester hydrolyzed. It is evident that this is not an all-or-none phenomenon – the L-5178Y cells are capable of steryl ester hydrolysis – the difference lies in the amount of ester hydrolyzed in each cell type.

We have conducted similar experiments employing cholesteryl palmitate-1-^{14}C. In this case the fatty acid is labeled, allowing us to follow its utilization by the cell (Table V). In the case of L-cells, a substantial percent of fatty acid is released from the cholesteryl ester, most of which can be recovered in the cellular phospholipid fraction. All other lipids, including free sterol, are also labeled, indicating that a portion of the fatty acid is converted to acetate,

Table V. Distribution of carbon-14 liberated by tissue culture cells from cholesteryl-1-^{14}C palmitate

	Carbon-14 distribution	
	L-cells %	L-5178Y cells %
Steryl ester	49.5	89.2
Triglycerides	8.0	1.9
Free fatty acids	8.2	0.9
Sterol	1.9	0.2
Diglycerides	5.0	2.1
Monoglycerides + phospholipids	27.7	5.6
CO_2	trace	trace
Defatted residue	trace	trace

which in turn is resynthesized into sterol. Small but detectable amounts of ^{14}C can be recovered as CO_2, and small amounts are present in the defatted cellular residue. A similar situation exists in the case of the L-5178Y cells, though far less ester is hydrolyzed. All L-5178Y lipid fractions contained ^{14}C as did CO_2 and defatted residue.

In another set of experiments we have attempted to determine the cellular location of the cholesteryl esters in L-5178Y cells. In these studies we have ruptured cells through the use of N_2 decompression. This material was then fractioned by differential centrifugation into debris, mitochondrial, microsomal, and supernant fractions, after which the free and ester sterol content of each fraction was determined. The results are presented in Table VI. The free sterol is equally distributed between the cellular debris on one hand and the mitochondria and microsomes on the other. This is not the case, however, with the esters of L-5178Y cells. Eighty percent of the cholesteryl ester is recovered in the cell debris fraction. Thus the free and esterified sterol are not proportionally distributed in this cell line, and the data indicate that the excess esters accumulate in the cell fraction consisting of membrane and

Table VI. Distribution of sterol in cellular fractions of L-5178Y cells

Cell fraction	Sterol distribution %		
	total	free	ester
Debris	71	51	82
Microsomes	14	25	8
Mitochondria	11	17	8
Supernatant	4	7	2

nuclear fragments and other material which pellet under low g conditions. The complete composition of this fraction is presently under investigation.

Discussion

The data obtained in this study indicate that the cholesteryl esters present in L-5178Y cells are derived primarily from the serum in which the cells are cultivated. The comparative studies conducted on L-cells and on L-5178Y cells indicate that the higher ester content of the latter cell line is due to the decreased amount of cholesteryl ester hydrolyzed by the L-5178Y cells.

The effect of serum on the cellular ester content of L-5178Y cells can be interpreted as being a reflection of two separate phenomena. First, the cholesteryl ester content of the cells is influenced by the type of serum in which the cells are grown; secondly, when the lipids of the L-5178Y cells are compared to those of other tissue culture cells grown in the same serum, the L-5178Y cholesteryl ester content is higher. These data also suggest that the effect of fetal calf serum on steryl ester content in L-5178Y cells is not related to the rate of hydrolysis of esters since growth in either horse or fetal calf serum did not affect the percentage of free ^{14}C sterol recovered in the cells. One possible explanation of this serum effect would be that cells take up steryl esters from fetal calf serum at much lower rates than from horse or calf sera due to some as yet unknown factor which influences ester uptake. In L-5178Y cells, with less hydrolytic activity, a rate-limiting step would be in cholesteryl ester hydrolysis. Variation in the amount of ester taken up by these cells from different sera would then be reflected in the cellular steryl ester content. In L-cells, on the other hand, the rate-limiting step would be the amount of cholesteryl ester taken up from the sera. Therefore, no excess esters would accumulate in the cells since esters taken up from the serum, regardless of uptake rate, would be hydrolyzed to free sterol and fatty acid. Studies are continuing to gain additional information on the relationship between cholesteryl ester uptake and hydrolysis in tissue culture cells.

Summary

Factors affecting cholesterol ester metabolism have been studied in the L-5178Y tissue culture line. Data obtained in this study indicate that this cell line accumulates cholesteryl esters from the serum in which the cells are cultivated and that this accumulation is a reflection of reduced cholesteryl ester hydrolysis by these cells when compared to other tissue culture cell lines. Studies have been conducted to determine the fate of the fatty acids liberated by the cells upon hydrolysis of cholesteryl esters. Additional experiments were performed to determine the intracellular location of free and esterified sterol.

References

MACKENZIE, C.G.; MACKENZIE, J.B. and REISS, O.K.: Regulation of cell lipid metabolism and accumulation. III. The lipid content of mammalian cells and the response to the lipogenic activity of rabbit serum. Exp. Cell Res. *36:* 533–547 (1964).

ROTHBLAT, G.H.; HARTZELL, R.W., Jr.; MIAHLE, H. and KRITCHEVSKY, D.: Cholesterol metabolism in tissue culture cells. Eds. G. H. ROTHBLAT and D. KRITCHEVSKY. Wistar Symp. Monograph No. 6, pp. 129–146 (Wistar Press, Philadelphia 1967).

ROTHBLAT, G. H.; HARTZELL, R.W., Jr.; MIAHLE, H. and KRITCHEVSKY, D.: The uptake of cholesterol by L-5178Y tissue culture cells: Studies with free cholesterol. Biochim. Biophys. Acta. *116:* 133–145 (1966).

Authors' address: G. H. Rothblat, Ph. D.; R. Hartzell, B. A.; H. Mialhe, B. A. and D. Kritchevsky, Ph. D. The Wistar Institute of Anatomy and Biology, *Philadelphia, Penn.* (USA).

Institute for Cardiovascular Research, Prague-Krč, Czechoslovakia
Director: Prof. JAN BROD

Study of Factors Affecting Arterial Enzyme Activities in Man and Some Animals

T. ZEMPLÉNYI, O. MRHOVÁ, D. URBANOVÁ and
M. KOHOUT, Prague-Krč

In our previous work [3–5] we investigated some regulatory factors of vascular enzyme activity. We observed, for example, that sex hormones have a clearcut effect on many arterial enzymic activities as demonstrated by sex-differences, by changes in gonadectomized animals and following the administration of sex hormones. Feeding animals with different types of 'atherogenic' or 'thrombogenic' diets also induces clearcut changes in vascular enzymic activities.

In this communication studies of some local metabolic factors in the pathogenesis of atherosclerosis will be presented.

It is well known that different arteries and also different segments of the same artery show a considerable variation in susceptibility to atherosclerosis. For example, the extent and severity of atherosclerosis is higher in the abdominal than in the ascending aorta in humans and most animals and birds. These and similar disparities provide a unique opportunity for investigation of 'local' factors in atherogenesis.

Fig. 1 summarizes results in the aortas of young Rhesus macaques. The activity of malate dehydrogenase and β-glucuronidase is significantly higher in the ascending than in the abdominal segments of the same aortas and a similar trend in the activity of succinate dehydrogenase can be observed. The activities of ATPase, 5'-nucleotidase, carboxylic esterase and of the glycolytic enzyme lactic dehydrogenase are higher in the ascending aorta.

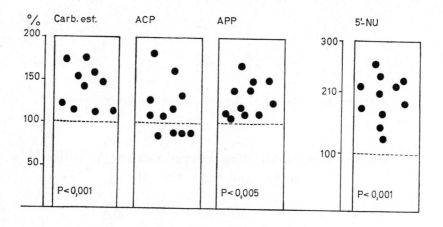
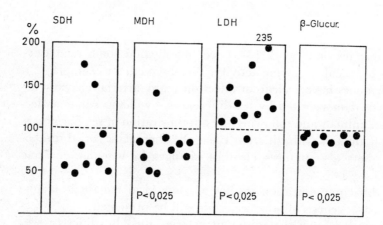

Fig. 1. Enzyme activities in ascending aortas of Macacus Rhesus. Activity of abdominal segments from the same aortas put at 100%.

Without going into details, Figure 2 shows that similar findings, in particular concerning Krebs cycle enzymes and phosphomonoesterases, were observed in investigations of enzymic activities in the aortas of very young calves and ducks. In addition, we observed in chicken aortas also a disparity in the phosphomonoesterase activities of ascending and abdominal aortic segments.

In contrast to the above results there is a different pattern of distribution in enzymic activities in arteries or arterial segments of older humans. As shown in the last three rows of Figure 2 this is

	Tricarboxylic acid cycle enzymes	Phosphomonoesterase II and/or I
Abdominal as compared with asc. aorta in 30 days old chickens	↗	↘
Abdominal as compared with asc. aorta in 50-60 days old ducks	↗	→
Abdominal as compared with asc. aorta in 30 days old calves	↗	→
Abdominal as compared with asc. aorta in young monkeys (Macaca mulatta)	↗	→
Abdominal as compared with asc. aorta in 6 months old pigs	↘	↘
Abdominal aorta as compared with asc. aorta in older humans	↘	↗
Femoral artery as compared with brachial artery in older humans	↘	↗
Thoracic aorta as compared with pulmonary artery in older humans	↘	↗

Fig. 2.

characterized by a lower activity of Krebs cycle enzymes and a higher activity of phosphomonoesterases in the artery or arterial segment more susceptible to atherosclerosis. A similar trend (row 5) could be observed for Krebs cycle enzymes in the aortic segments of older pigs.

However, these results were obtained from intact segments of aortas of older humans and the findings in young subjects and particularly in children are therefore of particular significance. Comparison of the enzymic activities in the ascending with the abdominal aortas shows (Table I) that the ratio of the mean activity for malate dehydrogenase and phosphomonoesterase II undergoes a striking change during the first years of life. The relative activity of the former is somewhat higher in the abdominal aorta during early childhood but it decreases very quickly to about 50% in relation to the ascending aorta. Quite the opposite is true for the activity of phosphomonoesterase II.

Table I. Mean values of asc./abd. activity ratios in intact aortas of children and young people

Enzyme	Age (years)		
	0–1	2–20	30–42
Malate dehydrogenase	0.9062	1.2548	1.917
Lactate dehydrogenase	1.2387	1.2070	1.196
Phosphomonoesterase II	0.9955	0.7311	0.7792
ATP-ase	0.9520	0.9683	0.7320
5'-nucleotidase	0.9045	0.9516	1.00

The above age-linked changes of the enzymic pattern in the arteries of humans and probably animals and birds seem to reflect a general phenomenon. Our comparative histological, histochemical and biochemical studies indicate [5] that the activity of most of the enzymes investigated is confined to muscle cells. This is obviously the reason for the higher activity for example of Krebs cycle enzymes in the abdominal aortas of ducks, chickens or pigs where there are clearcut differences in histological structure in the ascending and abdominal aortic segments. This is doubtless also the reason for differences in the other species, although the differences in anatomical structure are not so striking.

In agreement with many others [1, 2] we must suppose that muscle cells are more susceptible to injuries of daily life and this leads to their decreased enzymic activity. Hypoxia, caused for example by vascular intimal thickening may have a fundamental role in these events, since it is known that intimal thickening occurs in the abdominal aorta, coronary and iliac arteries and in other predilection sites of atherosclerotic lesions.

It is reasonable to expect that the decrease in enzyme activity, in particular that of Krebs cycle enzymes, leads to a lower production of macroergic phosphate compounds, lower synthesis of proteins – including enzymes – and initiates a vicious circle leading finally to the deterioration of all defence mechanisms of the vessel wall against further injuries and against the accumulation of lipids as well. It may be added that the increased activity of some enzymes, in particular that of phosphomonoesterases, seems to reflect injury to the vessel wall and reaction of its connective tissue [3–5] *.

* All the data on enzyme activities were calculated on the basis of either the desoxyribonucleic acid content of tissue or the protein content of aortic extracts. For technical reasons it was not possible to include all the slides projected at the symposium into this paper.

May I add that it is our belief that the problem of atherogenesis is neither the question of lipids only nor of vascular metabolism only but a question of a proper balance of both these factors.

References

1. ADAMS, C.W.M.: Arteriosclerosis in man, other mammals and birds. Biol. Rev. *39:* 372–423 (1964).
2. CONSTANTINIDES, P.: Experimental atherosclerosis (Elsevier, New York 1965).
3. ZEMPLÉNYI, T.: The lipolytic activity of blood and tissues and problems of atherosclerosis. Adv. lip. Res. *2:* 235–293 (1964).
4. ZEMPLÉNYI, T.; LOJDA, Z. and MRHOVÁ, O.: Enzymes of the vascular wall in experimental atherosclerosis in the rabbit; in Sandler and Bourne's atherosclerosis and its origin, p. 459–509 (Academic Press, New York 1963).
5. ZEMPLÉNYI, T.; MRHOVÁ, O.; URBANOVÁ, D. and LOJDA, Z.: Comparative aspects of vascular enzymes. Acta Zool. Path. Antverp. *39:* 45–68 (1966).

Authors' address: Doc. T. Zemplényi, M. D.; O. Mrhová, Chem. Eng., D. Urbanová, M. D. and M. Kohout, RN Dr., Inst. Cardiovascular Research, *Prague-Krč* (Czechoslovakia).

Discussion Session 6

H. ENGELBERG (Beverly Hills, Cal.): It is interesting that plasma lipemia has been shown by various workers to decrease tissue oxygen tension. Thus increased plasma lipids contribute to tissue hypoxia which Dr. ROBERTSON has shown initiates a fat-accumulating process in aortic endothelial cells grown in tissue culture. In 1955 we first showed that the clearing of lipemia by heparin injection resulted in an increase of oxygen consumption in atherosclerotic humans who had a subnormal total oxygen consumption, and suggested that this occurred because of the removal of lipid intimal films following triglyceride lipolysis after heparin. The effect of hypoxia on endothelial cells, and of plasma lipemia on tissue oxygenation, suggest that increased plasma lipid levels may contribute to atherogenesis via hypoxia entirely apart from the question of lipoprotein infiltration into the vascular wall. I would like to ask Dr. ROBERTSON if the hypoxia levels he used in his tissue culture work are comparable to those observed in human studies of interstitial tissue oxygen tension following alimentary lipemia.

A.L. ROBERTSON (Cleveland, Ohio): As indicated in the last slide, it is possible that a variety of hemodynamic factors that alter perfusion rates of the arterial intima such as changes in blood laminal flow, platelet and/or fibrin deposition, and intima hyperplasia may contribute to local hypoxia and, indirectly, by blocking protein and phospholipid synthesis, to atherogenesis. The effects of heparin sulfate and other strong anionic substances (Angiology, *12*, 525–534, 1961) seem to occur by affecting cell membrane surface changes and rate of incorporation of serum lipoprotein by intimacytes. Pinocytosis, resulting from alimentary hyperlipemia still requires the presence of other local tissue changes, such as those mentioned before, to explain the characteristic spotty distribution of atheromatous lesions in man and experimental animals.

D. KRITCHEVSKY (Philadelphia, Pa.): We (CRISTOFALO and KRITCHEVSKY, 1965, 1966; KRITCHEVSKY and HOWARD, 1966) have been carrying out biochemical studies with a human diploid cell strain (WI-38) which retains many of the characteristics of normal cells *in situ*, e.g. a stable diploid karyotype, a limited life span and non-malignant growth *in vivo*. The glycolytic rate of these cells resembles that of other serially cultured cells and their respiratory and glycolytic capacity is relatively constant over their life span. This is also true of their respiratory capacity. There is a significant increase in total lipid content of the older WI-38 cells regardless of whether total lipid is calculated on a dry weight (25 vs. 19 mg lipid/100 mg dry wt.) or cellular (15 vs. 6.5 mg lipid/10^8 cells) basis. Phospholipids account for 70% of the total lipid in both young and old cells. In the young cells lecithin is the predominate phospholipid (53% of total phospholipid). In older cells the amount of lecithin rises to 70% with a concomitant fall in the amount of phosphatidyl ethanolamine (12% in young cells and 7% in old cells) and phosphatidyl inositol (15% in young cells and 8% in old cells). The major component of the neutral lipid in WI-38 cells is cholesterol (12.6% of total lipid in young cells and 13.1% of total lipid in old cells). The free/ester cholesterol ratio is about 3:1 in both young and old WI-38 cells.

References

1. CRISTOFALO, V.J. and KRITCHEVSKY, D.: Growth and glycolysis in the human diploid cell strain WI-38. Proc. Soc. exp. Biol. N.Y. *118*: pp. 1109–1113 (1965).
2. CRISTOFALO, V.J. and KRITCHEVSKY, D.: Respiration and glycolysis in the human diploid cell strain WI-38 J. cell. comp. Physiol. *67*: pp. 125–132 (1966).
3. KRITCHEVSKY, D. and HOWARD, B.V.: The lipids of human diploid cell strain WI-38. Ann. Med. exp. Fenn. (In press).

P. CONSTANTINIDES (Vancouver, B.C.): I would like to congratulate Dr. MCMILLAN on his very important contribution to the solution of the problem of the mechanism of cell proliferation in arterial lesions.

We have made similar findings in injured arteries of rats that were given three microcuries of tritiated thymidine per gram body weight twenty hours before sacrifice. We found labelling of nuclei in endothelial and smooth muscle cells around the foci of injury.

I would like to ask Dr. McMillan whether he has had the opportunity to make time studies, and if so, whether he thinks the proliferation gradient is from the endothelium downwards or the other way around.

G. McMillan (Montreal, Que.): No, Dr. Constantinides, we have not yet made time studies that would let me answer your question.

P. Constantinides (Vancouver, B.C.): I tend to agree with Dr. Robertson's statement that lipid is not carried through the endothelial cytoplasm by ordinary pinocytosis. We have followed intravenously injected yolk particles in cells of rat and found no membranous envelope around them. However, these were large micellar particles, approximately 1,000 Å in diameter and I wonder whether Dr. Robertson has any relevant information on smaller particles. Turning to another point, I do not think that cells in atherosclerotic lesions can be sharply separated into cells that phagocytose lipids only and cells that produce collagen only. One sees so often cells that do many jobs at the same time in these lesions that it is very hard to put them in pigeonholes.

A. L. Robertson (Cleveland, Ohio): The observations reported this morning were made with micelles of serum beta-lipoproteins measuring 1200 to 600 Å in diameter. If these micelles were tagged with ferritin or colloidal metals in preparation for electron microscopic studies, pinocytosis occurred. That pinocytosis was due to the tagging substance rather than the lipoprotein micelles was further proven by control studies with ferritin and colloidal metals alone.

In reference to the second question, functional rather than morphological criteria were used to identify hyman intimacytes in two major cell types: atherophils and fibrophils. We have until now been unable to demonstrate transformation of one cell type into the other in cultures. I fully agree with Dr. Constantinides that atherophils may be multipotential and represent a stem cell that may, under appropriate environmental conditions, evolve into cells resembling modified smooth muscle cells, macrophages, endothelial and even blood monocytes. This was not the case for fibrophils which in their behavior resembled connective tissue fibroblasts. Identification of such cell differences in routine tissue sections without culture could not be easily made. I would also like to mention that similar behavior on serum lipoprotein incorporation and cell type we have been able to demonstrate so far only in a nonhuman primate, the squirrel monkey and in swine arteries but not in other laboratory animals commonly used in the inductions of experimental atherosclerosis.

P. Constantinides (Vancouver, B.C.): Dr. Mead, I wonder how we can marry your finding of much unsaturated fatty acid in plasma membranes with the Dawson-Danielli model of these membranes.

Unsaturated fatty acids are supposed to be electron-dense after osmic acid fixation; yet according to the Dawson-Danielli model the lipids reside in the middle leaflet of the plasma membrane, which is the electron-lucent leaflet.

J. F. Mead (Los Angeles, Cal.): Since the polysaturated fatty acids are present in the membrane in the form of phospholipids and since it has been shown, by most workers, that the osmium is located in the polar portion of the phospholipid complex, I do not believe any difficulty exists, despite the known reduction of osmic acid by unsaturated centers. Incidentally, I do believe that the Dawson-Danielli proposal in its original form needs modification.

A. Scanu (Chicago, Ill.): This question is directed to the interesting and provocative finding of Dr. L. Robertson that serum lipoproteins are split at the membrane surface so that only the lipid moiety is able to penetrate into the cell. Have you any idea, Dr. ROBERTSON, of the mechanism whereby the splitting takes place?

Also, what are the effects of inhibitors of protein synthesis (actinomycin D, puromycin, etc.) in the permeability of serum lipoproteins in your system?

A. L. ROBERTSON (Cleveland, Ohio): Our data suggests that the splitting of serum β-lipoproteins occurs at the cell surface in the presence of adenosinetriphosphate and by exchange with cell plasma membrane lipids which are incorporated into the cell cytoplasm.

Because we were discussing at this time the effects of hypoxia on lipoprotein incorporation by human intimacytes, time did not allow to mention the effects of other protein synthesis inhibitors in this phenomena. We have also found that puromycin and actinomycin D acting on 50 S ribosomes or messanger RNA will also induce transformation of micellar intracellular lipids into globular fat.

Our present explanation of these effects is that similarly to hypoxia, by inhibiting availability of protein for micellar coating, these compounds induce formation of lipid vacuoles and eventually foam cell formation.

A. SCANU (Chicago, Ill.): Dr. HOLLANDER on which basis have you applied the ultracentrifugal technique described for serum lipoproteins, to separate lipoproteins from the intima of the aorta? Have you made any attempt to correlated lipoproteins separated from aorta and serum on the basis of the properties of their protein moieties, i.e., on an immuno-chemical basis?

W. HOLLANDER (Boston, Mass.): We have not correlated the lipoprotein composition of the aorta and plasma in the same individual. However, the composition of the high density lipoproteins (d 1.063–1.210) in the normal and diseased aorta appears to be similar to that of normal plasma. The aortic high density lipoproteins also have the electrophoretic mobility of a plasma alpha-lipoprotein. The low density lipoproteins (d < 1.063) in the plaque contain a higher percentage of cholesterol and a smaller percentage of triglyceride and protein than the low density fractions of normal plasma. In general, the extractable content and composition of the aortic low density lipoproteins vary directly with the lipid content and composition of the aortic intima. We have been able to demonstrate that the aortic low density lipoproteins migrate electrophoretically as a beta-lipoprotein and with immunodiffusion techniques that the aortic low density lipoproteins behave immunologically like plasma beta-lipoproteins.

P. ALAUPOVIC (Oklahoma City, Okla.): Dr. ROBERTS, what evidence do you have for the suggestion that serum lipoproteins are split into protein and lipid moieties and that under these circumstances protein accumulates at the surface of the cell and lipid penetrates into the cell in the micellar form?

A. L. ROBERTSON (Cleveland, Ohio): The evidence we have on the splitting of serum β-lipoproteins at cell surface level was obtained by studies including (1) rates of incorporation of double-labeled lipoproteins by cell membrane ghosts and microsomal-cell sap fractions, (2) cytochemical demonstration of ATPase activity on cell membrane surface and (3) electron microscopic evidence of protein complexes on cell plasma membrane.

This data was presented at the Council on Atherosclerosis of the American Heart Association (Circulation *28:* 664; 1963, J. Cell Biology *23:* 18; 1964) and published as: Intracellular incorporation of plasma lipoproteins by arterial intima in relation to early stages of intravascular thrombosis; in: Biophysical mechanisms in vascular homeostasis and intravascular thrombosis, p. 267–273, ed. PH. N. SAWYER (Appleton-Century-Crofts, New York 1965).

N. KALANT (Montreal, Que.): We have fed a high-fat diet containing *no* propylthiouracil to rats and rabbits and measured the various acid polysaccharides in the isolated intima from the aorta. Initially, before there are demonstrable plaques or chemically measurable increases in intimal lipids, we found a marked decrease in chondroitin sulfates with no change in hyaluronic acid or heparin sulfate. Later, when lipid accumulation occurs, we found in the plaque areas a two-fold increase in the heparin sulfate and chondroitin sulfates, again with no change in the hyaluronic acid. The turnover rate of these sulfated polysaccharides was also considerably increased in the plaque area. The direction of changes in concentration of the polysaccharides described by Dr. HOLLANDER is the same as that which we found. I wonder whether Dr. HOLLANDER would agree that this data on polysaccharides should not be disregarded or attributed entirely to hypothyroidism but that both his and our results are compatible with the concept expressed earlier in the meeting that the initial step in lipid deposition is a decrease in the intimal chondroitin sulfates, permitting increased permeability of the intima to plasma lipoproteins and that the subsequent increase in sulfated polysaccharides may be due to co-precipitation with lipoproteins derived from the plasma.

W. HOLLANDER (Boston, Mass.): Your observations are extremely interesting and lend additional support to the concept that a disturbance in acid mucopolysaccharide metabolism precedes the accumulation of lipids in the atherosclerotic plaque. I agree with you that the accumulation of cholesterol in the arteries may be due to an increased permeability of the intima to plasma lipoproteins. However, I am not certain from our own data whether the changes in the acid mucopolysaccharides are primary or secondary to the deposition of the lipids.

W.A. THOMAS (Albany, N.Y.): I would like to correct a comment just made by Dr. HOLLANDER regarding our paper. The so-called 'mild' diet that we used did not contain thiouracil. Nonetheless we demonstrated that protein synthesis by the aorta was markedly depressed (to $50\% \pm$ of control values). Of course, when we added thiouracil to the diet the depressant effect was even greater.

Gonadal Hormones, Blood Lipids and Ischemic Heart Disease

R. H. Furman, P. Alaupovic, R. H. Bradford and
R. P. Howard, Oklahoma City, Okla.

Epidemiologic Considerations

It is well known that the clinical manifestations of coronary atherosclerosis are much more common in men than in women, particularly among subjects between 35 and 60 years of age. The ratio of male to female deaths from coronary artery disease in the United States reaches a maximum value during the fifth decade of life; thereafter it diminishes, approaching unity in the 70th and later years (Fig. 1). This decrease of the male/female ratio of coronary deaths, occurring as it does in the fifth decade, is popularly ascribed to a presumed marked increase in coronary heart disease mortality in women as a result of the menopause (Oliver [1]). Careful examination of mortality data provides no evidence for an effect of the menopause on the death rate from arteriosclerotic heart disease in women.

Although mortality rates from coronary artery disease increase exponentially in both men and women (Fig. 2) and are always higher in men, it is important to note that the factor by which the mortality rate in men grows each year begins to diminish during the fifth decade, i.e., the curve is noted to 'flatten' somewhat and the male mortality rate shows a smaller exponential growth rate after age 45. This diminishing rate of increase of coronary artery disease mortality among men is such that, beyond the 45th year of age, the growth of

Ratios of male to female death rates for coronary disease and hypertensive disease, whites and nonwhites, by age, United States 1955

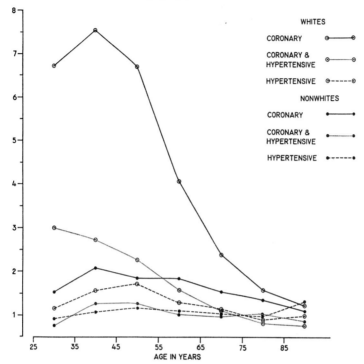

Fig. 1. Graph of values of the ratios of male/female mortality due to arteriosclerotic heart disease and hypertension in white and non-white populations in the U.S. in 1955. A high male/female death rate ratio is largely restricted to the white population. The ratios are derived from age-specific death rates (number of deaths in a specific age group during any year divided by the average population of the group at risk that year). Data for the year 1955 were chosen because in that year the National Vital Statistics Division of the National Center for Health Statistics and the National Heart Institute of the United States Public Health Service coded multiple (up to five) causes of death. Thus, mortality from hypertension with or without arteriosclerotic heart disease could be evaluated.

The authors are greatly indebted to Dean E. KRUEGER of the Biometrics Research Branch of the National Heart Institute, U.S.P.H.S., for expert assistance in preparing the data presented in this Figure and in Figure 2.

coronary artery disease mortality among women is greater than it is among men and the mortality rate curves approach one another.

The slowing of the rate at which coronary artery disease mortality increases with age in men is undoubtedly due to shrinkage of the pool of male heart attack candidates resulting from the many deaths among relatively young, coronary disease-vulnerable men.

Death rates for coronary heart disease
White and non-white population, by age and sex: United States 1955.

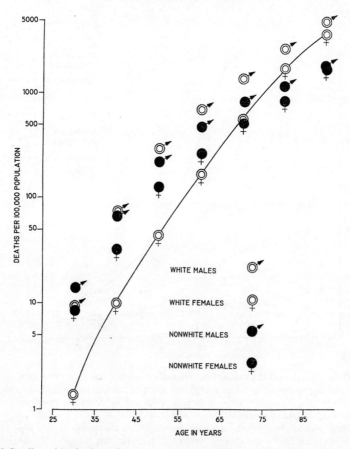

Fig. 2. Semilogarithmic plots of coronary artery disease mortality (#420: arteriosclerotic heart disease, including coronary disease, W.H.O. Manual of the International Statistical Classification of Diseases, Injuries and Causes of Death) in white and non-white U.S. population 1955. Data are age-specific death rates. The flattening of the curves depicting mortality in white men and non-white men and women is readily apparent in contrast to the relatively straight-line character of the curve depicting mortality in white women.

The data depicted in Figure 2 represent semilogarithmic plots of mortality as a function of age and thus the vertical distance (difference) between points indicates the log of their ratio: log male/female mortality = log male mortality − log female mortality. Thus it can be seen that, after age 45, the diminishing male/female coronary artery disease mortality ratio is due to a diminished acceleration of male mortality with age. This is further substantiated by the fact that the curve depicting female mortality is virtually a straight line (Figure 2). There is no evidence of a change in the steadily accelerating mortality rate with age in women which could be ascribed to the menopause. Additional evidence that the death rate from arteriosclerotic heart disease among white women in the United States is not influenced by the menopause was offered recently by Tracy [2] who derived the following equation expressing mortality as a function of age: $\log R = kA - mA^2 + c$, in which R = death rate, A = age and c, k and m are empirical constants, and found that these constants were the same after the age of the menopause as before. In other words, pre- and postmenopausal states were identical in respect to their influence on mortality from arteriosclerotic heart disease.

If the menopause signaled an increased rate of death from coronary heart disease among women, the mortality rate curve for women (Figure 2) would show a greater rise at some interval following the menopause. The menopause occurs commonly between the ages of 45 and 54. Approximately 25% of women cease menstruating before the age of 45 while only 2% continue to menstruate after the age of 55 (Millot and Daux [3]). If one assumes that 5–10 years must elapse before loss of ovarian function might be reflected in a significant increase in clinical or anatomical manifestations in coronary atherosclerosis, evidence of menopause-related increase in coronary artery disease mortality should be apparent in women by age 55 or 60. The curves depicting mortality rates for women in Figure 2 do not suggest any such menopause-related increase.

That the mortality data presented in Figure 2 can be relied upon to reveal an influence of the menopause on diseases which are known to be influenced by ovarian function is attested to by the fact that data from the same source indicate a marked decrease in mortality rate from breast carcinoma following the menopause.

The higher coronary artery disease mortality of men is most evident in an 'affluent society', i.e., among populations enjoying what questionably might be called a 'high standard of living'. In 1950 coronary heart disease mortality in the 45 to 54 age group showed a male/female ratio of 5/1 in the United States, while in Italy it was 2/1 and in Japan only slightly greater than unity.

The sex differential in coronary artery disease mortality is very much smaller in the American Negro (Master, Dack and Jaffe [4]) and virtually disappears when hypertension is also listed as a cause of death (Fig. 1). The existence of earlier and more severe coronary atherosclerosis in Negro women in comparison with white women is reflected in the observation that the average age at the time of the first myocardial infarction in Negro women is ten years less than that of white women (56 years versus 66.5 years) (KEIL and McVAY, Jr. [5]).

World Health Organization statistics for 1961 indicate that the male/female ratios of deaths from arteriosclerotic and degenerative heart disease in Finland and in the Netherlands were 10/1 and 9/1 respectively, but these peak ratios occurred during the *fourth* decade, well in *advance* of the menopause. The male/female mortality rate ratio in both countries dropped to approximately 6/1 by the 45th year and continued to decline, reaching unity around the 80th year, as in the United States.

In view of these many observations it seems unlikely that the diminishing male/female coronary artery disease mortality ratio in middle aged and older individuals in the U.S. white population bears a causal relationship to the menopause.

The absence of a 'menopause effect' on the sex differential in coronary heart disease mortality casts considerable doubt on the widely-held belief that estrogens are responsible for the lower mortality rate from this disorder in women. However, ovarian hormone secretion does not cease completely after the menopause and the adrenal cortex is an additional source of estrogens. Thus the possibility exists, however remote it may seem, that, if estrogens are indeed protective against coronary artery disease, the low level of estrogen production in the postmenopausal period may be of continued prophylactic importance. The effects of estrogens on serum lipids and lipoproteins (discussed below) do not of themselves readily suggest any clear mechanism for such prophylaxis, if such indeed exists.

The smaller sex differential in the U.S. Negro and in populations in which the way of life is, in many ways, simpler than that of the white American, indicates that the differential is not based on sex alone but involves an additional factor or factors as yet unrecognized.

Sex differences in the anatomical (autopsy) evidence of coronary atherosclerosis are less striking than are the clinical manifestations (LOBER [6]). It is also of interest that, in white persons under 50 years of age, myocardial infarction is more likely to be fatal in men than it is in women (GOODALE, THOMAS and O'NEAL [7]), a phenomenon not observed in nonwhites (THOMAS, BLACHE and LEE [8]). These observations raise the possibility that thrombogenic factors or vulnerability to fatal arrhythmias secondary to myocardial ischemia may be of relevance to the sex differential in coronary heart disease mortality.

The onset of the decline in the sex differential in the U.S. population in the middle of the fifth decade, and even earlier in Finland and the Netherlands, as the acceleration of the male mortality rate becomes less marked, emphasizes clearly the high decree of vulnerability of men in 'early middle age' to fatal heart attacks. It would seem more logical to presume that men are *predisposed* to coronary artery disease than to assume that women are relatively '*protected*'. Do androgens in some way provide the basis for the vulnerability of the male to coronary artery disease? The effects of androgens on serum lipids and lipoproteins (discussed below) are suggestive of such an influence, but the matter remains largely conjectural.

Gonadal Hormones and Serum Lipids

The lower mortality rate due to coronary artery disease observed in women is popularly ascribed to a 'hypocholesterolemic' and/or 'prophylactic' effect of estrogen. Androgens, on the other hand, are not generally regarded as 'atherogenic' or 'hypercholesterolemic', although either or both views are occasionally encountered.

Sex differences do exist in respect to serum lipids and lipoproteins, and gonadal hormones predictably alter serum lipid and lipoprotein patterns (FURMAN, ALAUPOVIC and HOWARD [9]) although the significance for atherogenesis of these sex differences in lipid transport patterns, and of the changes induced by the administration of gonadal hormones, is by no means clear. Differences in lipid patterns are evident between men and women, and between younger (premenopausal) and older (postmenopausal) women. Younger women have less cholesterol circulating in the form of very low density (VLD) lipoproteins (d <1.006 g/ml) than do older women and men, and lower serum triglyceride levels (FURMAN, HOWARD, LAKSHMI and NORCIA [10]). The lipid content of the lipoprotein fractions increases with age in women and serum cholesterol, phospholipid and triglyceride levels are higher. The serum lipid and lipoprotein patterns of men and older women have in common higher triglyceride levels and a higher content of lipid in VLD-lipoproteins than do younger women.

Estrogen administration does *not* change the serum lipid and lipoprotein pattern of hypogonadal women to one resembling that of the nonpregnant premenopausal female. Estrogen administration increases the lipid content of a- and VLD-lipoproteins (Figures 3a, b), resulting in higher serum phospholipid levels (a-lipoproteins) and triglyceride levels (VLD-lipoproteins) and lower cholesterol/phospholipid ratios [9]. The serum cholesterol level response to estrogen administration varies, depending on the magnitude and direction of

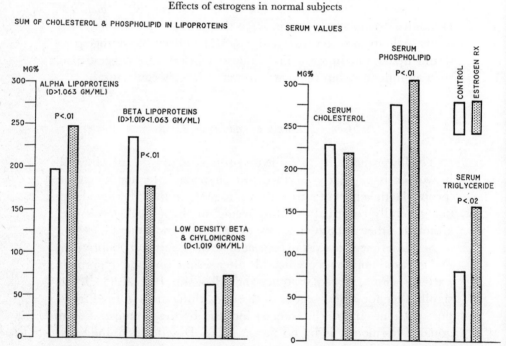

Fig. 3a. Effects of estrogen administration on serum lipids and lipoproteins. (Mean values derived from studies in seven estrogen-deficient, normolipemic women administered various estrogens.) Estrogen increases the lipid content of a-lipoprotein (mainly phospholipid), usually diminishes that of β-lipoprotein and increases serum triglyceride (which circulates principally as d < 1.019 g/ml lipoprotein). The increase in a-lipoprotein phospholipid accounts for the significant increase in serum phospholipid levels and the resultant fall in cholesterol/phospholipid ratios. The increase in a-lipoprotein cholesterol is approximately equal to the decrease in β-lipoprotein cholesterol, thus the net change in serum cholesterol levels is not statistically significant. (Transitional values excluded.)

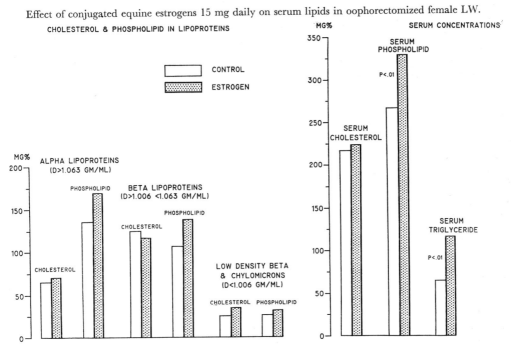

Fig. 3b. Effects of conjugated equine estrogens (Premarin®, Ayerst Laboratories, New York) 15 mg/day on serum and lipoprotein lipids in an oophorectomized woman (LW). The increased content of phospholipid in α- and β-lipoproteins and the increased content of triglyceride in VLD (d<1.006 g/ml) lipoproteins resulted in statistically significant increase of serum phospholipid and triglyceride levels in this subject. (Estrogen administered 196 days. Transitional values excluded.)

change in the cholesterol content of β-lipoproteins. Increased serum triglyceride and serum phospholipid levels during estrogen administration are reliable indicators of estrogen effect. (A detailed study of estrogen effects on serum lipids and lipoproteins in normal subjects and in subjects with various hyperlipidemic syndromes has recently appeared [9].)

Gonadal Hormones and Lipoprotein Composition

Quantification of serum lipoproteins usually involves fractionation of serum by physical means, such as preparative ultracentrifugation or electrophoresis on various media, following which the lipid content of each of the isolated fractions is determined. Lipoprotein 'concentrations' are said to increase or decrease according to whether the lipid content of any isolated fraction increases or decreases. Thus the increased content of cholesterol and phospholipid in the d 1.063–1.210 g/ml serum fraction occurring during estrogen administration is taken as evidence that estrogen increases a-lipoprotein concentrations. Analyses of the protein content of carefully isolated and washed lipoprotein fractions have provided additional information regarding the effects of estrogen administration (ALAUPOVIC, HOWARD and FURMAN [11], FURMAN, ALAUPOVIC and HOWARD [9]). Estrogen diminishes the cholesterol content of purified a-lipoproteins relative to the protein content and similar but smaller changes usually occur in β-lipoproteins. The lipid/protein ratio of VLD-lipoproteins increases due mainly to an increased content of triglyceride (Figure 4). Inasmuch as the cholesterol content of the d 1.063–1.210 g/ml (a-lipoprotein) fraction of serum is increased when estrogens are

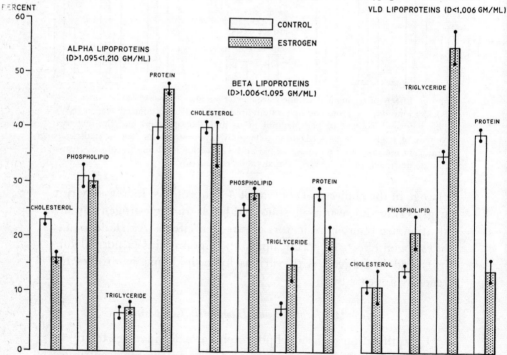

Fig. 4. Effects of estrone sulfate 2.5 mg/day on the composition of purified lipoproteins in a hypogonadal woman (MM). The principal relative changes are (1) decreased a-lipoprotein cholesterol, increased protein (2) increased triglyceride in β- and VLD-lipoproteins, decreased protein. (Transitional values excluded. Sera were obtained for study only after serum lipid values had remained stable over periods of several weeks. Period of sodium estrone sulfate administration, 156 days.)

administered, while the cholesterol/protein ratio is diminished, it would appear that estrogens increase the concentration of α-lipoprotein 'particles' while diminishing the cholesterol content of the individual α-lipoprotein particles. Alternatively, estrogen-induced increase of cholesterol 'binding' by α-lipoprotein-protein (apolipoprotein A) could explain the compositional changes observed in α-lipoprotein.

The possible significance for atherogenesis of such estrogen-induced changes in the concentration and composition of α-lipoproteins and low density lipoprotein particles is not clear. It is evident, however, that apolipoprotein A is importantly involved in the transport and metabolism of triglyceride-rich very low density lipoproteins (FURMAN, SANBAR, ALAUPOVIC, BRADFORD and HOWARD [12]). It is possible that estrogen stimulates production of apolipoprotein A and thereby enhances transport of cholesterol in the form of α-lipoproteins and transport of triglyceride in the form of VLD-lipoproteins. Estrogen-induced increases in the concentration of serum triglycerides, i.e., triglyceride-rich VLD-lipoproteins, may be of little significance for atherogenesis if the triglyceride-rich lipoprotein particles are very large (i.e., above the upper limit of an as yet undetermined range of particle sizes critical to atherogenesis) and contain relatively little cholesterol, as previously noted (GUSTAFSON, ALAUPOVIC and FURMAN [13]).

The effect of androgens on serum lipid and lipoprotein patterns is generally opposite to that noted during estrogen administration (Figures 5a, b). Androgens diminish the lipid content of α- and VLD-lipoprotein fractions of serum, whereas these amounts are increased by estrogen. Although the effect of androgens on β-lipoproteins is not constant, β-lipoproteins are increased more often than decreased (FURMAN and HOWARD [14]). Since androgens lower the concentration of phospholipid-rich α-lipoproteins, serum phospholipid levels diminish relatively more than do serum cholesterol levels, and C/P ratios increase. An important effect of androgen administration on serum lipids and lipoproteins is the decrease in the concentration of serum triglycerides (VLD-lipoproteins), an effect which is most evident in subjects with hyperglyceridemia ('hyperlipemia') in whom androgen may bring about substantial and sustained lowering of serum lipid levels [9].

As might be anticipated, the effect of androgens on the composition of 'purified' α-lipoprotein particles appears to be opposite to that of estrogens. Inasmuch as the cholesterol content of the d 1.063–

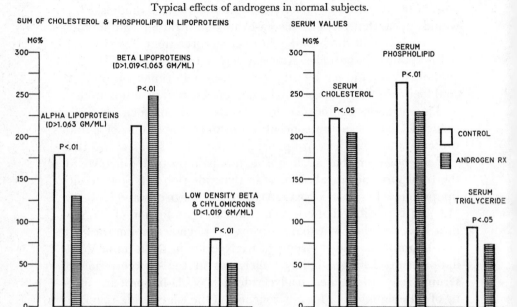

Fig. 5a. Typical effects of androgen administration on serum lipids and lipoproteins. (Mean values in five normolipemic men treated with methyltestosterone.) In contrast to estrogens, androgens diminish the lipid content of a-lipoprotein (principally phospholipid) and serum triglyceride levels. (Serum triglycerides circulate principally as $d<1.019$ g/ml lipoproteins.) Beta-lipoprotein cholesterol and phospholipid content usually increase. Because the changes in the cholesterol content of a- and β-lipoprotein cholesterol are divergent, the net change in serum cholesterol is relatively slight. Serum phospholipid levels diminish significantly (a-lipoprotein has high phospholipid content) and the serum cholesterol/phospholipid ratio increases. (Transitional values excluded.)

1.210 g/ml (a-lipoprotein) fraction is decreased by androgen administration, while the cholesterol/protein ratio is increased, it would appear that androgens decrease the concentration of a-lipoprotein particles in the serum while increasing the cholesterol content of the individual 'particles'. Studies in dogs in the authors' laboratory (CRIDER, BRADFORD and FURMAN [15]) have indicated that methyltestosterone administration diminishes the incorporation of C^{14}-lysine into apolipoprotein A. Androgen-induced reduction in apolipoprotein A synthesis would diminish the transport of triglyceride in the blood, where it exists mostly in the form of VLD-lipoprotein.

The parallelism between the changes in the concentrations of a- and VLD-lipoproteins (i.e., they vary together), resulting from either estrogen or androgen administration, is of considerable interest and directs our attention again to the likelihood that a-lipoprotein

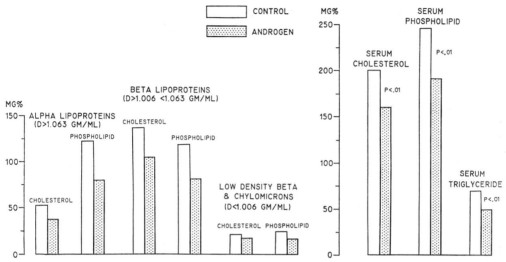

Fig. 5b. Effect of methyltestosterone, 75 mg/day on serum and lipoprotein lipids in a hypogonadal man (ADB). The lipid content of all three lipoprotein fractions was diminished by treatment, accounting for a statistically significant decrease in serum cholesterol, phospholipid and triglyceride levels. A decrease in β-lipoprotein lipid during androgen administration is observed less often than an increase or no change. (Methyltestosterone administered 313 days. Transitional values excluded.)

protein (apolipoprotein A) plays an important role in the metabolism of serum triglycerides and their transport as VLD-lipoproteins.

In 1964 reports appeared of an effect of estrogen on the relative amounts of $C^{14}O_2$ produced in human subjects administered C^{14}-glucose labeled in the C-1 or C-6 position (GORDON and GOLDBERG [16]). It was stated that ovarian activity during the normal menstrual cycle, or the administration of 'natural estrogens' (GORDON [17]), but not stilbestrol, to estrogen-deficient women, enhanced $C^{14}O_2$ production from glucose-1-C^{14} over that produced from glucose-6-C^{14}. This was interpreted as a stimulating effect of estrogen on the activity of the hexose monophosphate 'shunt', on the assumption that the ratio of glucose-1-C^{14} to glucose-6-C^{14} oxidation provides a measure of shunt activity.

These reports suggested to us the possibility that the increased serum triglyceride levels noted following estrogen administration might be, in part at least, a manifestation of increased lipogenesis

Effects of methyl testosterone 75 mg daily on the percent composition of serum lipoproteins in hypogondal male ADB.

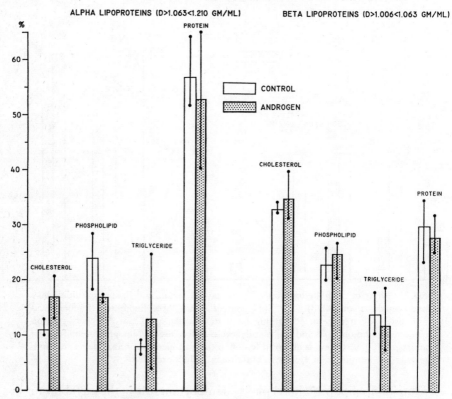

Fig. 6. Effects of methyltestosterone administration, 75 mg/day, on the composition of 'purified' α- and β-lipoproteins in hypogonadal male (same study as depicted in figure 5b). Compare with Figure 4. The principal relative changes are increased cholesterol and triglyceride, diminished phospholipid and protein in α-lipoprotein. Little change was noted in the relative amounts of lipid and protein in β-lipoprotein. (No analyses of VLD-lipoproteins were undertaken in this study. Transitional values excluded. Sera were obtained for study only after serum lipid values had remained stable over periods of several weeks. Period of methyltestosterone administration, 313 days.)

from carbohydrate secondary to estrogen stimulation of hexose monophosphate shunt activity. Accordingly, we undertook studies of the effect of estrogen administration to normal and hypogonadal subjects on the oxidation of glucose-1-C^{14} and glucose-6-C^{14}. This was accomplished by comparing the per cent of administered C^{14} radioactivity recovered as $C^{14}O_2$ during 60 and 150 minute periods following intravenous glucose-1-C^{14} or glucose-6-C^{14} administration. The results are tabulated in Table I.

Table I

	No. of studies	Ratio, $C^{14}O_2$ from glucose-1-C^{14} to $C^{14}O_2$ from glucose-6-C^{14}*	
		60 min	150 min
Premenopausal			
1. K.A. (no treatment)	1	2.21	1.66
2. L.H. (no treatment)	1	1.00	0.77
3. J.McL. (no treatment)	1	1.27	1.41
4. F.S. (no treatment)	1	1.42	1.15
Hypogonadal			
1. V.B.			
Control mean of	5	0.83	0.70
Estrone sulfate (2.5 mg/day)	1	1.35	1.33
Estrone sulfate (25 mg/day)	1	1.06	0.99
Diethylstilbestrol (10 mg/day)	1	0.92	0.42
Diethylstilbestrol (20 mg/day)	1	0.77	0.70
Control	1	2.18	2.72
2. G.L.			
Control mean of	3	1.30	1.15
Premarin® (10 mg/day) mean of	4	1.26	0.96
Premarin® (20 mg/day) mean of	3	1.20	0.94
Postmenopausal			
1. M.B.			
Control (menopausal) mean of	2	1.50	1.19
Premarin® (5 mg/day)	1	1.05	0.96
Premarin® (10 mg-day)	1	0.92	0.73
Estradiol benzoate (1 mg IM 4×weekly)	1	1.98	1.29
2. K.L.			
Premarin® (10 mg/day) mean of	2	1.08	0.82
Control	1	1.65	1.30
Estradiol benzoate (1 mg IM 4×weekly)	1	1.41	1.16
3. L.W.			
Premarin® (5 mg/day) mean of	2	1.93	1.16
Control	1	1.34	1.20
Male			
1. P.G.			
Control	1	1.38	1.12
Estradiol benzoate (1 mg IM 4×weekly)	1	1.11	1.11
2. R.B.	1	1.64	1.14

* Ratio of percent of injected C^{14} radioactivity recovered as $C^{14}O_2$ following glucose-1-C^{14} injection to that recovered following glucose-6-C^{14} injection.

The data in Table I fail to confirm the observations of GORDON and GOLDBERG and do not suggest that estrogen enhances the glucose-1-C^{14}/glucose-6-$C^{14}CO_2$ ratio, or that the glucose-1-C^{14}/glucose-6-C^{14} ratio differs between premenopausal women and hypogonadal women, or that men differ from women in this respect.

Discussion

The changes induced by gonadal hormones in serum lipid and lipoprotein patterns, and in the composition of individual lipoprotein particles, are indeed striking, yet their significance in the genesis of human atherosclerosis is obscure. There are as yet no studies of the effect of estrogens on the formation of apolipoprotein A. If estrogens were found to enhance apolipoprotein A synthesis or availability, such an effect would enhance serum lipoprotein stability and attenuate atherogenesis.

If estrogens afford some degree of protection against the development of atherosclerosis, or death from myocardial infarction, such protection would appear to be manifest mainly in association with some factor or factors inherent in a so-called 'affluent' society and largely limited to its white members. In view of the absence of a 'menopause effect' on coronary artery disease mortality in white women in the U.S., if one assumes that estrogens play a 'protective' role in this disease, very small amounts of hormone must suffice.

If, alternatively, one takes the position that men are coronary disease-prone, the possible etiologic role of androgen is suggested by experimental studies [15] indicating that testosterone inhibits synthesis of apolipoprotein A, the protein moiety of a-lipoproteins and a component of VLD (d <1.006 g/ml) lipoproteins [18]. The characteristic lowering of a- and VLD-lipoprotein concentrations and the changes in their composition during testosterone administration probably are a result of diminished availability of apolipoprotein A. Acknowledging the primary importance of the serum lipids, particularly their physical state, in atherogenesis, one might be persuaded that androgen-induced diminution of apolipoprotein A availability, and the resultant diminution of serum lipoprotein stability, would favor atherogenesis.

In conclusion, one can say only that there is no ready explanation for the lower mortality rate and incidence of clinical manifestations in respect to atherosclerotic heart disease among women, principally white women in populations 'enjoying' a high standard of living. It can be said with confidence that the diminishing ratio of male/female death rates which characterizes mortality from arteriosclerotic heart disease beginning in the middle of the fifth decade of life is due to a slowing of the rate of acceleration of mortality in men because of the decrease in the size of the pool of heart attack candidates which is a consequence of the very high coronary artery disease mortality among relatively young men.

There is no evidence that the menopause *per se* contributes significantly to the decrease in the ratio of male/female mortality rates which is first apparent in the fifth decade.

It is not evident how the effects of gonadal hormones on serum lipids and lipoproteins can account for the lower attack rate and mortality rate noted in white women. It is possible, however, that by influencing apolipoprotein A synthesis, especially under certain dietary or other environmental conditions, these hormones may be responsible for differences in the inherent stability of the serum lipoprotein transport system, which, over several decades of life, result in a significant sex difference in the manifestations of the disease.

Finally, one must keep in mind that genetic factors may endow men with a vulnerability to atherogenesis which is greater than that of women. There are several diseases, exclusive of those involving the genito-urinary system, in which a sex difference is evident, although the disorders do not appear to be influenced by, or dependent on, gonadal function. Gout is a classic example. Like atherosclerosis, however, these disorders remain an etiologic enigma.

References

1. OLIVER, M.F.: In: Pathogenesis and treatment of occlusive vascular disease (Lippincott, Philadelphia 1960).
2. TRACY, R.E.: That estrogens may have nothing to do with coronary heart deaths. Fed. Proc. 25, I: 665 (1966).
3. MILLOT, J. and DAUX, J.L.: Influence de la ménopause précoce, naturelle ou chirurgicale sur le déclenchement des coronarites. Arch. Mal. Cœur 52: 297 (1959).
4. MASTER, A.M.; DACK, J. and JAFFE, H.L.: Age, sex, and hypertension in myocardial infarction due to coronary occlusion. Arch. intern. Med. 64: 767 (1939).
5. KEIL, P.G. and McVAY, L.V., Jr.: A comparative study of myocardial infarction in white and Negro races. Circulation 13: 712 (1956).
6. LOBER, P.H.: Pathogenesis of coronary sclerosis. Arch. Path. 55: 357 (1953).

7. GOODALE, F.; THOMAS, W.A. and O'NEAL, R.M.: Myocardial infarction in women. Arch. Path. *69:* 599 (1960).
8. THOMAS, W.A.; BLACHE, J.O. and LEE, K.T.: Race and the incidence of acute myocardial infarction. Arch. intern. Med. *100:* 423 (1957).
9. FURMAN, R.H.; ALAUPOVIC, P. and HOWARD, R.P.: Effects of androgens and estrogens on serum lipids and the composition and concentration of serum lipoproteins in normolipemic and hyperlipidemic states. Progr. biochem. Pharmacol. *2:* 215–249 (Karger, Basel/New York 1967).
10. FURMAN, R.H.; HOWARD, R.P.; LAKSHMI, K. and NORCIA, L.N.: The serum lipids and lipoproteins in normal and hyperlipidemic subjects as determined by preparative ultracentrifugation. Effects of dietary and therapeutic measures. Changes induced by *in vitro* exposure of serum to sonic forces. Amer. J. clin. Nutr. *9:* 73, (1961).
11. ALAUPOVIC, P.; HOWARD, R.P. and FURMAN, R.H.: Effects of estrogens and androgens on the α- and β-lipoprotein composition in human subjects. Circulation *28:* 647 (1963).
12. FURMAN, R.H.; SANBAR, S.S.; ALAUPOVIC, P.; BRADFORD, R.H. and HOWARD, R.P.: Studies of the metabolism of radioiodinated human serum alpha lipoprotein in normal and hyperlipidemic subjects. J. lab. clin. Med. *63:* 193 (1964).
13. GUSTAFSON, A.; ALAUPOVIC, P. and FURMAN, R.H.: Studies of the composition and structure of serum lipoproteins: Isolation, purification and characterization of very low density lipoproteins of human serum. Biochemistry *4:* 596 (1965).
14. FURMAN, R.H. and HOWARD, R.P.: Influence of gonadal hormones on serum lipids and lipoproteins: Studies in normal and hypogonadal subjects. Ann. intern. Med. *47:* 969 (1957).
15. CRIDER, Q.E.; BRADFORD, R.H. and FURMAN, R.H.: Effect of methyltestosterone on the protein moiety of canine serum alpha-lipoproteins. Circulation *32:* 8 (1965).
16. GORDON, E.S. and GOLDBERG, M.: Carbon-14 studies of energy metabolism in various thyroid states. Metabolism *13:* 591 (1964).
17. GORDON, E.S.: Carbon-14 studies of glucose and fatty acid oxidation in human subjects. In: Sym. Resp. Pattern Analysis Intermediary Metab. Study, p. 43 (1964).
18. GUSTAFSON, A.; ALAUPOVIC, P. and FURMAN, R.H.: Studies of the composition and structure of serum lipoproteins: Separation and characterization of phospholipid-protein residues obtained by partial delipidization of very low density lipoproteins of human serum. Biochemistry *5:* 632 (1966).

Authors' address: R. H. Furman, M.D.; P. Alaupovic, Ph.D.; R.H. Bradford, Ph.D., M.D., and R.P. Howard, M.D., Oklahoma Medical Research Foundation, *Oklahoma City, Okla.* (USA).

Biologic Half-Life of Lipoprotein Triglyceride in the Normal and Diabetic Rabbit

R. L. Hirsch, New York, N.Y.

The biologic half-life of lipoprotein triglycerides (TG) in the plasma of the rabbit was measured in control and diabetic rabbits. Rabbits were rendered diabetic by the intravenous administration of alloxan monohydrate. Lipoprotein triglycerides were injected as hyperlipemic plasma obtained from donor rabbits rendered hyperlipemic by the parenteral administration of either alloxan (ADH plasma) or an alkaline aqueous extract of hog pituitary glands (PE plasma). The plasma was injected at a constant rate for 30–80 min. Plasma TG concentration was measured at intervals during the infusion and for 2 h after the infusion was abruptly stopped. Plots of plasma TG concentration *vs* time had the characteristics of first order kinetics. The biologic half-life obtained from the decay portion of the curves ranged from 20–25 min for ADH plasma and 18–40 min for PE plasma when control animals were infused. The half-life was independent of rate of input, total TG infused and the height of the plasma TG concentration reached by the end of the infusion. In contrast the half-life of ADH plasma infused into diabetic rabbits ranged from 44–150 minutes. The volume of distribution of the injected TG was shown to be the same in control and diabetic rabbits. Regular insulin was given to diabetic rabbits, either as a single injection of 5–40 units, or in daily doses from 3 days to several weeks prior to the start of the infusion. In each group, the half-life remained elevated in about $1/2$ of the animals.

Author's address: R. L. Hirsch, New York Blood Center, 310 East 67 Street, *New York, N. Y.* (USA).

Effects of Gonadal Hormones on Cholesterol Metabolism in Rat

S. Mukherjee, S. Gupta and A. Bhose, Calcutta

The effects of exogenously administered sex hormones on serum cholesterol level in man and experimental animals have featured prominently in many recent studies but the exact mechanism by which these effects are brought about remained speculative to a large extent. Using intact, gonadectomized and hormone-supplemented rats, maintained on fat-free, cholesterol and fat-fed diets, an attempt has been made to study the effects of gonadal hormones on the regulation of rates of biosynthesis and degradation of cholesterol in rat liver and their inter-relationship with cholesterol content of serum. Both liver slices and liver microsomes in presence of dialyzed soluble enzymes have been employed to determine the rate of incorporation of acetate-1-C^{14} into cholesterol and liver mitochondrial preparations have been used to study the degradation of 4-C^{14}-cholesterol to bile acids.

Females have normally higher rates of synthesis than males on all experimental diets. Gonadectomy of both sexes resulted in decreased biosynthesis of cholesterol and hormone replacement reversed the effects of gonadectomy. Higher rates of synthesis observed in ovariectomized rats given estrogen supplements suggest a probable return to normal condition of the animal rather than a stimulatory effect of estrogen on cholesterol biosynthesis. Moreover, if estrogen is administered to intact males marked inhibition of synthesis is obtained. In cholesterol-fed animals, the hormonal influences are masked to a great extent by inhibition of sterol synthesis by exogenously administered cholesterol. Synthesis is also considerably depressed in the absence of any dietary fat.

Conversion of cholesterol to bile acids is significantly higher in females as compared to males, whether or not fat or cholesterol is present. The catabolism of cholesterol is however maximum when cholesterol is present in the diet of the animals. Orchidectomy results in appreciable increase in bile acid formation, the effect being more pronounced among cholesterol-fed rats. Hormone replacement therapy does not however reverse the effects in gonadectomized males. Ovariectomy, on the other hand, causes significant lowering of cholesterol catabolism and estrogen supplementation counteracts this effect. Further, catabolic rates in orchidectomized rats are comparable to those in intact females; similarly, estrogen-treated males have rates of degradation alike those of intact females.

Our studies further reveal that serum cholesterol level in rats cannot be satisfactorily explained merely in terms of hepatic synthesis and breakdown of cholesterol. Evidences of mobilization of cholesterol from liver by estrogen treatment may point to similar redistribution of sterol in other tissues following hormone withdrawal or its supplementation.

Authors' address: S. Mukherjee, D. Sc.; S. Gupta, M. Sc. and A. Bhose, Laboratories of Lipid Research Department of Applied Chemistry, University of Calcutta, *Calcutta* (India).

From the Cardiovascular Institute, and the Division of Cardiovascular Disease, Department of Medicine, Michael Reese Hospital and Medical Center, Chicago, Illinois.

Estrogens and Atherosclerosis*

Ruth Pick, G.B. Clarke and L.N. Katz, Chicago, Ill.

Some degree of incidence of arteriosclerosis has been shown to be present in almost all animal species, from fishes to primates [9]. This state is so universal in animals living under such diverse conditions that it would seem that merely having an arterial system is all that is necessary for the development of arteriosclerotic changes. However, these changes are usually so slight that they do not interfere with the blood supply of the organs and, therefore, do not lead to organic disease. Also, these slight changes show no consistent pattern as far as sex incidence is concerned, but they do show an increase with advancing age of the animals.

In 20th century man arterial disease, particularly atherosclerosis of the coronary arteries has become, as is well-known, a major health problem leading to disability and premature death in middle-age, in most, if not all, countries of Western civilization, living on luxurious diets rich in calories, particularly calories from animal fats [2]. In these countries a definite pattern of a sex-difference is apparent, where women before the menopause are much less prone to develop coronary heart disease than men of the same age. These observations have given rise to extensive clinical and animal experimental studies, on the role that estrogens may have on this accelerated form of the disease.

In our laboratory, we have been studying the effect of estrogens on cholesterol-oil diet-induced plasma lipid changes and aortic and coronary atherosclerosis for more than a decade. These experiments, which have become more elaborate with the introduction of new techniques, have given us many leads and yet we are not even

* Supported by grants from the National Heart Institute (HE-06375), Public Health Service and the American Heart Association.

close to a true elucidation of the mechanisms by which estrogens protect the coronary arteries in cockerels from cholesterol-oil induced atherosclerosis. We have used several different natural, semisynthetic and synthetic estrogens, and so far all of them that were protective were also feminizing. Several estrogen-like compounds with low feminizing activity which we tested neither protected cockerels against coronary atherosclerosis nor produced the usual well-known plasma lipid changes, namely, elevation of phospholipids and therefore a reduction of the cholesterol/phospholipid (C/P) ratio to normal. For this reason, our routine procedure in studying estrogen effects in the laboratory has been to use conjugated equine estrogens (Premarin) in the drinking water of the birds.

The usual effect of the combination of cholesterol-oil feeding and estrogen treatment on plasma cholesterol, C/P ratio, thoracic and coronary atherosclerosis is shown in Figure 1. Figure 2 shows

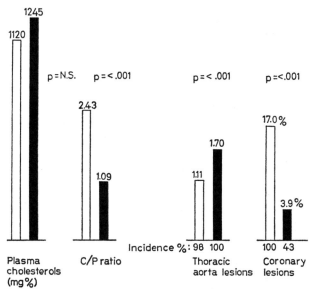

Fig. 1. Estrogen prophylaxis of coronary atherosclerosis in cholesterol-fed cockerels. *Open bars* are 1 CO controls; *solid bars*, 1 CO Premarin, 25 mg/chick/day.
1CO = 1% cholesterol and 5% cottonseed oil.
C/P ratio = $\frac{\text{cholesterol}}{\text{phospholipid}}$ ratio.
Incidence % = % of birds showing atheroma. The # on top of the bars for thoracic lesions is the mean severity of lesions graded grossly on a scale of 0 (no lesion) to 4+ (most severe lesions). The % on top of the bars for coronary lesions denotes vessels showing atheroma as % of all vessels counted in one section from each of two blocks of tissue (mean for group).

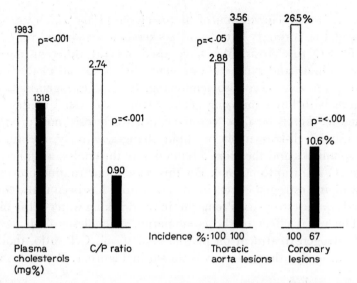

Fig. 2. Estrogen-induced reversal of coronary atherosclerosis in cholesterol-fed cockerels. *Open bars* are 1 CO for 12 weeks; *solid bars*, 1 CO for 12 weeks with estrogen added in last 5 weeks. Estrogens = Premarin, 25 mg/bird/day in the drinking water. Other symbols as in Figure 1.

the decrease of the C/P ratio and regression of coronary lesions in birds on a continuous atherogenic diet for 12 weeks with estrogens added for the last 5 weeks. The protective effect of estrogens on coronary atherosclerosis can be slightly but significantly counteracted by making the cockerels hypothyroid with thiouracil (Fig. 3), suggesting that normal thyroid function is involved in the estrogen-effect on coronary atherosclerosis. Also, as seen in Figure 4, 'blocking' the RES system with India ink decreases estrogen protection. Similar results are obtained, as seen in Figure 5, when either heparin or dicumarol is added to the diet containing estrogens. The effect with heparin is statistically significant, while that with dicumarol, even though in the same direction, does not reach statistical significance [5, 6, 7, 8].

Recently, we have become concerned with the effect of different doses of Premarin on plasma, liver and bile lipid composition, as well as on aortic and coronary atherosclerosis [1]. In these experiments all animals were fed the standard diet of chick-starter mash supplemented with 1% cholesterol and 5% cottonseed oil. Four dosage levels of Premarin were used, 12.5, 25, 50 and 75 mg/bird/

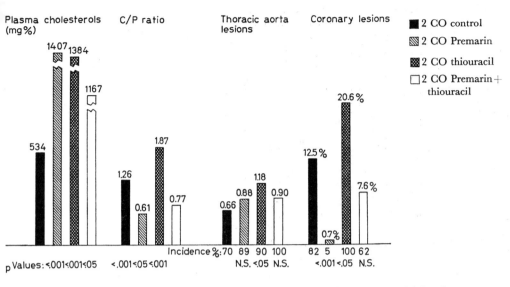

Fig. 3. Thiouracil counteraction of estrogen anti-atherogenesis in cholesterol-fed cockerels. Thiouracil: 0.1–0.2% in diet. Other symbols as in Figure 1. From PICK, R. *et al.*, Circulat. Res. *5:* 510–514, 1957 [8].

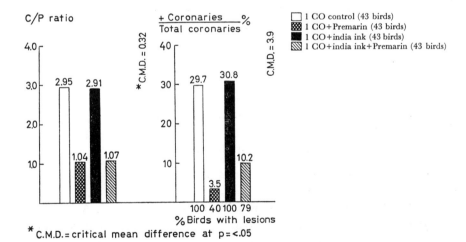

Fig. 4. Effect of estrogens and India ink given alone or in combination on plasma C/P ratio and coronary atherosclerosis in cholesterol-oil fed cockerels. India ink = 5 mg/100 gm of body weight/bird given one week before the start of the experiment, with the same dose repeated two weeks after the start of the experiment. Other symbols as in Figure 1. From PICK, R. *et al.*, Acta Cardiol. *21:* 133–144, 1966 [5].

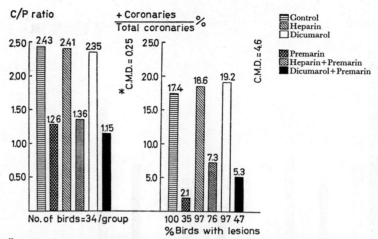

Fig. 5. Effect of heparin, dicumarol and estrogen alone or in combination on plasma C/P ratio and coronary atherogenesis in cholesterol-oil fed cockerels. Heparin = Depot heparin (Upjohn Co.) intramuscularly, 5–10 mg 5 days/week. Dicumarol (Abbott Laboratories) orally 30 mg/bird/day. Other symbols as in Figure 1. From PICK, R. *et al.*, J. Atheroscler. Res. *7:* 11–16 (1967) [6].

day. The plasma, bile and liver cholesterol levels, digitonin precipitable fecal sterols and plasma phospholipids were determined by the usual methods. Plasma fatty acid composition was determined by gas liquid chromatography.

The changes from the control noted in various parameters were classified as:

1. No effect,
2. an effect which is not dose-dependent,
3. a dose-dependent effect with a maximum at one or two dose levels,
4. a diphasic, dose-dependent effect.

Only the most significant results of these dose-relationships will be shown.

The effect on atherogenesis is illustrated in Figure 6. As can be seen, increasing doses of Premarin had a dose-dependent augmentatory effect on the protective action on the coronary arteries. Thoracic aorta lesions show a diphasic effect – an increase at the two lower doses (our usual dose level) with a decrease at higher dosages. Abdominal aorta lesions show the same trend, but are not significantly different from the control.

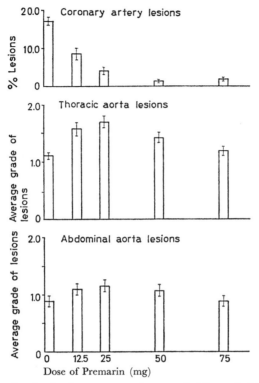

Fig. 6. Effect of various levels of estrogens on atherogenesis. The vertical lines in the bars are the standard error of the mean.

Figure 7 shows the effect on plasma and bile lipids. Increasing doses of Premarin showed a diphasic response on plasma cholesterol and bile cholesterol – but the effects in plasma and bile were in opposite directions. Liver cholesterol also showed a diphasic response parallel to the pattern of the bile cholesterol. The plasma phospholipids show the well-known effect of an increasing rise dependent on dose; and the C/P ratio, the concomitant fall.

The effects of estrogen dosage on plasma unsaturated fatty acids are seen in Figure 8. As can be seen, arachidonic acid showed a diphasic dose-dependent effect, a statistically not significant elevation at the lower dosages, a significant decrease at the two higher dosages. Linolenic and linoleic acids decreased significantly but the decrease is not dose-dependent.

It is possible that further analysis may reveal other, perhaps more complex interactions among the lipids in ester form in the plasma

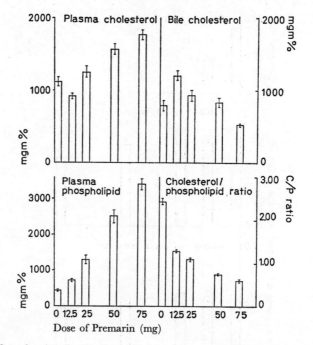

Fig. 7. Effect of various levels of estrogens on plasma and bile lipids. Symbols as in Figure 6.

and the several vascular beds. It would be valuable to know exactly in what form these fatty acids are in the plasma as far as their combination with glycerides, cholesterol and phospholipids are concerned, as well as their quantitative distribution among the several classes of lipoproteins.

The significance of these lipid changes in atherogenesis is not clear at the present time. They may only be part of the estrogen effect. Data are also available indicating a direct effect of estrogens on the tissues themselves. It is apparent from our studies that the action of estrogens is complex and the effect depends, at least, on the blood vessel examined, the animal species used, and the nature and amount of hormone employed.

Results from human studies testing the efficacy of estrogens on survival time of patients with proven coronary heart disease are contradictory. It appears that in two studies using conjugated equine estrogens a beneficial effect was observed [3, 10], while in studies using ethinyl estradiol no benefit of this treatment was noted [4]. This discrepancy is unexplained and deserves further investigation.

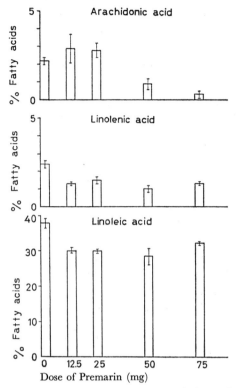

Fig. 8. Effect of various levels of estrogens on unsaturated plasma fatty acids. Symbols as in Figure 6.

Summary

In cockerels estrogens have profound and complex effects on atherogenesis in different vascular beds. They also have significant effects on lipid metabolism as demonstrated in their effect on plasma, liver and bile lipids.

References

1. CLARKE, G.B.; PICK, R.; JOHNSON, P. and KATZ, L.N.: Effect of estrogen dosage upon plasma, liver and bile lipids in cholesterol-fed cockerels. Circulat. Res. *19:* 564–570 (1966).
2. LILIENFELD, A.M.; ACHESON, R.; EPSTEIN, F.; GEARING, F.R.; GIBSON, T.C.; HIGGINS, I.T.T.; KLAINER, L.M.; SACKETT, D.L.; SCHUMAN, L.M.; SCHWEITZER, M.D.; STALLONES, R.A.; WHITE, K.L. and WINKELSTEIN, W., Jr.: Epidemiology of cardiovascular disease. The Heart and Circulation, vol. I/Research, pp. 220–225 (Federation of American Societies for Experimental Biology, Washington 1964).

3. Marmorston, J.; Moore, F.J.; Hopkins, C.E.; Kuzma, O.T. and Weiner, J.: Clinical studies of long-term estrogen therapy in men with myocardial infarction. Proc. Soc. exp. Biol. Med. *110:* 400–408 (1962).
4. Oliver, M.F. and Boyd, G.S.: Influence of reduction of serum lipids on prognosis of coronary heart disease. A five-year study using estrogen. Lancet *ii:* 499–505 (1961).
5. Pick, R.; Johnson, P.J.; Kakita, C. and Katz, L.N.: The role of the reticuloendothelial system (RES) on blood lipid alterations and atherogenesis in cholesterol-fed and estrogen treated cockerels. Acta Cardiol. *21:* 133–144 (1966).
6. Pick, R.; Kakita, C. and Katz, L.N.: The effect of heparin and bishydroxycoumarin (Dicumarol) on cholesterol-induced atherosclerosis in cockerels. J. Atheroscler. Res. *7:* 11–16 (1967).
7. Pick, R.; Stamler, J.; Rodbard, S. and Katz, L.N.: The inhibition of coronary atherosclerosis by estrogens in cholesterol-fed chicks. Circulation *6:* 276–280 (1952a). – Estrogen-induced regression of coronary atherosclerosis in cholesterol-fed chicks. Circulation *6:* 858–861 (1952b).
8. Pick, R.; Stamler, J. and Katz, L.N.: Effects of hypothyroidism on estrogen-induced inhibition of coronary atherogenesis in cholesterol-fed cockerels. Circulat. Res. *5:* 510–514 (1957).
9. Roberts, J.C. and Straus, R.: Comparative Atherosclerosis (Hoeber, New York 1965).
10. Stamler, J.; Pick, R.; Katz, L.N.; Pick, A.; Kaplan, B.; Berkson, D. and Century, D.: Effectiveness of estrogens for therapy of myocardial infarction in middle-age men. J. Amer. Med. Ass., *183:* 632–638 (1963).

Authors' address: Ruth Pick, M.D.; G.B. Clarke, Ph.D. and L.N. Katz, M.D. Cardiovascular Institute and Division of Cardiovascular Disease, Dept. of Medicine, Michael Reese Hospital and Medical Center, *Chicago, Ill.* (USA).

From the Institut de Biochimie Clinique, University of Geneva, Switzerland

Laboratory Animals with Spontaneous Diabetes and/or Obesity: Suggested Suitability for the Study of Spontaneous Atherosclerosis*

A.E. Renold, A.E. Gonet, W. Stauffacher and
B. Jeanrenaud, Geneva

The increasing prevalence of diabetes mellitus in man has come under intensive scrutiny in recent years, and genetic analysis suggests that upward of 5% of the population carries the predisposition to that disorder in a form likely to lead to its expression, while between 25 and 30% carry the predisposition in a non-expressed form, probably as a recessive gene. Epidemiologic studies have also demonstrated repeatedly that a close association exists between diabetes mellitus and obesity, either condition being much more frequent, in any given population, in the presence of the other. Finally, there can be no doubt but what the incidence of vascular disease in general, and of coronary atherosclerosis in particular, is greatly increased among diabetics, and also among obese subjects. Although these are all acknowledged and well-established facts, surprisingly little is known about the true and detailed nature of the links connecting these major diseases of civilized mankind.

The purpose of this paper is that of drawing attention to the fact that the combination of diabetes mellitus and obesity, and perhaps also of atherosclerosis, is not limited to man but is similarly found in laboratory animals. Since such syndromes in animals can be studied more readily than their human counterparts, in terms of tissue metabolism, overall metabolism, and genetics, it is clear that these syndromes deserve close attention and might well reward

* Supported in part by grants-in-aid from the Fonds National de la Recherche Scientifique (Grant 3618); The Fondation Emil Barell pour le développement des recherches médico-scientifiques; and through Diabetes Foundation Inc., Boston, Mass.

the investigator with better insight into the general problem of pathogenetic mechanisms involved in diabetes mellitus, in obesity, and in atherosclerosis.

As of the present, we are aware of 11 syndromes (Table I) in mice and other small rodents, syndromes characterized by diabetes mellitus and/or obesity. In some, such as in the chinese hamster (Cricetulus griseus), diabetes mellitus is the primary anomaly, and obesity is rarely, if ever, seen. In the majority of the syndromes, however, hyperglycemia is associated with a greater or lesser degree of obesity, and also with an increase, rather than a decrease of insulin-secreting cells in the islets of Langerhans, and also in circulating insulin. This had led us, and others, to assume that in that type of diabetic syndrome in laboratory mammals, the primary event is not a decrease in the insulinogenic capacity but rather decreased insulin effectiveness.

Table I. Spontaneous diabetes and/or obesity in small rodents maintained on standard laboratory diets. RENOLD et al. [17]

Mice	Hyper-glycemia	Obesity	Pancreatic β-cells	ILA or IRI[1] of serum
1. Yellow – O.[2] [1]	+	+	+	
2. O.-H.[2], Bar Harbor [2–4]	++	++	++	++
3. O.-H., Edinburgh [5]	+	++		
4. O., New Zealand [6]	+	++	++	++
5. K.K., Japan [7]	++	+	+	
6. Diabetic, Bar Harbor [8]	+++	+		
7. $C_3Hf \times I$, F-1, Wellesley [9]	+	+	+++	++
8. Spiny mice [10, 11]	++	++	+++	++
Hamsters and rats				
9. Chinese hamster [12, 13]	+++	(+)		
10. Sand rat [14, 15]	++	+	±	±
11. 'Fatty' rat, Waltham [16]	(+)	++		

[1] ILA = Insulin-like activity; IRI = Immunologically reactive insulin.
[2] O. = Obese; H. = Hyperglyaemic.

As an example of the type of metabolic studies to which these syndromes lend themselves, mention will only be made here of a study carried out by STAUFFACHER, in our laboratory, designed to test the hypothesis that congenitally diabetic and obese mice exhibit a relatively greater resistance to insulin of muscle than of adipose tissue, thereby explaining the disposal of a greater proportion of glucose and other substrates into fat [18, 19]. In this study, insulin was injected intraperitoneally in very small amounts, that is in amounts insufficient to affect overall body metabolism, but sufficient to alter the metabolism of tissues lining the peritoneal cavity, a method first introduced by RAFAELSEN and coll. [20]. Since both a muscle (the diaphragm) and a representative adipose tissue (epididymal adipose tissue) have a large peritoneal surface, it is possible to test simultaneously in any one animal the effect of the small intraperitoneally injected amount of insulin on both these tissues. A tracer amount of glucose, uniformly labelled with ^{14}C, is injected together with the insulin. The indices used for estimating the metabolic activity of the two tissues were those of incorporation of radioactivity from glucose into glycogen, in the diaphragm, and into total lipid, in epididymal adipose tissue. The conditions were so chosen that, in the presence of even minute concentrations of insulin, equal amounts of radioactivity were incorporated into diaphragm glycogen and epididymal adipose tissue lipid. Under these conditions, therefore, the ratio of counts in diaphragm to counts in adipose tissue was unity or, in the presence of higher concentrations of insulin, greater than unity.

The data shown in Table II clearly demonstrate the shift in this ratio, and therefore in the relative sensitivity of diaphragm and adipose tissue to insulin, in animals with genetically determined obesity. It should first be noted that in animals with another type

Table II. Diaphragm to adipose tissue ratio in three types of experimental obesity. Age of all animals: 8 weeks. STAUFFACHER et al. [18, 19]

Insulin dose (μU/ml)	New Zealand Obese	Bar Harbor O.-H.	Goldthioglucose Obese
0	0.03±0.01* (5)	0.01±0.003 (6)	0.08±0.04 (4)
300	0.03±0.01 (5)	0.01±0.003 (6)	0.98±0.27 (5)
3000	0.34±0.07 (5)	0.04±0.01 (5)	1.60±0.14 (6)
10000	1.47±0.25 (6)	0.06±0.03 (4)	2.00±0.2 (7)
30000	—	0.08±0.02 (5)	—

* Mean ± S.E.M. Number of animals in parentheses.

of obesity, purely secondary to over-eating, i.e. in goldthioglucose-obese mice (third column in Table II), the relative sensitivity of diaphragm and adipose tissue remained normal, with ratios of diaphragm to adipose tissue radioactivity at unity or well above it in the presence of all concentrations of insulin tested. In the genetically obese and hyperglycemic, however, the ratios remained well below unity for most concentrations of insulin tested, and the two strains differed, furthermore, in the severity of the anomaly so demonstrated. Whereas the 'resistance' to insulin of the diaphragm could be overcome, in the New Zealand obese animals, by concentrations between 3000 and 10000 μU of insulin per ml, even 30000 μU of insulin per ml barely increased the uptake of glucose into the diaphragm of the obese hyperglycemic animals of the Bar Harbor strain!

As yet relatively little is known about vascular disease in animals with spontaneous diabetes and/or obesity, although GONET, in our laboratory, has accumulated evidence on this point. He has found clear-cut evidence of both atherosclerosis and of arteriolosclerosis in diabetic spiny mice, as clearly seen in the example shown in Fig. 1, which shows an early atheromatous plaque in a coronary artery of a spiny mouse (Acomys cahirinus, No. 8 in Table I). There is evidence of fatty infiltration of the intima, with an as yet non-organized thrombus, with numerous lipophages, and with hypertrophy of the muscular layer. It should be stressed that the animal from which the section shown in Fig. 1 was obtained was not kept on any type of atherogenic diet, but on a normal laboratory diet. Spiny mice with diabetes also exhibit interesting changes of capillary basement membranes [11], which show some resemblance to the type of basement membrane lesion often seen in human diabetes.

Among the possible pathogenetic factors of atheromatosis, much interest has always centered on the lipid composition of blood plasma. It is perhaps of interest to note here that spiny mice with hyperglycemia often develop severe hyperlipemia and the type of fatty infiltration of liver, kidney and muscle characteristic of the syndrome which has been termed, by CARLSON, the lipid mobilization syndrome [22]. An example of the fatty liver seen in such animals is shown in Fig. 2, which is stained with Best's carmine, and which clearly indicates both the accumulation of lipid around the central veins and the peristence of the glycogen deposition in the portal areas.

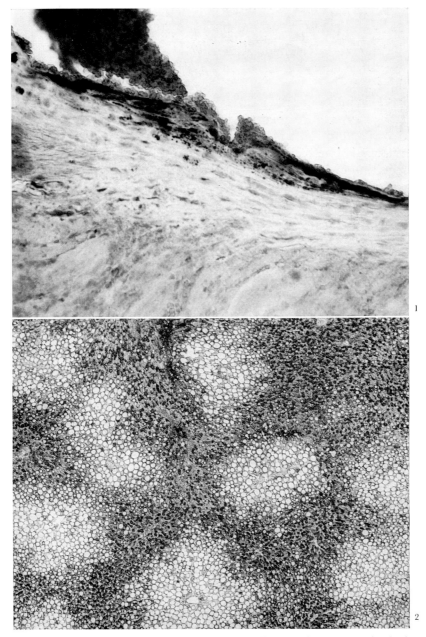

Fig.1. Atheromatous lesion of a coronary artery in a diabetic spiny mouse maintained on normal laboratory diet. Stained with Sudan III (Enlarged 350×).

Fig.2. Fatty infiltration of the liver in a diabetic spiny mouse maintained on normal laboratory diet. Stained with Best's carmine (Enlarged 52×).

Summary and Conclusions

Among small laboratory animals, 11 syndromes are now known which are characterized by hyperglycemia and/or obesity. All of these have been found by chance, not by systematic screening of large populations of laboratory animals. It is therefore likely that many more syndromes could be found. It is suggested that recognition of mutants with metabolic anomalies and diseases in laboratory animals is one of the major frontiers of medical research today, requiring major effort and organisation. The data collected by the authors suggest that strains exhibiting a diabetes and/or obesity might prove to be fruitful objects of study in the expoloration of the pathogenesis of atherosclerosis.

References

In October, 1966, the first international Symposium on spontaneous diabetes in animals will be held in Kalamazoo, Michigan. All of the presented papers will be published in the March issue of Diabetologia.

1. WEITZE, W.: Hereditary adiposity in mice and the cause of this anomaly (Store Nordiske Videnskabsboghandel, Kopenhagen 1940).
2. MAYER, J.: The obese hyperglycemic syndrome of mice as an example of 'metabolic' obesity. Amer. J. clin. Nutr. *8:* 712 (1960).
3. CHRISTOPHE, J.: Le syndrome récessif obésité-hyperglycémie de la souris. Ses relations possibles avec le diabète gras humain. Bull. Acad. roy. Méd. Belg. *5:* 309 (1965).
4. HELLMAN, B.: Studies in obese-hyperglycemic mice. Ann. N.Y. Acad. Sci. *131:* 541 (1965).
5. FALCONER, D.S. and ISAACSON, J.H.: Adipose, a new inherited obesity of the mouse. J. Hered. *50:* 6 (1959).
6. SNEYD, J.G.T.: Pancreatic and serum insulin in the New Zealand strain of obese mice. J. Endocrin. *28:* 163 (1964).
7. NAKAMURA, M.: A diabetic strain of the mouse. Proc. Japan-Acad. *38:* 348 (1942).
8. HUMMEL, K.P.: Personal communication.
9. LIKE, A.A.; STEINKE, J.; JONES, E.E. and CAHILL, Jr., G.F.: Pancreatic studies in mice with spontaneous diabetes mellitus. Amer. J. Path. *46:* 621 (1965).
10. GONET, A.E.; STAUFFACHER, W.; PICTET, R. and RENOLD, A.E.: Obesity and diabetes mellitus with striking congenital hyperplasia of the islets of Langerhans in spiny mice (Acomys Cahirinus). Diabetologia *1:* 162 (1966).
11. PICTET, R. et GONET, A.E.: Cellules mixtes (exocrines et endocrines) dans le pancréas de la souris à piquants Acomys Cahirinus. C.R. Acad. Sci. (Paris) *262:* 1123 (1966).
12. MEIER, H. and YERGANIAN, G.A.: Spontaneous hereditary diabetes mellitus in Chinese hamster. I. Pathological findings. Proc. Soc. exper. Biol. Med. *100:* 810 (1959).
13. YERGANIAN, G.A.: Spontaneous diabetes mellitus in the Chinese hamster, Cricetulus griseus. IV. Genetic aspects. Ciba Found. Coll. Endocrinology *15:* 25 (1964).

14. SCHMIDT-NIELSEN, K.; HAINES, H.B. and HACKEL, D.B.: Diabetes mellitus in the sand rat induced by standard laboratory diets. Science *143:* 689 (1964).
15. HACKEL, D.B.; FROHMAN, L.; MIKAT, E.; LEBOVITZ, H.E.; SCHMIDT-NIELSEN, K. and KINNEY, T.D.: Effect of diet on the glucose tolerance and plasma insulin levels of the sand rat (Psammomys obesus). Diabetes *15:* 105 (1966).
16. ZUCKER, T.F. and ZUCKER, L.M.: Hereditary obesity in the rat associated with high serum fat and cholesterol. Proc. Soc. exp. Biol. Med. *110:* 165 (1962).
17. RENOLD, A.E.; STAUFFACHER, W. and GONET, A.E., in Advances in Metabolic Disorders, vol. 3 (Academic Press New York, in press).
18. STAUFFACHER, W.; CROFFORD, O.B.; JEANRENAUD, B. and RENOLD, A.E.: Comparative studies of muscle and adipose tissue metabolism in lean and obese mice. Ann. N.Y. Acad. Sci. *131:* 528 (1965).
19. STAUFFACHER, W.; JEANRENAUD, B. et RENOLD, A.E.: Métabolisme du glucose dans le tissue adipeux et le muscle des animaux présentant une obésité par hyperphagie. Acta clin. belg. (in press).
20. RAFAELSEN, O.; LAURIS, V. and RENOLD, A.E.: Localized intraperitoneal action of insulin on rat diaphragm and epididymal adipose tissue *in vivo*. Diabetes *14:* 19 (1965).
21. GONET, A.E.: Unpublished observations.
22. CARLSON, L.A.: Recent advances in the metabolism of plasma lipids. Progr. biochem. Pharmacol. *4:* pp. 170–178 (Karger, Basel/New York 1967).

Authors' address: A. E. Renold, M.D.; A. E. Gonet, M.D.; W. Stauffacher, M.D. and B. Jeanrenaud, M.D., Institut de Biochimie Clinique, University of Geneva, *1211 Geneva 4* (Switzerland).

From the Research Laboratory, the Memorial Hospital, Worcester, Massachusetts

Survival after Cerebral Thrombosis

R.W. ROBINSON, Worcester, Mass.

A study of survival after a first attack of cerebral thrombosis in 921 patients first observed between 1947 through 1956 was completed. This allowed an analysis of survival ranging from 8 to 15 years, with an average of 12 years. The immediate mortality was 21%. Of the 725 patients who survived the initial attack 94% were traced in 1964. 17% died in the first year. 50% were dead in 3.7 years. The 5 year mortality was 63%. At 15 years there were still 2% surviving in contrast to 47% in the age and sex matched United States white population. 85% of the deaths were due to vascular disease.

148 patients with cerebral thrombosis first observed between 1957 and 1962 were submitted to the estrogen study group: 88 treatment and 66 control cases. The male patients received 5 mg and the women 15 mg stilbesterol daily. Definite estrogenic lipid effects were produced by stilbesterol in these patients. Life table analysis shows that resistance to death is the same in the first year. As the survival time increases the estrogen treated group experiences a somewhat lower mortality rate. After treatment is started, age is of no consequence. Segregation of the data according to sex, indicates that the treatment of females is definitely more successful. At 5, 6 and 7 years, the difference between the groups was twice the standard error. In males estrogen produced no significant change in survival over the control group. Women had a 3 to 1 increased rate of survival at 5, 6 and 7 years as compared to the controls, based on the actual data (chi square test).

Author's address: R.W. Robinson, Research Laboratory, The Memorial Hospital, *Worcester, Mass.* (USA).

Institute of Pharmacology, University of Milan, Milan, Italy

On the Effect of Norepinephrine and Nicotinic Acid on Free Fatty Acid Transport and Incorporation in Tissue Lipids

A. Sólyom and L. Puglisi, Milan

In short-term *in vivo* experiments we have investigated the effects of norepinephrine (NE) and nicotinic acid (NAc) on the incorporation of palmitate-1-^{14}C into liver and plasma lipids. In order to obtain information on how the incorporation of free fatty acids (FFA) into tissue lipids is related to FFA mobilization, four groups of rats were treated with: I. saline, II. NAc, III. NE or IV. NE+NAc, 20 min before the intravenous injection of labeled palmitate. The animals were killed 5, 20 or 40 min after palmitate injection.

The difference in plasma FFA levels between the experimental groups was constant during the period investigated. NAc significantly reduced and NE increased plasma FFA concentrations, while treatment with NE+NAc resulted in values similar to those of controls. No change was found in liver and plasma triglyceride concentrations.

In the liver the total lipid radioactivity did not differ significantly in the four experimental groups. After 5 min palmitate was already incorporated, in great part, into phospholipids (PL) and triglycerides (TG). The PL radioactivity was nearly the same in all groups at 5 min. However, the results obtained at 20 and 40 min showed an inverse correlation with the levels of plasma FFA. NAc treatment caused higher and NE lower values than those obtained in control or NE+NAc treated animals. On the other hand, there was a direct correlation between FFA mobilization and incorporation into liver TG, which could be observed 5 minutes after palmitate injection. The highest TG radioactivity was found after NE, the lowest after NAc, while NE+NAc treatment resulted, like the controls, in intermediate values. The PL:TG radioactivity ratios in the liver were closely and inversely related to FFA mobilization.

In plasma there was no close relationship between TG radioactivity and FFA concentrations. Not only NAc, but also NE and NE+NAc treatment caused lower TG radioactivity than that of controls. These results, as well as the decreasing TG radioactivity in the liver after the first 5 minutes, which is due possibly to TG transport from liver to plasma, suggest an effect of NE on TG metabolism independent of its effect on FFA mobilization.

It has been concluded that FFA mobilization has a direct effect on TG synthesis in the liver and that reducing plasma FFA levels may result in decreasing liver and plasma TG concentrations without affecting directly PL synthesis in the liver.

Authors' address: A. Sólyom, M.D. and L. Puglisi, Ph. D., Institute of Pharmacology, University of Milan, *Milan* (Italy).

Discussion Session 7

L.A. CARLSON (Stockholm): It was very interesting to hear the story of Dr. RENOLD's mice. Apparently the pathologic-anatomic picture we have is very similar to the one we see within 24 hours after increasing the mobilization of FFA in dogs about 5 times. To what extent and for how a long time have your mice had increased FFA mobilization?

A.E. RENOLD (Geneva): Our observations have been limited so far to animals killed two to six weeks after the onset of diabetes with catabolism. Preliminary findings suggest that at least the fatty liver may be seen much earlier.

J. STAMLER (Chicago, Ill.): Dr. FURMAN has presented a most intriguing and though provoking analysis of the U.S. coronary heart disease mortality data on men and women. Based on it, he has posed significant questions concerning the concept that estrogens play a fundamental role in accounting for the male-female difference. I would like to sound a note of caution. As useful as the mortality data are, they have their limitations. These result from the nature of the information put on the death certificate by physicians, and from the method of coding that information. In view of the limited precision of mortality statistics, great caution is indicated in their interpretation.

With respect to the mortality of U.S. white males and females for the year 1960, the following are the rates for arteriosclerotic heart disease for white males ages 35–39, 40–44, 45–49 and 50–54 respectively: 50.1, 124.2, 254.9 and 462.6 per 100,000. The rate for the men age 40–44 is 2.5 times that for the men 35–39. The rate for the men age 45–49 is 2.1 times that of the men age 40–44. The rate for the men age 50–54 is 1.8 times that for the men 45–49. The corresponding rates for white females age 35–39, 40–44, 45–49 and 50–54 were: 6.5, 19.4, 40.8 and 85.6 per 100,000. The rate for the women age 40–44 was 3.0 times that for the women 35–39. The rate for the women 45–49 was 2.1 that of the women age 40–44. The rate for the women 50–54 was 2.1 times that of the women 45–49. The sex differentials are sizeable for all these age groups, being 7.7, 6.4 and 5.4 respectively. They are slightly lower in the older than in the younger age groups, because the female mortality rate is increasing more steeply than the male, or (vice versa) the male rate is increasing less markedly than the female.

My main doubt and uncertainty relates to the fact that the women 40–44 had a threefold increase compared with the women 35–39, whereas the men of similar age had a 2.5-fold increase. For the sake of argument, might it not be inferred that the greater increase in rate for the women, compared with the men – in the crucial age period at which many women experience the menopause – reflects the slackening off of the protective effect of estrogenic hormones? I hasten to add, however, that the difference between a 3.0 and a 2.5-fold increase in rate is small. I find it difficult to draw any clean theoretical conclusion with respect to sex hormones from these data, but – to repeat – is it not possible to infer that the more marked gradient for the females could be attributed to the decrease estrogenic protection?

Time does not permit an extended discussion of the varied patterns of the sex differential in U.S. Negroes, and in populations of the economically less developed countries, compared with U.S. whites. Most of our thoughts on the possible reasons for these phenomena are in the literature – cf. STAMLER, J.J. nat. med. Ass., N.Y. *50:* p. 161 (1958); Postgrad. Med. *25:* pp. 610, 685 (1959); J. chronic Dis. *12:* pp. 440, 456, 464 (1960); Geriatrics, *16:* p. 382 (1961); Amer. J. Cardiol. *10:* p. 319 (1962); *Atherosclerosis and Its Origin*, p. 231 (Academic Press, New York 1963); *Preventive Cardiology* (Grune and Stratton, New York 1966).

Suffice it here to note that the most intriguing phenomenon is the decidedly higher coronary mortality rates for middle-aged Negro women, compared with white, resulting in a lower Negro sex differential. Again, it is appropriate first to emphasize the need for caution. There are serious problems of validity with regard to Negro mortality rates. Thus, the data on age of deceased Negroes are subject to significant error. Moreover, an inordinately high percent of Negro deaths attributed to coronary disease are signed out not by physicians who render medical care, but by coroners' physicians or medical examiners.

With these uncertainties, it is not possible at this juncture to state to what extent the apparently high coronary mortality rates in Negro women are due to such phenomena as the much greater prevalence in Negro women compared with white of hypertension, or severe obesity or premature oophorectomy. In all likelihood, every one of these factors plays a role. Further work is needed in this area, especially prospective living population studies. In any case, I do not believe that one can utilize these data to arrive at any conclusion with respect to estrogens and their role, or lack thereof, in accounting for the sex differential. These data on race differences in the sex ratio of coronary mortality pose important questions, but cannot answer them: If estrogens play a significant role in the sex differential, why is the differential less for Negroes than for whites? If estrogens do not play a significant role, what accounts for the sex differential, and again why is it less for Negroes than whites? Perhaps I am weak in logical inference at this juncture, but I cannot see how these mortality data clarify the whole problem.

With respect to data on the sex differential that are more precise than the mortality rates, limited statistics are available from prospective epidemiologic studies, particularly the one at Framingham, Massachusetts. Ten year incidence data of new coronary heart disease, by age and sex, were presented by Dr. THOMAS R. DAWBER and his colleagues in 1963 (Trans. Assoc. Life Ins. Med. Dirs. of Amer., 47: 70, 1964). For the groups originally age 30–39, 40–49 and 50–59 respectively, the ten-year incidence rates for men were 38, 96 and 160 per 1,000. For women, they were 3, 31 and 86. These groups – originally age 30–39, 40–49 and 50–59 respectively – averaged age 40, 50 and 60 over the ten years of follow-up. Thus, the group averaging 40 and the group averaging 50 afford a neat comparison of the transition women undergo as they evolve from high level ovarian estrogen secretion to the low level of the menopause. For these two groups of women, the ten year rates of new coronary disease were 3 and 31 respectively, i.e. the women of average age 50 had rates about ten times higher than those average age 40. For men they were 38 and 96 respectively, i.e. a 2.5-fold difference. Thus, this age-related increase in rate was four times greater for women than for men. This certainly is suggestive of a 'break in the curve'. Again, these data could easily be interpreted to indicate that with removal of the protective effect of estrogens, American women – exposed generally to the same environmental conditions (diet, sedentary living habit, smoking, etc.) as men in terms of risk of atherosclerotic coronary heart disease – leap forward in their incidence rate with removal of estrogens, and proceed headlong down the path toward catching up with men. Again, I want to emphasize that I am not impressed with the use of statistics of this kind to document this particular argument. I believe the question at issue has to be solved in another way, e.g. through animal experimentation, and through studying special population groups, as Dr. FURMAN himself has done in his neat investigations of eunuchs, and as others have done in studies on oophorectomized women, men treated with estrogens, etc.

I would conclude this overly long disquisition by expressing appreciation to Dr. FURMAN for having stimulated us to a new look at these data and their possible meaning. I apologize for reviewing my old thoughts at the onset of this new look. I must confess that I am as yet wedded to the well-documented thesis that estrogens are chiefly responsible for the sex differential. But this is certainly not a closed question.

R. H. FURMAN (Oklahoma City, Okla.): I am sure none of us suffers any delusion regarding the 'limited-precision' of death certificates in epidemiologic studies. Such criticism is applicable to virtually all epidemiologic data. On the other hand, it is not difficult to ascertain that death has resulted from arteriosclerotic heart disease. If it were 'under diagnosed', i.e., 'under-coded' on death certificates (as it may well be in respect to women), it would tend to exaggerate rather than minimize sex differentials in mortality rates. The data which I presented were prepared by the National Vital Statistics Division of the National Center for Health Statistics and the National Heart Institute of the United States Public Health Service. They were made available to me by Dean E. KRUEGER of the Biometrics Branch of the National Heart Institute. As a measure of their validity, it may be pointed out that these mortality data are regarded as reliable when applied to a disease about which there exists no challenge concerning gonadal influence: the declining mortality rate from breast carcinoma in women following the menopause is clearly evident in these same mortality figures. Finally, the changes noted in the United States mortality data in respect to the male/female mortality rate ratios are not unique to the United States. I have no objection to Dr. STAMLER's sounding a note of 'caution', but these mortality data hardly need 'interpretation', they speak for themselves.

The mortality data for 1960 which Dr. STAMLER has cited reveal exactly the same exponential relationships of mortality and age as shown in Figure 2 in which 1966 mortality data are used. Dr. STAMLER has cited data only up to age 54, but mortality in the older age groups shows further decrease in the magnitude of the sex differential.

Mortality rates increase exponentially each year in both men and women, i.e., there is an 'acceleration' of mortality rate in both sexes. This 'acceleration' of mortality rate, however, diminishes in men beginning sometime in the fifth decade. (This is undoubtedly due, as already mentioned, to decrease in the size of the male population at risk because of the high vulnerability of young men to fatal heart attacks.) It must be kept in mind that the mortality rate is *always accelerating*, but in women there is a steady acceleration without diminution, although the mortality rate *per se* is always less in women than it is in men. (An occasional exception to this is noted in the tenth decade.) These phenomena – diminution of the acceleration of mortality rate in men beginning in the fifth decade and the constant, steady acceleration of the rate in women – account for the decreasing sex ratio which begins at this age. Additionally, the lack of any change in the slope of the female mortality rate curve in any temporal relationship to the time of the menopause places the onus of proof on him who says the menopause increases vulnerability of women to death from arteriosclerotic heart disease. Thus, I would state, in reply to his question, that one can not infer that the higher mortality rates from arteriosclerotic heart disease among older women are the result of decreased estrogenic protection.

Perhaps the chief difficulty confronting many who deal with factors responsible for the changing sex differential is their tendency to consider coronary heart disease death rates as arithmetic rather than exponential functions. The need for the latter approach is self evident upon inspection of Figure 2.

Dr. STAMLER raises the question of the validity of mortality rates derived from death certificates again in respect to the Negro. The high coronary artery disease mortality rate in the Negro in the U.S.A. is a medical dictum and the very much smaller sex differential in the Negro cannot be explained by a greater prevalence of hypertension among Negro women, inasmuch as the sex difference remains extremely small when death associated with hypertension is excluded. Furthermore, a small or negligible sex differential is not peculiar to the U.S. Negro population inasmuch as the sex differential is small also among populations whose diets and way of life are relatively austere. The questions of obesity and the presumed 'greater occurrence of premature oophorectomy in Negro women compared to white' seem speculative and, assuming that they were factors of significance, which seems unlikely, could hardly be expected to efface the sex differential to such a large degree in the Negro.

Dr. STAMLER echos the question we have asked many times, 'if estrogens play a significant role in the sex differential, why is the differential less for Negroes than for whites?' One must also add the question, why is the sex differential so small or lacking in populations enjoying less 'affluence'? Dr. STAMLER is not 'weak in logical inference at this juncture', but is having difficulty addressing himself to a phenomenon which troubles him because it is antithetical to his long-espoused assumption that estrogens are indeed responsible for lower arteriosclerosis vulnerability in white women. These mortality data obviously do not 'clarify the whole problem' but, as epidemiologic data so often do, they define for us a question which must be answered by more study and research.

Finally, Dr. STAMLER addresses himself to the question of *de novo* heart disease, i.e., the incidence of heart disease, the occurrence of 'new coronary heart disease'. The data which Dr. STAMLER cites from the Framingham study exhibit an exponential relationship between age and incidence, just as is the case with mortality rate data, with a tendency of the curve for men to flatten during the fifth decade. I cannot agree that there is any suggestion of a 'break in the curve' when plotted in this fashion.

Dr. STAMLER confesses, in closing, that he remains 'wedded' to the thesis that estrogens are chiefly responsible for the sex differential. It is clear that the thesis is by no means 'well-documented' and that the wedding has taken place without benefit of clergy.

D. BRUNNER (Jaffa): In our observations on certain population groups we could not find a difference in the average cholesterol, triglyceride and alpha-cholesterol percentage levels of women 45–50 years old and women 55–65 years old. In the first group the total cholesterol in 161 women was 240.0 ± 27 mg% triglycerides 109.1 mg% ± 30.5 and alpha-cholesterol percentage $24.7 \pm 3.9\%$. In the older age group total cholesterol was 248.9 mg%, triglyceride 106.1 ± 29.0 mg% and alpha-cholesterol percentage $25.7 \pm 4.1\%$.

It seems, that all those differences are environmental phenomena and not an obligatory biological development.

By the way, I think that the prognosis of women who have suffered myocardial infarction is worse than in men but the incidence rate is much smaller.

J. J. GROEN (Jerusalem): With reference to the papers by Drs. FURMAN and PICK, I would like to take up again the question why coronary heart disease is several times more frequent in middleaged males than in middleaged females. We can theoretically put forward three hypotheses to explain the difference: a chromosal, a hormonal and a cultural hypothesis. The fact that this difference is frequency between the sexes is only seen in the so-called western populations is more marked in the American white, whereas it is much less in the coloured, makes a biological explanation by a difference in the chromosomes seems unlikely. There are also only minute differences in hormone levels and excretion between males and females of the so-called 'western' and 'eastern' cultures. The differences in the serum lipids between western males and females (in alpha-lipoproteins-lipids mainly) seem equally insufficient to explain the marked differences in prevalence of coronary heart disease between the western sexes. By exclusion, it seems therefore as if certain ways of life, be it through physical activity or freedom of certain interhuman stresses, which affect the western male rather than the western female could be responsible. I am well aware of the methodological difficulties that we shall encounter in our approach to this cultural aspect of the problem but I feel that we have to look into this direction for a further elucidation.

L. W. KINSELL (Oakland, Cal.): In view of Dr. HIRSCH's findings with infused glyceride-containing lipoprotein I wonder if Dr. GROSS would agree that no studies on triglyceride production and utilization are interpretable (in the intact human subject) unless one has available *in each subject* data on the removal rate of infused specific lipoprotein?

R. G. GOULD (Palo Alto, Cal.): We performed such direct infusion studies in two men and two dogs using glycerol-labeled $S_f > 20$ lipoproteins. In dogs the turnover rates of plasma $S_f > 20$ triglyceride obtained from these studies agreed very well with those calculated from the specific activity-time curves of liver and plasma triglycerides following injection of labeled glycerol during the same study. This constitutes the validation of these calculations and of the kinetic model upon which they are based.

In man, since the slopes of disappearance of triglyceride from the infused labeled lipoproteins were parallel to those observed after the injection of labeled glycerol, the fractional (as well as absolute) turnover rates of plasma $S_f > 20$ triglyceride can be obtained directly from the slope of decline of specific activity of this lipid after injection of labeled glycerol.

These results point out the need for performing such studies in each species or under each set of experimental conditions, in order to validate the calculations made under these conditions, but I do not agree that they are necessary in each individual.

S. GARATTINI (Milan): I have two remarks. The first one tends to add support to the hypothesis that drugs such as nicotinic acid and 3.5 dimethylpyrazole interfere with the 3.5 cyclic AMP system. It is possible to stimulate lipolysis *in vitro* with dibutyril 3.5 cyclic AMP, a derivative slowly metabolized by phosphodiesterase-nicotinic acid and 3-carboxyl-dimethylpyrazole lose their capacity to inhibit lipolysis in these experimental conditions. The other point refers to the fact that in fasted animals the increase of plasma FFA does not depend on an activation of the cyclic AMP system. Yet nicotinic acid and dimethylpyrazole are very effective in lowering this increased level of plasma FFA. These drugs are active at doses several times lower than the ones required to affect the increased plasma FFA induced, for instance, by catecholamines. It may be possible that they have different mechanisms in relation to the dose level.

R. PAOLETTI (Milan): A 3.5 AMP-dibutyryl derivative is not greatly metabolised by phosphodiesterase and therefore nicotinic acid, which according to Krishna activates the phosphodiesterase, is not active against hypersecretion of FFA by this substance. Regarding the second remark, I must say that no definite evidence has been so far presented whether 3.5 AMP is involved in the lipase stimulated by fasting or not. As a possible tool I would like to suggest the use of fish adipose tissue. In fishes, adipose tissue lipase is activated by fasting but not by norepinephrine or ACTH, according to the recent work of my former colleague Dr. T. FARKAS. Addition *in vitro* of 3.5 AMP dibutyryl derivative to such adipose tissue should indicate if the lipase is activated or not by the cyclic 3.5 AMP system.

O.J. POLLAK (Dover, Del.): Study of hibernating animals – neglected to this date – may prove rewarding.

Primates

Department of Pathology, University of Chicago

Recent Progress in Studies of Experimental Primate Atherosclerosis*

R.W. WISSLER, Chicago, Ill.

The purpose of this presentation is two-fold: First, to try to give some perspective as to the present application of the subhuman primate in the study of experimental atherosclerosis and second, to present briefly the results of three recent studies in this laboratory which indicate that the Rhesus monkey may offer particular advantages in the study of the pathogenesis of this disease process.

It is important to recognize the special importance of the experimental approach in the study of atherosclerosis. More than almost any other major malady of man, it is not possible to study this disease adequately in the human subject. Three of the reasons for this difficulty are listed here:

1. Atherosclerosis develops gradually during many years. The scientist and physician often cannot follow the development of the disease process in people in a practical period of time and for the individual subject it is difficult to decide when the disease starts.

2. The fact that people keep changing their diets as well as their physical activity and that they often live in many variable environments makes it quite difficult to plan and execute controlled experiments.

3. Not only is it difficult to know when the disease starts but it is almost impossible to evaluate the severity of the disease in the individual living human subject until he or she develops a clinical manifestation of atherosclerosis.

* This work was supported by U.S.P.H.S. Grant 5 RO1 HE-06894.

From these facts it is evident that experimental models are necessary if one is to do meaningful studies which will clarify the pathogenesis of atherosclerosis and which will permit the evaluation of prevention or therapy in the diseased artery.

Much has been learned from studies in the rabbit, fowl, rat, dog, etc., but each of these models leaves something to be desired. However, two experimental models have been developed recently which seem to have greater similarity to the disease in man. These are the dietary production of atherosclerosis in swine and in the Rhesus monkey.

In this paper and indeed in this section of the Symposium attention will be focused on studies in the monkey and his relatives. Thus far, the Rhesus, the Cebus, the Squirrel, the Woolly, the Baboon, and the Chimpanzee have been used with varying degrees of success [2–4, 7–19, 21–28].

Figure 1 represents an oversimplified summary of the relative susceptibility of these families of subhuman primates to spontaneous atherosclerosis and to the experimentally induced disease, realizing, of course, that not nearly all of the possible methods of producing the disease have been treated.

Subfamilies of primates studied	Spontaneous disease	Experimentally produced disease
Baboon	++	++
Cebus	+	+++
Chimpanzee	+	+
Rhesus	+	++++
Squirrel	+++	+++
Wooley	+	++

Fig. 1. Primate susceptibility to atherosclerosis.

The Squirrel monkey [13, 14] is particularly susceptible to 'spontaneous' disease and to a lesser extent so is the Baboon [6–9, 12, 17, 18]. Other representatives listed seem to develop only fatty streaks or very minor plaques in their native environment. On the other hand, the studies of MANN and ANDRUS [10] as well as the extensive studies of TAYLOR, COX et al. [2, 3, 19–23] have established that the Rhesus monkey can readily develop very severe disease when maintained on a diet rich in cholesterol and relatively saturated lipids.

The following list summarizes the main observations reported by TAYLOR, COX and co-workers in a comprehensive study utilizing the Rhesus monkey fed a low-fat commercial monkey ration supplemented by butter and cholesterol [2, 3, 19–23]:

1. Little or no atheromatous disease was found in controls.
2. A high fat, high cholesterol ration supported sustained hyperlipemia and atherosclerosis, especially if given as a mixed diet once a day.
3. In general there was definite correlation between blood lipid levels and atherogenesis.
4. A few individual animals appeared to be resistant to diet-induced hyperlipemia and to arterial atherogenesis.
5. Severe disease resulted from prolonged feeding of the stock ration enriched by butterfat and cholesterol.

a) Aorta, coronaries, cerebral as well as other peripheral arteries were involved.
b) Myocardial infarction and gangrene of the leg have developed following arterial occlusion.

Some of the apparent advantages of the Rhesus monkey and some of its disadvantages as demonstrated by the studies of TAYLOR, Cox and co-workers [19–23] can be listed as follows:

Advantages

1. Normal blood lipids and the levels of cholesterolemia resulting in atherogenesis are similar to those in man.
2. The lesions show a striking resemblance to those in man.
3. The arteries involved and the 'clinical' catastrophies are similar to those in man.

Disadvantages

1. Natural protection of the female has not been observed.
2. The Rhesus is expensive and relatively difficult to handle.
3. It is also relatively difficult to obtain uniform, disease-free animals.

The advantages appear to us to far outweigh the disadvantages and furthermore we believe that most of the disadvantages can be overcome with increased attention to collecting and breeding and with development of better facilities for controlled experiments in which careful dietary regulation and metabolic manipulation are possible.

In the next few paragraphs a brief and simplified summary will be given of three recent experiments in this laboratory, each of which illustrates utilization of the Rhesus monkey in studies of the pathogenesis of atherosclerosis.

The first of these studies which will be summarized is the most recent. The experiments were very simple in design (Fig. 2) and the

Experimental groups	Number of animals	Experimental diet and treatment	
		Diet	Dose
I	8	12.5% of butterfat – 12.5% of coconut oil 2% cholesterol all in a commercial lowfat primate ration	400 mg of meprobamate per 100 g of diet 3 times weekly
II	8		—
III	8		20 mg of ephedrine SO$_4$ per 100 g of diet 3 times weekly

Fig. 2. Diet and drugs fed to Rhesus monkeys for a nine-month period.

results were largely unexpected. Following the lead of Dr. TAYLOR and his group [19–22] Rhesus monkeys were fed a balanced low fat commercial primate ration[1] enriched with 25% fat and 2% cholesterol [26]. The experiments differed in design from the ones reported by TAYLOR, COX, etc. in that the food fat was a 50:50 mixture of coconut oil and butterfat instead of butterfat alone. The reasons for including coconut oil have been reported previously [24]. Another lead reported earlier was also followed [29] in that $1/3$ of the animals received meprobamate and $1/3$ of the animals received ephedrine sulfate. The unexpected findings were that almost all of these animals achieved very high blood lipid levels and at autopsy, after only 40 weeks, they showed a remarkable amount of atherosclerosis whether one judges by surface area involved (Fig. 3) or by the combined evaluation of surface area and severity (Fig. 4).

Blood vessels	Group		
	I Meprobamate	II Controls	III Ephedrine
Aorta	30	49	35
Coronaries	30	34	27

Fig. 3. Surface area of aortic lesions in Rhesus monkeys fed meprobamate and ephedrine SO$_4$ for 40 weeks.

[1] Rockland low fat primate ration obtained from: Tecklad, Inc., Monmouth, Illinois.

Gross evaluation	I Meprobamate	II Controls	III Ephedrine SO$_4$
Estimated intimal area involved (%)	52	57	64

Fig. 4. Combined incidence and severity of atherosclerotic lesions in Rhesus monkeys fed butter and coconut oil for 40 weeks.

Figure 5 simply illustrates the gross characteristics of these lesions in the abdominal aorta and Figure 6 in the thoracic aorta of a different animal. Microscopically the aortic lesions varied from simple fatty deposition in pre-existing arterial cells (Figure 7) to fatty lesions with considerable intimal proliferation (Figure 8) to quite

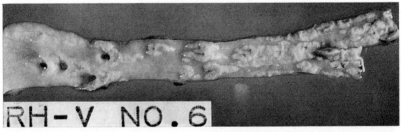

Fig. 5. Gross photograph of abdominal aorta in one of the animals (No. 6) fed the high fat, high cholesterol diet, with no meprobamate or ephedrine. Severe fatty plaques with fibrous 'caps' are located above aortic bifurcation. There are also faintly visible fatty 'streaks' below the ostia of the renal arteries and above the superior mesenteric and celiac arteries.

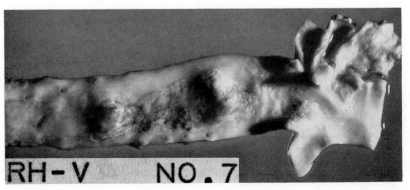

Fig. 6. Gross photograph of the thoracic aorta of another animal (No. 7) in the same diet group as above. The lesions involve the majority of the thoracic aortic intima as well as the great vessels of the neck. There are soft fatty plaques with focal ulceration, fatty plaques with fibrous caps and fatty streaks.

Fig. 7. Photomicrograph of Oil Red 0 stained section of a lesion in the thoracic aorta. This animal (No. 5) was fed the high fat (50:50 butterfat and coconut oil) high cholesterol diet. Much of the dark staining lipid is in pre-existing arterial medial cells. Only a little evidence of intimal proliferation is seen. Oil Red 0. Magnification, 153.

Fig. 8. Photomicrograph of thoracic aorta of another animal (No. 4) from the same group. This fatty lesion is characterized by thickened intima with abundant fatty deposition in the intima as well as the underlying media. The dark material represents lipid, except that in the extreme right part of the media where a few flecks of calcium are present. Note occasional cholesterol clefts in the thickened intima. Oil Red 0 stain. Magnification, 100.

severe lesions with these characteristics of fat and fibrosis plus intense calcification, cholesterol slits and necrosis of the plaque (Fig. 9). Standard samples of the coronary arteries revealed well

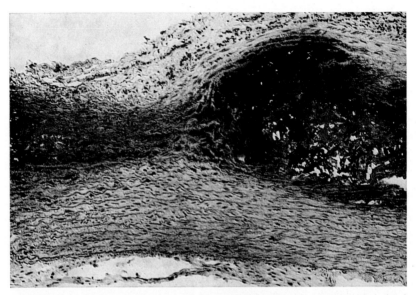

Fig. 9. Photomicrograph of thoracic aorta of animal No. 4. Besides considerable intimal proliferation, fibrosis, intense calcification, cholesterol slits and necrosis of the plaque are also present. Oil Red 0 stain. Magnification, 100.

developed plaques in about 50% of the animals. Some of the coronary arteries were extensively involved with stenosis and calcification (Figure 10).

Double tracer techniques were used in these studies and the lipid analyses of the aorta and other issues are still being completed. Some of these results will be reported by Dr. GETZ at the Autumn Meeting of the Council on Arteriosclerosis of the American Heart Association in New York [5]. At present it appears that the type of lipid being deposited in these lesions, especially the types of fatty acids incorporated in the cholesterol ester, may help to determine the cellular damage and the fibrous proliferation which occurs in the artery wall.

Other recent studies in this laboratory have demonstrated that the Rhesus monkey can be utilized:

1. To study the effects of 'table-prepared' human diets on the development of atherosclerosis [27] and

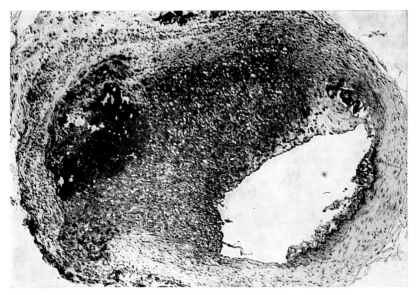

Fig. 10. Photomicrograph of a cross section of the left main coronary artery of animal No. 4. The lumen of this artery is narrowed by the fat filled thickened intima. The deeply buried necrotic center is prominent and stains positively with both Oil Red 0 and the Von Kossa stains. The atheromatous lesion is accompanied by marked thinning of the media and by accumulation of inflammatory cells in the adventitia. Oil Red 0 stain. Magnification, 100.

2. To furnish lipoproteins for donor experiments in which the major classes of serum lipoproteins ultracentrifugally fractionated from these Rhesus monkeys which have been fed human diets have been used to produce model intramural aortic lesions in normal recipient Rhesus monkeys [30].

The major results of these two types of approaches will be summarized very briefly. The human table-prepared diets were carefully planned by Mrs. DOROTHEA TURNER who is the Chief Medical Nutritionist at the University of Chicago Hospitals. She formulated these rations so that they would simulate the Average American diet on the one hand, and the Prudent diet recommended jointly by the Committee on Nutrition and the Council on Arteriosclerosis of the American Heart Association (low in cholesterol, hard fats, calories and refined sugar [1, 31]. The basic composition of the two rations is summarized in Figure 11. As compared to the Average diet the prudent ration contained little or no eggs, fatty meat and butterfat.

	American average		American prudent	
	Dry wt. %	Consumed g/day	Dry wt. %	Consumed g/day
Fat	22.6	26.9	18.0	14.0
Protein	25.0	29.8	31.5	24.4
Carbohydrate	45.6	54.3	44.5	34.5
Ash	2.2	2.6	2.5	1.9
Fiber	3.2	3.8	3.3	2.6
Total solids	98.6	117.4	99.8	77.4
Water g/100 g/solids	101.4		150.0	
Daily calories/monkey		560		360
Daily food consump. (g)		260.6		223.8

Fig. 11. Diet composition (average of 6 analyses) and food consumption.

Fatty acid patterns of the lipids extracted from each of the diets are shown in Figure 12. It is evident that there were more short chained fatty acids in the average diet and that there was more linoleic acids in the prudent diet. Diet consumption was consistently complete and weight gains were constant and consistent with the calories consumed.

Figure 13 shows the terminal average values for the three monkeys of each group at 107 weeks for serum cholesterol, phospho-

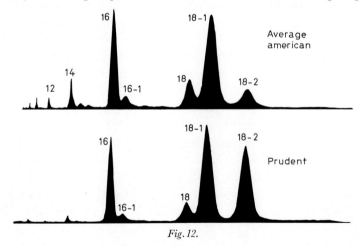

Fig. 12.

Diet	Cholesterol	Phospholipids	Total lipids
American average	384	427	816
American prudent	202	228	452

Fig. 13. The results obtained at the end of the experiment from the serum in Rhesus monkeys fed 'Average' and 'Prudent' American diet.

lipid and total lipid. The contrasts in these averages are evident and the values appear to be in the range one might expect to observe in man. Figure 14 shows the contrasts in fatty acids of the low density lipoprotein cholesterol esters fractionated from the pooled serum of the two groups. Note that these ester fatty acids reflect the fatty acid patterns of the diets.

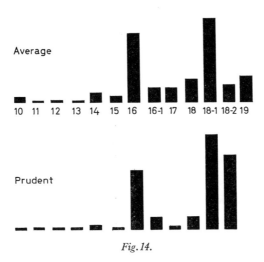

Fig. 14.

At autopsy a careful evaluation was made of the severity of aortic atheromatous change in each monkey, both grossly and microscopically, relying on standard sections taken from predetermined areas.

Although space does not permit illustration of these lesions, Table 1 reflects the contrast in lesions in the aorta, both in terms of surface area involved and in terms of incidence and severity at the microscopic level.

This experiment utilized too few animals to be definitive and it is now being repeated with additional animals in each group. The results reported here do indicate however, that it is feasible to study human dietary factors using the Rhesus monkey and that the usual 'table prepared' human diet results in hyperlipemia and atherogenesis in the Rhesus monkey.

A major original purpose of this study was to use the ultracentrifugally separated lipoproteins from these animals to investigate the reaction of the normal aortic media to each lipoprotein fraction

Table I. Frequency and severity of aortic lesions in Rhesus monkeys fed 'Average' and 'Prudent' American diets for 107 weeks

Gross evaluation		'Average' American diet	'Prudent' American diet
Estimated intimal surface area involved (%)		95, 70, 5	10, 10, 15
Major microscopic findings	Std. aortic sections	'Average' American diet	'Prudent' American diet
1. Lipid deposition in the intima	1A	3/3 a, a, c	2/3 a, a
	2A	3/3 a, a, c	2/3 b, b
	3A	2/3 b, c	2/3 a, b
2. Lipid deposition in both the intima and innermost part of the media	1A	1/3 c	0/3
	2A	1/3 a	0/3
	3A	2/3 b, c	1/3 a
3. Swelling and focal proliferation of intimal cells	1A	1/3 b	0/3
	2A	1/3 a	0/3
	3A	1/3 a	1/3 a
4. Degeneration and increase of collagen in the media	1A	3/3 a, a, b	1/3 a
	2A	2/3 a, a	0/3
	3A	3/3 a, a, a	0/3

a = mild; b = moderate; c = mild.

from the monkeys fed the 'Average' and the 'Prudent' diets. Figure 15 illustrates the distribution of the model lesions which were produced in the first of three experiments which we have now performed. As the figure indicates, we transferred ultracentrifugally separated lipoprotein fractions to the media of recipient animals as small intramurally injected depots using a 30-gauge needle in two experiments and a French-made 'non-traumatic' jet type of syringe [1] in another. Several animals were injected simultaneously and they were then killed at intervals and the tatoo-identified lesions were sampled for light and electronmicroscopy.

Table II gives a brief and over-simplified summary of the differences in lipid disappearance which we have observed in these model lesions. Alpha 1 or high density lipoprotein disappeared rapidly and showed no particular effect on the aorta no matter

[1] Dermajet Akra, Brevete, SGDG, Pau, France.

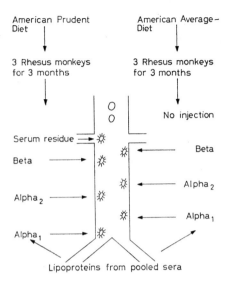

Fig. 15.

Table II. Visible* aortic lipid in smooth-muscle cells resulting from injecting lipoproteins intramurally

Lipoproteins and residue	Intervals after injection (days)			
	2	4	7	9
Alpha 1 (HD)				
average	++	+	0	0
prudent	++	+	0	0
Beta (LD)				
average	++	+++	+++	++
prudent	++	++	++	+
Alpha 2 (VLD)				
average	0	+	+	0
prudent	0	+	+	0
Defatted protein				
average	0	0	0	0
prudent	0	0	0	0

* Combined oil red 0 and electron microscopic evaluation.
0 = No lipid visible.
+ = A few cells with sparse lipid droplets.
++ = Moderate number of cells with medium number of lipid droplets.
+++ = Many cells with many lipid droplets.

whether the lipoprotein came from the 'Average' or the 'Prudent' fed group. Beta or low density lipoprotein from the 'Average' fed group appeared to be retained longer in the vessel wall and it appeared to do more damage to the smooth muscle cells at the ultrastructural level than did any other fraction. Alpha 2 or very low density lipoprotein has not yet been studied adequately because we have not had sufficient quantities to permit injection of comparable quantities, but we hope to have better data soon. As of now there appears no particular difference in the fate of the very low density lipoproteins from the two groups. The subnatant defatted protein is apparently not traumatic to the artery wall.

Electronmicroscopy of the seven day model lesions resulting when the beta lipoprotein from the 'Average American Diet' fed monkeys was injected, indicated that the smooth muscle cells of the media reacted to the injection with hyperplasia of the Golgi apparatus and production of many free ribosomes. Mitochondria appeared relatively less numerous than usual [30].

These three studies are being amplified and the results will be reported in more detail soon.

The results of these experiments and of those previously reported by MANN and ANDRUS [10] and by TAYLOR, COX et al. [2, 3, 19–23] as well as those reported by others [4, 6, 8, 10–17, 24, 25, 28] and those to be reported in the following pages indicate that the nonhuman primate is an appropriate animal for experimental studies of man's most menacing disease. The current status of the nonhuman primate as a suitable animal for this type of study has also been summarized recently in the publication 'Comparative Atherosclerosis' [17]. TAYLOR et al. [20] as well as HOWARD ET AL. [7] give almost identical reasons for utilizing the nonhuman primates for the study of atherosclerosis. A paraphrase of the latter points seems to be a fitting conclusion for this introductory paper. Briefly stated, subhuman primates are useful for the study of atherogenesis because:

a) Their general anatomical resemblance to man is obvious and advantageous.
b) In the larger genera the arteries permit anatomical and chemical studies.
c) Their atherosclerotic lesions resemble those found in Homo Sapiens.
d) Their lipid metabolism appears to be similar to that in man.

References

1. Consumption of Food in the United States, 1909–1952 Agriculture, Economic Research Service, Economic Statistical Analysis Division, Washington, D.C.
2. Cox, G.E.; Taylor, C.B.; Cox, L.G. and Counts, M.A.: Atherosclerosis in Rhesus monkeys. I. Hypercholesteremia induced by dietary fat and cholesterol. Arch. Path. 66: 32–52 (1958).
3. Cox, G.E.; Trueheart, R.E.; Kaplan, J. and Taylor, C.B.: Atherosclerosis in Rhesus monkeys. IV. Repair of arterial injury – an important secondary atherogenic factor. Arch. Path. 76: 166–176 (1963).
4. Fox, H.: Arteriosclerosis in lower mammals and birds: its relation to the disease in man. In 'Arteriosclerosis', Ed. by E.V. Cowdry, pp. 153–193 (Macmillan, New York 1933).
5. Getz, G.S.; Wissler, R.W.; Hughes, R.H. and Miller, L.H.: The composition and phospholipid synthesis in Rhesus monkey aortas. (In press).
6. Gilbert, C. and Gilman, J.: Structural modifications in the coronary artery of the Baboon (Papio Ursinus) with special reference to age and endocrine status. Sth. afr. J. med. Sci. 25: 59–70 (1960).
7. Howard, A.N.; Gresham, G.A.; Richards, C. and Bowyer, D.E.: Serum proteins, lipoproteins, and lipids in baboons given normal and atherogenic diets. In 'The Baboon in Medical Research', Ed. by H. Vagtborg, pp. 283–299 (University of Texas Press 1965).
8. Lindsay, S. and Chaikoff, J.L.: Naturally occurring arteriosclerosis in animals: A comparison with experimentally induced lesions. In 'Atherosclerosis and its Origin', Ed. by Maurice Sandler and G.H. Bourne, pp. 349–437 (Academic Press, New York 1963).
9. Lindsay, S. and Chaikoff, J.L.: Naturally occurring arteriosclerosis in nonhuman primates. J. Atheroscler. Res. 6: 36–61 (1966).
10. Mann, G.V. and Andrus, S.B.: Xanthomatosis and atherosclerosis produced by diet in an adult Rhesus monkey. J. lab. clin. Med. 48: 533–550 (1956).
11. Mann, G.V.; Andrus, S.B.; McNally, A. and Stare, F.J.: Experimental atherosclerosis in Cebus monkeys. J. exp. Med. 98: 195–218 (1953).
12. McGill, H.C., Jr.; Strong, J.P.; Holman, R.L. and Werthessen, N.T.: Arterial lesions in the Kenya baboon. Circulat. Res. 8: 670–679 (1960).
13. Middleton, C.C.; Clarkson, T.B. and Lofland, H.B.: Atherosclerosis in the Squirrel monkey. Naturally occurring lesions of the aorta and coronary arteries. Arch. Path. 78: 16–23 (1964).
14. Middleton, C.C.; Clarkson, T.B.; Lofland, H.B. and Prichard, R.W.: Naturally occurring atherosclerosis in the Squirrel monkey. Circulation 28: 665–666 (1963).
15. Portman, O.W. and Andrus, S.B.: Comparative evaluation of three species of New World monkeys for studies of dietary factors, tissues, lipids and atherogenesis. J. Nutr. 87: 429–438 (1965).
16. Portman, O.W. and Sinisterra, L.: Dietary fat and hypercholesteremia in the Cebus monkey. J. exp. Med. 106: 727–742 (1957).
17. Roberts, T.C., Jr.; Straus, R. and Cooper, M.S.: Comparative Atherosclerosis (Harper and Row, 1965).
18. Stare, F.J.; Andrus, S.B. and Portman, O.W.: Primates in Medical Research with special references to New World monkey. Proc. of a conference on research with primates. Ed. by D.E. Pickering, pp. 59–66 (Tektronix Foundation Beaverton, Washington 1963).

19. TAYLOR, C.B.; COX, G.E.; COUNTS, M. and YOGI, N.: Fatal myocardial infarction in the Rhesus monkey with diet-induced hypercholesteremia. Amer. J. Path. *35:* 674 (1959).
20. TAYLOR, C.B.; COX, G.E.; MANALO-ESTRELLA, P. and SOUTHWORTH, J.: Atherosclerosis in Rhesus monkeys. II. Arterial lesions associated with hypercholesteremia induced by dietary fat and cholesterol. Arch. Path. *74:* 16–34 (1962).
21. TAYLOR, C.B.; MANALO-ESTRELLA, P. and COX, G.E.: Atherosclerosis in Rhesus monkeys. V. Marked diet-induced hypercholesteremia with xanthomatosis and severe atherosclerosis. Arch. Path. *76:* 239–249 (1963).
22. TAYLOR, C.B.; PATTON, D.C. and COX, G.C.: Atherosclerosis in Rhesus monkeys. VI. Fatal myocardial infarction in a monkey fed fat and cholesterol. Arch. Path. *76:* 404–412 (1963).
23. TAYLOR, C.B.; TRUEHEART, R.E. and COX, G.C.: Atherosclerosis in Rhesus monkeys. III. The role of increased thickness of arterial wall in atherogenesis. Arch. Path. *76:* 14–28 (1963).
24. WISSLER, R.W.; FRAZIER, L.E.; HUGHES, R.H. and RASMUSSEN, R.A.: Atherogenesis in the Cebus monkey. I. A comparison of three food fats under controlled dietary conditions. Arch. Path. *74:* 312–322 (1962).
25. WISSLER, R.W.; FRAZIER, L.E.; HUGHES, R.H. and RASMUSSEN, R.A.: The development of atheromatous disease in Cebus monkeys using saturated and unsaturated fats. Fed. Proc. *19:* 17 (1960).
26. WISSLER, R.W.; GETZ, G.S.; VESSELINOVITCH, D.; FRAZIER, L.E. and HUGHES, R.H.: Acute severe experimental atherosclerosis in Rhesus monkeys. Fed. Proc. *25:* 597 (1966).
27. WISSLER, R.W.; HUGHES, R.H.; FRAZIER, L.E.; GETZ, G.S. and TURNER, D.: Aortic lesions and blood lipids in Rhesus monkeys fed 'table prepared' human diets. Circulation *32:* 220, Suppl. 2 (1965).
28. WISSLER, R.W.; HUGHES, R.H.; FRAZIER, L.E. and RASMUSSEN, R.A.: Effects of feeding fats with varying fatty acid composition on the blood and tissue lipids of Cebus monkeys. Circulation *22:* 833 (1960).
29. WISSLER, R.W.; HUGHES, R.H.; FRAZIER, L.E. and RASMUSSEN, R.A.: Factors influencing development of aortic fatty plaques in Cebus monkeys maintained on constant diet. Circulation *26:* 673–674 (1962).
30. WISSLER, R.W.; TRACY, R.E.; MOLNAR, Z.; RACKER, D.; MANCINI, M. and HUGHES, R.H.: Lipoprotein induced model lesions in the aortic media of Rhesus monkeys. Fed. Proc. *23:* 101 (1964).
31. ZUKEL, M.C.: Diet management and therapy. Amer. J. clin. Nutr. *16:* 270–276 (1965).

Author's address: R. W. Wissler, M. D., University of Chicago, Dept. of Pathology, 950 East 59th Street, *Chicago, Ill.* (USA).

Department of Nutrition, Harvard School of Public Health, Boston, Massachusetts, and Yerkes Laboratories of Primate Biology of Emory University[1], Orange Park, Florida

Comparative Studies of Spontaneous and Experimental Atherosclerosis in Primates
II. Lesions in Chimpanzees Including Myocardial Infarction and Cerebral Aneurysms[2]

S. B. ANDRUS[3], Boston, Mass., O. W. PORTMAN[4], Beaverton, Oregon, and A. J. RIOPELLE[5], Covington, La.

(With colour plate I)

Introduction

From the comparative study of spontaneous vascular lesions in New World monkeys, it is apparent that there are distinct species differences (ANDRUS and PORTMAN, 1966). This applies to the incidence of both gross aortic lipid deposition (demonstrated by Sudan staining) as well as microscopically identifiable atherosclerosis. Thus of 5 species, comprising 73 individuals shot and prepared in the Llanos of Colombia, South America, mean aortic sudanophilia (expressed as % of surface area involved) ranged from

---

[1] Now incorporated in the Regional Primate Research Center at Emory University, Atlanta, Georgia.

[2] Supported in part by grants-in-aid from the National Institutes of Health (H-04208; HE-00136; AM-05628; HE-10098), the John A. Hartford Memorial Fund, and the Fund for Research and Teaching, Department of Nutrition, Harvard School of Public Health.

[3] Assistant Professor of Pathology, Department of Nutrition, Harvard School of Public Health and Department of Pathology, Harvard Medical School, Boston, Massachusetts.

[4] Senior Scientist, Oregon Regional Primate Research Center, Beaverton, Oregon; Professor of Biochemistry, University of Oregon Medical School, Portland, Oregon; Established Investigator of the American Heart Association during the period of this research.

[5] Director, Delta Regional Primate Research Center, Covington, Louisiana.

0.1% in Cebus to 9% in Saimiri. Lagothrix, Allouatta, and Callicebus showed intermediate values of 0.6, 0.6, and 1.3% respectively. Only Lagothrix, however, demonstrated definitive atherosclerosis or appreciable gross fibrotic lesions, less than 1% of the aortic surface being involved with the latter change. The bulk of the sudanophilic material in all 5 species proved to be extracellular in location.

Similarly when 3 representative species were subjected to the same series of experimental high-fat diets containing cholesterol, species differences became more clear-cut (PORTMAN and ANDRUS, 1965). On control diets both Saimiri and Lagothrix had serum cholesterol levels distinctly higher than Cebus. In spite of the fact that all 3 species responded to the experimental diets with comparable elevations of serum cholesterol levels, the associated aortic sudanophilia showed a pronounced spectrum of reactivity which reflected the above differences seen in the jungle specimens, i.e., Saimiri > Lagothrix > Cebus. In terms of atherosclerosis per se, Saimiri and Lagothrix appeared comparably involved while Cebus were completely free of the definitive lesion. Under more severe experimental conditions, primarily with higher dietary levels of cholesterol, Cebus will respond with definable atherosclerosis (MANN et al., 1953).

The present study of induced atherosclerosis in chimpanzees represents an extension of the above comparative study in New World species. Insofar as possible, the same experimental design was followed in both studies. The chief difference between the two was the duration of the experiments. Because of the prevalence and relative severity of spontaneous lesions in the chimpanzee[1], this experiment was continued for 12 months in contrast to the 6 months' duration of the previous study.

The histologic characteristics of atherosclerosis in the above species including the present chimpanzee material have recently been described and a concept of pathogenesis formulated (ANDRUS and PORTMAN, 1966). The distinction between elastica-oriented extracellular lipid and the definitive lesion of atherosclerosis was emphasized, the former often representing the bulk of gross aortic sudanophilia, especially in lesions without surface elevation. It was felt that the definitive lesion frequently follows the former, particularly when the extracellular lipid occurs in conjunction with the

[1] ANDRUS, unpublished data.

spontaneous musculo-elastic change in the intima. The fate of the bulk of the elastica-oriented extracellular lipid, especially when unassociated with the musculo-elastic change, is unknown. It was also pointed out that despite the species differences in the incidence of definitive atherosclerosis, the lesions when present are very similar, from species to species, whether spontaneous or induced.

Methods

This experiment was conducted at the Yerkes Laboratories of Primate Biology in Orange Park, Florida, 5 of the animals having been born and raised in that colony. Following weaning, these animals had subsisted on the Yerkes natural diet (Table I). The remaining 9 animals came from various sources and their dietary past history is unknown, these individuals being identified by approximate ages in Table II. The majority of the animals were sexually mature (8–24 years) and consisted of 4 males (mean weight 51.2 kg) and 10 females (mean weight 43.8 kg) (Table II). Ten of the animals were placed on the high-fat diets, corn oil or coconut oil, with and without 0.5% cholesterol (approximately 0.1 g/100 kcal), (Table I). Because of palatability problems, the sucrose used as source of carbohydrate in the study of New World monkeys was here replaced by corn starch. Diet groups included 2 or 3 animals per group. Prior to the definitive diets the animals were observed for 2–3 months while eating the Yerkes natural diet and for an additional 6–12 months while eating a semi-purified control diet (Table I). Four controls were included, these animals receiving the Yerkes natural diet. At monthly intervals blood was drawn for determinations of serum cholesterol, beta lipoproteins, and hematologic values and blood pressure measurements were made. After 12 months on the experimental diets, the animals were anesthetized and in addition to the above procedures, electrocardiographic tracings and chest X-rays were taken. One animal was sacrificed after 6 months, following which the decision was made to extend the experimental time to 12 months in view of the extent of spontaneous lesions. Following exsanguination, complete autopsies were performed and a number of fixatives and staining procedures used. As indicated previously (ANDRUS and PORTMAN, 1966), Zenker's fixation and the use of longitudinal sections of arteries were found to be particularly helpful in delineating the vascular smooth muscle cells present in the lesions[1]. Aortic sudanophilia expressed as % of surface area involvement was measured by the combined use of planimetry and transparent grids, the latter measurements being made under a dissecting microscope. The specimens had been stained with Sudan IV in 70% ethanol following fixation of the opened specimen in 10% formalin or Zenker's acetic solution. Sudanophilic lesions in the coronary arteries and great vessels (innominate, carotid, subclavian, mesenteric, renal, and common iliac arteries) were estimated visually on a scale of 1–4+. Lesions of cerebral arteries, which were all more or less comparable in size, were simply enumerated. Aortic specimens were examined from 26 additional control chimpanzees. Of these 13 were complete specimens and satisfactory for accurate quantitation of lesions: 8 males (mean weight 35.8 kg) and 5 females (mean weight 26.2 kg). Data on X-rays, electrocardiographic tracings, blood pressure, hematology, and tissue lipids will be published elsewhere.

[1] In the absence of Zenker's fixation two techniques have been helpful in suggesting the identity of these cells in the atherosclerotic lesion: viz. the Verhoeff-van Gieson and the periodic acid-Schiff stains. The latter demonstrates rather intense staining of the intimal stroma adjacent to the muscle cell cytoplasm (HAUST et al., 1960).

Results

Diet acceptance and weight gain. During the preliminary attempts to feed a semi-purified diet to chimpanzees, sucrose was used as the source of carbohydrate. Although initially accepted, this diet was rejected by many animals within weeks. Consistent diet acceptance was maintained by changing the carbohydrate source to corn starch and allowing the animals a carrot daily. Eight of the 10 chimpanzees on the high-fat diets gained large amounts of weight during the year's duration of the experiment; the other two animals lost only small amounts of weight. Although the majority of the animals were adults and could have been expected to maintain a nearly constant weight on the Yerkes natural diet, the animals on the high-fat diets gained an average of 17.2% of their initial weight (range: —6.2% to +34.1%). The low residue semi-purified diet had a much higher caloric density than the Yerkes natural diet or presumably than the diet of the chimpanzee in its native habitat. Insofar as possible, the caloric intakes of the semi-purified diets were maintained at the levels used at the Yerkes Laboratory in the feeding of the natural diet. By this practice an adult chimpanzee received approximately 3,000 cal. per day. There was little wastage regardless of the nature of the diet.

Mean serum cholesterol concentrations. The Yerkes natural diet and the low corn oil semi-purified control diets used during initial diet testing were not compared critically as to serum cholesterol response. Some representative serum cholesterol values associated with these diets are given in Table III. It would appear that the low corn oil semi-purified diet based on corn starch produced somewhat lower serum cholesterol levels than the other two control diets[1]. One might expect the natural diet to be associated with cholesterol levels comparable to the starch-based semi-purified diet. The natural diet differs, however, from the latter in several respects, i.e., the presence of small amounts of both animal fat and cholesterol (Table I). There were no differences in serum cholesterol response between the sexes (Table III). Serum cholesterol values for the individual experimental animals are given in Table II. In view of the small numbers of animals involved, any conclusions

---

[1] Such a carbohydrate effect has been described in rats (PORTMAN *et al.*, 1956a) and in Cebus monkeys (PORTMAN *et al.*, 1956b), and in various other species including man. The subject has recently been reviewed by McGANDY *et al.* (1966).

Table I. Dietary lipids, control and experimental

Diet designation	Calories from corn oil %	Calories from coconut oil %	Cholesterol g/100 kcal
Natural[1]			0.001
Control[2]	16		
Corn[3]	45		
Corn +	45		0.1
Coconut		45	
Coconut +		45	0.1

[1] The natural diet was a varied mixture of grains, fruits, vegetables, and milk, etc. Analysis of the detailed dietary history indicated a composition by weight of 20% protein, 69% carbohydrate, and 9.8% fat, which supplied 2.6% of the calories as butterfat and 17.5% of the calories as vegetable fat.

[2] The composition of the semi-purified control diet was (in %): 'vitamin-free' casein, 25.0; corn oil, 8.0; corn starch, 62.3 (equal quantities of sucrose were substituted for corn starch in the sucrose variant of this diet); salts IV (HEGSTED, 1941), 4.0; choline, 0.5; para-aminobenzoic acid, 0.1; inositol, 0.1.

The following ingredients were added (in mg/kg of diet): thiamine, 10; riboflavin, 10; pyridoxine, 10; niacin, 49; calcium pantothenate, 30; folic acid, 1; biotin, 1; vitamin A acetate, 12,500 units; crystalline calciferol, 1,000 units. Ascorbic acid was supplied at 100 mg/day.

[3] The composition of the 4 experimental diets was (in %): 'vitamin-free' casein, 21.5; fat as indicated, 25.2; corn starch, 48.7. All other ingredients were comparable to those in the control diet.

concerning the effects of the experimental diets must be tentative. It would appear that there are no differences in serum cholesterol levels of animals eating the high corn oil experimental diet and those eating the comparable low corn oil control diet, i.e., based on corn starch. Adding cholesterol to the high corn oil diet or substituting coconut oil for corn oil caused an elevation of serum cholesterol, the two effects appearing roughly comparable in extent. These effects were additive and thus the animals receiving coconut oil and cholesterol showed the most pronounced changes.

Lipoprotein changes. Table III shows the serum lipoprotein concentrations of chimpanzees fed the Yerkes natural diet and the various semi-purified diets. The alterations in lipoproteins associated with the various diets parallel the changes in serum cholesterol concentrations. In general, lipoprotein response of individual animals was more consistent than the serum cholesterol changes.

Gross aortic lesions. Three basic lesions were seen in the aortas following Sudan staining. Diffuse, poorly defined and pale orange

Table II. Mean serum cholesterol levels and vascular lesions in 4 control and 10 experimental chimpanzees after 12 months of diet feeding

Sex-No.	Age years-months	Final weight kg	Diet	Mean serum[1] cholesterol mg %	Aortic lesions[2] lipid[3] %	Aortic lesions[2] fibrous %	Other arterial lesions[4] coronary arteries	Other arterial lesions[4] great vessels	cerebral arteries	Myocardial lesions	Cerebral aneurysms No.
F-130	13-6	47.7	natural	139	2.3	0.1	0	tr.	23		
F-160	9-6	28.8	natural	167	6.0	<0.1	0	1+	24		
F-218	12±	49.0	natural	177	26.0	0	0	1+	3		
F-164	7-7	30.9	natural	199	19.8	0	0	tr.	15		
F-288	13±	59.5	corn	160	9.4	0.2	0	1-2+	0		
F-264	12±	57.5	corn	180	7.5	0.2	0	2-3+	5		
M-215	15±	44.7	coconut	290	10.0	0.1	tr.	–	–		
M-189	10±	48.5	coconut	305	40.9	<0.1	4+	3+	12		
M-191	18±	63.1	corn+	150	33.4	0.2	2+	3+	19	focal necroses	
F-94[5]	21-7	45.3	corn+	270	17.5	0.2	0	1-2+	–	focal fibrosis	
F-220	21±	44.6	corn+	280	28.9	0.8	0	2-3+	8		
F-208	10±	34.6	coconut+	310	38.6	0	tr.	1+	4		
F-138	18±	40.2	coconut+	420	75.7	0	tr.	1-2+	11		1
M-59	23-9	48.5	coconut+	600	47.7	0	1+	4+	17	fibrosis	4

[1] Values represent the means of 12 monthly determinations. Variance analysis: effect of diet, F = 6.25, p = 0.01; type of fat, F = 5.32, p = 0.05; presence of cholesterol, F = 22.0, p<0.01.
[2] Figures represent % aortic surface area involvement.
[3] Variance analysis: effect of diet, F = 17.6, p<0.01; type of fat, F = 9.99, p = 0.02; presence of cholesterol, F = 10.26, p = 0.02.
[4] Coronary arteries and great vessels are graded visually on a 1-4+ basis. Values for cerebral arteries are the actual numbers of lesions seen.
[5] This animal was sacrificed after 6 months of diet feeding.

Table III. Mean serum cholesterol and beta lipoprotein concentrations from chimpanzees maintained on the various control and experimental diets (mg %) [1]

Diet	Number of animals	Serum cholesterol	Lipoproteins, Sf classe [2]				
			0–12	12–20	20–35	35–100	100–400
Natural [3]							
males	25	213.0	253	12	3	3	0
females	28	219.0	252	11	3	4	2
Control (purified)							
sucrose	14	214.2 [5]	276	15	6	10	1
corn starch	11	163.5	265	11	4	10	1
Corn [4]	3	174.0	219	9	3	1	2
Coconut	2	297.5	435	34	5	4	1
Corn +	3	233.3	412	16	3	4	0
Coconut +	3	443.3	567	66	7	3	1

[1] The values for the natural diet were determined from the means of 2–5 values for each animal. Values for the experimental diets were derived from the means of determinations made monthly in the case of serum cholesterol and at 6 and 12 months after beginning experiments in the case of lipoproteins.

[2] The flotational classes of lipoproteins are those described by GOFMAN et al. (1949). Variance analysis: Sf 0–12, effect of diet, $F = 12.81$, $p<0.01$; type of fat, $F = 2.01$, $p>0.05$; presence of cholesterol, $F = 16.93$, $p<0.01$. Sf 12–20, effect of diet, $F = 25.00$, $p<0.01$; type of fat, $F = 3.98$, $p>0.05$; presence of cholesterol, $F = 7.25$ $p\ 0.05–0.01$.

[3] In view of the large numbers of animals, data for males and females are given separately. In the other diet groups in the absence of sex-oriented differences, data from the sexes are pooled.

[4] One animal of this group is not represented in Table II.

[5] $t = 3.57$, for the difference in the serum cholesterol levels of the sucrose and corn starch groups, $p<0.005$. For variance analysis of serum cholesterol values, see Table II.

---

staining areas were frequently seen, especially in the arch and thoracic aorta. Such changes were increased by dietary cholesterol, but because of the impossibility of accurate and reproducible measurements these values have not been included in Table II. Measurements of the other two types of aortic change are recorded therein, expressed as % of the surface area involved. By far the commonest lesion was focal, discrete, rather intense red sudanophilia, frequently associated with some elevation of the intima. Because of the difficulty of judging minimal elevation of the intima and because it was apparent that microscopically well-defined atherosclerotic lesions can exist without gross intimal elevation, no attempt was made to quantitate such lesions separately.

All but one of the 17 control aortas showed discrete sudanophilia. The mean % surface area involvement of this group was 5.94%, though in one mature female this value reached 26%

(Fig. 3). Although the females showed somewhat more sudanophilia than males, 8.6 and 3.0% respectively, these differences were not statistically significant. No significant correlation could be demonstrated between the extent of the spontaneous lesions and age of the animal. The only completely negative aorta, however, was that of an infant of 6.7 kg. Fig. 1 shows diagrammatic profiles and values for % area involvement of the various segments of the aorta. Among the control animals the arch and abdominal portions were about equally involved while the thoracic segment was least affected. Within the thoracic segment, the upper portion was generally more involved than the lower.

The results of the dietary manipulations on aortic sudanophilia were comparable to the effects on serum cholesterol levels (Table II). Thus, there was little difference between the control (natural) diet and the high corn oil experimental diet. Adding cholesterol to the latter or substituting coconut oil produced roughly equal increases in sudanophilia and these increases were cumulative in the group receiving both factors. The one animal sacrificed after 6 months on the experimental diet, No. 94, showed less sudanophilia than the other 2 animals on the same diet. With increased severity of lesions, there appeared a shift in the segmental aortic profiles (Fig. 1), with preponderant involvement frequently being in the thoracic or abdominal segments. This effect was even more apparent on simple inspection, particularly in the abdominal aorta, as here the sudanophilic foci tended to be more sharply circumscribed than elsewhere, more frequently elevated, and the staining intensity greater. All of these attributes were exaggerated by the more severe diets.

The third type of lesion consisted of elevated fibrous-appearing plaques with or without peripheral sudanophilia. These were seen occasionally in the arch or thoracic aorta, but most commonly in the abdominal segment. These lesions were relatively uncommon and inconspicuous, being seen in only 6 of the 17 control animals, with a mean % area involvement of 0.2%. In one mature female such involvement approached 2% of the aortic surface area. These lesions bore no apparent relationship to age of animal or extent of sudanophilia. The experimental diets had no appreciable effect on these lesions (Table II).

Gross configuration of certain lesions. The experimental animals invariably showed a pair of striking matching sudanophilic lesions

	Control		Experimental			
	Male	Female	Corn	Coconut	Corn +	Coconut +
Number of animals	8	9	2	2	3	3
Mean % sudanophilia — Segment 1	4.2	11.8	10.0	15.5	32.0	40.3
Segment 2	1.6	5.6	3.0	32.0	15.7	59.0
Segment 3	4.2	11.9	18.5	24.5	44.0	57.7
Total	3.0	8.6	8.5	25.5	26.3	54.3

Fig. 1. Aortic segmental profiles of 17 control and 10 experimental chimpanzees. Sudanophilic involvement of the arch, thoracic, adominal portions, and of the whole aorta are expressed in % surface area and are designated as segments 1, 2, 3 and total respectively. The values for any one animal in the lower portion of the figure have been connected by lines; males ———, females ------. The upper portion represents mean sudanophilia of the various segments and of the whole aorta for the various experimental and control groups.

of the lesser and greater curvatures of the aortic arch, just proximal to the apex. These were commonly seen, though less well developed, in control animals, especially lesions of the lesser curvature. The lesions of the lesser curvature were more discrete and intensely stained than those from the greater curvature and more frequently elevated. They were linear and oriented parallel to the direction of blood flow. Lesions from the greater curvature were generally larger, though paler stained, flatter and more diffuse, consisting of a fern leaf or sun burst pattern of fine radiating lines, frequently with an uninvolved central portion (Fig. 4). The configuration of these lesions strongly suggested the action of hydrodynamic forces. Under high magnification this impression was reinforced by the finding of minute punctate foci of intense sudan staining, again arranged linearly. These latter changes were often prominent in lesions about the ostia of large branch arteries. In general, such ostia were sites of predilection for lesions as were the insertion of the ductus arteriosus and the bifurcation of the abdominal aorta. In the case of periostial lesions of the descending aorta, the changes were generally found distal to the ostium and frequently with a non-sudanophilic central region. Under magnification sudanophilic lesions from various sites occasionally showed minute, sharply circumscribed, clear nodules with little or no sudan staining (Fig. 5).

Other gross vascular lesions. The mural and valvular endocardium of the left ventricle showed minimal to moderate amounts of patchy sudanophilia, especially in the experimental animals.

Coronary artery lesions, which were not seen in control animals or in the two receiving the high corn oil diet, were found in 6 of the 8 animals receiving coconut oil and/or cholesterol (Table II). Both the incidence and severity of the lesions appeared more marked among males than females. The anterior descending branch of the left coronary artery was by far the most severely involved segment (Fig. 5). These lesions when marked were strongly sudanophilic and elevated and in at least one instance would appear to have distinctly compromised the vessel lumen.

Quite similar lesions were found in the great vessels, in descending order of involvement subclavian > carotid > iliac > abdominal arteries. These lesions, which were quite minimal in control animals were aggravated by the experimental diets (Table II).

Lesions were seen in the cerebral arteries of the 4 control animals submitted to complete autopsy. These lesions were so common and extensive that no effect could be associated with the experimental diets, with the possible exception of aneurysm formation (Table II). The uncomplicated lesions consisted of small sudanophilic foci, frequently slightly elevated, occasionally with gray fibrotic centers, and almost invariably at points of branching. Vertebral and basilar arteries and the Circle of Willis were about equally involved, the middle and anterior cerebral arteries less so. Aneurysms were seen in the middle cerebral arteries of two animals receiving the high coconut oil diet with cholesterol. These lesions varied from 1–3 mm in diameter. In one male 4 out of 17 discrete foci of sudanophilia were associated with aneurysms, these being located bilaterally and symmetrically so. One female with 11 sudanophilic lesions showed a single aneurysm. In all instances these lesions were at points of origin of smaller branches and generally in the area of the acute angle thus formed. Focal sudanophilia was present in the generally paper-thin walls of these structures, and several showed marked focal fibrous thickening of the wall.

Lesion correlations. There was a moderately high degree of correlation between mean serum cholesterol levels and aortic sudanophilia (Fig. 2), r = 0.7048 and p 0.001–0.01. Aortic sudanophilia appeared to correlate by rank (Spearman) with the visual grading of coronary artery and great vessel lesions, giving p values of 0.01–0.02 and 0.02–0.05 respectively. Aortic sudanophilia did not correlate with cerebral artery sudanophilia and indeed the latter did not correlate with any other determinant including serum cholesterol.

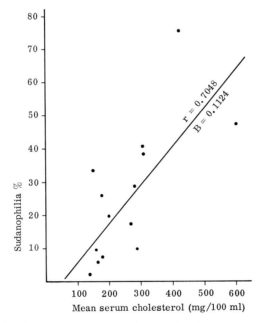

Fig. 2. Correlation between mean serum cholesterol values and percentages of the aortic endothelial surfaces involved with discrete sudanophilic lesions. Ten experimental chimpanzees fed the various high-fat diets and 4 control animals fed the natural diet are represented. p 0.001–0.01.

Microscopic findings – general. The histologic details of the vascular lesions of the present animals have been presented in some detail as part of the comparative study mentioned above (ANDRUS and PORTMAN, 1966). A brief summary of these findings will be given followed by a description of certain aspects of the present material not previously described.

In the chimpanzee, as in the New World monkeys, it was concluded that the simplest and presumably the earliest form of lipid deposition consisted of fine extracellular lipid oriented along the internal elastic membrane and adjacent elastic structures. Following this, lipid appeared intracellularly, first seen in the smooth muscle cells (Fig. 11), followed closely by its appearance in other coarsely and finely vacuolated cells. With the advent of increased intimal stroma, both elastic and collagenous, classical atherosclerosis was definable.

Two spontaneous changes in the muscular and elastic components of the arterial wall, which are common in the New World monkeys, were particularly prominent in chimpanzees. One was the well documented musculo-elastic change in the intima (GROSS et al., 1934), seen in the thoracic and abdominal aorta and in many of the smaller muscular arteries (Figs 6, 7 and 18). The other, virtually undocumented, was the presence, in the aortic arch and occasionally in the proximal portions of the arch vessels,

Fig. 3. Spontaneous atherosclerosis in the lower abdominal aorta of one of the additional control animals, a female 19 years old. Though not apparent in this photograph, most of the sudanophilic areas were elevated. Lesions can be seen extending into both common iliac arteries (aortic bifurcation at right of photograph). Discrete sudanophilia involved 21% of the abdominal segment and 26% of the whole aorta. Formalin fixation, Sudan.

Fig. 4. Ascending portion of the aortic arch of experimental animal No. 59, fed coconut oil and cholesterol. Along the lesser curvature (right of photograph) is a discrete elevated sudanophilic plaque, the long axis of which is parallel to the direction of blood flow. Opposite this lesion and along the greater curvature is a larger, more diffuse sudanophilic area, less elevated than the above, and made up of streaks radiating outward from a central clear area. A fine punctate dusting of more intense sudanophilia is superimposed on the paler areas. See microscopic preparation of same lesion (Fig. 13 and 14). Formalin fixation, Sudan.

Fig. 5. Right and left coronary arteries of experimental animal No. 189, fed coconut oil. The anterior descending branch of the left coronary artery and its branches are most severely involved. The left circumflex branch (right of photograph) shows a number of pale, frequently elevated nodules. The right coronary artery (left of photograph) is least involved. Formalin fixation, Sudan.

Fig. 6. Longitudinal section of internal carotid artery of experimental animal No. 191 fed corn oil and cholesterol, showing an intimal musculo-elastic layer. The linear cytoplasmic masses lying between the endothelium and the well-defined internal elastic membrane are those of longitudinally oriented smooth muscle cells, in contrast to the circularly oriented muscle of the underlying media seen here in cross section. The increase in intimal elastica in very close association with the smooth muscle cells is not demonstrated. Compare with Fig. 7. Zenker's fixation, Mallory's aniline blue[1].

[1] ANDRUS and PORTMAN (1966), reprinted by permission of the Zoological Society, London.

Progress in Biochemical Pharmacology, vol. 4 Plate I

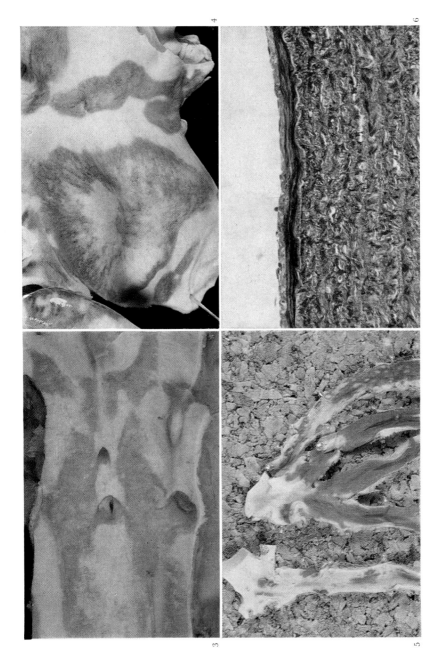

Andrus/Portman/Riopelle   S. Karger Publishers, Basle/New York

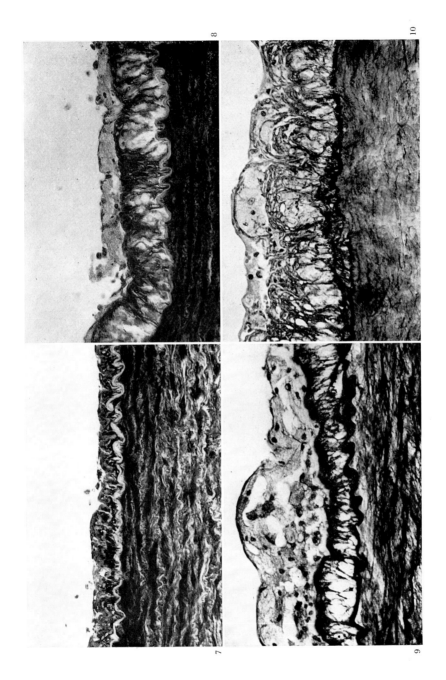

Fig. 7. Cross section of same artery illustrated in Fig. 6. The intimal smooth muscle cells are here seen in cross section while those of the media are in longitudinal profile. Zenker's fixation, Mallory's aniline blue.

Fig. 8. Cross section of same artery as in Figs. 6 and 7, cut at a different level. Lipid-containing foam cells surmount a well developed musculo-elastic layer. The latter also contains lipid both extracellularly and within smooth muscle cells (not apparent here). See Figs. 9, 10, and 11. Zenker's fixation, Mallory's aniline blue.

Fig. 9. Cross section of internal carotid artery from experimental animal No. 189, fed coconut oil. This lesion is similar to that in Fig. 8 and demonstrates the reduplication of the internal membrane associated with the musculo-elastic lesion. Lipid again is present in the latter as well as in the overlying foam cells. Formalin fixation, Verhoeff-van Gieson.

Fig. 10. Cross section of a coronary artery of same animal as in Fig. 9. This lesion is similar to those in Figs. 8 and 9 though the increasing amounts of lipid at the base of the plaque have expanded the musculo-elastic layer into a series of arcades arching inward from the internal elastic membrane. Zenker's fixation, Verhoeff-van Gieson.

Fig. 11. Longitudinal section of internal carotid artery of experimental animal No. 59 fed coconut oil and cholesterol. The intimal musculo-elastic layer contains demonstrable lipid both extracellularly and within the vacuolated smooth muscle cells, at least 6 of which are present in this field. When this lesion was viewed in cross section it closely resembled that in Fig. 8 except for the absence of surface foam cells. Zenker's fixation, Mallory's aniline blue.

Fig. 12. Tangentially cut section from the lesser curvature of the ascending portion of the aortic arch from one of the additional control animals. Immediately beneath the endothelium is some condensation of elastica but no definitive internal elastic membrane. Below this and extending well into the media is a loose textured area where the smooth muscle cells, each invested with elastica, are largely longitudinal in orientation in contrast to more circularly oriented muscle cells of the underlying areas. This close association of smooth muscle cells and elastica is characteristic of their appearance in the apical stroma of maturing intimal plaques. Increased collagen and extracellular lipid can be demonstrated in the loose textured area. The lipid present was grossly sudanophilic. Formalin fixation, Verhoeff-van Gieson.

Fig. 13. Cross section of the lesion from the lesser curvature of the aortic arch illustrated in Figs. 4 and 14. The longitudinal smooth muscle cells of the inner media are here seen in cross section. The overlying intimal foam cells are separated from the medial longitudinal muscle by some condensation of elastica but no true internal elastic membrane. Compare with Fig. 14. Formalin fixation, Verhoeff-van Gieson.

Fig. 14. Frozen section of same lesion illustrated in Figs. 4 and 13. Lipid is seen within intimal foam cells and medial longitudinal smooth muscle. Formalin fixation, Sudan IV-hematoxylin.

Fig. 15. Frozen section of the edge of a pale fibrous plaque from the aortic arch of one of the additional control animals, a 16 year old male. Lipid is restricted to the deeper parts of the plaque and is seen as a fine dusting of the stroma (A). The darker masses (B), represent foci of calcification. Formalin fixation, Sudan IV-hematoxylin.

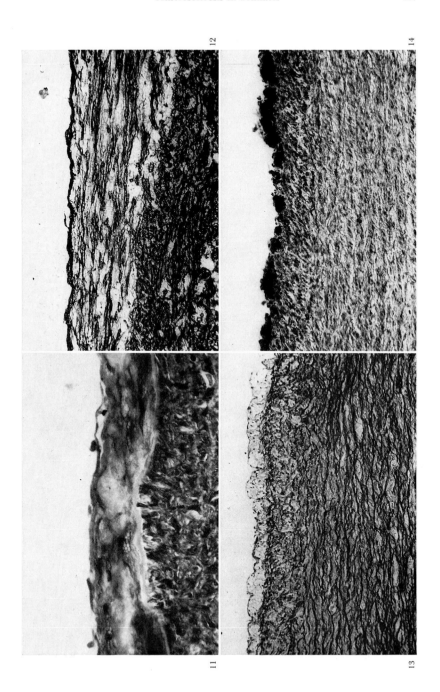

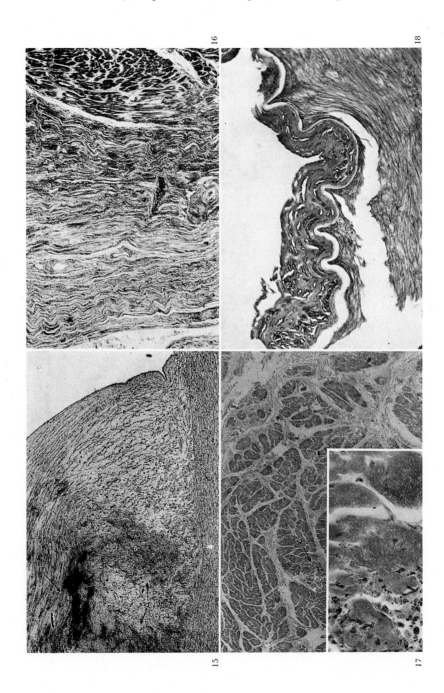

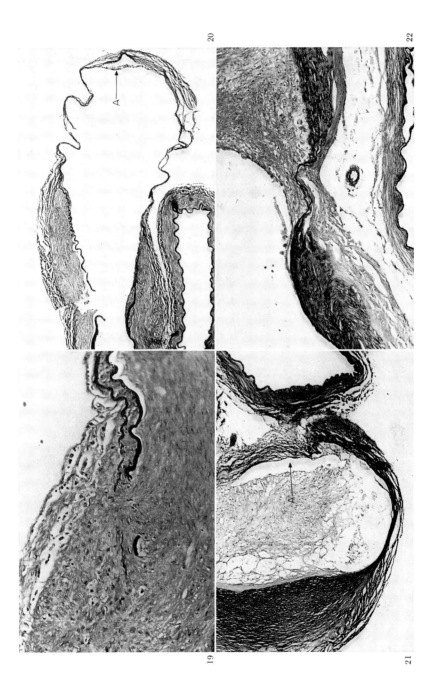

of large amounts of longitudinally oriented medial smooth muscle. The absence of a discrete internal elastic membrane in this region added to its bizarre appearance (Fig. 12). It was reminiscent of the intermediate layer of human coronary arteries described by Gross et al. (1934). These two phenomena have much in common despite obvious differences in configuration. Both changes are interpreted as resulting from hydrodynamic factors and both appear

---

Fig. 16. Myocardium of animal No. 59, fed coconut oil and cholesterol. Over a wide area the striated muscle has been replaced by dense acellular collagen. Normal appearing muscle is seen only at the right of the illustration. Small foci of muscle cell loss with inflammation were seen elsewhere. Zenker's fixation, Mallory's aniline blue.

Fig. 17. Myocardium of animal No. 191, fed corn oil and cholesterol. There is a delicate focal scarring of the myocardium as well as small foci of necrosis (A). The latter, seen in the insert, shows granular necrotic muscle cells and acute inflammation. Zenker's fixation, Giemsa.

Fig. 18. Basilar artery of experimental animal No. 288, fed corn oil, showing the edge of a musculo-elastic lesion in which lipid could not be demonstrated. The location of smooth muscle cells is indicated by the intense periodic acid-Schiff staining (dark linear markings) of the interstitium immediately adjacent to the muscle cell cytoplasm. Much of the grey appearing interstitium actually consists of unstained elastica and smaller amounts of alcian blue staining ground substance. Formalin fixation, alcian blue-periodic acid-Schiff.

Fig. 19. Basilar artery of experimental animal No. 189, fed coconut oil, showing transition between musculo-elastic change (right of photograph) and an atherosclerotic plaque. The latter shows disruption of the basal elastic structures and an extensive component of smooth muscle cells which are in direct continuity with those of the media. Formalin fixation, Verhoeff-van Gieson.

Fig. 20. Right middle cerebral artery of experimental animal No. 191, fed corn oil and cholesterol, showing an aneurysmal dilatation. There is marked attenuation of the wall which over wide areas consists solely of a thin collagenous membrane. In the depths of the aneurysm, at A, a thin layer of foam cells underlies the endothelium which is detached at this point. Formalin fixation, Verhoeff-van Gieson.

Fig. 21. Left middle cerebral artery of animal in Fig. 20, showing a second aneurysm. The lumen is all but obliterated by an extensive double-layered plaque, only a slit-like lumen remaining (A). The base of the lesion is made up of dense collagen, while the apical portions consist of a more delicate, cellular stroma rich in lipid, particularly in the deeper regions. The lipid occurs largely in extracellular pools and in scattered foam cells. Fine elastic fibrils are not apparent at this magnification. Compare with Fig. 22. Formalin fixation, Verhoeff-van Gieson.

Fig. 22. Higher magnification (cut from a deeper level) of the lesion in Fig. 21, showing transition between normal artery wall (left) and aneurysm with the edge of the overlying atherosclerotic plaque. The dense collagenous base appears to contain many smooth muscle cells as indicated by their intensely stained (periodic acid-Schiff) sheathes. Much of the fine fibrillar stroma of the apex is also stained, though more weakly. Formalin fixation, periodic acid-Schiff-hematoxylin.

susceptible to lipid deposition. As elsewhere extracellular elastica-oriented lipid appeared to precede intracellular lipid within smooth muscle cells. The rather dramatic gross lesions of the lesser and greater curvatures of the aortic arch described above (Fig. 4) represent lipid deposition as seen in the area of medial longitudinal muscle. The frankly atherosclerotic plaques that are frequently superimposed on both lesions differed in appearance. In the aortic arch the greater participation of lipid bearing *medial* smooth muscle cells formed a much deeper base to the lesion than that seen in the intimal musculo-elastic based lesion. The configuration of the lipid deposition in these two types of lesion suggested that the internal elastic membrane acts as a barrier to lipid penetration from the intima. Figs. 6–11 illustrate the intimal musculo-elastic lesion and a sequence of stages of superimposed atherosclerosis, while Figs. 12–14 demonstrate the comparable processes in the aortic arch. It should be pointed out that these are early lesions without appreciable increase in intimal stroma. The rate of maturation of the atherosclerotic process in the arch appeared rather indolent compared with that based on the musculo-elastic lesion seen elsewhere. This was a striking feature in that involvement of the arch would appear to be generally the earliest gross change seen in experimental animals, particularly in Cebus monkeys[1]. In view of the marked architectural similarities between the musculo-elastic phenomena and the atherosclerotic lesions and in view of various experimental relationships, the question was raised whether the latter lesion ever exists per se without being superimposed on the former.

Certain additional details bear repetition: the many apparent transitional forms between lipid-containing identifiable smooth muscle cells and classical foam cells; the intimate association of smooth muscle cells and the stromal elements, both elastic and collagenous; and the virtual absence of fibroblasts. It was concluded that the vascular smooth muscle cell is a multipotential cell and capable of supplying both the cellular and fibrous stromal elements of the atherosclerotic plaque. It should be noted that definitive atherosclerotic lesions were seen in the 4 control animals and in all of the experimental animals. Intimal plaques from the abdominal aorta appeared more mature than those from the thoracic, primarily

[1] ANDRUS, unpublished data.

due to larger amounts of stroma. Mature lesions were most marked in those animals receiving coconut oil and/or cholesterol. On the other hand, instances where *early* lipid deposition appeared to be clearly associated with musculo-elastic lesions were more prevalent among the controls, presumably because in the experimental animals many such foci had developed into the definitive lesion. The one experimental animal in whom definitive atherosclerosis was least marked (microscopic traces only) was also the one animal in whom aortic musculo-elastic change could not be demonstrated. And yet this individual had responded to the coconut oil and cholesterol diet with high serum cholesterol levels and with the most extensive aortic surface sudanophilia seen (75%). Such findings emphasize the potential significance of the intimal musculo-elastic lesion as a forerunner of atherosclerosis.

Senescent changes. Late sequellae and senescent changes in atherosclerotic plaques were not treated in the previous comparative study. Such changes were more prominent in spontaneous lesions than in the relatively short-term lesions produced experimentally. They have been most pronounced in our chimpanzee control material, probably due to the undoubtedly greater age of these animals. Hemorrhage within intimal plaques, or mural thrombosis overlying plaques has been seen, though rarely, in our experimental lesions. Hemosiderin or other evidence of previous hemorrhage was uncommon. The senescent plaques most commonly showed increasing density of the collagenous stroma with relative attenuation of the musculo-elastic elements. Smooth muscle cells may persist or disappear, in the latter case leaving empty lacunae outlined by intense periodic acid-Schiff staining of the stroma (Fig. 22). Lipid may persist within smooth muscle cells at the base of the lesion or occasionally in foam cells, but was characteristically extracellular and finely dispersed within the collagen (Fig. 15). Solid crystals of cholesterol may be seen extracellularly. Calcification, which is rare in even senescent lesions in New World monkeys, was not uncommon, within the internal elastic membrane or the collagenous interstitium (Fig. 15). Variable metachromasia could still be demonstrated by alcian blue or toluidine blue.

Coronary artery lesions. Musculo-elastic lesions of the coronary arteries were seen in all 14 animals. Among New World species there seemed to be a rough correlation between species size and prevalence of this lesion[1]. Thus, 67% of Allouatta demonstrated the lesion while no such changes were seen in Callicebus or Saimiri; Cebus and Lagothrix showed intermediate values. Of all 77 New World specimens, 33% of males and 42% of females showed the lesion. In chimpanzees the lesion was commoner in the epicardial branches of the coronary arteries than in the intramural branches. In the former location in terms of circumferential location it appeared more prominently on that portion adjacent to the myocardium. There was no correlation between extent of the process in the coronary arteries and that in the cerebral arteries or aorta. Nor could any relationship be seen within the coronary arteries between the extent of musculo-elastic change and atherosclerosis. There was a suggestion that the basic musculo-elastic change, irrespective of lipid deposition therein, was more prominent in those animals receiving cholesterol. The internal elastic membrane underlying the lesions may show wide areas of discontinuity. In some instances the musculo-elastic proliferation appeared external to the internal elastic membrane. The process of lipid deposition and intimal plaque formation appeared no different in the coronary arteries than that occurring elsewhere (Fig. 10). In three of the animals, well-defined lesions were readily demonstrable histologically, in two of these the process being extensive. The third animal, No. 59, whose epicardial arteries had been graded grossly as 1+ showed, however, an exceptionally extensive musculoelastic change in the intramural branches, and extensive lipid deposition was superimposed on the latter lesion.

Myocardial changes. The above animal, No. 59, with the marked intramural coronary artery changes had shown a gross area of diffuse scarring of the anterior wall of the left ventricle, 2–3 cm in diameter. This was densely collagenous with loss of striated muscle over wide areas (Fig. 16). Elsewhere were seen small foci of muscle loss with infiltrates of polymorphonuclear cells, lymphocytes, and macrophages containing eosinophilic debris. No necrotic muscle was apparent. A second animal, No. 191, showed focal fibrosis as well as focal necrosis associated with acute inflammation (Fig. 17). The gross lesions in the coronary arteries of this animal had been graded as 2+. These myocardial changes are consistent with anoxic damage though admittedly the amount of coronary artery disease

[1] ANDRUS, unpublished data.

in at least one of these cases, No. 191, would hardly seem appropriate. No myocardial parasites were seen. A third animal, No. 94, whose coronary arteries were completely free of lesions, showed minute scattered foci of myocardial fibrosis.

Cerebral artery lesions. The thin-walled cerebral arteries commonly showed several prominent features; musculo-elastic intimal lesions, with varying amounts of lipid, and well-defined atherosclerotic plaques (Figs. 18 and 19). These lesions resembled comparable lesions from other sites, though with the formalin fixed material with which we worked, smooth muscle cell identification was more difficult while metachromasia was more apparent with alcian blue staining. Most striking, however, were areas of discontinuity of the medial smooth muscle at points of origin of branching vessels. The five aneurysms present were similarly located and in each instance showed lipid deposition within the wall. The lipid bearing foci varied from a thin lining layer of foam cells to a thick intimal plaque (Figs. 20 and 21). The lumen of the aneurysm seen in Fig. 21 was almost obliterated by what appeared to be an extensive atherosclerotic lesion. The base of this plaque consisted of dense collagen in which foci of intense staining by the periodic acid-Schiff reaction suggested the presence of large numbers of smooth muscle cells (Fig. 22). Overlying this was a bulky apex of delicate fibrillar stroma rich in lipid, largely extracellular. A number of spindle-shaped cells were seen, the cytoplasm of which was outlined by fine fibrils of elastica and periodic acid-Schiff positive collagen. Though generally bipolar, some cells appeared tripolar and appeared to be fibroblasts. The presence of small numbers of capillary-like lacunae was suggested. Where uninvolved with lipid bearing lesions, aneurysm walls consisted of an attenuated sheet of collagen with remnants of internal elastic membrane (Fig. 20). At actual mouths of the aneurysms one on occasion saw either atherosclerotic intimal plaques or musculo-elastic lesions (Fig. 20). Hypertension was not present.

Discussion

Discussion of various aspects of the histopathology and pathogenesis of lesions has been included above. Discussion here will be largely limited to the broader comparative aspects of the study and features unique to the chimpanzee.

The chimpanzee would appear to show decidedly more extensive spontaneous aortic sudanophilia and definitive atherosclerosis as well as musculo-elastic intimal change than any of the 5 New World species examined. The occurrence of spontaneous cerebral lesions in laboratory animals is in our experience unique and the extent of these lesions striking. These statements apply to animals born in the Yerkes colony as well as those whose past dietary histories are unknown. The serum cholesterol levels of chimpanzees responded to the dietary manipulations in much the same way as did those of the New World species. Thus, coconut oil alone was comparable to corn oil plus cholesterol, whereas coconut oil plus cholesterol produced the most pronounced changes. Similarly, the effects of the experimental diets produced quantitative changes in aortic sudanophilia in the chimpanzees comparable to those in the 2 more reactive monkey species, Saimiri and Lagothrix. Because of the difference in duration of the two experiments, it is difficult to quantitate the relative degrees of reactivity of these three species. The impression is gained, however, that the chimpanzee is probably more reactive than the other species not only in terms of the increase in aortic sudanophilia but also in the extent of definitive lesions.

Experimentally induced coronary atherosclerosis we have seen only in the Cebus[1] and the Rhesus (MANN and ANDRUS, 1956). In both instances the animals received high dietary levels of cholesterol, at least a ten-fold increase over what was used here. Three of the experimental chimpanzees showed appreciable amounts of coronary artery atherosclerosis, the most marked example being in an animal fed coconut oil without cholesterol. Two animals showed acute focal myocardial changes consistent with ischemic damage and in one of these a grossly visable area of scarring would appear to represent a healed infarct.

Despite a distinct effect of the experimental diets on vascular lesions elsewhere, the incidence of atherosclerosis in the cerebral arteries was unaffected by diet. Histologically these lesions showed a wide spectrum of apparent ages from minimal lipid deposition in musculo-elastic lesions to senescent lesions that probably antedated the experiment. We have seen experimentally induced cerebral artery lesions only in the Rhesus (MANN and ANDRUS, 1956). It is of note that the apical portion of the double-layered plaque in

---

[1] ANDRUS and MANN, unpublished data.

the cerebral aneurysm of the above animal (No. 191, Figs. 21 and 22), is one of the rare examples where we have seen intimal fibroblastic cells in the atherosclerotic process. One other instance was in experimental atherosclerosis in the Rhesus monkey (MANN and ANDRUS, 1956), in which such cells were associated with vascularization of the lesions. The apical portions of the present lesion may well represent organization of an old thrombus, though fibrin and hemosiderin were not apparent and the presence of capillaries could not be definitely established.

Despite the apparent lack of correlation between the extent of uncomplicated cerebral atherosclerosis and the experimental diets, the occurrence of cerebral aneurysms did appear so related. The two animals bearing aneurysms were from the most atherogenic diet group (coconut oil plus cholesterol) and were the two most reactive experimental animals both in terms of serum cholesterol elevation and aortic sudanophilia. We have not seen cerebral aneurysms occurring spontaneously in any sub-human primates or indeed in any species other than man. In humans these lesions are uncommon being seen in 1–5% of autopsies (ALPERS, 1965). HASSLER (1965), however, in studying microaneurysms with microdissection techniques has found an incidence of 17%. The chimpanzee lesions involved the middle cerebral arteries which is the site of predilection of HASSLER's microaneurysms. Although the present lesions measured no more than 1–3 mm in greatest diameter, they were not microstructures in relation to the size of chimpanzee cerebral arteries, being generally larger than the parent vessel.

FORBUS (1930) first noted the potential significance of medial muscular defects in the pathogenesis of cerebral aneurysms. GLYNN (1940) emphasized the importance of the integrity of the internal elastic membrane in preventing dilatation of the arterial wall. Most authors now agree that deficiencies at both levels of the arterial wall are important pathogenic factors, i.e., medial defects, presumably developmental, and changes in the internal elastic membrane, acquired or degenerative. CARMICHAEL (1950) has implicated atherosclerosis as the source of the internal elastic membrane changes. An example of such change is seen in Fig. 19, where in some areas there is complete disappearance of this structure. It would also appear that the splitting and reduplication of the internal elastic membrane associated with the simple musculo-elastic change might well predispose to aneurysm formation when coincident with a medial defect. Both SMITH and WINDSOR (1961) and HASSLER (1965) have described comparable lesions at bifurcations of cerebral arteries, though the former authors do not mention the presence of smooth muscle cells. In both instances the authors implicate these intimal changes in aneurysm formation, though by different postulated mechanisms. In any case, in chimpanzees as in humans, points of branching of cerebral arteries appear to be critical areas and frequently demonstrate one or more of the following changes: medial muscular defects, intimal musculo-elastic lesions, and atherosclerosis. All three of these alterations would appear pertinent to any discussion of the origin of the present aneurysms. Hypertension, the physiologic factor frequently considered in the etiology of this lesion, could not be demonstrated.

Summary

As part of a series of comparative studies of atherosclerosis in non-human primates, 10 chimpanzees were maintained for one year on a series of semi-purified high fat diets (45% of calories from fat). The two variables were the type of fat (corn or coconut oils) and the presence or absence of cholesterol (0.5%). There were 2 or 3 animals per diet group. Four control animals were fed Yerkes Laboratory natural diet. Insofar as possible conditions were similar to those of a comparable study carried out earlier in 3 New World species of monkeys. Monthly serum cholesterol determinations were performed. At termination the animals were sacrificed and complete autopsies performed. Various tissue and serum lipids were measured as well as accurate quantitation of gross aortic sudanophilic lesions. With few exceptions animals were sexually mature (8–24 years), consisting of 4 males (mean weight 51.2 kg) and 10 females (mean weight 43.8 kg). Vascular specimens were available from 13 additional control animals, 8 males (mean weight 35.8 kg) and 5 females (mean weight 26.2 kg).

Discrete aortic sudanophilia was present in all control animals except one infant. Mean % surface area involvement of all 17 control aortae was 5.9%, though individuals showed levels of 20–25%. There was no significant correlation between age of animal and extent of lesions, nor were the differences between the sexes significant. There were no differences between the control animals and those receiving the corn oil diet without cholesterol.

Adding cholesterol to the latter diet or substituting coconut for corn oil caused a distinct elevation of serum cholesterol and an increase in aortic sudanophilia. These effects were additive and thus the animals receiving coconut fat and cholesterol showed the most pronounced changes. There was a high degree of correlation between mean serum cholesterol and total discrete aortic sudanophilia, $r = 0.7048$, $p\ 0.001-0.01$. Mean serum cholesterol correlated to a lesser degree with coronary artery lesions, and not at all with lesions of the cerebral arteries. Both from the gross and microscopic data, progression of the aortic lesions appeared to be associated with preponderant involvement of the lower portions of that structure as compared with the arch.

Histologically spontaneous and induced lesions appeared identical, though senescent changes were more prominent in the former. Adding coconut oil and/or cholesterol to the diet increased the extent of atherosclerosis. The lesions were indistinguishable from those seen in New World monkeys previously studied, Cebus, Wooly, and Squirrel. Spontaneous musculo-elastic lesions were common, and as in the latter species this change appeared to precede and even predispose the intima to lipid deposition. Longitudinal medial smooth muscle is particularly prominent in the aortic arch of chimpanzees and is frequently the site of lipid accumulation. Progression of such changes into classical atherosclerosis however appears indolent in contrast to rates of maturation of lesions elsewhere. Despite the preponderance of females, appreciable amounts of coronary atherosclerosis were seen only in 3 males, these animals having received cholesterol and/or coconut oil. Two of these animals showed myocardial lesions consistent with ischemic damage, both active and healed. Cerebral atherosclerosis was seen commonly in both control and experimental animals and in 2 of the latter was associated with aneurysm formation.

Acknowledgement

The authors gratefully acknowledge the technical assistance of Mr. THOMAS P. FAHERTY and CHARLES M. ROGERS, Ph.D.

References

ALPERS, B.J.: Aneurysms of the circle of Willis; in FIELDS' and SAHS' Intracranial aneurysms and subarachnoid hemorrhage, pp. 5–21 (Thomas, Springfield, Ill. 1965).

Andrus, S.B. and Portman, O.W.: Comparative studies of spontaneous and experimental atherosclerosis in primates. Some recent advances in comparative medicine, pp. 161–177 (Academic Press, London 1966).
Carmichael, R.: The pathogenesis of non-inflammatory cerebral aneurysms. J. Path. Bact. 62: 1–19 (1950).
Forbus, W.D.: On the origin of miliary aneurysms of the superficial cerebral arteries. Bull. Johns Hopk. Hosp. 47: 239–284 (1930).
Glynn, L.E.: Medial defects in the circle of Willis and their relation to aneurysm formation. J. Path. Bact. 51: 213–222 (1940).
Gofman, J.W.; Lindgren, F.T. and Elliott, H.: Ultracentrifugal studies of lipoproteins of human serum. J. biol. Chem. 179: 973–979 (1949).
Gross, L.; Epstein, E.Z. and Kugel, M.A.: Histology of the coronary arteries and their branches in the human heart. Amer. J. Path. 10: 253–273 (1934).
Hassler, O.: On the etiology of intracranial aneurysms; in Fields' and Sahs' Intracranial aneurysms and subarachnoid hemorrhage, pp. 25–37 (Thomas, Springfield 1965).
Haust, M.D.; More, R.H. and Movat, H.Z.: The role of smooth muscle cells in the fibrogenesis of arteriosclerosis. Amer. J. Path. 37: 377–389 (1960).
Hegsted, D.M.; Mills, R.C.; Elvehjem, C.A. and Hart, E.B.: Choline in the nutrition of chicks. J. biol. Chem. 138: 459–466 (1941).
Mann, G.V.; Andrus, S.B.; McNally, A. and Stare, F.J.: Experimental atherosclerosis in Cebus monkeys. J. exp. Med. 98: 195–218 (1953).
Mann, G.V. and Andrus, S.B.: Xanthomatosis and atherosclerosis produced by diet in an adult Rhesus monkey. J. lab. clin. Med. 48: 533–550 (1956).
McGandy, R.B.; Hegsted, D.M.; Myers, M.L. and Stare, F.J.: Dietary carbohydrate and serum cholesterol levels in man. Amer. J. clin. Nutr. 18: 237–242 (1966).
Portman, O.W.; Lowry, E.Y. and Bruno, D.: Effect of dietary carbohydrate on experimentally induced hypercholesteremia and hyperbetalipoproteinemia in rats. Proc. Soc. exp. Biol. Med. 91: 321–323 (1956a).
Portman, O.W.; Hegsted, D.M.; Stare, F.J.; Bruno, D.; Murphy, R. and Sinisterra, L.: Effect of the level and type of dietary fat on the metabolism of cholesterol and beta lipoproteins in the Cebus monkey. J. exp. Med. 104: 817–828 (1956b).
Portman, O.W. and Andrus, S.B.: Comparative evaluation of three species of New World monkeys for studies of dietary factors, tissue lipids, and atherogenesis. J. Nutr. 87: 429–438 (1965).
Smith, D.E. and Windsor, R.B.: Embryologic and pathogenic aspects of the development of cerebral saccular aneurysms; in Fields' Pathogenesis and treatment of cerebrovascular disease, pp. 367–386 (Thomas, Springfield, Ill. 1961).

Authors' addresses: Dr. S.B. Andrus, Assistant Professor of Pathology, Department of Nutrition, Harvard School of Public Health, *Boston, Mass.*; Dr. O.W. Portman, Senior Scientist, Oregon Regional Primate Research Center, *Beaverton, Oregon*, and Dr. A.J. Riopelle, Director, Delta Regional Primate Research Center, *Covington, La.* (USA).

From the Department of Laboratory Animal Medicine of The Bowman Gray School of Medicine of Wake Forest College, Winston-Salem, North Carolina, USA

Pathologic Characteristics of Atherosclerosis in New World Monkeys*

T. B. Clarkson, B. C. Bullock and N. D. M. Lehner, Winston-Salem, N.C.

Monkeys occupy an important place among laboratory animals used for studies of atherosclerosis principally because they share with human beings taxonomic classification as *Primates*. The atherosclerotic lesions that occur naturally in nonhuman primates are most often fatty streaks, complicated lesions occurring very infrequently. Emphasis in our laboratories has been primarily on the New World species of monkeys because of their size, availability, and susceptibility to atherosclerosis both in their natural environment and as a result of manipulation in the laboratory. The purpose of this paper is to present some of our observations on atherosclerosis of squirrel, Cebus, and spider monkeys.

The serum cholesterol levels of recently trapped squirrel, Cebus and spider monkeys have been studied by St. Clair et al. [8]. These data are summarized in Table I. Spider monkeys (*Ateles* sp) have somewhat higher serum cholesterol levels and transport a higher percent of cholesterol in the beta lipoprotein fraction than do squirrel monkeys (*S. sciureus*) or Cebus monkeys (*C. albifrons*).

Naturally occurring atherosclerosis is common among wild caught mature squirrel monkeys (*Saimiri sciureus*) maintained in the laboratory for 3–24 months, Middleton et al. [6]. About 90% of mature laboratory acclimated squirrel monkeys have fatty streaks of the aorta visible after gross staining with Sudan IV; while about 10% have small raised aortic plaques visible without gross staining.

* Supported in part by grants from the U.S.P.H.S., National Institutes of Health (FR-00180, FR-00236 and HE-04371).

Table I. Serum cholesterol levels of some new world monkeys[1]

Species	Number	Total serum cholesterol (mg%)[2]	Alpha-lipoprotein cholesterol (mg%)[2]	Beta-lipoprotein cholesterol (mg%)[2]
Saimiri sciureus	220	105 ± 2.1	30 ± 0.8	74 ± 0.2
Cebus albifrons	57	90 ± 3.6	21 ± 1.7	67 ± 3.2
Ateles sp	29	130 ± 7.0	12 ± 0.7	118 ± 6.8

[1] Adapted from data of St. Clair et al., Fed. Proc. 25, No. 2: 388 (1966).
[2] Mean values ± standard error of the mean.

Fatty streaks appear to be more common in thoracic aorta, while fibrous plaques are seen more commonly in the abdominal aorta (Fig. 1).

The aortic lesions occurring naturally in squirrel monkeys consist microscopically of intra- and extracellular accumulations of Sudan IV positive material associated with varying numbers of spindle cells. The spindle cells are usually seen in greatest number between the lumen of the vessel and the main collection of lipid.

Fig. 1. Sudan IV stained aorta from an adult female squirrel monkey. There is diffuse fatty streaking of the thoracic portion and a raised plaque near the iliac bifurcation.

Fraying and splitting of the internal elastic lamina is frequently seen in the central portion of lesions.

Coronary artery atherosclerosis occurs in about 30% of mature laboratory acclimated squirrel monkeys and most often affects the small intramyocardial arteries. No occlusive lesions of coronary arteries have been seen in squirrel monkeys.

The progression of atherosclerosis among squirrel monkeys would appear to be age related. Atherosclerotic lesions are much more common among mature than juvenile squirrel monkeys. Squirrel monkeys examined shortly after capture appear to have less extensive atherosclerosis than those maintained for varying lengths of time in the laboratory. This would seem to be particularly true in the case of coronary artery atherosclerosis which seems to be rare among free living squirrel monkeys. Squirrel monkeys maintained in the laboratory usually have serum cholesterol levels of about 200 mg% which would appear to be a hypercholesteremic level as compared with the levels (105 mg%) maintained by these monkeys in their native environment. This increase in serum cholesterol level has been studied by MacNintch et al. [4]. The serum cholesterol increase took place over a 55 day period during which the animals were held at a compound in South America and shipped by air to our research farm. The monkeys were held in a relatively quiet environment at our research farm for about 100 days. During this interval serum cholesterol values gradually declined. When the monkeys were moved to the animal facility at the medical center the change of environment apparently caused another increase in serum cholesterol levels. The fluctuations seen in serum cholesterol levels were almost entirely in the beta lipoprotein fraction.

The extent and severity of aortic and coronary artery atherosclerosis of squirrel monkeys has been increased by feeding semi-synthetic diets to which cholesterol (0.5%) had been added, Middleton et al. [7]. In this experiment the effect of the level of dietary protein and cholesterol on atherosclerosis in squirrel monkeys was studied using a 2×2 factorial design. The diets were either high (25%) or low (9%) in protein, high in fat (25%) and fed with or without the cholesterol supplement which was calculated to be 0.5% of the diet by weight. The cholesterol supplement increased serum cholesterol levels in these animals whether they were fed the high or low level of protein. The degree of aggravation of aortic or coronary atherosclerosis was greater among animals fed high levels of protein. The high protein-cholesterol group had 44% of thoracic (Fig. 2) and 21% of abdominal aortic intimal surface involved with sudanophilic lesions compared with 18 and 4% respectively for the low protein-cholesterol group. Coronary artery atherosclerosis (Fig. 3) was about twice as extensive in those animals fed high protein and cholesterol as was observed in the low protein-cholesterol-fed groups.

Naturally occurring atherosclerosis of *Cebus albifrons* monkeys appears to be rare. In our own autopsy series we have seen almost no arterial disease among animals of this group. *Cebus albifrons* monkeys are susceptible to dietary cholesterol-induced atherosclerosis. Mann et al. [5] produced atherosclerosis in Cebus

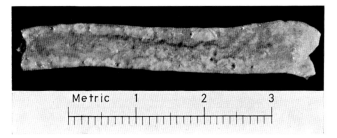

Fig. 2. Unstained thoracic aorta from an adult male squirrel monkey that had been fed a semi-synthetic diet containing 25% protein, 25% fat and 0.5% cholesterol for 1 year. (2.3×)

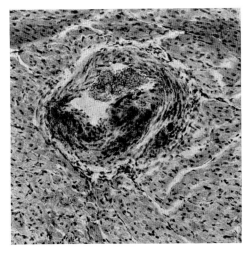

Fig. 3. Sudan IV and hematoxylin stained frozen section of the left ventricular myocardium from an adult male squirrel monkey fed the 'high protein with cholesterol' diet. A lipid containing plaque fills approximately 40% of the lumen of the intramyocardial artery. (110×)

monkeys by feeding diets high in cholesterol (5%) and low in sulfur containing amino acids. WISSLER et al. [9] also produced aortic atherosclerosis in *Cebus albifrons* monkeys by dietary means. Their semi-synthetic diets contained 25% dietary fats of different types (butter fat, coconut oil, and corn oil), 25% protein from casein and cholesterol content adjusted to 0.5%. One of four monkeys which received butter fat and one of four receiving coconut oil developed gross aortic atherosclerosis. Monkeys fed corn oil did not develop atherosclerotic lesions. No coronary artery lesions were seen in any of the animals.

Age differences in the response of *Cebus albifrons* monkeys to dietary induced atherosclerosis have recently been described by CLARKSON et al. [1]. Old or very young monkeys were fed a high-protein, high-fat diet either with or without a supplement of 0.5% cholesterol by weight. The serum cholesterol levels of the cholesterol-fed young and old animals increased similarly during the first 3 to 4 months of the experiment; however, for the remainder of the experiment (9 months) the young monkeys returned to near control levels while the older animals remained at a definite hypercholesterolemic level. During the period in which there were wide differences in the serum cholesterol level of the young and old group, LOFLAND et al. [3] studied the disappearance of intravenously administered radioactive cholesterol from the serum of the animals. Young cholesterol-fed animals were found to remove the labeled cholesterol from the blood significantly faster than old animals.

The effect of the atherogenic diet on the prevalence of aortic, coronary artery, and carotid artery atherosclerosis of the young and old *Cebus albifrons* monkeys is summarized in Table II. The

Table II. Effect of cholesterol feeding on the prevalence of atherosclerosis of young and old *Cebus Albifrons* monkeys[1]

Group	Diet	Aortic atherosclerosis	Coronary artery atherosclerosis	Carotid bifurcation atherosclerosis
Young	Control	1/6	1/6	0
Young	Cholesterol-fed	5/5	3/5	1/5
Old	Control	2/5	2/5	0
Old	Cholesterol-fed	3/3	3/3	3/3

[1] Expressed as number with gross atherosclerosis over the number in the group.

raised aortic atherosclerotic plaques occurred primarily in the arch (Fig. 4). All of the old cholesterol-fed monkeys had plaques at the carotid artery bifurcation (Fig. 5); while only one of the young cholesterol-fed monkeys had lesions at this site. The coronary artery lesions seen in some of the old cholesterol-fed monkeys were extensive (Fig. 6). Although major differences in prevalence of lesions were seen only at the carotid bifurcation, differences in extent of the lesions at these sites were striking. The effect of cholesterol feeding on the extent of atherosclerosis of the aorta, coronary artery, and carotid artery bifurcation is summarized in Table III. The

Fig. 4. Sudan IV stained aortic arch from an adult *Cebus albifrons* fed a 0.5% cholesterol diet for 1 year. (Sudan IV, 2.3×)

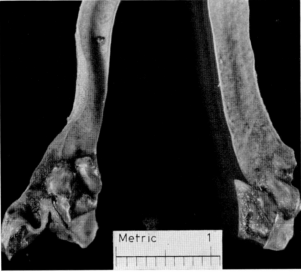

Fig. 5. Raised plaques at the carotid artery bifurcations from the monkey whose aortic arch is shown in Fig. 4. Stained with Sudan IV (2.3×)

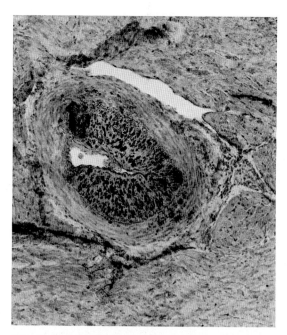

Fig.6. Sudan IV and hematoxylin stained frozen section from the interventricular septum of the same Cebus monkey. (110×)

Table III. Effect of cholesterol feeding on atherosclerosis of young and old *Cebus albifrons* monkeys[1]

Group	Diet	Aortic atherosc. index[2]	Coronary artery atherosc. index[3]	Carotid bifurc. atherosc. score[4]
Young	Control	0.08±0.08	0.45±0.45	0
Young	Cholesterol-fed	6.50±3.5	4.50±2.6	0.2±0.2
Old	Control	0.12±0.09	0.86±0.68	0
Old	Cholesterol-fed	66.60±18.6	37.80±9.0	2.6±2.5

[1] All values are expressed as means for the group, followed by the standard error of the mean.
[2] % of aortic intimal surface that was sudanophilic.
[3] % of arteries seen with atherosclerosis in 15 step sections of left ventricle and septum.
[4] Based on an arbitrary scale of severity, 0–3.

striking age difference susceptibility of these monkeys to the dietary induced disease is evident.

Preliminary experiences with spider monkeys (*Ateles* sp) have indicated the value of these animals in atherosclerosis research.

Fatty streaks and raised intimal plaques (Fig. 7) have been found in a small necropsy series of mature spider monkeys. Coronary artery atherosclerosis has also been seen in these animals. The lesions seen to date have been small, but FINLAYSON [2] has reported fairly severe left coronary artery atherosclerosis in a female

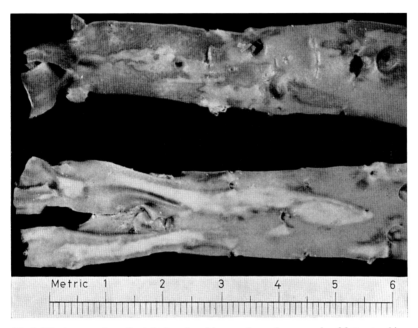

Fig. 7. Distal aortas from 2 adult female spider monkeys. An example of fatty streaking is shown at the top and of raised fibrous plaques at the bottom of the photograph. (Sudan IV, 2.4×)

spider monkey that had been in captivity for 13 years. It is of interest that spider monkeys appear to transport a higher percentage of cholesterol in the beta lipoprotein fraction than do other New World monkeys.

Summary

Naturally occurring atherosclerosis is common among squirrel monkeys (*Saimiri sciureus*). The serum cholesterol level of laboratory confined monkeys is higher than that of recently trapped animals. The prevalence and extent of atherosclerosis appears to increase with laboratory confinement.

Cebus albifrons develop very little atherosclerosis naturally although they are susceptible to dietary induced disease, old monkeys being more susceptible than young ones.

Limited experience with spider monkeys (*Ateles* sp) suggests that these animals may have considerable naturally occurring atherosclerosis.

References

1. CLARKSON, T.B.; MIDDLETON, C.C.; LOFLAND, H.B. and BULLOCK, B.C.: Effect of age on cholesterol-induced atherosclerosis in *Cebus* monkeys. Fed. Proc. *24:* 310 (1965).
2. FINLAYSON, R.: Spontaneous arterial disease in exotic animals. J. Zool. *147:* 239–343 (1965).
3. LOFLAND, H.B.; CLARKSON, T.B. and MIDDLETON, C.C.: Cholesterol turnover in *Cebus* monkeys. Fed. Proc. *23:* 2: 497 (1964).
4. MACNINTCH, J.E.; MIDDLETON, C.C.; CLARKSON, T.B.; ST. CLAIR, R.W. and LOFLAND, H.B.: The effects of changing environment on serum cholesterol levels in squirrel monkeys. Fed. Proc. *25,* 2: 388 (1966).
5. MANN, G.V.; ANDRUS, S.B.; MCNALLY, A. and STARE, F.J.: Experimental atherosclerosis in Cebus monkeys. J. Exp. Med. *98:* 195–218 (1953).
6. MIDDLETON, C.C.; CLARKSON, T.B.; LOFLAND, H.B. and PRICHARD, R.W.: Atherosclerosis in the Squirrel monkey. Arch. Path. *78:* 16–23 (1964).
7. MIDDLETON, C.C. and LOFLAND, H.B.: Aggravation of atherosclerosis in squirrel monkeys (*Saimiri sciureus*) by diet. Fed. Proc. *24,* 2: 311 (1965).
8. ST. CLAIR, R.W.; MIDDLETON, C.C.; CLARKSON, T.B. and LOFLAND, H.B.: Serum lipids, lipoproteins and atherosclerosis in New World primates. Fed. Proc. *25,* 2: 388 (1966).
9. WISSLER, R.W.; FRAZIER, L.E.; HUGHES, R.H. and RASMUSSEN, R.A.: Atherogenesis in Cebus Monkey: I. Comparison of three food fats under controlled dietary conditions. Arch. Path. *74:* 312–322 (1962).

Authors' address: Department of Laboratory Animal Medicine, Bowman Gray School of Medicine, *Winston-Salem, North Carolina* (USA).

Department of Pathology, Queen's University and the Kingston General Hospital, Kingston, Ontario, Canada

Electron Microscopic and Immuno-Histochemical Studies of Fatty Streaks in Human Aorta*

M. Daria Haust, London, Ont.

(With colour plate II)

Introduction

Previous light microscopic studies showed that certain forms of atherosclerotic lesions in men contained proteinaceous material in which fibrin was often demonstrated by special stains. The presence of fibrin indicated that this material was largely derived from blood; it was either insuded into the intima from the lumen [1–4] or deposited upon it in the form of thrombi [5–10]. It was difficult to assess, however, whether other blood constituents also participated in the development and progression of atherosclerotic lesions. The recently available antibody techniques provided means for specific demonstration of various blood proteins in these lesions at light [11–18] and electron microscopic [19] levels.

It was recognized on the basis of light microscopic studies [9] that fibrous atherosclerotic plaques were characterized by the presence of cells which had all the morphologic features of smooth muscle cells. Moreover, these cells appeared to be associated with fibrogenesis in the fibrous plaques [20, 21]. It was, however, not possible to determine whether in fatty streaks and dots the cells containing fat were of similar nature. With the improved techniques it became possible to adequately preserve tissues obtained at necropsy for electron microscopy and thus to re-examine the human

---

* Supported by grants-in-aid from the Medical Research Council of Canada (MA-1037 and MA-1250), and from the Ontario Heart Foundation, Toronto, Ontario, Canada.

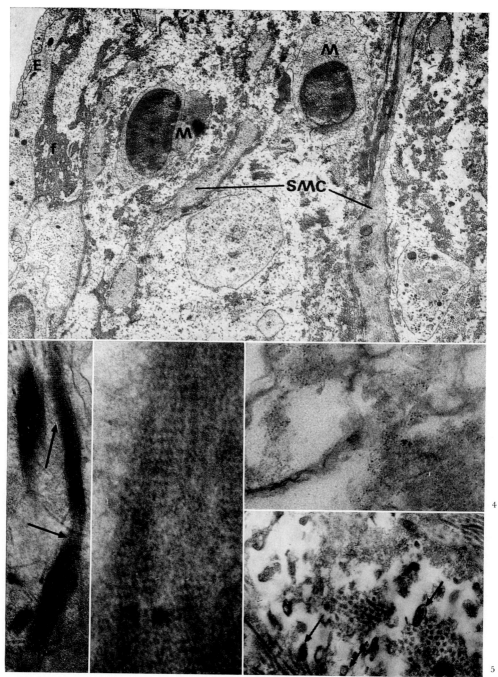

Figs. 1, 2, 3, 4 and 5.

lesions with special reference to the cellular components of fatty streaks [22–24].

The present report is a resumé of the results of investigations carried out in our laboratories during the past six years[1] and undertaken to elucidate the above problems.

Material and Methods

Antisera directed against human fibrinogen (using Cohn fraction I of human plasma obtained from Nutritional Biochemicals Corporation, Cleveland, Ohio), serum albumin (source as above), and alpha-1 and alpha-2 lipoproteins (obtained according to the method of HAVEL et al. [25]) were produced, purified and characterized by methods described previously [15]. For the fluorescent antibody technique, the gamma globulin fraction, or the 7S of the gamma globulin of the antifibrinogen, anti-albumin, anti-alpha-1 and anti-alpha-2 lipoproteins were conjugated with fluorescein isothiocyanate by a modified COONS's method [26]. Measures adopted to reduce the nonspecific staining, control procedures, preparation of tissues and sections for the staining with the specific conjugate were described previously [15]. For the demonstration of albumin and alpha-1 lipoproteins it was necessary to fix the sections for 15–20 min in absolute alcohol either prior to, or following the staining with the conjugate.

Methods for tissue preparation and staining for electron microscopy were described elsewhere [19, 24]. Conjugation of antifibrinogen with ferritin for electron microscopy was carried out by a modified method of SINGER and SCHICK [27].

[1] This investigation was carried out in collaboration with Dr. Robert H. More, Professor and Head of the Department; Drs. J. U. Balis, J. C. Wyllie and J. H. Choi participated in various phases of this work.

Fig. 1. An 'early' fatty aortic streak. The intact endothelial layer (E) is in some areas narrow. The subendothelial intima contains mononuclear cells (M) and smooth muscle cells (SMC). Electron-dense fibrillar and granular material (f) and round but irregular bodies (to right in the photograph) are present in the widened intercellular space. Note the paucity of collagen and elastica in the area. X 6,500.

Fig. 2. On higher magnification clumps of the electron-dense extracellular fibrillar material exhibit periodicity (arrows). X 31,500.

Fig. 3. The axial periodicity of the electron-dense extracellular material similar to that seen in Figs. 1 and 2, ranges from 200 to 220 Å, and is thus characteristic of fibrin. X 151,200.

Fig. 4. The granular and fibrillar material in the intercellular space is labelled when the ferritin-conjugated antibody technique is employed. Note that the interstices and some non-descript structures are not labelled. Cryostat-cut tissue; uranyl acetate only. X 64,000.

Fig. 5. Higher magnification of an area similar to that seen at right in Fig. 1. Irregularly shaped, partly osmiophilic bodies (arrows) are intermingled with 'dissolving' and fragmented connective tissue elements. X 31,500.

Results

1. Electron microscopic observations. In fatty streaks the endothelial cells contained at times fat, and often were partly separated from each other by fibrillar or granular electron-dense material (Fig. 1). Similar material was present in the widened, subendothelial intercellular spaces; some of it exhibited an axial periodicity (Fig. 2). Under higher power the latter was found to be characteristic of fibrin (200–220 Å) (Fig. 3). It was possible to demonstrate with the ferritin-conjugated antibody technique that some of the electron-dense extracellular material not exhibiting the typical periodicity, was also fibrin (Fig. 4). In addition, numerous bodies varying in size, shape and electron density (Fig. 5) were present in the extracellular space; many of these contained myelin figures. These bodies were interpreted as being precipitated fats, as they corresponded to the fine dispersed extracellular fat of light microscopy. Collagen fibrils were often fragmented and permeated by the fibrillar and granular material, and the elastic tissue components appeared in various stages of 'dissolution' [28].

Two definite types of cells were present in the fatty streaks. One, with all the characteristic features of a smooth muscle cell (SMC) [23, 24], was more numerous in the 'early' lesion. Various numbers of fat droplets were present in many of these SMC's (Figs. 6 and 7). When the fat droplets occupied much of the cytoplasm, the SMC assumed the appearance of a foam cell (Fig. 7); it was termed

---

Fig. 6. A smooth muscle cell from a more advanced lesion than that seen in Fig. 1, contains fusing fat droplets (F). These droplets have an irregular outline and are surrounded by free ribosomes. The basement membrane (BM) enveloping the smooth muscle cell is thickened. Fibrin-like material (f) and dense bodies (b) similar to those seen in Fig. 5 are in close proximity to the BM. X 13,000.

Fig. 7. A myogenic foam cell from a well developed fatty streak. The cytoplasm is filled and distended with numerous fat droplets, but the cell is still recognizable as a smooth muscle by the presence of the enveloping basement membrane. X 8,750.

Fig. 8. Complex lipid inclusions and lysosomes as well as membrane bound vacuoles are present in a macrophage from a fatty streak. Numerous small smooth vesicles and mitochondria are also present in the 'loose' cytoplasm. X 18,000.

Fig. 9. Two foam cells derived from macrophages contain numerous vacuoles and complex lipid inclusions; an eccentric nucleus (N) is present in one cell and in the other there are myelin figures (arrow). X 5,390.

Unless specified otherwise all electron micrographs (Figs. 1–9) were prepared from blocks of tissue fixed in glutaraldehyde and post-fixed in osmic acid, and sections stained with uranyl acetate followed by lead citrate.

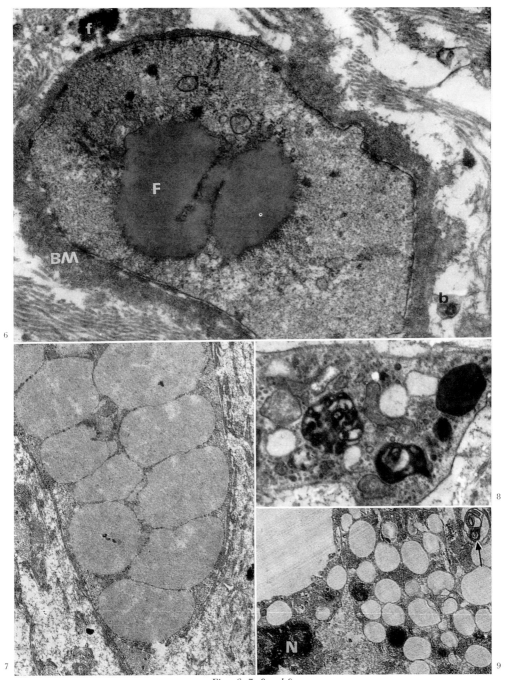

Figs. 6, 7, 8 and 9.

myogenic foam cell [23, 24] to indicate its origin. The basement membrane of a fat-containing SMC was often broadened and extracellular fat bodies as well as material resembling fibrin were occasionally observed in its vicinity (Fig. 6). In advanced lesions myogenic foam cells were often disintegrating, and their fat-containing fragments were observed in the extracellular space [24]. The second cell type in the fatty streaks appeared somewhat later in the development of the lesion. When free of the intracellular fat, it had some features of a lymphocyte and some of a macrophage (Fig. 1). With the progression of the lesion fat, presumably phagocytosed, appeared in the cytoplasm of these cells (Fig. 8). This process culminated in the formation of the second type of a foam cell (Fig. 9). Fat droplets of complex composition, lysosomal bodies (Fig. 8) and myelin figures (Fig. 9) characterized these (non-myogenic) foam cells.

2. Fluorescent immuno-histochemical studies. In the early fatty streaks fibrin was distributed throughout the lesion in finely dispersed, filamentous form (Fig. 10), whereas clumps of fibrin were a conspicuous feature of more advanced lesions (Fig. 11). In the latter, fibrin was often associated with fat accumulations (Fig. 11). No

Fig. 10. Early fatty streak. Section stained with fluorescent antifibrinogen. Fine strands of fibrin are distributed throughout the lesion. Note the paucity of connective tissue in the area of the lesion. X 250.

Fig. 11. Advanced fatty streak. Section stained with fluorescent antifibrinogen. Fibrin is seen in clumps and bands. It often surrounds pools of fat, here represented by white spots. X 250.

Fig. 12. Advanced fatty streak, similar to that represented in Fig. 11. The section was treated with unconjugated antifibrinogen prior to staining with fluorescein conjugated antifibrinogen. Specific staining is abolished. X 250.

Fig. 13. An early fatty streak stained with fluorescent anti-albumin. The albumin permeates diffusely the lesion situated in the upper two-thirds of the intima. In addition, small focal (intracellular) accumulations of albumin can be detected. X 150.

Fig. 14. Fatty streak stained with fluorescent anti-alpha 1 lipoprotein. The specific staining is observed only in a narrow superficial zone of the lesion. X 150.

Fig. 15. Fatty streak stained with fluorescent anti-alpha 2 lipoprotein. Note that as compared with Fig. 14, the specific staining is rather diffuse and it reaches almost the level of the intimal elastic lamina. Some specific staining is present also in small foci (intracellular). X 150.

Figures 10–15 were prepared from cryostat-cut sections of fatty streaks stained with the appropriate fluorescein isothiocyanate-conjugated antibody and examined with ultraviolet light. The specific staining is indicated by yellow-green fluorescence. Autofluorescence is exhibited variously by collagen (dark blue), elastica (silver-blue) and lipids (dark-blue, yellow and white).

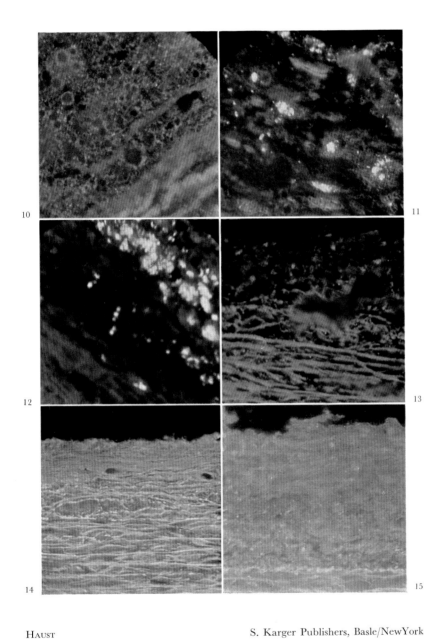

fibrin was detected in control (normal) intimal areas. The specific staining was always abolished with appropriate control procedures not only for fibrin (Fig. 12) but also for other proteins tested. Fibrin was present in all 50 fatty streaks examined.

Albumin was demonstrated in normal intimas in the superficial layer and in small amounts only. In the fatty streaks it permeated into the deep layers of the lesions reaching at times the musculo-elastic layer above the internal elastic membrane (Fig. 13). It was present in a very finely dispersed from in the extracellular space and in addition, it could be at times demonstrated in the endothelial and intimal cells (presumably macrophages) (Fig. 13). Albumin never assumed the filamentous or clumped form characteristic of fibrin (Figs. 10 and 11). Of the 50 fatty streaks examined, twenty showed strong specific staining, twenty-two were moderately positive and eight were negative.

Alpha-1 lipoproteins were present in approximately four-fifths of the fifty fatty streaks examined. The specific fluorescence was confined to the more superficial layers of the lesion (Fig. 14) and was observed in some cells as well as extracellularly. Normal intimal areas of only older individuals exhibited the specific fluorescence in the superficial layers, but the staining was inconsistent and less intense than that of fatty streaks.

Alpha-2 lipoproteins were distributed throughout fatty streaks, and often reached the level immediately adjacent to the internal elastic lamina (Fig. 15). The specific staining was intense and present intra- and extracellularly. The number of lesions stained, and the staining of normal intimal areas were comparable with findings obtained for alpha-1 lipoproteins (see above).

Summary and Conclusions

The results of the studies carried out in our laboratories during the past six years utilizing immuno-histochemical techniques show that fibrin, albumin and alpha-1, and alpha-2 lipoproteins are present in fatty streaks of human aorta. Normal controls did not contain fibrin and the remaining proteins were present in small amounts; their distribution was different from that observed in the lesions. Our results are largely in agreement with the work of other investigators [11–14]. The discrepancy existing between the findings with respect ot the albumin may be attributed to a slight alteration in procedures. In our studies slides were fixed in alcohol prior to, or following the staining with conjugated anti-albumin, whereas no fixation was used in the studies of other investigators [14], and no albumin was detected in these.

The electron microscopic studies show that the smooth muscle cell is involved in the development and progression of the fatty streaks as it accumulates fat and gives rise to the myogenic foam cell. The other foam cell in these lesions is derived from a mononuclear cell. These results are largely in agreement with those of others [29–31].

It is concluded on the basis of these and other studies that plasma proteins enter the intima during the process of atherosclerosis and participate in the formation of fatty streaks of human aorta. The progression and the outcome of these lesions may well depend upon the ability of the intimal cells (smooth muscle cells, mononuclears) to handle this 'influx' into the arterial wall.

References

1. RÖSSLE, R.: Über die serösen Entzündungen der Organe. Virchow's Arch. path. Anat. *311:* 252–284 (1944).
2. MOVAT, H.Z.; HAUST, M.D. and MORE, R.H.: The morphologic elements in the early lesions of arteriosclerosis. Amer.J. Path. *35:* 93–101 (1959).
3. HAUST, M.D. and MORE, R.H.: Morphologic evidence and significance of permeation in the genesis of arteriosclerosis. Circulation *16:* 496 (1957).
4. MORE, R.H. and HAUST, M.D.: Encrustation and permeation of blood proteins in the genesis of arteriosclerosis. Amer.J. Path. *33:* 593 (1957).
5. ROKITANSKY, C.: A manual of pathological anatomy. Day, G.E. (trans.). The Sydenham Society, London *4:* 272 (1852).
6. CLARK, E.; GRAEF, I. and CHASIS, H.: Thrombosis of the aorta and coronary arteries, with special reference to the 'fibrinoid' lesions. Arch. Path. *22:* 183–212 (1936).
7. DUGUID, J.B.: Thrombosis as a factor in the pathogenesis of aortic atherosclerosis. J. Path. Bact. *60:* 57–61 (1948).
8. CRAWFORD, T. and LEVENE, C.I.: The incorporation of fibrin in the aortic intima. J. Path. Bact. *64:* 523–528 (1952).
9. HAUST, M.D.; MORE, R.H. and MOVAT, H.Z.: The mechanism of fibrosis in arteriosclerosis. Amer.J. Path. *35:* 265–273 (1959).
10. MORE, R.H. and HAUST, M.D.: Atherogenesis and plasma constituents. Amer.J. Path. *38:* 527–537 (1961).
11. WISSLER, R.W. and KAO, V.: Immunohistochemical studies of the human aorta. Fed. Proc. *21:* 95 (1962).
12. WATTS, H.F.: Role of lipoproteins in the formation of atherosclerotic lesions. In: Evolution of the atherosclerotic plaque. p.98–113 (The University of Chicago Press, Chicago and London 1964).
13. WOOLF, N. and CRAWFORD, T.: Fatty streaks in the aortic intima studied by an immuno-histochemical technique. J. Path. Bact. *80:* 405–408 (1960).

14. KAO, V.C.K. and WISSLER, R.W.: A study of the immuno-histochemical localization of serum lipoproteins and other plasma proteins in human atherosclerotic lesions. Exp. Mol. Path. *4:* 465–479 (1965).
15. HAUST, M.D.; WYLLIE, J.C. and MORE, R.H.: Atherogenesis and plasma constituents I. Demonstration of fibrin in white plaques by the fluorescent antibody technique. Amer.J.Path. *44:* 255–267 (1964).
16. WYLLIE, J.C.; MORE, R.H. and HAUST, M.D.: Demonstration of fibrin in yellow aortic streaks by the fluorescent antibody technique. J.Path.Bact. *88:* 335–338 (1964).
17. HAUST, M.D.; CHOI, J.; WYLLIE, J.C. and MORE, R.H.: Demonstration of albumin in human atherosclerotic lesions by the fluorescent antibody techniques. Circulation *32,* 11, 17 (1965).
18. CHOI, J.H.: Demonstration of lipoproteins in human atherosclerotic lesions by the fluorescent antibody technique. Fed.Proc. *25:* 665 (1966).
19. HAUST, M.D.; WYLLIE, J.C. and MORE, R.H.: Electron microscopy of fibrin in human atherosclerotic lesions. Immuno-histochemical and morphologic identification. Exp.Mol.Path. *4:* 205–216 (1965).
20. HAUST, M.D. and MORE, R.H.: New functional aspects of smooth muscle cells. Fed.Proc. *17:* 440 (1958).
21. HAUST, M.D.; MORE, R.H. and MOVAT, H.Z.: The role of smooth muscle cells in the fibrogenesis of arteriosclerosis. Amer.J.Path. *37:* 377–389 (1960).
22. HAUST, M.D.; MORE, R.H. and BALIS, J.U.: Electron microscopic study of intimal lipid accumulations in the human aorta and their pathogenesis. Circulation *26:* 656 (1962).
23. HAUST, M.D. and MORE, R.H.: Significance of the smooth muscle cell in atherogenesis. In: Evolution of the atherosclerotic plaque. p.51–63 (The University of Chicago Press, Chicago and London 1963).
24. BALIS, J.U.; HAUST, M.D. and MORE, R.H.: Electron microscopic studies in human atherosclerosis. Cellular elements in aortic fatty streaks. Exp.Mol.Path. *3:* 511–525 (1964).
25. HAVEL, R.J.; EDER, H.A. and BRAGDON, J.H.: The distribution and chemical composition of ultracentrifugally separated lipoproteins in human serum. J.clin. Invest. *34:* 1345–1353 (1955).
26. GOLDSTEIN, G.; SLIZYS, I.S. and CHASE, M.W.: Studies on fluorescent antibody staining. I. Non-specific fluorescence with fluorescein-coupled sheep anti-rabbit globulins. J.exp.Med. *114:* 89–110 (1961).
27. SINGER, S.J. and SCHICK, A.F.: The properties of specific stains for electron microscopy prepared by the conjugation of antibody molecules to ferritin. J.Biophys. biochem.Cytol. *9:* 519–537 (1961).
28. HAUST, M.D.; MORE, R.H.; BENCOSME, S. and BALIS, J.U.: Electron microscopic studies in human atherosclerosis. Extra-cellular elements in aortic dots and streaks. Exp.Mol.Path. *6:* 300–313 (1967).
29. GEER, J.C.; MCGILL, H.C., Jr. and STRONG, J.P.: The fine structure of human atherosclerotic lesions. Amer.J.Path. *38:* 263–287 (1961).
30. MCGILL, H.C., Jr. and GEER, J.C.: The human lesion, fine structure. In: Evolution of the atherosclerotic plaque. p.65–76 (The University of Chicago Press, Chicago and London 1964).
31. GEER, J.C. and GUIDRY, M.A.: Cholesteryl ester composition and morphology of human normal intima and fatty streaks. Exp.Mol.Path. *3:* 485–499 (1964).

Author's address: M. Daria Haust, M.D., Department of Pathology, Health Sciences Centre, University of Western Ontario, *London, Ontario* (Canada).

From the Department of Pathology, University of Cambridge and the Donner Laboratory, University of California, Berkeley

Aortic and Coronary Atherosclerosis in Baboons

A.N. Howard, G.A. Gresham, D.E. Bowyer, Cambridge and F.T. Lindgren, Berkeley, Cal.

The use of baboons in the field of atherosclerosis would seem to offer many advantages. As McGill and his co-workers have shown, they develop aortic atherosclerosis in the wild state [1]. They are omnivorous, readily eat experimental diet rich in fat and are a convenient size for the repeated taking of blood for the study of plasma biochemistry [2]. We have found them particularly hardy and less prone to death from infection than other non-human primates. Over the last four years we have studied a large number of baboons freshly captured from Africa or kept on a variety of stock and experimental diets for long periods. The following account reviews briefly our results and gives some recent unpublished data.

Atherosclerosis in Wild Baboons

A study was made of some 23 male and female adult baboons captured during an expedition made to Kenya by members of the Southwest Foundation for Research and Education, San Antonio, Texas [3]. Some 14 baboons had aortic atherosclerosis visible after macroscopic Sudan staining and 8 had fibrous plaques. Microscopically, the lesions consisted of slight or conspicuous intimal elevations including sudanophil lipid, elastic fibres and smooth muscle cells.

In addition, 50 young baboons 1–2 years of age, were examined from the Lister Institute, London after being kept for short periods of 3–6 months on a laboratory stock diet. None of these animals showed disease, and we conclude that below the age of 4–6 years it is extremely rare to find aortic atherosclerosis. These studies confirm the classic observation of McGill's that wild adult but not young baboons show disease [1].

Experimental Diets

Juvenile baboons (1–4 years) kept on a laboratory rat cake diet supplemented with vitamin C tablets thrive extremely well and even after 2–3 years, no arterial disease is evident. However, supplementation of the laboratory stock diet with 15% dried egg yolk, 15% butter or 2% cholesterol, 20% butter (Table 1) produces hypercholesterolaemia in 3–4 months and extensive aortic lesions after 18 months [3].

Table I. Plasma lipids in baboons given atherogenic diets for 1½–2 years, mg/100 ml

Diet		Cholesterol in diet %	Cholesterol	Plasma triglyceride	Phospholipid	$\alpha:\beta$ Lipoprotein
Egg yolk	15%	0.55	218	46	326	0.52
Butter	15%					
Cholesterol	2%	2.06	302	42	367	0.59
Butter	20%					
Semi synthetic, beef fat	20%	0.02	128	72	260	1.0
Control		0.05	98	38	214	2.0

In some species, such as the rabbit, it is not necessary to give cholesterol in the diet, since hypercholesterolaemia can be achieved by feeding a semisynthetic diet containing all known essential nutrients [4]. Seven baboons were given a semisynthetic diet containing 20% beef tallow for 9 months. Values of cholesterol were 90, 102, 105, 120, 135, 150 and 195 mg/100 ml. It would seem that at least two of these animals were above the normal range (which is 70–135 mg/100 ml). Although baboons may behave in a similar way to the rabbit, the response is very much less and the system would not seem suitable for the routine production of atherosclerotic lesions.

It was therefore concluded that the most effective way of producing atherosclerosis was by using a diet containing 15% egg yolk and 15% butter, since the cholesterol content (0.5%) was not grossly in excess of that consumed by man.

Pathological Studies

Two types of lesion were seen macroscopically corresponding to the fatty streak and early fibrous plaque in man. Fatty streaks were particularly common around the origins of branches leading off the aorta. Fibrous plaques were common in the mid aortic region and at the bifurcation of the abdominal aorta (Fig. 1).

When examined histologically the fatty streaks showed an accumulation of extracellular and intracellular sudanophilic lipid, together with abundant intimal elastic fibrils; material could be seen giving the staining reaction of mucopolysaccharide by Hale's method or Alcian Blue. Fibrous plaques appeared histologically

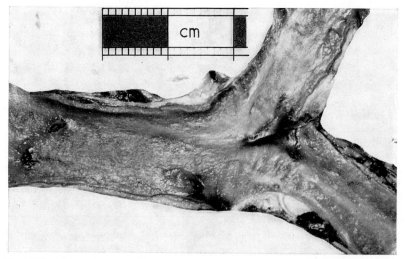

Fig. 1. Fibro-fatty plaque at bifurcation of aorta in a mature baboon. Sudan ×4.

very similar to the fatty streaks except that the lesions were much more elevated and the amount of fibrous material in relation to the lipid was greater (Fig. 2). There appeared no difference between the histology of those lesions seen in adult wild baboons and those produced experimentally in young baboons. As described elsewhere, lesions bore a very close resemblance to the early human fatty streak and fibrous plaque [5]. For this reason the production of atherosclerosis in this species would seem to be of considerable interest. In view of the occurrence of atherosclerosis in the wild baboon, on its natural diet, it is concluded that the function of the atherogenic diet is purely to hasten the onset of the disease and may not be its basic cause.

When animals on an atherogenic diet are injected *in vivo* with Evans blue dye (1 mg/kg in a 5% aqueous solution) a few hours before death, the dye is taken up at the sites where lesions normally occur, namely, near the orifices of the emergent vessels from the

Fig. 2. Large fibrous plaque in baboon aorta showing elastic fibres in intimal thickening. Weigerts – resorcin – fuchsin ×400.

aorta and at the bifurcation in the abdominal aorta. This preferential taking up of dye might indicate an increase in protein permeability at the sites where atherosclerosis occurs. Similar results have recently been obtained by MUSTARD *et al.* in pigs [6].

Coronary Artery Lesions

Like McGILL *et al.* [1], we find that coronary atherosclerosis in wild baboons is rare, and have found the disease in only one animal given the egg yolk and butter diet described above. Until recently it appeared that the production of coronary lesions in this species was extremely difficult. In the course of the routine examination of baboon hearts obtained from a number of laboratories in England, it was found that 5 animals out of a group of 18 baboons received from a laboratory in Scotland (A.D. Little and Co. Ltd.) had a considerable amount of intimal thickening in the coronary arteries (Fig. 3). The type of disease produced was fibro-fatty, rather than the fat-filled foam cell type lesions obtained previously. Further investigation revealed that these animals had been fed on a ordinary laboratory stock diet which had been autoclaved before cubing. In serial determinations over a six month period there was a gradual rise of plasma cholesterol (Fig. 4). Although this may partially

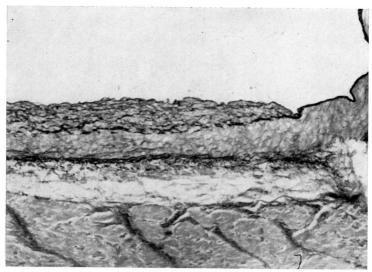

Fig. 3. Fibrous intimal thickening in coronary artery of baboon given an autoclaved laboratory diet. Weigerts – resorcin – fuchsin ×130.

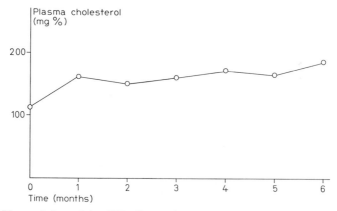

Fig. 4. Plasma cholesterol (mg/100 ml) over six months in six young baboons (2–3 years) given an autoclaved laboratory animal diet (containing soya bean meal, meat and bone meal, white fish meal, wheat, salt and vitamin mixture).

explain the results, it is difficult to see why on other hypercholesterolaemic diets no coronary lesions should have been produced. The heating process may destroy a factor which normally keeps the plasma cholesterol at low levels. Other work in the rabbit has shown that hypercholesterolaemia produced by feeding a semisynthetic diet can be prevented by feeding soya bean meal [4]. It is known that the factor present is destroyed by heat and a deficiency of the same substance may be causing the hypercholesterolaemia in these animals [7]. Other vitamins such as riboflavin and vitamin E may also be destroyed and contribute to the condition.

Biochemical Studies

The chief biochemical change in the plasma of animals given atherogenic diets, is an increase in plasma cholesterol but not triglycerides and an increase in beta lipoproteins (Table II). Using the analytical ultracentrifuge it has been shown that the S_f 0–20 cholesterol rich species but not the S_f 20–400 triglyceride rich species were elevated [3]. It would therefore appear that if it were possible to postulate an atherogenic lipoprotein this would be the S_f 0–20 species. Long term studies have shown that after two years on the atherogenic diets, the alpha lipoprotein (HDL) concentration also rises and the alpha:beta ratio returns to normal or is reversed [8].

Table II. Plasma lipoproteins in baboons (mg/100 ml)

Diet	No. of animals	S_f 0–12 *	S_f 12–20	S_f 20–10^6
15% egg yolk 15% butter	6	308±86	12±10	1±1
Controls	4	116±22	2±3	0

* Difference between means significant P <0.01.

Changes in the plasma fatty acid composition of the lipids were small and have been reported extensively elsewhere [9]. In the cholesterol esters, the chief change was an increase in oleic acid and a decrease in linoleic acid. Such a finding is common in the plasma of all species given an atherogenic diet, since the feeding of cholesterol stimulates the production of cholesterol oleate. BOWYER et al. [10] have pointed out the possible atherogenic effect of large plasma concentrations of cholesterol oleate which can be metabolised at only a small rate by those species which are susceptible to atherosclerosis. Such an hypothesis, which has so far not been investigated in baboons, is likely to stimulate interest in the metabolism of radioactively labelled plasma lipids in perfused baboon arteries.

Conclusions

It is possible to produce aortic atherosclerosis in baboons when fed cholesterol-containing diets over a period of 18 months. The lesions obtained bear a close resemblance to the early human fatty streak and fibrous plaque and are similar to those seen in adult animals captured in the wild state. Hypercholesterolaemia and coronary artery lesions were found in young baboons given an autoclaved stock diet for 6 months; the occurrence of such lesions with other hypercholesterolaemic-producing diets is small. The chief biochemical abnormality in the plasma of animals given the atherogenic diet is an increase in the S_f 0–20 cholesterol-rich low density lipoproteins. The increase in plasma cholesterol oleate may be important in the production of atherosclerotic lesions since the baboon aorta may have a limited capacity to metabolise this lipid.

Acknowledgements

This work was supported in part by grants from the National Institutes of Health, Bethesda P.H.S. Grant No. H6300 and the British Heart Foundation.

References

1. McGill, H.C.; Strong, J.P.; Holman, R.L. and Werthessen, M.T.: Arterial lesions in the Kenya baboon. Circulat. Res. *8:* 670 (1960).
2. Howard, A.N. and Gresham, G.A.: The care and use of baboons in the laboratory. Symposia of the zool. Soc. Lond., No. 17, 1966 on Some Recent Developments in Comparative Medicine, p. 75–89 (Academic Press, London and New York).
3. Howard, A.N.; Gresham, G.A.; Hales, C.N.; Lindgren, F.T. and Katzberg, A.A.: Atherosclerosis in baboons: pathological and biochemical studies. 2nd International Symposium on the baboon and its use as an experimental animal. San Antonio, Texas 1965 (Texas University Press 1966).
4. Howard, A.N.; Gresham, G.A.; Jones, D. and Jennings, I.W.: The prevention of rabbit atherosclerosis by soya bean meal. J. Atheroscler. Res. *5:* 330–337 (1965).
5. Howard, A.N.; Gresham, G.A.; Richards, C. and Bowyer, D.E.: Plasma proteins lipoproteins and lipids in baboons given normal and atherogenic diets. Symposium, The baboon and its use as an experimental animal. 1965 (Texas University Press 1965).
6. Mustard, J.F.; Glynn, M.F.; Nishizawa, E.E.; Packham, M.A. and Rowsell, H.C.: Recent advances in blood coagulation factors and thrombosis. Progr. biochem. Pharmacol. vol. 4 (Karger, Basel/New York 1967).
7. Gresham, G.A. and Howard, A.N.: Recent advances in the pathology of atherosclerosis in animals other than primates. Progr. biochem. Pharmacol. vol. 4 (Karger, Basel/New York 1967).
8. Peeters, H. and Blaton, V. (personal communication).
9. Gresham, G.A.; Howard, A.N.; McQueen, J. and Bowyer, D.E.: Atherosclerosis in Primates. Brit. J. exp. Path. *46:* 194–203 (1965).
10. Bowyer, D.E.; Howard, A.N.; Gresham, G.A. with Bates, D. and Palmer, B.V.: Aortic perfusion in experimental animals. A system for the study of lipid synthesis and accumulation. Progr. biochem. Pharmacol. vol. 4 (Karger, Basel/New York 1967).

Authors' address: A.N. Howard, G.A. Gresham and D.E. Bowyer, Dept. of Pathology, University of Cambridge, *Cambridge* (England) and F.T. Lindgren, The Donner Laboratory, University of California, *Berkeely, Cal.* (USA).

Department of Pathology, Albany Medical College
Albany, New York

Metabolic Studies of Protein Synthesis in Aortas of Monkeys Fed Atherogenic Diets

W. A. Thomas, K. T. Lee and D. N. Kim, Albany, N. Y.

We have produced extensive atherosclerosis in Rhesus monkeys using variations of dietary regimens previously described by Taylor [1], Wissler [2] and others. Figure 1 shows an atheroma in the aorta of one of our monkeys.

We have selected for combined metabolic and ultrastructural studies subgroups of monkeys fed mild atherogenic diets with atherosclerosis at an early stage, when most lesions are in a non-necrotic phase [3–5]. Figure 2 shows a typical non-necrotic atherosclerotic lesion in one of these monkeys.

Table I gives basic information on the subgroups to be presented. With the relatively mild diets used, lesions at 8 months were extensive but not very thick. At 16 months they were much thicker but only a few had become necrotic. The high mortality was largely because more severe diets were given for the first few weeks.

Figure 3 gives quantitative chemical data. The aortic weights, RNA and protein contents were increased somewhat by 16 months. However, the most striking observation was that the DNA content had doubled. Since the amount of DNA per diploid cells is the same in all cells in any given species these figures suggest the number of cells per aorta has doubled (if we disregard polyploidy). Hence, any observations expressed on a per cell basis will appear quite different than if expressed on a per weight basis.

Figure 4 illustrates data on protein synthesis at least insofar as it can be determined by ^{14}C-glycine incorporation into TCA pre-

This work was supported by USPHS Grant H-7155

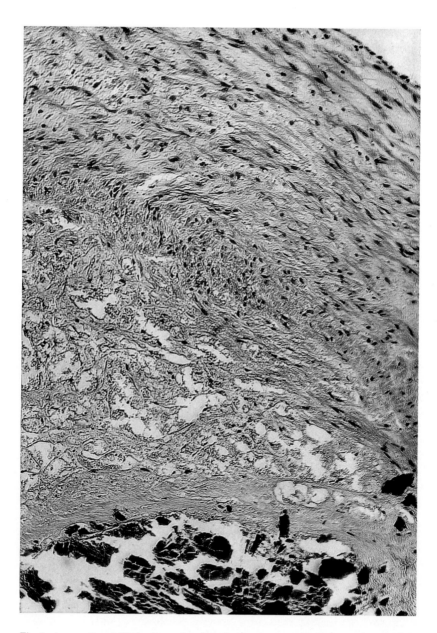

Fig. 1. A necrotic calcified atherosclerotic lesion in the aorta of a monkey fed a severe atherogenic diet for 2 years. H and E stain 190 ×.

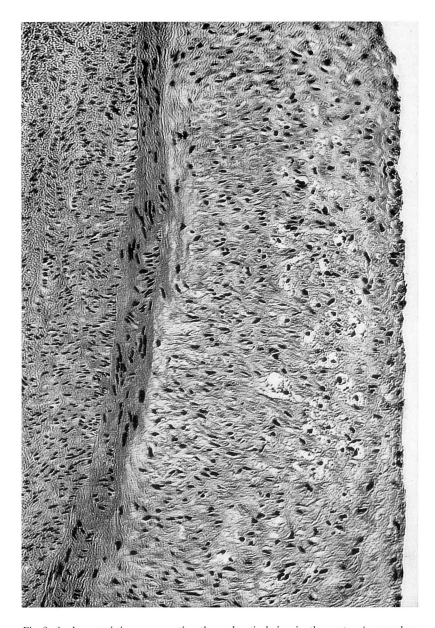

Fig. 2. A characteristic non-necrotic atherosclerotic lesion in the aorta of a monkey. H and E stain 240 ×.

Table I. General information pertaining to experimental monkeys

			Stock group	8-month group	16-month group
Number at outset			8	8	10
Number of survivors			8	8	5
Average body wt. at outset, gm			2,140	2,690	1,960
Average body wt. of survivors, gm			2,910	3,110	2,680
Severity of non-necrotic atherosclerotic lesion	Number with	2+	1/8	8/8	5/5
		3+	0/8	4/8	5/5
		4+	0/8	1/8	4/5

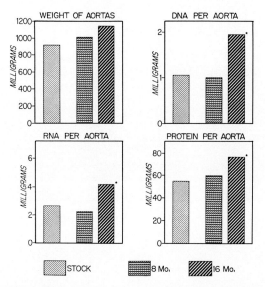

Fig. 3. Weight, DNA, RNA, and protein content of aortas of experimental monkeys in the 16-month group were significantly greater than those in the stock group. No significant difference was observed between the 8-month and stock group. (Courtesy of Academic Press.)

cipitable proteins. (Note that these data must be interpreted with caution since apparent rates of incorporation in aorta could have been affected significantly by transport of labeled proteins made in the liver and elsewhere in the body.) Incorporation rates expressed per aorta or by weight are greatly increased at 16 months. However, to us the most pertinent observation is that incorporation rates per unit of DNA (and presumably per cell) are not increased.

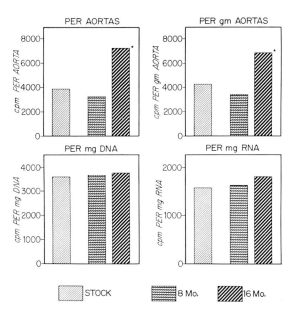

Fig. 4. ¹⁴C-glycine incorporation into total proteins of aortas of experimental monkeys. Apparent rates of incorporation per whole aorta and per gram of aorta were significantly greater in the 16-month group than in the stock group but that per mg DNA was similar in the two groups. However, see text for another possible explanation. (Courtesy of Academic Press.)

Since we are dealing here with averages for all of the cells in the aorta these results could have been obtained if each of the cells in the atherogenic groups was functioning at the same rate as the corresponding cell in the control group. However, the same result could also have been obtained if some cells in the atherogenic group were functioning at a higher rate than in control but with these counter balanced by others functioning at a lower rate.

Electron microscopy observations suggest that the latter explanation is the correct one. Because of space limitations we are presenting here only a stylized drawing of smooth muscle cells from controls and atherogenic diet-fed animals. Fig. 4a represents a normal smooth muscle cell in the media and Fig. 4b a well differentiated smooth muscle intimal atherosclerotic lesion. Cells like these were found both in controls and in atherogenic diet-fed monkeys. However, in the atherogenic diet-fed animals most of the cells in the intimal lesions were smooth muscle cells that had one of the two forms shown in figures 4c and 4d. This cell in Fig. 4c shows only

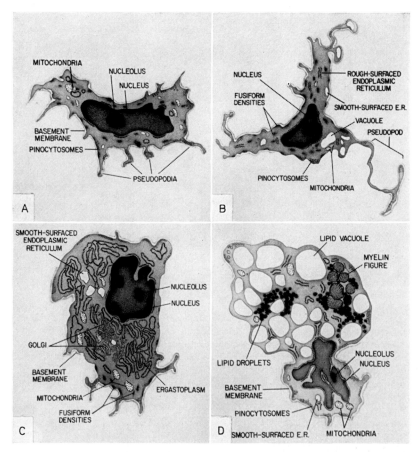

Fig. 5. A – Normal medial smooth muscle cell. B – 'modified' smooth muscle cell in lesion. C – Fibroblast-like smooth muscle cell in lesion. D – Foamy smooth muscle cell in lesion.

slight differentiation, has a great deal of rough-surfaced endoplasmic reticulum and presumably is very active in protein synthesis. The cell in Fig. 4d, which has been photographically reduced in size, was greatly distended by lipid droplets and by what are probably degenerative products of cell organelles. It contains relatively little rough-surfaced endoplasmic reticulum and is perhaps not as active in protein synthesis as a normal smooth muscle cell.

If we can conclude that if the cells appearing very active in electron micrographs are in fact very active *in vivo*, normal values for ^{14}C-glycine incorporation per cell could have been obtained only if some of the cells in the lesions or in the media or both are hypoactive. If we use the same reasoning that we applied to results from rats (presented in another section), we can conclude that the hypoactive cells probably received some form of chemical injury that is being reflected in their ability to produce protein.

One possible explanation for the results observed is that the direct effect of the diet has been to stimulate hyperactivity and that secondary injury to some cells has occurred as a result of overstimulation. A more likely explanation is that the direct effect of the diet has been to injure cells with hyperactive appearing cells representing a reparative reaction to the injury.

References

1. TAYLOR, C.B.; COX, G.E., MANALO-ESTRELLA, P., and SOUTHWORTH, J.: Atherosclerosis in Rhesus monkeys. II. Arterial lesions associated with hypercholesterolemia induced by dietary fat and cholesterol. Arch. Path. 74: 16–34 (1962).
2. WISSLER, R.W.; GETZ, G.S., VESSELINOVITCH, D. and FRAZIER, L.E.: Acute severe experimental atherosclerosis in Rhesus monkeys. Fed. Proc. 25: 597 (1966).
3. LEE, K.T.; JONES, R.; KIM, D.N.; FLORENTIN, R.; COULSTON, F. and THOMAS, W.A.: Exp. Mol. Pathol. 3: 108–123 (1966).
4. SCOTT, R.F.; COULSTON, F.; DAOUD, A. and JONES, R.: Atherosclerosis in monkeys: Light and electron microscopy features. Exp. Mol. Pathol. (1967).
5. SCOTT, R.F.; MORRISON, E.S.; KROMS, M.; NAM, S.C. and ZUMBO, O.: Atherosclerosis in rhesus monkeys: Lipid chemistry features. Exp. Mol. Pathol. (1967).

Authors' address: W. A. Thomas, M.D., K.T. Lee, M.D. and D.N. Kim, M.D., Dept. of Pathology, Albany Medical College, *Albany, N.Y.* (USA).

Discussion Session 8

R. WISSLER (Chicago, Ill.): I want to compliment Dr. HAUST on her beautiful presentation and I want to try to clarify the slight differences in her published results and ours, especially as to the presence of serum albumin in the lesions. I believe that you have routinely fixed your tissues in alcohol, have you not, while we work entirely with unfixed artery blocks and sections. We believe that almost any plasma protein may pass through the artery wall, and could probably be expected to be found free in small quantities. Our results would suggest that low density lipoproteins and fibrin are the only two substances 'trapped' by the artery wall, i.e. bound so firmly that the usual washing procedures used in fluorescent antibody studies will not remove them from the unfixed section.

DARIA HAUST (London, Ont.): Yes, Dr. WISSLER, we have to fix the cryostat-cut sections in absolute alcohol for 15–20 minutes either prior to or following the staining with the conjugated anti-albumin and before the washing procedures. The same applies to alpha-1 (high density) lipoproteins. Sections used for the staining with conjugated anti-fibrinogen and anti-alpha-2 (low density) lipoproteins did not require fixation.

R. WISSLER (Chicago, Ill.): I want to reemphasize that we should continue to look for irritating substances in the aortic wall. I think they will prove to be important in ascertaining why some lesions progress and others do not. This is different from the simple analysis of the various components of the lesions. It involves testing many possible substances directly. There may be many such chemicals which cause the proliferation of cells over the area of lipid deposition like the ones we discovered which were characteristic of the plaques produced in the Cebus monkey by coconut oil, cf. Arch. Path. *74:* p. 312 (1962).

H. MALMROS (Lund): If we feed commercial pellets soaked in cholesterol-ether to rabbits for a year, it is possible to induce lesions containing cholesterol not only in the atherosclerotic plaques but also in other tissues, e.g. the cornea.

In a slide you can see such a case of arcus cornea. After some more months a fibrous membrane covered the cornea and the animal became completely blind (Fig. 1).

When we examined this fibrous membrane in the microscope we found it loaded with crystals of cholesterol-esters. It is obvious that the crystals act as a foreign substance inducing a fibrous reaction of the collagen (Fig. 2).

R. H. FURMAN (Oklahoma City, Okla.): Two comments: First, a suggestion to Dr. CLARKSON. It has been reported that in animals chronically treated with corticosteroids areas of necrosis may appear in the myocardium which seemed to be about as large as those which you showed in your animals. It is my recollection that these areas of necrosis are ascribable to potassium deficiency. It is conceivable that Dr. CLARKSON's animals were chronically stressed and, as such, were subjecting themselves to some degree of hyperadrenalcorticosteroidism. One might consider the possibility that the lesions in the myocardium in these animals may have arisen in part as a result of such a phenomenon. The second comment relates to the use of coconut oil. The C-8 and C-10 fatty acids present in coconut oil are ingested as triglycerides and hydrolyzed in the gut to free fatty acids and, probably, monoglycerides, in which form they enter the mucosa. Most of the C-8 and C-10 (and shorter chain) fatty acids go directly to the liver via the portal system where they undergo oxidation to CO_2 or to ketone bodies, or undergo chain lengthening by the addition of two carbon fragments, ultimately to form longer chain fatty acids characteristic of the species. I think one is likely to be disappointed if one looks for C-8 and C-10 fatty acids in the plasma or in the lesions of these animals fed with coconut oil, except for those C-8 or C-10 fatty acids that enter the mucosa as monoglycerides and are subsequently re-esterified to triglyceride with endogenous, longer chain fatty acids.

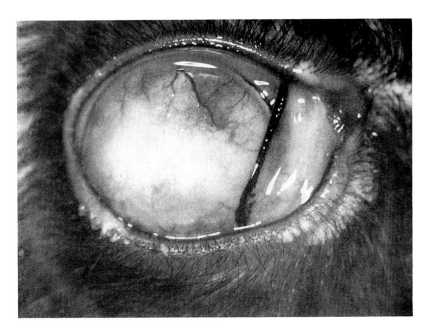

Fig. 1.

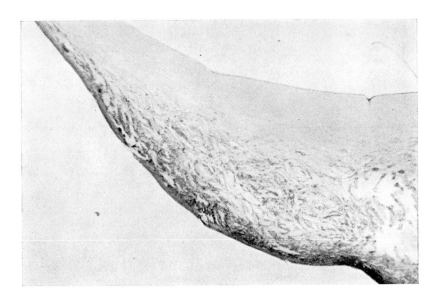

Fig. 2.

L. H. KRUT (Cape Town, S. Africa): I would like to comment on the point raised by Dr. CONSTANTINIDES. I think it is important to point out that cholesterol is not a very stable substance. A group at Hammersmith Hospital have shown that sunlight will degrade cholesterol. Since some people also bake their cholesterol before feeding, it is appropriate to comment that heating cholesterol can produce many breakdown products which are easily demonstrable on thin-layer plates. I don't know the significance of this but it may be relevant to keep in mind when preparing experimental diets. I would also like to make some comments on Dr. WISSLER's question about the possible role and nature of a lipid 'irritant' in the development of the atheromatous lesion. We believe that the presence of crystalline cholesterol in the arterial intima is responsible for the development of the atheromatous plaque. It was shown some years ago that crystalline cholesterol will produce a granulomatous reaction in connective tissue. More pertinently, the intima of the aorta has been shown to respond to such material by a fibrous reaction. BYERS and FRIEDMAN demonstrated that crystals of cholesterol acetate left in contact with the intimal surface of the abdominal aorta in the rat gives rise to a massive overgrowth of fibrous tissue which completely surrounds this crystalline mass.

J. L. BEAUMONT (Paris): Adding to the remark of Dr. KRUT, may I say that (with J. R. CLAUDE) we have found that simple exposure to daylight is able to produce changes in the cholesterol molecule. This 'photolysis' produces many derivatives beginning with 7 hydroxycholesterol. When the entire extraction and analysis procedure was run in the dark we found only cholesterol in the blood and also in the normal and diseased aortas of men and animals. No derivatives were detectable in these conditions.

L. H. KRUT (Cape Town, S. Africa): I would like to corroborate Dr. FURMAN's comments with respect to short chain dietary fatty acids being reflected in the blood lipids. This, however, does not hold for those fatty acids from about C 10 and higher which are abundant in coconut fat. These will be reflected in serum lipid fractions. In particular, the cholesterol ester fatty acids will take on the dietary fatty acid pattern as shown by the Rockefeller group in about 1959.

I would also like to comment on the speculative role of high cholesterol oleate being atherogenic. It is extremely difficult to draw inferences from fatty acid proportions relating to the situation in man. I would like to point out that the South African Bantu, who has a remarkable degree of immunity to coronary heart disease, show a fatty acid pattern in their plasma cholesterol esters which is higher in oleate and lower in linoleate when compared to that in white subjects with established coronary heart disease. The pattern in the Bantu resembles closely that pattern suggested by Dr. HOWARD as being atherogenic.

A. N. HOWARD (Cambridge): Since the total cholesterol level of the Bantu is low, the absolute circulating level of cholesterol oleate will also be low, even if it is elevated as a percentage of the total fatty acid.

L. H. KRUT (Cape Town, S. Africa): The fatty acid pattern I referred to applies to the beta lipoprotein cholesterol esters. I would also like to add, by way of illustration, that fatty acid proportions are difficult to relate to this disease process. We have found that, in human serum lipids, C 18:2 and C 18:1 are inversely related. Further, a rise in cholesterol is paralleled by a rise in cholesterol linoleate. This would serve to further emphasise the points I raised earlier in regard to the reciprocal relations between C 18:2 and C 18:1 in relation to cholesterol oleate being atherogenic.

D. E. BOWYER (Cambridge): I think Dr. KRUT has missed the point. I made the hypothesis that cholesterol oleate would accumulate intimally if there was an increased amount in the plasma or because of a lipoprotein instability it was more readily dissociated from the plasma lipoproteins or if there was a low cholesteryl oleate hydrolase activity in the arterial wall. Both Bantu and Whites may have similar ratios of plasma cholesteryl oleate to linoleate, but what distinguishes them with respect to susceptibility to atherosclerosis may be the cholesteryl oleate hydrolase of the arterial wall.

C. B. TAYLOR (Evanston, Ill.): I would like to ask one question directed to Dr. CLARKSON, Dr. HOWARD and Dr. MCGILL. In the free living animals that they have studied was there any sex relationship to this spontaneous disease? I ask this because recently I have reviewed a paper by LIKAR and others (In press, A.M.A. Arch. Path.) which shows that milk cows have a great deal more hypercholesterolimia and more vascular disease than male cows of equal age. Dr. W. CONNOR, of the University of Iowa tells me that in the human female lactation is associated with a marked increase of serum cholesterol.

T. CLARKSON (Winston-Salem, N.C.): Our group, in collaboration with Dr. MCGILL have completed an autopsy study of 220 squirrel monkeys in South America. Females had significantly more fatty streaking of the aorta than males. Female squirrel monkeys bear one young per year and lactate for about 3 months.

H. MCGILL (New Orleans, La.): Fatty streaks in baboons are about equal between the sexes. In humans, females consistently have slightly more fatty streaks than males.

A. N. HOWARD (Cambridge): Our own studies on baboons agree with those of Dr. MCGILL.

L. W. KINSELL (Oakland, Cal.): Regarding cholesterol esters, some years ago we extracted atheromata from arteries of patients who had been on high unsaturated fat intake for some time before their death. Cholesterol esters obtained by simple lipid-solvent extraction had large amounts of linoleate. Solution of calcific material (from the atheroma) and re-extraction with lipid solvent gave cholesterol esters with a very high stearate content. Just what this means is by no means clear, since none of us knows as yet the fate and function of cholesterol esters. Perhaps simple matters relating to solubility play some part, i.e. cholesteryl stearate, being relatively much more insoluble than cholesteryl linoleate tends to be deposited in and to remain in the intima. We believe it also suggests that 'newer' portions of the atheroma are potentially reversible, since they exchange readily, whereas 'older' portions here ceased to be a part of the freely miscible pool.

R. H. FURMAN (Oklahoma City, Okla.): I should like to get back to the cholesterol ester question. A recent study by BLOMSTRAND (BLOMSTRAND, R.; GURTLER, J. and WERNER, B.: Intestinal absorption and esterification of ^{14}C-labeled fatty acids in man. J. clin. Invest. 44: pp. 1766–1777 [196[5]] demonstrated that cholesterol is preferentially esterified with oleic acid during the process of lymph cholesterol formation in human subjects. I would think, in view of this observation, that in simple experiments involving cholesterol feeding without an overwhelming amount of any one type of fatty acid in the diet, one would expect that, if the fed cholesterol deposited in lesions, the predominant cholesterol ester would be cholesterol oleate. One would have to assume, of course, that cholesterol is preferentially esterified with oleate in these animals as well as in the human.

D. E. BOWYER (Cambridge): To return to the problem of cholesterol ester accumulation in early fatty streaks and in old alteromatous lesions in man. The total cholesterol ester content of early streaks rises, but the one which increases most is cholesterol oleate. In the later atheromatous lesion there is also an increase in cholesterol esters, but the composition is like that of the plasma cholesterol esters, suggesting an indiscriminate deposition, and not alteration of composition by transacylation or hydrolysis.

E. SMITH (London): Returning to the question of cholesterol oleate, I have already published work showing that there is a marked difference in composition in the cholesterol esters of fatty of streaks and fibrous plaques. In fatty streaks there is a very high proportion of cholesterol oleate, whereas in fibrous plaques the cholesterol ester composition is close to that of serum.

In our recent work we have found that the high cholesterol oleate pattern is invariably associated with lipid contained in fat filled cells, whereas the serum-type cholesterol esters are associated with extracellular lipid. If one thinks that fatty streaks are not the precursors of fibrous plaques, it would be reasonable to suggest that the plaque lipid is derived from serum. But if one postulates that the extracellular lipid originates from fat filled cells which break down, then there must be some hydrolytic or transesterifying system present.

H. MALMROS (Lund): Even if the cholesterol esters contain much oleate, I don't think the oleic acid, as such, has any atherogenic effect. We have fed 8% trioleate to rabbits for one year without any serious complication. The trioleate level in the different serum lipid factions was very high and also the trioleate concentration of subcutaneous fat. We didn't, however, find any atherosclerotic lesions in the aorta or in any other artery.

A. N. HOWARD (Cambridge): The absolute concentration of plasma cholesterol oleate is probably the determining factor. Cholesterol esters in rabbits given oleic acid are usually quite low.

H. McGILL (New Orleans, La.): Our experiments with baboons provide additional information pertaining to the problem of fatty acids esterified to cholesterol. Aortic fatty streaks occurred in animals receiving both fat mixtures, each mixture having different proportions of fatty acids with regard to saturation, chain length, and other characteristics. Depot fat, liver, and aortic intimal cholesterol esters contained fatty acids that closely paralleled the fatty acids in the diet. We found no difference histologically or electron microscopically between fatty streaks in animals receiving different fat mixtures. These findings suggest that the composition of cholesterol ester fatty acids in fatty streaks is influenced greatly by the fatty acid composition of the diet, and that the fatty acid composition of the cholesterol esters in the lesions does not influence greatly the morphologic characteristics of the lesions. It is, of course, possible that the fatty acid composition could influence the progression of the lesions and possibly their conversion into fibrous plaques.

In connection with the suspected association of fatty streaks with pregnancy and lactation, we have compared fatty streaks in young women dying of complications of pregnancy and the puerperial period, and have found no more extensive fatty streaks in these women than in women of similar age and geographic origin dying of other causes. Therefore, in the human, I think it is probably not true that pregnancy and lactation are associated with increased arterial lipid deposition.

A. N. HOWARD (Cambridge): Did you analyse intima or the whole artery of the baboons?

H. McGILL (New Orleans, La.): Yes, the analyses of cholesterol esters were based on intimal tissue alone.

C. B. TAYLOR (Evanston, Ill.): LIKAR and others (in press A.M.A. Arch. Path.) have studied older dairy cows at autopsy. They have described, with convincing illustrations, quite striking 'fatty' atherosclerotic lesions. Interestingly many of their serum cholesterol levels ranged from about 250 to 300 mg%.

R. PAOLETTI (Milan): An additional factor which might induce the well known increase in plasma lipids during pregnancy may well be PPP secreted by the human placenta and which shows a considerable lipolytic activity on adipose tissue.

P. CONSTANTINIDES (Vancouver, B.C.): Dr. ANDRUS, I was intrigued by the aneurysms in the cerebral arteries of your monkeys. Did you have any occasion to study the blood pressure of these animals during the experiment? It is interesting that years ago, certain lipids were associated with blood pressure effects.

S. ANDRUS (Boston, Mass.): Blood pressure data in the experimental chimpanzees plus a large number of control animals is currently being analyzed. Contrary to our original impressions, it would appear that the animals bearing aneurysms were not hypertensive when compared with large numbers of control animals. I should add that there seen to be certain difficulties in collecting such data in chimpanzees.

P. CONSTANTINIDES (Vancouver, B.C.): With regard to Dr. HOWARD's autoclaved diets it reminds me of ALTSCHUL's baked cholesterol diet cakes that were much more atherogenic than ordinary diet. Perhaps the trace amount of cholesterol in the stock diet is activated somehow in the same sense.

A.N. HOWARD (Cambridge): Most laboratory stock diets contain soya bean meal and the protective soya bean factor present would be destroyed by heat.

A.N. HOWARD (Cambridge): In Dr. TAYLOR's monkey which developed myocardial infarction was death due to coronary thrombosis or atherosclerotic debris and constriction?

C.B. TAYLOR (Evanston, Ill.): As reported previously (A.M.A. Arch. Path. *76:* 404, 1963) one coronary occlusion had the definite characteristics of a recent occlusive thrombus. The other five occlusions were old and usually recanalized. There was no evidence of plugging by atherosclerotic debris only.

D. BRUNNER (Jaffa): In your studies it is demonstrated that the animals after 16 weeks of atherogenic diet had a significant increase in cholesterol-lipoproteins, but in the next 47 weeks without change in their diet a certain lowering of the lipoprotein levels occured.

I am asking this question because in our nutrition-study, where Oriental Yemenite Jews received for 7 months a western diet, high in calories, fat and proteins, we observed a similar 'rebound syndrome'. Dr. WISSLER, can you explain it?

R. WISSLER (Chicago, Ill.): No, I have no ready explanation for this phenomenon but I expect that enzyme activation take place which help the animal metabolize lipids and cholesterol more readily. All I can say is that this phenomenon has been observed by many experiments in many animal species under many experimental conditions, specially in the rat and the monkey by us and by Dr. PICK in the cockerel. If any one has a more complete answer to Prof. BRUNNER's question, I hope they will step forward and answer the question.

D. KRITCHEVSKY (Philadelphia, Pa.): I would like to comment in relation to Dr. McGILL's statements concerning the apparent difficulty in inducing atherosclerosis in the baboon as compared to other types of primates. If we look at the two mammalian species which seem quite resistant to cholesterol induced atherosclerosis, namely the dog and the rat, we find that they have very high levels of taurine conjugated bile acids and practically no glycine conjugates; the rabbit, on the other hand, is said to lack the enzyme required for synthesis of taurocholanic acids and has only the glycine conjugates. If we examine the taurine/glycine ratio in the bile salts in monkeys, we find that the susceptible Rhesus monkey has a low taurine/glycine ratio while the somewhat resistant Cebus monkey has a high taurine conjugate content. In man, the taurine/glycine ratio is 1:3, whereas our preliminary findings in the baboon indicated a ratio of about 1:1. Many workers have tried to feed taurine to rabbits, but have found no effect on either serum cholesterol levels or atherosclerosis. If we now consider the fact that one of the initial chemically evident lesions in the atherosclerotic artery is the accumulation of ester cholesterol, and recall that Dr. ZEMPLÉNYI has shown a differential rate of lipolysis by aortas of different species, that Dr. BOWYER has shown a differential hydrolysis of different cholesterol esters by rabbit aorta, that Dr. PATELSKI has indicated an increased lipolysis by aorta slices in the presence of bile conjugates, and that Dr. VAHOUNY has recently reported that taurine conjugate bile acids protect the integrity of lipases, we may come out with the following hypothesis: That the taurocholanic acids in some way protect the integrity of the aortic cholesterol ester hydrolase and if we attempt to correlate biliary taurine/glycine ratios with susceptibility to atherosclerosis, the place to look may be in their effect on aortic metabolism.

R. H. FURMAN (Oklahoma City, Okla.): I would like to return briefly to the question posed by Dr. TAYLOR and remind the audience that estrogens produce hypertriglyceridemia whether they are of exogenous origin or endogenous, e.g. the increase in estrogens occurring during the course of pregnancy. Clinicians in the audience will recognize the term 'hyperlipemia of pregnancy', occurring in the third trimester in human subjects. Lipemia occurs during the course of pregnancy in many vertebrates. There is a tremendous outpouring of estrogens from the placenta during the course of pregnancy in mammals and in a healthy colony of primates where females are bringing forth young more or less continuously, such gravid females are likely to be hypertriglyceridemic in comparison with nonpregnant females or male animals. Under these circumstances one would anticipate that the breeding females would exhibit increased numbers of fatty streaks.

S. ANDRUS (Boston, Mass.): I should like to point out that in contrast to Dr. McGILL's data, we feel that we have demonstrated a distinct difference between coconut and corn oil. To date four species have been subjected to the same basic regime i.e. corn oil with and without cholesterol and coconut oil with and without cholesterol. The species are chimpanzee, cebus, lagothrix, and saimiri. In all four, coconut oil has consistently been more effective than corn oil in elevating the serum cholesterol either with or without cholesterol. Cholesterol added to either diet, augmented the effect of the fat, and the two effects i.e. the fat effect and the cholesterol effect, were additive. Thus, corn oil plus cholesterol was equivalent to coconut oil alone, and coconut oil plus cholesterol produced the most marked changes. When available, as in three of these species, data on low density beta lipoproteins showed a similar response. In three of these species, i.e. chimpanzee, lagothrix, and saimiri, the increase in aortic sudanophilia paralleled the serum responses to the dietary manipulations, coconut oil alone again being as effective as corn oil plus cholesterol. The fourth species, cebus, did not respond to this relatively mild dietary regime and would appear to be a less reactive species.

L. W. KINSELL (Oakland, Cal.): One of the problems in the evaluation of cholesterol metabolism as it relates to atherosclerosis has been the great differences that exist between different species. To date one can have no assurance that findings in any experimental animal really have interpretable meaning when transposed to the human. Most primates are herbivorous. Consequently, their pattern of cholesterol absorption may have much more in common with the rabbit than with man, since some recent studies have shown that man apparently has very limited ability to absorb dietary cholesterol.

With regard to Dr. FURMAN's comments, we have a few data dealing with the formation of different types of cholesterol esters in man as evidenced by the cholesterol ester fatty acid pattern in chyle following the ingestion and absorption of certain specific fats. These data do not support very strongly the concept of preferential formation of cholesterol oleate. We suggest the need for more studies in *human* subjects dealing with chyle rather than plasma if one hopes to evaluate relationship between diet and plasma lipid composition.

J. PATELSKI (Poznan): Considering factors leading to cholesterol ester accumulation in the aortic wall, I would like to add that an enzymatic esterification of cholesterol by fatty acids has also been demonstrated in our laboratory in Poznan in 1964. The pH optimum of the enzyme activity is different (below 7.0) from that of cholesterol ester hydrolase. An adaptive increase in cholesterol esterifying activity could take place, in recent atherosclerosis at least, as is the case for the aortic lipase. However, this seems to be of small importance in advanced disease where a decrease in synthetic processes (except for trans-acyl reactions, perhaps) can be expected because of the reduced delivery of energy rich bonds in the aortic wall. Therefore, the enhancement in cholesterol ester accumulation seems to be more closely related to the deposition of serum cholesterol esters and to the decreased activity of cholesterol ester hydrolase in the aortic wall.

H. PEETERS (Brugge): Cholesterol esters do not exist in abstract but only in relation to a protein to which they are attached. We have shown the existence of cholesterol esters with given fatty acid ratios for alpha and for beta lipoproteins. In the aortic lesions there must also exist lipoproteins with a given protein to cholesterol ester ratio and a given fatty acid pattern. Diet influences the overall fatty acid pattern of the cholesterol esters but nevertheless cholesterol esters behave differently when linked with alpha protein, beta protein or with the protein inside the atheromatic lesion.

J. J. GROEN (Jerusalem): I would like to ask whether any of the speakers has examined in his atherosclerotic hypercholesterolaemic monkeys the effect of such factors as lack of physical activity or stress or frustration? Furthermore, has anybody seen in these animals, a myocardial infarction or is this just as rare in them as in atherosclerotic non-primates? You will understand how interested we, who try to study these problems in the human, are in your results.

C. B. TAYLOR (Evanston, Ill.): Our studies and those of others where animals were caged were necessarily subjected to lack of physical activity, stress (particularly during capture for studies) and to frustration, since they were individually housed and subject to some visiting by caretakers, etc. No control feeding experiments with these factors excluded have been done to my knowledge. I reported one case of fatal myocardial infarction in a rhesus monkey with diet induced hypercholesteraemia and coronary atherosclerosis with thrombosis (TAYLOR et al. A.M.A. Arch. Path., 1965).

D. KRITCHEVSKY (Philadelphia, Pa.): We (KRITCHEVSKY et al. 1965; SHAPIRO et al., 1965, 1966) have carried out a number of studies related to the biosynthesis and transport of cholesterol in baboons. The manner of transport of exogenous and endogenous free and ester cholesterol by serum alpha and beta lipoproteins and the exchange of cholesterol between various lipoprotein fractions has been studied. The data suggest that: (a) endogenous and exogenous cholesterol are transported by the serum alpha and beta lipoproteins in an indiscriminate fashion; (b) free and ester cholesterol (of endogenous or exogenous origin) are incorporated into and exchanged between the lipoprotein fractions, both *in vivo* and *in vitro;* and (c) the rates of biosynthesis and turnover of cholesterol in the baboon are comparable to those observed in man.

References

1. KRITCHEVSKY, D.; WERTHESSEN, N. T. and SHAPIRO, I. L.: Studies on the biosynthesis of lipids in the baboon: Biosynthesis and transport of cholesterol. Clin. chim. Acta *11:* pp. 44–52 (1965).
2. SHAPIRO, I. L. and KRITCHEVSKY, D.: Preliminary studies on the biosynthesis and transport of cholesterol in the baboon. In: The Baboon in Medical Research, pp. 339–352. (U. Texas Press, Austin 1965).
3. SHAPIRO, I. L.; JASTREMSKY, J. A.; EGGEN, D. A. and KRITCHEVSKY, D.: Transport of cholesterol in the baboon. In: The Baboon in Medical Research, Vol. II (In press, 1966).

Non Primates

From the Department of Pathology, University of Cambridge, England

Recent Advances in the Pathology of Atherosclerosis in Animals other than Primates

G.A. Gresham and A.N. Howard, Cambridge

Introduction

Atherosclerosis occurs in a wide variety of creatures and can also be induced or aggravated by numerous experimental procedures. Since it is impossible to encompass the many facets of research in this field, this short review will deal primarily with those species which are receiving considerable attention at the present time, namely the bird, pig, rat and rabbit. The first two are chosen because together with primates they were considered recently to be the most useful animals for atherosclerosis research [1] and the latter because much useful research has come from their study despite the often quoted difficulties in their use.

The Rat in Atherosclerosis Research

Extreme measures are often needed to produce atherosclerosis in the rat and involves feeding a diet containing fat, cholesterol and cholic acid [2, 3]. The observation that the rat responds differently to diets rich in saturated and in unsaturated fatty acids provides a provoking problem. Thus rats given a diet containing butter or cocoa butter develop thrombosis, those given a diet with peanut oil develop atherosclerosis (Fig. 1). Earlier work indicated the need for thiouracil as a hypercholesterolaemic agent but recently it has been found not to be essential for the production of both types of

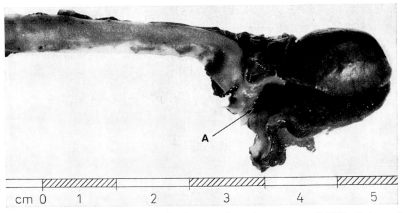

Fig. 1. Atherosclerotic lesion in proximal aorta of rat given peanut oil 40%, cholesterol 5% and cholic acid 2% for 4 months. Lesion at A. Sudan ×10.

lesions. RENAUD and ALLARD [4] found that this substance could be omitted providing the protein content of the diet was reduced to 10% and later showed that the precipitating factor was probably a methionine deficiency [5]. Also using a diet containing the normal amount of protein, the same result can be achieved with a more saturated fat such as cocoa butter. Likewise, in our own laboratory, thiouracil has been shown not to be essential in the peanut oil diet. Over a period of 4–6 months, a high proportion of rats given 40% peanut oil, 5% cholesterol and 2% cholic acid develop fibrous plaques in the proximal aorta.

Current work has been directed towards a comprehensive study of the biochemistry and physiology of the two groups of animals, given the butter and peanut oil diets, with the hope of finding a difference which was causally related to the production of the lesions.

There have been several studies of the coagulation factors in the plasma of such experimental animals but no clear-cut differences have been obtained [6, 7, 8]. In both groups there is a large increase in all the coagulation factors and animals appear to be in a state of hypercoagulability. Insufficient attention, has, however, been paid to the properties of the blood platelets in this system and it is likely that such studies may cast considerable light on the different mechanisms involved. In favour of this view are the enzymatic studies of ZEMPLENYI and MRHOVA [9]. He showed a greatly increased content of adenyl pyrophosphatase in the aortas of the butter fed but not peanut oil fed rats (Fig. 2). This enzyme which

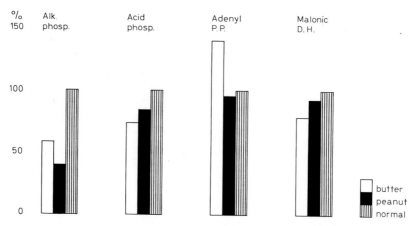

Fig.2. Results obtained by ZEMPLENYI and MRHOVÁ: Aortic enzymes in the aortae of rats given thrombogenic (40% butter, 5% cholesterol, 2% cholic acid and 0.3% thiouracil) and atherogenic (40% peanut oil replacing butter) diets (9). Normal control rats taken as 100% activity.
Alk Phosp. = alkaline phosphatase, Acid Phosp. = acid phosphatase, Adenyl PP = adenyl pyro phosphatase, Malonic DH = malonic dehydrogenase.

splits ATP to ADP may be responsible for increased platelet aggregation and thrombosis. Any damage to the vessel wall would be likely to liberate more ADP which is well known to promote platelet aggregation *in vivo* and *in vitro*. Studies with platelets, however, are difficult because of the intense lipaemia in rats given these diets; this obscures any measurement of platelet aggregation.

Determination of the plasma lipids in two groups of rats suggested an association between the elevated levels of blood triglycerides and thrombosis [10]. In a continuation of these studies, LINDGREN and co-workers have recently examined the distribution of both the high density (HDL) and low density lipoproteins (LDL) [11] as shown in Table I. In the butter fed rats there is 2½ fold increase in the S_f 0–20 lipoproteins and a 30% increase in the S_f 20–100 lipoproteins compared with peanut oil fed animals. The latter change explains the difference in triglyceride composition because this fraction is normally rich in triglycerides. It could be argued that this particular fraction is thrombogenic. However, the greatest difference was in the HDL which were completely absent in the peanut oil fed rats. This finding is interesting in view of the known interconversion of HDL to LDL [12] and the commonly observed reduction of HDL in hyperbetalipoproteinaemia. It is

Table I. Plasma lipoprotein concentrations (mg/100 ml) in rats given thrombogenic and atherogenic diets

Diet*	Lesion	Low density S_f			High density S_f
		0–20	20–400	400–10⁵	
Butter	thrombosis	1130	4040	273	90
Peanut oil	atherosclerosis	447	3190	610	0
Control		11	77	4	208

* Peanut oil and butter diets contain 40% fat, 5% cholesterol, 2% cholic acid, 0.3% thiouracil.

surprising, however, that this effect is greatest in the peanut oil group which have lower LDL. High density lipoproteins may be of more importance in the development of atherosclerosis than has previously been realised and they may play a special role in the transport of lipids in the arterial wall.

The Rabbit in Atherosclerosis Research

Since fibrous atherosclerosis has been produced in rabbits by feeding semisynthetic diets [13] and also by intermittent feeding of cholesterol [14], workers are less sceptical about using this animal for the production of atherosclerosis. Even the foam cell lesions, which were first produced by ANITSCHKOW by feeding cholesterol in oil, are a useful model for the study of lipid accumulation in the arterial endothelium. This is, after all, one of the most important problems in human atherosclerosis research because lipid is the major factor causing coronary artery obstruction.

Experimental methods which involve the feeding of large quantities of cholesterol can be justly criticized since the levels used are often greatly in excess of those encountered in human nutrition and particularly, as in man, most of the body cholesterol is endogenously synthezised and is not of dietary origin. A considerable advance in experimental pathology was made when it was discovered that rabbits given a semisynthetic diet containing all known essential nutrients acquire a moderate hypercholesterolaemia in about 8 weeks (400–600 mg/100 ml) and atherosclerosis of the aorta and coronary arteries in about 12–16 weeks [13, 15, 16]. The inclusion of fat in the diet is not essential but the lesions are more severe when beef fat or other highly saturated fats such as hydrogenated coconut oil are given. The disease is unrelated to dietary cholesterol and is not caused by essential fatty acid deficiency. It seems likely, however, that the diet is deficient in some other essential nutrient required by the rabbit for maintaining the vascular system in a healthy condition. Evidence to support this view has recently been obtained since it is found that the inclusion of commercial extracted soya bean meal in the diet prevents the disease [17]. This suggests that it may contain a protective factor similar to that present in the rabbit's natural diet. Recently it has also been possible to isolate concentrates from soya bean meal containing a heat labile substance which will prevent hypercholesterolaemia in rabbits given such diets [18]. Such a compound which has hitherto been undescribed would seem to qualify for inclusion among the list of vitamins required by the rabbit.

Birds in Atherosclerosis Research

There have been a number of notable modern advances in the use of birds for experimentation. Perhaps the most important is the discovery of the genetic factor in predisposing species to spontaneous atherosclerosis. Thus the Carneau pigeon is a superb example of a highly susceptible breed [19]. However, explanation of the biochemical and physiological lesions which lead to the development of the disease in this particular breed of bird is still being sought. Another feature of current research is the paradoxical effect of hormones in birds compared with mammals [20]. The ability of estrogens to promote hypercholesterolaemia and atherosclerosis in birds provides a provoking problem for further research in the important field of hormonal effects.

Certain breeds of turkey develop aortic atherosclerosis [21, 22, 23]; this is seen most often in the region of the coeliac axis and along the posterior aorta extending into the iliac arteries. Lesions in the anterior aorta differ in structure from those in the posterior part of the vessel; the latter are composed largely of smooth muscle together with some lipid, and, more closely resemble the human disease than the anterior lesions. Various experiments have suggested that the tendency to develop posterior aortic lesions is related to the speed of growth, since little or no disease can be seen in small animals whose growth had been inhibited by dietary means. However, recent experiments have shown that if these small animals are allowed to remain on the diet for a longer period then they eventually develop lesions. Three groups of turkeys were placed on a diet containing different levels of protein such that the time to reach 14 lb body weight varied from 14–20 weeks (Table II). Atherosclerosis was just as severe in the slow growing group. It is therefore concluded that the severity of the disease depends on the actual size of the bird rather than on the speed of growth.

Table II. Effect of speed of growth on turkey atherosclerosis in birds killed when 14 lb body weight

Group	Protein content of diet*	Mean weeks to reach 14 lb	Number of birds	Grades of atherosclerosis 0 1 2 3
1	40%	14	16	4 5 2 5
2	20%	16	12	3 3 0 6
3	10%	20	11	1 2 2 6

% area of disease in posterior aorta.
$0 = $ Nil; $1 = 0\text{--}1\%$; $2 = 1\text{--}5\%$; $3 = >5\%$.
* Diet contained soya bean meal, wheat meal, grass meal, yeast, dicalcium phosphate, limestone, salt and vitamin mix.
Since no animal products were used, the diet was cholesterol free.

The cause of atherosclerosis in turkeys is still obscure but epidemiological studies are proceeding in the hope that such experiments may show up an environmental or genetic factor. For instance, a study of wild turkeys from Texas and Brazil show that anterior aortic lesions are as frequent as those that occur in artificially fed birds but posterior lesions are less frequent. Histologically the lesions seen in the two different segments of the aorta are quite different (Figs. 3, 4); those in the anterior aortic segment consisting chiefly of cartilage, collagen with occasional calcium deposits. Those in the posterior section were composed chiefly of muscle with some elastin. A survey of several hundred birds examined in Cambridge also show that there is no correlation between the occurrence of lesions in the two different sites. For instance in females posterior lesions are almost invariably absent up to the age of 16 weeks in contrast to a high frequency in males [24]. This would suggest that the etiology of the lesions in the two different parts of the aorta may be different.

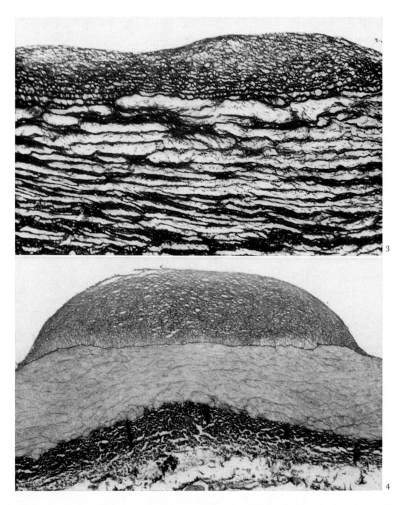

Fig. 3. Atherosclerotic lesion in anterior aorta of turkey. Weigerts resorcin fuchsin ×100.

Fig. 4. Atherosclerotic lesion in posterior aorta. Weigerts resorcin fuchsin ×55.

The Pig in Atherosclerosis Research

This animal suffers from the disadvantage of being expensive to buy and feed though the recent introduction of mini pigs to experimental atherosclerosis research has to some extent improved the situation [25]; but even these animals weigh up to 150 lb when mature.

The most notable work on this animal has been studies of pulsatile flow in extracorporeal shunts which has shown that platelet deposition corresponds closely to the distribution of aortic atherosclerotic lesions that occur spontaneously [26]. These results emphasise the important role of platelets not only as a precursor of thrombosis, but also as a potent factor in atherogenesis.

Epidemiological reports indicate that in man there is an inverse correlation between the occurrence of cardiovascular disease and the hardness of water supply, more disease being apparent in soft water districts [27, 28]. In order to study this phenomenon experimentally the effects of feeding hard and soft water to pigs in two different parts of Great Britain have been studied [29]. Abdominal aortae were obtained from two almost identical farms managed by the Pig Industry Development Association at Letchworth, England and Stirling, Scotland. The former has very hard water (275 mg/l carbonate), the latter soft water (42 mg/l carbonate). The diets of the pigs were the same and drawn from a central food supply. All pigs were killed at 200 lb body weight, the aortae were dissected out at the slaughter house, fixed in formalin and then stained macroscopically with Sudan dye according to the method of HOLMAN et al. [30]. Results were analysed according to centre, breed and sex. As shown in table III there was no statistical difference between either the number of pigs affected or the severity of disease between specimens obtained from the two centres. These results therefore provide no evidence to support the hypothesis that soft water increases severity of atherosclerosis. Differences were

Table III. Atherosclerosis in the abdominal aortae of pigs from two centres with different hardness of water. Soft (42 mg/L carbonate); hard (275 mg/L carbonate)

Centre	Water	Breed	Sex	No.	% with atherosclerosis	Average area affected sq/mm
Stirling	Soft	Large white	M	93	68.8	4.3
			F	95	62.1	3.5
		Landrace	M	47	74.4	7.6
			F	48	79.2	8.0
Letchworth	Hard	Large white	M	68	58.9	4.8
			F	73	52.0	2.8
		Landrace	M	36	83.3	12.6
			F	39	64.1	6.5

however observed between breeds; Landrace pigs being more severely affected than Large Whites. Also there was more disease in hogs (castrated males) than gilts. These results illustrate the importance of genetic and sex differences in the occurrence of the disease in pigs as in other species.

Acknowledgements

This work was supported in part by grants from the National Institutes of Health, Bethesda P.H.S. Grant No. H6300, the World Health Organisation and Eli Lilly, Indianapolis, (USA).

References

1. ROBERTS, J.C. and STRAUSS, R.: Comparative atherosclerosis (Harper and Row, New York 1965).
2. GRESHAM, G.A. and HOWARD, A.N.: The independent production of atherosclerosis and thrombosis in the rat. Brit.J.exp.Path. *41:* 395 (1960).
3. HOWARD, A.N. and GRESHAM, G.A.: The dietary induction of thrombosis and myocardial infarction. J.Atheroscler.Res. *4:* 40–56 (1964).
4. RENAUD, S. and ALLARD, C.: Thrombosis in connection with serum lipidic changes in the rat. Circulat.Res. *11:* 388 (1962).
5. RENAUD, S. and ALLARD, C.: Experimental pathology and dietary induced occlusive thrombosis in the rat. In: Methods and Achievements in Experimental Pathology, p.385. Ed. E. Bajusz and G. Jasmin (S. Karger, Basel/New York 1966).
6. DAVIDSON, E.; HOWARD, A.N. and GRESHAM, G.A.: The nature of the coagulation defect in rats fed diets which produce thrombosis or experimental atherosclerosis. Brit.J.exp.Path. *43:* 166 (1962).
7. TILLMAN, R.L.; O'NEAL, R.M.; THOMAS, W.A. and HIXON, B.B.: Butter, corn oil and fibrinolysis in rats. Circulat.Res. *8:* 423 (1960).
8. DAVIDSON, E.; HOWARD, A.N. and GRESHAM, G.A.: Blood coagulation studies in rats given diets which produce thombosis or atherosclerosis. Brit.J.exp.Path. *42:* 195 (1961).
9. ZEMPLENYI, T. and MRHOVÁ, O.: Arterial enzymes. In: Progress in Biochemical Pharmacology, vol.2 (Karger, Basel/New York 1967).
10. BIZZI, L.; HOWARD, A.N. and GRESHAM, G.A.: Plasma triglycerides and thrombosis in rats. Nature *197:* 195 (1963).
11. LINDGREN, F.T.; HOWARD, A.N. and GRESHAM, G.A.: In press.
12. ALAUPOVIC, P.: Recent advances in the metabolism of plasma lipoproteins: chemical aspects. In: Progress in Biochemical Pharmacology, vol.4 (Karger, Basel/New York 1967).
13. GRESHAM, G.A. and HOWARD, A.N.: Atherosclerosis produced by semisynthetic diet with no added cholesterol. Arch.Path. *74:* 1 (1962).
14. CONSTANTINIDES, P. and WYMAN, J.: Infarction and infarctoid necrosis in atherosclerotic rabbits. J.Atheroscler.Res. *2:* 285–305 (1962).
15. MALMROS, H. and WIGAND, G.: Atherosclerosis and deficiency of essential fatty acids. Lancet *2:* 749 (1959).
16. JONES, D.; GRESHAM, G.A. and HOWARD, A.N.: The effect of atromid on arterial disease induced by semisynthetic diet in the rabbit. J.Atheroscler.Res. *3:* 716–724 (1963).

17. HOWARD, A.N.; GRESHAM, G.A.; JONES, D. and JENNINGS, I.W.: The prevention of rabbit atherosclerosis by soya bean meal. J. Atheroscler. Res. 5: 330–337 (1965).
18. HOWARD, A.N. and GRESHAM, G.A.: In press.
19. PRICHARD, R.W.; CLARKSON, T.B.; LOFLAND, H.B., Jr.; GOODMAN, H.O.; HERNDON, C.N. and NETSKY, M.G.: Studies on the atherosclerotic pigeon. J. Amer. med. Ass. 179: 49 (1962).
20. KATZ, L.N.; NAGLE, R.; KAPLAN, P.; JOHNSON, P. and CENTURY, P.: The effect of Oestrogen on the concentration of bile cholesterol in the chicken. J. Atheroscler. Res. 3: 80 (1963).
21. CARNAGHAN, R.B.A.: Atheroma of the aorta associated with dissecting aneurysms in turkeys. Vet. Rec. 67: 568 (1955).
22. HOWARD, A.N. and GRESHAM, G.A.: Atherosclerosis in the Turkey in biological aspects of occlusive vascular disease. Ed. D.G. Chalmers and G.A. Gresham, p. 347 (Cambridge University Press 1964).
23. GRESHAM, G.A.; HOWARD, A.N. and JENNINGS, I.W.: Dietary fat and aortic atherosclerosis in the Turkey. J. Path. Bact. 85: 291 (1963).
24. GRESHAM, G.A.; HOWARD, A.N. and JENNINGS, I.W.: In press.
25. ZUGIBE, F.T.: Atherosclerosis in the miniature pig. In: Comparative Atherosclerosis (Harper and Row, New York 1965).
26. MUSTARD, J.F.; ROWSELL, H.C.; MURPHY, E.A. and DOWNIE, H.G.: Diet and thrombus formation: Quantitative studies using an extra corporeal circulation in pigs. J. clin. Invest. 42: 1783 (1963).
27. MORRIS, J.N.; CRAWFORD, M.D. and HEADY, J.A.: Hardness of local water – supplies and mortality from cardiovascular diseases in the county boroughs of England and Wales. Lancet 1: 860–862 (1961).
28. SCHROEDER, H.A.: Hardness of local water-supplies and mortality from cardiovascular disease. Lancet 1: 1171 (1961).
29. HOWARD, A.N.; JENNINGS, I.W. and GRESHAM, G.A.: Atherosclerosis in pigs obtained from two centres differing in hardness of water supply. Proc. Int. Soc. Geograph. Path. (Karger, Basel/New York 1967) (In preparation).
30. HOLMAN, R.L.; MCGILL, H.C.; STRONG, J.P. and GEER, J.C.: Techniques for studying atherosclerotic lesions. Lab. Invest. 7: 42–47 (1958).

Authors' address: G. A. Gresham and A. N. Howard, Dept. of Pathology, University of Cambridge, *Cambridge* (England).

Research Laboratories of the Pharmaceutical Department of CIBA Ltd., Basel

Mechanism of Atherogenesis in the Parakeet

R. Hess and W. Stäubli, Basel

Although experimental atherogenesis appears invariably to be associated with an elevation of serum lipid levels, the relationship between the exposure of the arterial wall to an excess of circulating lipoproteins and the cellular mechanism of lipid uptake remains obscure. Our own studies on the nature of the vascular response in atheroma formation were carried out in adult parakeets (Melopsittacus undulatus) of both sexes, i.e. in a species in which atheromatosis, particularly of the brachiocephalic aorta, is easily produced by adding cholesterol to the ordinary seed diet (Blohm et al., 1956; Finlayson and Hirchinson, 1961; Hess, 1963).

The result of feeding the parakeets a diet containing 2% of cholesterol for 3 months is given in Table I. An approximately five-fold increase in serum lipids was accompanied by an increase in aortic total lipids of the same magnitude. The main composition of tissue lipids (exemplified by the liver) became similar to that of serum. Deposition of lipid in the innermost layers of the aorta became

Table I. Distribution of lipids in serum and tissues of the parakeet*

Lipid content	Serum		Liver		Aorta	
	Normal	Chol.-fed	Normal	Chol.-fed	Normal	Chol.-fed
Total lipids (mg per ml/gm)	9.2	52.4	179.9	176.7	0.0146	0.0717
Total cholesterol %	20.2	17.9	1.4	16.0		
Triglyceride %	58.0	60.0	28.5	60.0		
Phospholipids %	26.2	18.0	65.9	24.0		

* Normal (untreated) birds as compared to animals fed a seed diet containing 2% cholesterol and 4% corn oil for 3 months.
Mean values of 3 to 4 animals of both sexes per group.

visible in the light microscope at about 12 days of feeding, and atheromas were present at 18 days.

Ultrastructurally, the aorta of the parakeet is characterised by a loose network of elastic lamellae, the interstices of which are filled with smooth muscle cells of a spider-like appearance that are embedded in a large extracellular space. The latter contains a weakly electron-dense and finely granular ground substance. The endothelium is of the usual appearance and it extends over a distinct basement membrane (HESS and STÄUBLI, 1963a).

The first visible stage in atheroma formation consisted of a progressive deposition of lipid-like material in the subendothelial space, accompanied by a breaking up of the basement membrane. Prominent cytoplasmic alterations of the endothelium (as seen, for instance, in the rat, HESS and STÄUBLI, 1963b) were not encountered. The newly deposited material consisted of globular particles and clusters of larger aggregates. The smallest units of these particles had a diameter of 23 to 120 mμ and, therefore, they fall into the range of lipoproteins of very low density (cf. SCANU, 1965). This material extended into deeper layers of the aortic wall as cholesterol feeding continued. Concurrently with the extracellular deposition of lipid-containing vesicles, the smooth muscle cells became progressively loaded with lipid vacuoles. As a result of this process the cells were transformed into lipophages and into typical foam cells. As the lesion developed, smooth muscle cells of deeper layers of the artery were transformed in the same way.

By means of repeated intravenous injection of electron-dense colloids (thorium dioxide, gold sulphide) in hyperlipidaemic birds, it was demonstrated that marker particles traversed the endothelium rather rapidly, apparently within cytoplasmic vacuoles of more than 100 mμ in diameter. These particles were deposited extracellularly in the subendothelial space together with the lipoprotein-like globules or vesicles. The metal-containing colloid was further taken up by smooth muscle cells and appeared to be incorporated in the same membrane-bounded vacuoles that contained the lipid-like inclusions (Fig. 1).

Endothelial transfer and uptake by smooth muscle cells of macromolecular substances derived from the circulation appears to depend initially on pinocytotic and phagocytotic mechanisms. The possibility exists that these processes require an enhanced formation of cytoplasmic membranes which, in turn, may be considered to derive from local synthesis (NEWMAN et al., 1961).

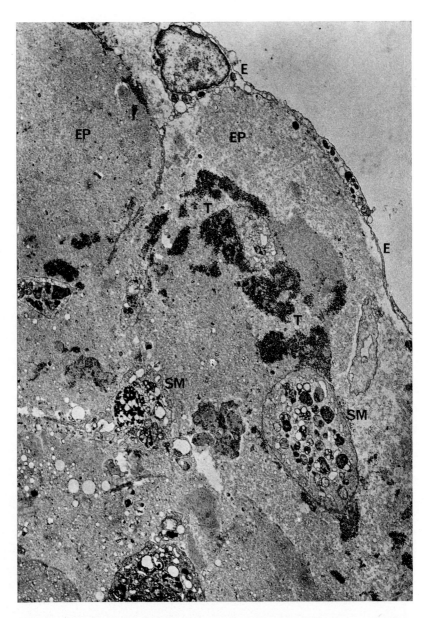

Fig. 1. Brachiocephalic aorta of a parakeet, at 18 days of cholesterol feeding and following two injections of ThO_2 (Thorotrast). The subendothelial space is filled up with extracellular particles (EP), possibly of lipoprotein nature, and with thorotrast particles (T). Lipid storing smooth muscle cells (SM) contain both types of particles, mainly concentrated in vacuoles. E = endothelium. ×5600.

References

BLOHM, T.R.; LERNER, L.J.; KARIYA, T. and WINJE, M.E.: Effects of cholesterol feeding in the shell parakeet (Melopsittacus undulatus). Circulation *14*: 497 (1956).

FINLAYSON, R. and HIRCHINSON, V.: Experimental atheroma in budgerigars. Nature *192*: 369–370 (1961).

HESS, R.: Cholesterin-induzierte Atheromatose in Melopsittacus undulatus. Path. Microbiol. *26*: 343–347 (1963).

HESS, R. and STÄUBLI, W.: Vergleichende histochemische und elektronenmikroskopische Untersuchungen von Aortaveränderungen bei experimenteller Lipoidose. Verh. Dtsch. Ges. Path. *47*: 369–372 (1963a).

HESS, R. and STÄUBLI, W.: The development of aortic lipidosis in the rat. A correlative histochemical and electron microscope study. Amer. J. Path. *43*: 301–335 (1963b).

NEWMAN, H.A.I.; McCANDLESS, E.L. and ZILVERSMIT, D.B.: The synthesis of C^{14}-lipids in rabbit atheromatous lesions. J. biol. Chem. *236*: 1264–1268 (1961).

SCANU, A.M.: Factors affecting lipoprotein metabolism. Adv. Lip. Res. *3*: 63–138 (1965).

Authors' address: R. Hess, M.D. Pharmaceutical Department, CIBA Ltd., *4000 Basel* (Switzerland).

The Wistar Institute of Anatomy and Biology and School of Veterinary Medicine
University of Pennsylvania, Philadelphia, Pennsylvania

Influence of Special Fats on Experimental Atherosclerosis in Rabbits

D. KRITCHEVSKY and SHIRLEY A. TEPPER, Philadelphia, Pa.

Introduction

In the course of our experiments involving the influence of the cholesterol vehicle in experimental atherosclerosis in the rabbit, we have observed that, generally, the atherogenicity of the diet is an inverse function of the iodine number of the oil used as the cholesterol vehicle. In using two oils of low iodine value, however, it was found that a diet using the more saturated oil (cocoa butter, iodine value, 35) was less atherogenic than one containing palm oil (iodine value, 53) (KRITCHEVSKY and TEPPER, 1965a). One of the major differences between these two fats lies in the spectrum of their saturated fatty acids. The saturated fatty acids of cocoa butter are primarily palmitic (23.4% of total fatty acids) and stearic (40.5%), whereas palmitic acid (45.4%) is the principal saturated fatty acid of palm oil. In view of these observations and of the fact that myristic and palmitic acids have been reported to be most hypercholesteremic in man (HEGSTED *et al.*, 1965), we devised an experiment to study the effects of individual saturated fatty acids upon cholesterol-induced atherosclerosis in rabbits.

Pure saturated triglycerides (trilaurin, trimyristin, tripalmitin and tristearin) were subjected to randomization with corn oil. Randomization is a process whereby two fats are interesterified in the presence of a strong base to yield a single mixed triglyceride whose fatty acid composition reflects the fatty acids of the two original fats. In order to obtain fats of equivalent saturation, the ratios of triglyceride/corn oil were 19/81, 20/80, 23/77 and 25/75 for trilaurin, trimyristin, tripalmitin and tristearin, respectively. The resulting fats all had an iodine value of 102 ± 2.

Experimental Procedure

Male rabbits of the Dutch belted strain (average starting weight 1.7–2.0 kg) were divided into groups of equal starting weight and were maintained for two months on a diet consisting of rabbit chow augmented with 2% cholesterol suspended in 6% of the appropriate fat. After eight weeks the rabbits were bled by venipuncture, killed and the liver and aorta removed. An aliquot of the liver was dissolved in 15% alcoholic KOH and the non-saponifiable material extracted with petroleum ether. Serum and liver were assayed for total cholesterol by the method of MANN (1961). Aortas were graded visually on a 0–4 basis following the system of DUFF and MCMILLAN (1949), the thoracic aorta and arch being graded separately. The visual grading of the aortas has been shown to correlate well with their cholesterol content (KRITCHEVSKY et al., 1961). The aortas were stripped, minced and extracted by stirring in chloroform-methanol 2:1 for 4 h. The chloroform extract was washed with water, dried over anhydrous sodium sulfate and assayed for free and ester cholesterol by the method of SPERRY and WEBB (1950).

The special fats were prepared by randomization of pure triglycerides (trilaurin, trimyristin, tripalmitin and tristearin) with corn oil. Ratios of triglyceride to corn oil used were calculated to yield fats of similar iodine value. The iodine values of the fats thus obtained were: trilaurin fat (L)-104.3; trimyristin fat (M)-100.3; tripalmitin fat (P)-100.0 and tristearin fat (S)-103.1. In addition to these groups, two additional groups of rabbits were used – one group was given cholesterol in corn oil which had been subjected to the randomization treatment, iodine value-128.3, (R) and another group was fed cholesterol suspended in 'native' corn oil, iodine value-126.1 (CO). The fatty acid composition of the various fats is given in Table I.

Results and Discussion

In all, three experiments were carried out. The results of these experiments are summarized in Table II. It is seen that the more

Table I. Fatty acid composition of rearranged fats

F.A.	Starting corn oil (G)	Rearr. corn oil (R)	Wistar corn oil (CO)	L (19/81)*	M (20/80)	P (23/77)	S (25/75)
Iodine value	130.6	128.3	126.1	104.3	100.3	100.0	103.1
12:0	—	—	—	19.0	0.3	—	0.4
14:0	0.1	0.1	—	0.4	18.2	0.3	0.2
16:0	11.6	11.4	11.0	9.5	10.0	30.9	12.3
16:1	0.3	0.3	0.3	0.1	0.2	0.3	0.2
18:0	1.9	2.2	1.6	1.7	1.6	2.6	23.4
18:1	26.5	26.2	29.2	20.8	21.3	20.0	20.4
18:2	57.4	58.2	54.9	46.7	47.0	44.1	41.9
18:3	1.3	0.9	1.5	0.9	0.7	0.9	0.6
20:0	—	0.5	1.2	0.3	0.3	0.2	0.3

* Ratio of triglyceride/corn oil used to obtain fats of equivalent saturation.

Table II. Influence of special fats on experimental atherosclerosis in rabbits (2% cholesterol in 6% fat; 8 weeks)

Fat	Survival	Weight gain (g)	Liver wt. (g)	Cholesterol liver (g/100 g)	Cholesterol serum (mg/100 ml)	Avg. atheromata arch.	Avg. atheromata thoracic
Expt. 1							
L	14/15	135	113.2	3.12	2297	2.1	1.3
M	12/15	106	104.6	2.56	2160	1.8	1.3
P	15/15	153	114.7	2.99	2045	2.2	1.3
S	15/15	231	111.2	2.64	2188	1.7	1.0
R	14/15	256	104.4	2.26	2412	1.6	1.0
CO	14/15	115	112.9	2.99	2474	1.5	1.2
Expt. 2							
L	11/12	272	120.9	6.18	2455	1.6	0.9
M	11/12	111	105.0	9.12	2523	1.7	1.3
P	12/12	268	114.4	6.46	2310	1.5	0.8
S	11/12	173	121.0	7.91	2152	1.7	0.9
R	10/12	203	111.2	7.43	2486	1.3	0.9
CO	12/12	249	123.1	9.30	3156	1.7	1.0
Expt. 3							
L	7/9	63	111.1	10.84	2315	2.4	1.3
M	8/9	254	109.7	10.60	1906	2.0	1.5
P	7/9	456	112.0	6.83	1961	2.6	1.7
S	6/9	155	93.0	4.71	1469	2.1	1.3
R	9/9	362	117.2	6.51	1554	1.7	1.2
CO	8/9	237	122.5	7.56	2075	1.9	1.3

saturated fats generally tended to produce more severe atheromata than either the rearranged or intact corn oil. Liver cholesterol levels were all grossly elevated with the lauric acid (L), and myristic acid (M) fats usually yielding the greatest elevations. Among the more saturated fats, atheromata tended to be more severe in the lauric (L) and palmitic (P) acid groups. The serum cholesterol levels were, however, almost always the highest in the groups fed corn oil.

In an earlier series of experiments (KRITCHEVSKY and TEPPER, 1965b), we observed that when cholesterol suspended in certain fats (medium chain triglyceride and tristearin) was fed to rabbits, their atheromata levels were not much higher than those observed in cholesterol-corn oil fed groups, but their serum cholesterol levels were considerably lower. In an effort to relate the serum cholesterol level to the atheromata, we suggested calculation of a 'relative atherogenic effect' (RAE) for each fat. If the serum cholesterol level is taken as an index of cholesterol absorption and the severity of atheromata is related to the cholesteremia, then the ratio of atheromata to serum cholesterol is an expression of the amount of circulating sterol which has been deposited in the aorta. Thus the RAE, which is calculated by dividing the average atheromata by grams of circulating cholesterol, is a value which relates serum cholesterol to atherogenesis. In Table III we present the average RAE for all the fats used in all three experiments. On this basis we see that the fat rich in palmitic acid was the most atherogenic when compared with corn oil. The difference between the average atheromata of the aortic arch in rabbits fed fat P or corn oil is significant at the 5% level of probability.

Table III. Average atheromata and relative atherogenic effect (average of 3 experiments)

Fat	Atheromata arch.	thoracic	Average atheromata	Average serum cholesterol mg/100 ml	RAE	RAE (CO = 1)
L	1.97±0.19*	1.14±0.13	1.56	2355	0.66	1.22
M	1.84±0.14	1.32±0.15	1.58	2223	0.70	1.30
P	2.04±0.10	1.26±0.10	1.65	2121	0.78	1.44
S	1.78±0.16	1.03±0.09	1.41	2041	0.69	1.28
R	1.52±0.14	1.00±0.13	1.26	2200	0.57	1.06
CO	1.67±0.16	1.12±0.13	1.40	2605	0.54	1.00

* Standard error.

The results of the measurement of the aortic cholesterol are presented in Table IV. The cholesterol content of all the aortas was elevated and the amount of ester cholesterol present in the aortas was high, ranging from 17 to 56%. The normal rabbit aorta contains practically no ester cholesterol. Severity of thoracic atheromata could not be correlated with ester cholesterol.

Table IV. Cholesterol content of thoracic aortas of rabbits fed cholesterol and special fats

Fat	Average aorta cholesterol, mg/100 g wet wt.					
	experiment 1		experiment 2		experiment 3	
	total	% ester	total	% ester	total	% ester
L	452	24.4	203	38.9	381	40.3
M	426	53.4	366	48.7	636	41.3
P	500	17.0	277	17.5	510	46.5
S	324	32.5	201	24.7	392	41.0
R	391	53.4	220	33.0	345	42.2
CO	429	55.5	210	26.4	394	34.1
Normal	50	4.7	—	—	—	—

Summary

Rabbits were maintained for 8 weeks on atherogenic diets consisting of chow augmented with 2% cholesterol and 6% fat. Four of the fats used were prepared by randomizing corn oil with pure trilaurin (L), trimyristin (M), tripalmitin (P) and tristearin (S) to yield materials of similar iodine value (102 ± 2) but differing in one predominate saturated fatty acid. Native corn oil (CO) and corn oil which had been subjected to the randomization process (R) were also used as cholesterol vehicle. In three experiments average atheromata were: L-1.56; M-1.58; P-1.65; S-1.41; R-1.26 and CO-1.40. Average serum cholesterol levels (mg/100 ml) were: L-2355; M-2223; P-2121; S-2041; R-2200 and CO-2605. Calculation of the relative atherogenic effect (average atheromata divided by average serum cholesterol in g/100 ml) showed that the four special fats were all considerably more atherogenic than either untreated or randomized corn oil. The fat rich in palmitic acid was the most atherogenic.

Acknowledgements

This work was supported, in part, by a grant (HE-03299) and a Research Career Award (5-K6-HE-734) from the National Heart Institute, USPHS. We are indebted to Dr. F.H. MATTSON, Proctor and Gamble Co., Cincinnati, Ohio for preparation of the rearranged fats and to Dr. V.K. BABAYAN, Drew Chemical Co., New York, N.Y. for the pure triglycerides used in their preparation.

References

1. DUFF, G.L. and McMILLAN, G.C.: The effect of alloxan diabetes on experimental cholesterol atherosclerosis in the rabbit. J.exp.Med. *89:* 611–630 (1949).
2. HEGSTED, D.M.; McGANDY, R.B.; MYERS, M.L. and STARE, F.J.: Dietary fatty acids and changes in serum cholesterol. Fed.Proc. *24:* 262 (1965).
3. KRITCHEVSKY, D.; MOYNIHAN, J.L.; LANGAN, J.; TEPPER, S.A. and SACHS, M.L.: Effects of D- and L-thyroxine and of D- and L-3,5,3^1-triiodothyronine on development and regression of experimental atherosclerosis in rabbits. J.Atheroscl.Res. *1:* 211–221 (1961).
4. KRITCHEVSKY, D. and TEPPER, S.A.: Cholesterol vehicle in experimental atherosclerosis. VII. Influence of naturally occurring saturated fats. Med. Pharmacol. exp. *12:* 315–320 (1965a).
5. KRITCHEVSKY, D. and TEPPER, S.A.: Cholesterol vehicle in experimental atherosclerosis VIII. Effect of a medium chain triglyceride (MCT). Exp.molec.Path. *4:* 489–499 (1965b).
6. MANN, G.V.: A method for measurement of cholesterol in blood serum. Clin. Chem. *7:* 275–284 (1961).
7. SPERRY, W.M. and WEBB, M.: A revision of the Schoenheimer-Sperry method for cholesterol determination. J.biol.Chem. *187:* 97–106 (1950).

Authors' address: Dr. David Kritchevsky and Shirley Tepper, Wistar Institute of Anatomy and Biology 36th and Spruce Street, *Philadelphia, Pa.* (USA).

The Wistar Institute of Anatomy and Biology, Philadelphia, Pennsylvania

Influence of Linolexamid on Experimental Atherosclerosis in Rabbits

D. KRITCHEVSKY and SHIRLEY A. TEPPER, Philadelphia, Pa.

Linolexamid (N-cyclohexyl linoleamide) has been reported to reduce serum cholesterol levels and atherosclerosis in cholesterol-fed rabbits (TOKI and NAKATANI, 1965). We have completed two experiments in which linolexamid was fed to male, Dutch-belted rabbits maintained on 2% cholesterol in 6% corn oil for two months. The drug was added to the diet at a level of 68.1 g/50 lb., based on an estimated food consumption of 100 g/rabbit daily, the daily ingested dose being 300 mg linolexamid. After 8 weeks, the rabbits were killed, serum and liver total cholesterol estimated by the method of MANN (1961), and severity of aortic atherosclerosis graded visually on a 0–4 scale. The results of the two experiments are summarized in Table I.

Table I. Influence of linolexamid (N-cyclohexyl linoleamide) on experimental atherosclerosis in rabbits*

Group	Survival rate	Weight gain (g)	Liver wt. (g)	Cholesterol liver (g/100 g)	Cholesterol serum (mg/100 ml)	Avg. atheromata arch.	Avg. atheromata thoracic
Exp. 1							
L**	7/8	202	89.6	4.06±0.77***	797±190	1.3	1.1
C	8/8	98	102.6	7.75±0.88	1497±313	1.9	1.4
Expt 2							
L	14/16	395	106.8	5.38±0.38	1876±237	1.5	1.0
C	14/16	305	107.3	8.68±0.70	2236±246	1.6	0.9

* All rabbits fed 2% cholesterol in 6% corn oil.
** L = linolexamid (300 mg/day); C = control.
*** Standard error.

It is evident that in the two experiments linolexamid lowered liver cholesterol significantly (p <0.01 in experiment 1 and p <0.001 in experiment 2). Serum cholesterol levels were also considerably lower in the linolexamid treated rabbits in both experiments, but differences were not significant below the 5% level. Atheromata were somewhat reduced in the first experiment and no different in the second. Average atheromata for the two experiments were: Linolexamid group – arch-1.5; thoracic aorta-1.0; and in the controls – arch-1.7; thoracic aorta-1.1. The sharp reductions observed in serum and liver cholesterol levels indicate that linolexamid is a compound that merits further study. Additional experiments are in progress in our laboratory.

Acknowledgement

We wish to thank Dr. K. Toki (Sumitomo Chemical Co., Ltd., Osaka, Japan) for a generous supply of linolexamid.

This investigation was supported in part by Public Health Service Research Grant No. HE 03299–0951 from the National Heart Institute. Dr. Kritchevsky is a recipient of Public Health Service Research Career Award No. K6–HE–734 from the National Heart Institute.

References

1. Mann, G.V.: A method for measurement of cholesterol in blood serum. Clin. Chem. 7: 275–284 (1961).
2. Toki, K. and Nakatani, H.: Effects of N-cyclohexyl linoleamide (Linolexamid) on experimental atherosclerosis in rabbits. Proc. Second International Symposium on Drugs Affecting Lipid Metabolism, 1965; in press.

Authors' address: Dr. David Kritchevsky and Shirley A. Tepper, Wistar Institute of Anatomy and Biology, 36th and Spruce Street, *Philadelphia, Pa.* (USA).

From the Medical Research Laboratory for the Study of Atherosclerosis, Lund and the
Department of Pathology, Allmänna Sjukhuset, Malmö

Induction of Atherosclerosis in Dogs by a Thiouracil-Free Semisynthetic Diet Containing Cholesterol and Hydrogenated Coconut Oil*

H. Malmros, Lund, and N. H. Sternby, Malmö

Since 1946 when Steiner and Kendall demonstrated atherosclerotic lesions in dogs fed thiouracil and cholesterol it has been generally thought that it is necessary to suppress thyroid function in order to produce high levels of serum cholesterol and arterial lesions. Cholesterol alone even in large amounts produced only a very moderate rise of the cholesterol level and no clear atherosclerosis. In these experiments Steiner and Kendall used 10 g crystalline cholesterol suspended in 40 ml cottonseed oil mixed in the dog food.

From these and similar experiments rather far reaching conclusions have been drawn. They have been quoted in support of the view that dietary cholesterol does not largely influence the serum-cholesterol level in man. It has been suggested that the 'resistence' of dogs to cholesterol might be due to an intrinsic high rate of cholesterol synthesis in the liver and that this synthesis would be rapidly suppressed after the absorption of dietary cholesterol.

However, in investigations on pet dogs we have often found relatively high levels of serum cholesterol (Table I).

Schiller and Berglund (1964) likewise frequently found hypercholesterolaemia in pet dogs whereas in a control group fed on commercial laboratory biscuits they observed markedly lower cholesterol levels. Since this seemed to suggest that the composition of the diet may influence the serum cholesterol level in dogs we

---

* These studies were supported by grants from the Medical Research Council and from the Swedish National Association against Heart- and Chest Diseases.

Table I. Total serum cholesterol in pet dogs

Dog No.	Total serum cholesterol mg/100 ml	Breed	Age year	Sex	Diagnosis
1	285	Chow-Chow	9	bitch	
2	152	New Foundland	2	bitch	
3	175	Golden Retriever	2,5	hound	
4	*316*	Boxer	10	hound	
5	171	Boxer	2	hound	
6	287	German Shepherd	2	hound	
7	254	Dachshound	3	bitch	
8	153	Poodle	1,5	hound	
9	283	Swedish Hound	2,5	bitch	
10	170	Boxer	2	hound	
11	*528*	Dachshound	2	hound	hypothyreosis
12	*327*	Boxer	4	hound	obesity
13	210	Poodle	10	hound	
14	246	Setter	5	hound	
15	*344*	Coccer Spaniel	9	hound	diabetes
16	*336*	German Shepherd	8	bitch	
17	175	Dachshound	4	hound	
18	*386*	Swedish Hound	2,5	bitch	
19	260	Boxer	1	bitch	
20	273	Dalmation	2,5	bitch	
21	151	Dachshound	3	bitch	
22	*338*	German Shepherd	10	bitch	

carried out a series of control experiments using a semisynthetic diet containing cholesterol and fat (Table II).

Table III and Figure 1 show that a strong hypercholesterolaemia can be produced also in dogs if the diet apart from cholesterol contains saturated fat in the form of hydrogenated coconut oil. Thyroid function need not then be suppressed with thiouracil to produce high serumcholesterol levels and these can rise to 1000 mg% or more.

Preliminary investigations have shown such a hypercholesterolaemia to be sufficiently high to produce a very marked atherosclerosis within a year (Fig. 2).

Even when the semisynthetic diet does not contain cholesterol but only hydrogenated coconut oil hypercholesterolaemia may appear with cholesterol levels as high as 500 mg%. Even in these cases microscopic examination, after one year, reveal some atherosclerotic changes in the coronary arteries. A diet containing hydrogenated coconut oil does not produce a marked rise in the serum triglyceride level in dogs as it does in rabbits.

Table II. Composition of experimental semi-synthetic diet in pellet form*

Substance	Percentage by weight	
Casein, purified	21	cholesterol 5%
Sucrose	30	
Fat	17	
Salt mixture	3	
Cellulose	12,5	
Choline chloride	0,4	
p-Aminobenzoic acid	0,1	
Inositol	0,1	
Aerosil (colloidal silicic acid)	12,5	
Talc	3,4	

Vitamins	mg/100 g ration
Ascorbic acid	400
Nicotinic acid	20
Calcium pantothenate	4
Alpha-tocopherol acetate	2
Riboflavin	2
Aneurine hydrochloride	2
Pyridoxine hydrochloride	2
Folic acid	1
Vitamin A-palmitate	0,6
K (Menadione sodium sulphate)	0,2
Biotin	0,06
D_2 (Calciferol)	0,004
B_{12} (Cyanocobalomin)	0,004

Salt mixture	mg/100 g
K_2HPO_4	645
$CaCO_3$	600
NaCl	335
$CaHPO_4 \times 2H_2O$	150
$MgSO_4$	99,7
$FeSO_4$	14,7
$MnSO_4 \times 4H_2O$	10,0
KI	1,6
$CuSO_4 \times 5H_2O$	0,6
$ZnCl_2$	0,5

* Tablets prepared by Apoteket Hjorten, Lund, Sweden.

Table III. Total serum cholesterol (C) and Triglycerides (T) in 9 dogs fed atherogenic diets

Dog No.		Before	1 month	2 months	3 months	5 months	6 months	7 months	8 months	10 months	11 months	12 months	13 months	Diet
1/65	C	248	472	542	399	392	451	572	488	373	393	495	374	hydrog. coconut oil
	T	52	187	52	86	137	109	148	122	185	95	207	91	
5/65	C	165	297	324	339	360	327	302	276		260	249		hydrog. coconut oil
	T	35	58	52	64	53	45	42	46		51			
2/65	C	212	723	725	841	1064	713	1172	1132	1091	1073	1030	775	hydrog. coconut oil +5% cholesterol
	T	49	108	63	46	82	50	93	78	97	60	73	44	
6/65	C	148	464	770	662	591	543	456	570		575			hydrog. coconut oil +5% cholesterol
	T	38	51	51	62	128	67	161	98		44			
7/66	C	136	652		677	755								hydrog. coconut oil +5% cholesterol
	T	22	63		105									
8/66	C	202	551		763	463								hydrog. coconut oil +5% cholesterol
	T	17	34		30									
9/66	C	172	344	1162										hydrog. coconut oil +5% cholesterol
	T	30	22											
3/65	C	220	264	277	324	316	312	365	412	395	414	655	438	dog pellets +cholesterol
	T	101	53	41	58	61	88	136	126	147	92	186	124	
4/65	C	246	217	334	289	291	322	359	318	366	341	251	300	dog pellets +cholesterol
	T	55	113	39	35	34	49	47	41	33	28	37	20	

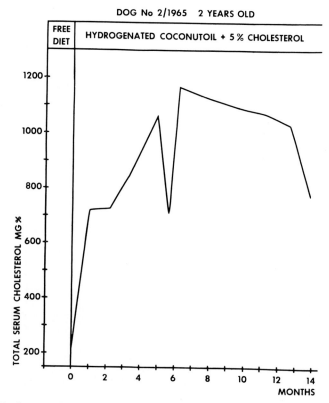

Fig.1. Total serum cholesterol in a dog fed 17% hydrogenated coconut oil+5% cholesterol in a semisynthetic diet for 1 year.

Fig.2. Atherosclerotic lesions in aorta.

To study the effect of cholesterol without any extra fat we used commercial dog pellets with cholesterol added to them. This was done by soaking the pellets in cholesterol dissolved in ether, which was left to evaporate. This diet also produced a certain rise in the serum cholesterol level with a few values as high as 400–600 mg %. Preliminary investigations have shown microscopic changes resembling an early atherosclerosis to appear within a year. These investigations make it apparent that:

(1) Dogs are not resistent to dietary cholesterol;

(2) they react to saturated fat (hydrogenated coconut oil) in their diet with a moderate hypercholesterolaemia even when there is no cholesterol in the diet;

(3) a combination of cholesterol and hydrogenated coconut oil causes a marked hypercholesterolaemia producing a pronounced atherosclerosis within one year;

(4) thyroid function need not be suppressed with thiouracil for cholesterol and saturated fat to exert this atherogenic effect.

References

STEINER, A. and KENDALL, F. E.: Atherosclerosis and arteriosclerosis in dogs following ingestion of cholesterol and thiouracil. Arch. Path. *42:* 433 (1946).

SCHILLER, I.; BERGLUND, N. E.; TERRY, J. R.; REICHLIN, R.; TRUEHART, R. E. and COX, G. E.: Hypercholesteremia in pet dogs. Arch. Path. *77:* 389 (1964).

MALMROS, H.: Vilken betydelse har födans kolesterolhalt för uppkomsten av atheroscleros. Näringsforskning *9:* 80 (1965).

MALMROS, H.: Importance of dietary cholesterol and fat in the pathogenesis of atherosclerosis. Paper read at the 4th International Congress of Dietetics, Stockholm 12–16 June, 1965.

Authors' address: Dr. Haqvin Malmros, Medical Research Laboratory for the Study of Atherosclerosis *Lund*, and Dr. Nils H. Sternby, Department of Pathology, Allmänna Sjukhuset, *Malmö* (Sweden).

Department of Pathology, Albany Medical College, Albany, New York

Aspects of Protein Metabolism in Rats Fed Atherogenic Diets*

W. A. Thomas, K. T. Lee, D. N. Kim, J. A. Sisson and
R. F. Scott, Albany, N.Y.

In this presentation we shall begin with a description of certain atherogenic diets for rats and resulting lesions. Secondly we shall describe metabolic studies related to protein synthesis by the liver in these rats. Thirdly, we shall present metabolic studies related to protein synthesis by the aorta.

Small proliferative atherosclerotic lesions characterized by smooth muscle cells develop eventually in practically all rats regardless of type of diet [1, 2]. Age of onset can be decreased, extent and severity of lesions increased and complications such as thrombosis induced, by various kinds of cholesterol-containing diets [3, 4]. Figure 1 illustrates an example of a lesion in a rat fed an atherogenic diet for 14 months. Figure 2 shows a smooth muscle cell which is the predominant cell in the lesion. The degree of response appears to be in large part determined by the severity of the diet although metabolic differences among species are also undoubtedly factors. With judiciously chosen diets the types of cells in the lesions are the same regardless of severity of the diet.

We have studied by biochemical means and by electron microscopy some of the metabolic effects of drastic and of relatively mild atherogenic diets on rats [5, 6, 7]. The period chosen for special study has been one in which a few small lesions are developing in 100% of rats fed atherogenic diets but with none in controls. We have carried out a few similar studies in Rhesus monkeys fed relatively mild diets and the results of these will be presented in the primate section [8].

* This work was supported by USPHS Grant H-7155.

Fig. 1a.

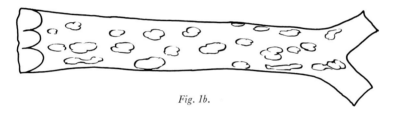

Fig. 1b.

Fig. 1a, b. An extensive intimal lesion composed largely of spindle cells from the aorta of a rat fed a severe atherogenic diet for 14 months is illustrated. H and E stain 175×. The diagram shows the extent of the lesions seen in the aorta. (Courtesy of Academic Press.)

All rats used for the experiments were weanlings and were fed the indicated diets 1–3 months. Fig. 3 shows a typical lesion and the typical location. Important ingredients in the 'mild' and 'severe' diets referred to in this report are shown in Table I. Rats fed the severe diets grew very little, were extremely debilitated and had a high mortality. Rats fed the mild diets grew well although not quite as well as controls, appeared healthy and virtually all survived.

By electron microscopy most of the cells in these lesions were either primitive smooth muscle cells or cells that were not differentiated enough to be identified in the pictures that were taken. An example is shown in Fig. 4. As judged by the extensive network of rough surfaced endoplasmic reticulum both the recognizable smooth muscle cells and cells not yet differentiated were very active in protein synthesis.

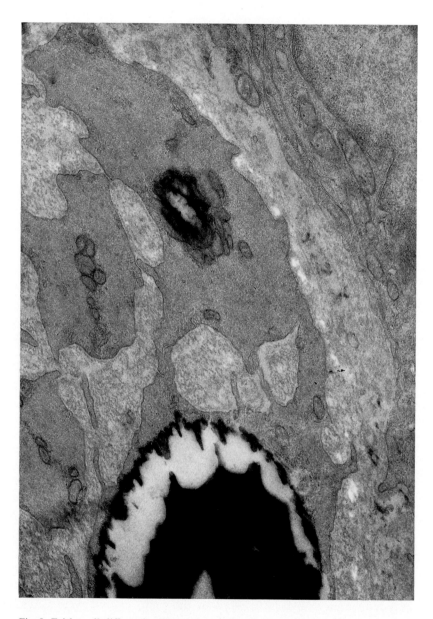

Fig. 2. Fairly well differentiated smooth muscle cells in a non-necrotic atherosclerotic lesion of the aorta of a rat fed a severe atherogenic diet for 14 months are illustrated. Lead hydroxide stain 22,000×. (Courtesy of Academic Press.)

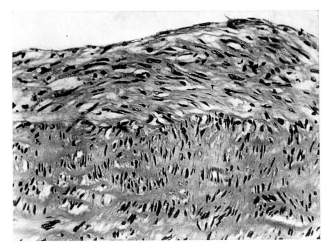

Fig. 3a.

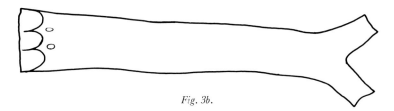

Fig. 3b.

Fig. 3a, b. A non-necrotic intimal lesion in the aorta of a rat fed a mild atherogenic diet for 3 months. The lesion is a multi-layered spindle cell lesion, with abundant intercellular material (H and E stain 225×). The diagram shows the characteristic location of lesions in rats fed atherogenic diets for only 3 months.

Table I. Diets (% by weight)

	'Severe'	'Mild'
Cholesterol	5	5
Sodium Cholate	2	2–3
Thiouracil	0.3	0
Fat	40	25–40
Casein	20	30–50
Sugar	+	+
Vitamins	+	+
Minerals	+	+

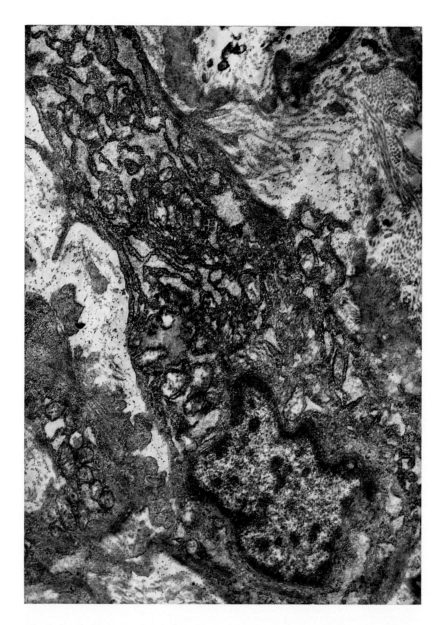

Fig. 4. A primitive smooth muscle cell seen in a non-necrotic atherosclerotic lesion in the aorta of rats fed a severe atherogenic diet for 3 months. Abundant dilated endoplasmic reticulum lined by numerous ribosomes suggesting active protein synthesis is present. Lead hydroxide stain 17,000×. (Courtesy of Academic Press.)

Our metabolic studies have centered about protein synthesis as an important pathway in the mainstream of cellular metabolism. We have studied the liver because of its overall importance in protein synthesis and the aorta because it is a prime site where the atherosclerosis occurs.

Metabolic Studies Related to Hepatic Cell Protein Synthesis of Rats Fed Atherogenic Diets

Rats fed atherogenic diets rapidly develop fibrinogen levels in plasma 2–3 × greater than corresponding control values (Figure 5). Total plasma proteins change only slightly. We have injected ^{14}C-glycine into rats fed severe diets for various periods and deter-

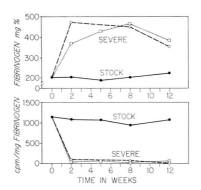

Fig. 5. Plasma fibrinogen levels and ^{14}C-glycine incorporation into fibrinogen *in vivo* for a period of 0–12 weeks in rats fed two slightly different severe atherogenic diets and a stock diet. The upper graph represents plasma fibrinogen levels expressed in milligrams per cent. The lower represents the radioactivity of the fibrinogen fraction expressed in counts per minute per milligram fibrinogen. Each point is the average of 6 determinations.

mined the relative amount of incorporation of isotope into fibrinogen after 1 h. We found incorporation rates to be approximately 1/15 of control values for fibrinogen, suggesting a very slow rate of synthesis. The elevated plasma fibrinogen levels must therefore be due to a decreased rate of degradation.

When we utilized an *in vitro* microsomal cell free system from the experimental rats, we found that the rate of incorporation of ^{14}C-leucine into microsomal protein was less than 1/2 control values whether expressed per unit of RNA or per unit protein

(Fig. 6). Examination of the liver by electron microscopy (Fig. 7 and 8) indicated that the controls had more ribosomes in rosettes or similar patterns that we associate with active protein synthesis than did rats fed atherogenic diets.

Also ultracentrifugal studies of hepatic ribosomes by sucrose gradient techniques indicated a much greater degree of aggregation of ribosomes into characteristic polysomal units in the control group than in those of rats fed the atherogenic diet. Studies of rapid (15 min) incorporation of ^{14}C-orotic acid into hepatic cytoplasmic RNA indicated a much lower rate in the atherogenic diet group.

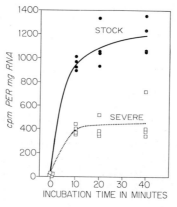

Fig. 6. ^{14}C-leucine incorporation into hepatic microsomal protein in rats fed a severe and a stock diet. The severe diet-fed group had low rates of leucine incorporation in the cell free system. Each point represents the mean of 4 determinations and each determination was made on a pool of 3 rats.

The excessive dispersal of ribosomes into small units as indicated by electron microscopy and sucrose gradient studies, and the depressed synthesis of rapidly labeled cytoplasmic RNA suggest (but do not prove) that one factor in the disturbed metabolism of the hepatic cells is a deficiency in synthesis of messenger RNA.

In other procedures we have determined by use of an amino acid analyzer the amino acids of the 105,000 g supernatant from hepatic homogenates after ^{14}C-glycine injection [7]. We found that the specific activity of the glycine in supernatant, which is largely cell sap, was as great or greater in the atherogenic diet-fed animals than in controls. This suggests that the labeled glycine was transported into the cell as readily as in controls. Measurements and comparisons of amounts of each of the amino acids indicate that

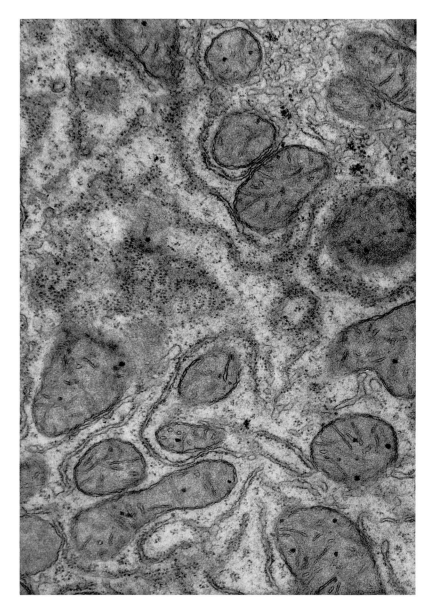

Fig. 7. An electron micrograph taken from the liver of a stock-fed rat shows a large area where the ergastoplasmic membranes have been grazed tangentially and presents an *en face* view of the characteristic rosette or spiral pattern of ribosomes attached to the membrane. Lead hydroxide stain 41,000 ×. (Courtesy of Academic Press.)

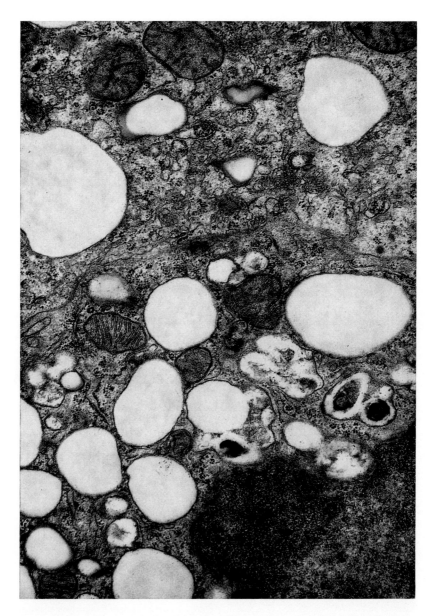

Fig. 8. Electron micrograph of the liver of a rat fed a severe atherogenic diet for 6 weeks. There are numerous lipid droplets in the cytoplasm and the usual parallel arrays of rough-surfaced endoplasmic reticulum often appear disorganized. Some ribosomes are still attached to the membrane but many are dispersed in the sap. Characteristic rosette or spiral patterns are much less common than in a stock-fed rat. Lead hydroxide stain 31,000 ×. (Courtesy of Academic Press.)

with all atherogenic diets studied threonine levels were ½ or less of control values. This was present even in groups that were fed relatively large amounts of protein. It is difficult to interpret this observation regarding threonine without further information. Potentially it could be of importance in various aspects of the regulating mechanism of the cell and it should be investigated further.

In any study of cellular metabolism energy transformations obviously play a central role. In our studies thus far only a small amount of information relating to bioenergetics has been obtained, and this has been limited to one set of experimental conditions [9]. In decapitated rats the total amount of hepatic ATP per mg of DNA was lower in the atherogenic diet group than in stock diet group after one month of feeding. The combined amount of ATP, ADP and AMP per mg of DNA was the same however in both groups suggesting that the amount of adenine available for these nucleotides was adequate at this period. In the atherogenic diet group the ATP/ADP ratios were significantly lower than in the stock-fed rats, suggesting greater utilization of hepatic ATP by the atherogenic diet group. At three months, when the livers of the atherogenic diet-fed rats are larger (partly because of fatty change) than the livers of the stock-fed group, the results are more difficult to interpret. Expressed per mg of DNA at three months the levels of ATP, ADP and AMP and the ATP/ADP ratio, are each significantly lower in rats fed the atherogenic diet than in stock-fed rats. This may mean that by three months there is deficient synthesis of the adenine nucleotides in the livers of the atherogenic diet-fed rats, although other interpretations are possible. For instance, one must consider that at three months the liver weights are much greater and the body weights much smaller in the atherogenic diet-fed group than in the stock-fed group. Injection of adenosine intraperitoneally into rats fed severe diets did not alter rates of fibrinogen synthesis as judged by ^{14}C-glycine incorporation into fibrinogen. These initial experiments suggest that further studies of bioenergetic systems in the liver, arterial walls and elsewhere would be useful in rats receiving atherogenic diets.

Metabolic Studies Related to Protein Synthesis by the Aorta

In *in vivo* studies related to protein synthesis in the aorta of rats fed some of the diets for 1, 2 or 3 months we injected ^{14}C-glycine intravenously and then sacrificed the rats one hour later. The aorta was removed leaving behind all or most of the small proximal portion containing the lesions and after a series of steps protein was separated and counts determined. An example of our results is shown in Fig. 9. Protein synthesis, at least as far as it can be determined by ^{14}C-glycine incorporation into TCA precipitable material appears markedly depressed per unit of weight or of RNA or DNA at 1, 2 and 3 months on diet. (Note that these data must be interpreted with caution since apparent rates of incorporation in aorta

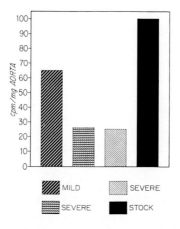

Fig. 9. ^{14}C-glycine incorporation into proteins of the aorta *in vivo* in rats fed a mild and two slightly different severe atherogenic diets and a stock diet for 4 weeks. Rates of incorporation in all three experimental diet groups were significantly lower than the stock group. However, the mild diet group showed much less depression than the severe diet groups.

could have been affected significantly by transport of labeled proteins made in the liver and elsewhere in the body.) Also, total amount of RNA per cell is reduced. These results indicate that the function of the cells not involved in the lesions is to some extent impaired and suggest that some form of chemical injury has occurred. The relationship between this manifestation of presumed injury to the cells of the wall of the aorta and the development of atherosclerotic lesions could be fortuitous. It could also be related as cause and effect with the lesions developing as reaction to injury.

The milder diets resulted in much less depression of protein synthesis than did the severe diets and yet virtually all rats had lesions by 3 months. Whether or not lesions could be produced with diets that produced no depression of protein synthesis we do not know. It seems unlikely since previous experience with very mild diets for rats containing only cholesterol suggest that no lesions are produced at least in a 3 month period.

In vitro studies have also been carried out with microsomes from aortas of rats fed severe diets for one month [6]. The results are shown in Fig. 10. The ability to incorporate ^{14}C-leucine into protein in the cell free system used has been impaired whether taken as per

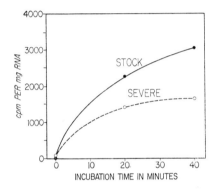

Fig. 10. ^{14}C-leucine incorporation into aortic microsomal protein in a cell free system in rats fed a severe atherogenic diet and a stock diet. The rate of incorporation in the severe diet group was much lower than the stock group.

unit of protein or RNA. Studies *in vivo* of rapid incorporation of ^{14}C-orotic acid (15 min) into total RNA of the cells indicated depressed synthesis in the aorta of those rats fed the atherogenic diet. Perhaps as suggested for the liver, synthesis of messenger RNA by the cells of the aorta has been impaired.

Summary

In summary, the results thus far obtained suggest that some form of cell injury has occurred in both livers and aortas of rats fed certain atherogenic diets. What relations if any the biochemical events described have to atherogenesis needs to be elucidated by more studies in depth of cell metabolism.

References

1. THOMAS, W.A.; JONES, R.; SCOTT, R.F.; MORRISON, E.; GOODALE, F. and IMAI, H.: Production of early atherosclerotic lesions in rats characterized by proliferation of 'modified smooth muscle cells'. Exp. Molec. Path., Suppl. *1:* 40–61 (1963).
2. THOMAS, W.A.; SCOTT, R. FOSTER; LEE, K.T.; DAOUD, A.S. and JONES, R.: Experimental atherosclerosis in the rat. Atlas 1964. Presented by W.A. THOMAS at Conference on Comparative Atherosclerosis. Amer. Heart Assoc., Beverly Hills, Calif., January 1964. Chp. in Comparative Atherosclerosis, ed. by ROBERTS and STRAUS, pp. 92–108 (Harper and Row, 1965).
3. SCOTT, R.F.; MORRISON, E.S.; THOMAS, W.A.; JONES, R. and NAM, S.C.: Short-term feeding of unsaturated vs saturated fat in the production of atherosclerosis in the rat. Exp. Molec. Path., Symposium *3/5:* 421–443 (1964).
4. THOMAS, W.A. and HARTROFT, W.S.: Myocardial infarction in rats fed diets containing high fat, cholesterol, thiouracil, and sodium cholate, Circulation *19:* 65–72 (1959).
5. KIM, D.N.; LEE, K.T.; JONES, R. and THOMAS, W.A.: Studies of fibrinogen synthesis in rats fed thrombogenic diets. Exp. Molec. Path. *5:* 61–82 (1966).
6. WORTMAN, B.; LEE, K.T.; KIM, D.N.; DAOUD, A.S. and THOMAS, W.A.: Studies related to protein synthesis in experimental animals fed atherogenic diets II. DNA, RNA and protein in aortas of rats fed atherogenic diets for 1 month. Exp. molec. Path., Suppl. *3:* 88–95 (1966).
7. SISSON, J.A.; KIM, D.N.; LEE, K.T.; FLORENTIN, R. and THOMAS, W.A.: Studies related to protein synthesis in experimental animals fed atherogenic diets III. Free amino acids in plasma and liver, and ^{14}C-glycine incorporation into plasma and structural proteins of rats. Exp. molec. Path., Suppl. *3:* 96–107 (1966).
8. LEE, K.T.; JONES, R.; KIM, D.N.; FLORENTIN, R., COULSTON, F. and THOMAS, W.A.: Studies related to protein synthesis in experimental animals fed atherogenic diets IV. ^{14}C-glycine incorporation into protein of aortas and electron microscopy of non-necrotic atherosclerotic lesions of monkeys fed atherogenic diets for 8 to 16 months. Exp. molec. Path., Suppl. *3:* 108–123 (1966).
9. SCOTT, R.F. and MORRISON, E.S.: Adenosine diphosphate/adenosine triphosphate ratios in the livers of rats fed high fat thrombogenic diets. Fed. Proc. *25:* 597 (1966).
10. KIM, D.N. and LEE, K.T.: Studies of plasma, aorta, liver and skeletal muscle protein synthesis in rats fed hypercholesterolemic thrombogenic diets. Fed. Proc. *25:* 597 (1966).

Authors' address: Dr. W.A. Thomas, Dr. K.T. Lee, Dr. D.N. Kim, Dr. J.A. Sisson and Dr. R.F. Scott, Department of Pathology, Albany Medical School, *Albany, N.Y.* (USA).

Discussion Session 9

G. A. GRESHAM (Cambridge): What is the structure of the membranes surrounding the lipid aggregates in the cell in the wall of the vessel that you have shown?

R. HESS (Basel): The accumulation of vacuoles represents an early and conspicious alteration of smooth muscle cells which are developing into lipophages. The vacuoles contain variable portions of osmiophilic lipid and the great majority of them are delimited by a single membrane. Non-membrane bound vacuoles (which are surrounded by cytoplasmic components such as myofilaments) are much less frequent, and they may represent a process of disintegration.

G. A. GRESHAM (Cambridge): What is the mechanism of action of coconut oil?

R. MALMROS (Lund): Coconut oil contains 40% lauric acid. Trilaurate is the most atherogenic agent we have tested. It also induced liver cirrhosis in a few animals.

P. CONSTANTINIDES (Vancouver, B.C.): I would like to make a general plea or comment. I think it is dangerous and paralysing to become species-bound, to look only for animal lesions that are structurally identical with the human lesions. We should be ready to study the basic process of atherogenesis in a wide range of species. We can learn about such basic things as lipid penetration, collagen formation, calcification, necrosis, thrombosis etc. wherever they occur. We can even learn from the deviations from the common human pattern. To give an example, we have learned a great deal about renal physiology by studying kidneys of some fishes that have *no* glomeruli.

T. CLARKSON (Winston-Salem, N.C.): I should like to comment on the statement just made by Dr. CONSTANTINIDES. In order to do meaningful metabolic experiments it is essential that one work with an animal model that has an atherosclerotic process with comparable stages to those observed in man. For example, one could not expect to find the same type of exchange between plaque and plasma in rabbits with 'fatty streaks' as compared with other animals that develop 'fibrous plaques'.

R. HESS (Basel): In general, the alterations of the arterial wall as seen in the present type of experimental atheromatosis are very similar in appearance to the human fatty streak (cf. report of M. D. HAUST at this symposium). However, plasma protein such as fibrin was not identified morphologically in the avian lesion.

RUTH PICK (Chicago, Ill.): Dr. GRESHAM, what is the underlying mechanism for the dissecting aneurysms? Is there always a plaque at this area and what pathology is found in the media?

G. A. GRESHAM (Cambridge): A plaque is always present at the site of rupture. I have not been alble to detect any medical changes.

R. PICK (Chicago, Ill.): Dr. KRITCHEVSKY, do you have any explanation why we find a difference between solid fats and oils on atherogenesis in chicks.

D. KRITCHEVSKY (Philadelphia, Pa.): The difference in the effect of various fats of atherogenesis is certainly not simple and must, among other things, depend upon the effect of these fats on cholesterol absorption. In experiments which we did with MCT, tristearin, and corn oil used as vehicles for cholesterol in the induction of atherosclerosis in rabbits we found that the average atheromata was the same in all three groups, but serum cholesterol levels in the corn oil group were 2 to 3 times higher than they were in the other groups. In an attempt to explain this we devised the idea of the relative atherogenetic effect (RAE) which is simply calculated by dividing the average atheromata by the average serum cholesterol. This gives us a measure of the amount of disposition of cholesterol as a function of the circulating sterol. When we apply this calculation to a number of our diets we always find the more saturated fat was the more atherogenic. (KRITCHEVSKY and TEPPER, Exp. Molec. Path. 4: 489 [1965].)

R. PICK (Chicago, Ill.): Dr. MALMROS, did you find any coronary lesions in the dogs on coconut oil and cholesterol?

H. MALMROS (Lund): Yes, in a dog fed hydrogenated coconut oil for a year we found atherosclerotic lesions on microscopic examination.

R. WISSLER (Chicago, Ill.): Dr. MALMROS, did you study the effects of hydrogenated fat on cholesterol synthesis by the dog liver? You may remember that in our 1962 article in the Archives of Pathology we reported that coconut oil seemed to stimulate hepatic synthesis of cholesterol in the Cebus monkey.

H. MALMROS (Lund): We have just started our investigations on dogs and have not yet been able to study the cholesterol synthesis in the liver. Dr. KRITCHEVSKY, in your paper you mentioned the atherogenic effect of corn oil. I don't think this expression is quite adequate because your experimental diet contained cholesterol. We have fed rabbits cholesterol-free, semisynthetic diet containing 8% corn oil and have not seen any atherosclerotic lesions at all.

D. KRITCHEVSKY (Philadelphia, Pa.): You are quite right. The expression is laboratory jargon for an enhancement of the atherogenic effect of cholesterol. Perhaps we need an oral shorthand notation for this effect which will be completely unambiguous.

R. L. HIRSCH (New York, N.Y.): I would like to comment on Dr. MALMROS' report of the production of atherosclerosis in dogs fed cholesterol alone, without the addition of thiouracil. Recently, Dr. ALFRED STEINER and I performed a similar experiment, feeding cholesterol as egg yolk powder for two years. No atherosclerosis developed, nor did the serum-cholesterol ever rise above 350 mg%.

In all of the dogs given cholesterol and thiouracil by STEINER and KENDALL, no dog ever developed atherosclerosis unless his serum cholesterol level rose to 600–800 mg% or greater and remained there for an appreciable length of time. It seems to me that Dr. MALMROS' success in producing atherosclerosis in his animals is due to some factor – the hydrogenated corn oil vehicle, or the synthetic diet, or a canine species difference – which enhanced the absorption of the cholesterol. This appears to be another piece of evidence to indicate that dietary cholesterol and hypercholesterolemia are directly and causally related to development of atherosclerosis.

O. J. POLLAK (Dover, Del.): A question to Dr. GRESHAM: Do the turkeys develop aneurysm? Recently, we have seen rabbits with a spontaneous abdominal aneurysm; one, with external rupture. We have also seen intraluminal hemorrhage from spontaneous calcific lesions in the ascending aorta. These were all 'control' rabbits.

The view of Dr. CONSTANTINIDES who urges study of lesions dissimilar to human and that of Dr. CLARKSON who urges study of human-like lesions are not exclusive of each other. To answer some questions, MALMROS' studies of lesions in dogs fed coconut oil, which develop slowly, over a year or two, are important. Such lesions can be used to advantage for study of progressive enzymatic changes. In other situations, rapid production of arterial alterations may be desirable. KRITCHEVSKY's study of detailed characterization of fats is most encouraging. Too often, we are exposed to generalizations: fats, proteins, and carbohydrates are mentioned as if there would be no differences between them.

G.A. GRESHAM (Cambridge): Certain breeds of turkey develop disserting aneurysms of the porteria aorta but there is no dilatation of the vessel prior to this event.

H. MALMROS (Lund): Dr. CONSTANTINIDES, in your excellent book on experimental atherosclerosis you have stated that you don't think it would be possible to induce a regression of well established atherosclerotic lesions. I would like to ask you if you have rried an experimental diet containing corn oil or some other linoleic-rich oil.

We have used such a diet in rabbit experiment and we think we have been able to induce a regression. VLES and coworkers have had the same experience with a linoleic rich diet in contrast to a diet containing very little fat.

D. KRITCHEVSKY (Philadelphia, Pa.): I would like to amplify what Dr. MALMROS has said concerning our experiments with rabbits which have been rendered atherogenic. Returning the animals to a diet containing unsaturated fat inhibits the exacerbation of the lesions if the diet contains no fat but with saturated fat, the lesions become much worse. Dr. VLES has reported the same thing. It should be pointed out that the animals are still hypocholesteremic even two months after cessation of the cholesterol-rich diet. Perhaps carrying them long enough on a saturated fat diet would result in regression of the lesions.

M. KRCILKOVÁ (Prague): Have you ever used dried cow milk in your experiment? We have found, that if we give cows milk formula, the total plasma lipids rise fast, non esterified fatty acids are low, in contrast to mothers milk. A total of 111 infants from Prague were included in the investigation, which was conducted from January to May 1966. Newborn to 24-month-old infants were subjected in addition to general clinical examination, biochemical investigation.

It was shown that total lipids, total cholesterol as well as the phospholipids are much lower in children than in normal adults. The serum lipids and lipoproteins show a tendency to increase towards the end of the first year. We summarized our results in 2 tables.

H. MALMROS (Lund): RAFSTEDT has studied the effect of milk on the serum-cholesterol in infants. He found a rapid increase in serum-cholesterol the first 4–5 days after the birth. If the child was fed on a fat-free formula diet instead of milk, the serum-cholesterol did not increase at all, or very little.

A.N. HOWARD (Cambridge): Dr. KRITCHEVSKY, have you analysed the fatty acid patterns of the cholesterol esters in these rabbits' aortas and plasmas?

D. KRITCHEVSKY (Philadelphia, Pa.): So far we have data on the fatty acids of the aorta total lipids and have observed a slight enrichment in the particular fatty acid fed. We have not yet analyzed the aorta and serum cholesteryl ester fatty acids but intend to do so.

G.A. GARCIA (Buenos Aires, Arg.): It is necessary then to make some points clear: (a) There is no doubt that the cholesterol increase is a very important element for diagnosis. (b) Cholesterol is very frequently high (>270 mg%) in young patients, sons of atherosclerotics, even when there are no clinic manifestations of disease. This is useful for the physician to look for preventive measures. (c) Usually, when there exists

Table I

Sex	Group age	Cholesterol (mg/100 ml)			Phospholipids (mg/100 ml)			Total plasma lipids (mg/100 ml)		
		n	x̄	s	n	x̄	s	n	x̄	s
Boys	0–24 hours	3	91.3	3.3	3	178.3	20.2	5	310.4	93.6
	1– 6 days	12	128.8	37.2	7	227.9	16.7	11	367.6	77.6
	1– 6 months	15	123.6	31.1	11	209.9	41.2	18	443.8	190.4
	7–12 months	12	160.8	26.9	7	212.0	23.0	11	406.1	97.8
	13–24 months	7	164.9	28.7	7	218.9	29.1	7	467.9	78.6
Girls	0–24 hours	10	106.7	26.8	8	178.4	20.9	11	324.4	69.3
	1– 6 days	9	146.7	35.1	5	237.6	36.7	9	399.9	89.0
	1– 6 months	13	152.4	23.7	12	222.8	27.6	16	491.6	108.9
	7–12 months	8	175.1	38.2	8	234.4	37.3	8	453.0	142.6
	13–24 months	7	192.3	30.4	4	227.2	34.2	7	584.4	128.7
Mixed	0–24 hours	13	103.2	24.4	11	178.4	20.7	16	320.0	78.0
	1– 6 days	21	136.5	37.4	12	231.9	27.3	20	382.2	84.5
	1– 6 months	28	137.0	31.4	23	216.7	35.4	34	466.3	159.2
	7–12 months	20	166.5	32.7	15	223.9	33.4	19	425.8	121.0
	13–24 months	14	178.6	32.6	11	221.9	31.3	14	526.1	121.6

Remarks: group 0–24 hours: before the intake of the first meal, group 1–6 days: breast milk, group 1–6 months, 7–12 months, 13–24 months: cow's milk formula.

Table II

Sex	Group age	Non esterified fatty acids (mg/100 ml)			Lipoprotein β/α index			Atherogenic index		
		n	x̄	s	n	x̄	s	n	x̄	s
Boys	0–24 hours	5	45.6	23.1	5	1.55	0.48	3	26.8	8.8
	1– 6 days	12	64.8	27.5	12	2.46	0.51	–	–	–
	1– 6 months	18	28.9	12.0	20	2.20	0.76	12	51.4	18.4
	7–12 months	11	33.0	18.3	12	2.70	0.97	7	49.8	17.0
	13–24 months	7	27.4	5.8	7	2.95	0.62	4	54.4	12.6
Girls	0–24 hours	11	46.8	21.0	11	1.79	0.38	3	29.6	3.0
	1– 6 days	9	62.8	28.8	9	2.74	0.67	–	–	–
	1– 6 months	17	28.7	21.2	18	2.05	0.50	11	59.9	14.6
	7–12 months	10	47.1	27.2	9	3.29	1.01	6	101.2	90.0
	13–24 months	7	22.6	13.0	8	3.02	0.83	5	67.5	13.9
Mixed	0–24 hours	16	46.4	21.9	16	1.71	0.43	6	28.2	6.7
	1– 6 days	21	63.9	28.1	21	2.58	0.60	–	–	–
	1– 6 months	35	28.8	17.1	38	2.13	0.65	23	55.4	16.7
	7–12 months	21	39.7	24.1	21	2.95	1.03	13	73.6	67.5
	13–24 months	14	25.0	10.3	15	2.97	0.74	9	61.7	14.8

Remarks: n = number of cases, x̄ = mean of values, s = standard deviation.

clinical manifestation of disease, the cholesterol comes down to an almost normal standard. (e) The action of lowering the cholesterol with drugs does not mean removal of atherosclerosis. (f) The fact that the cholesterol decreases by means of good hygiene and *diet* shows that there is favourable action on atherosclerosis.

And, finally, it is very interesting to mention the casual finding of WINITZ, GRAFF and SEEDMAN. These authors found out that with a diet composed of salts, vitamins, 18 aminoacids, 555 g of glucose and ethyl-linoleate as an essential lipid, the cholesterol came down in the tenth day in all of the patients under control (18), without taking in consideration the initial value.

C.B. TAYLOR (Evanston, Ill.): 1. Is the subdiaphragmatic portion of the aorta of the broad breasted bronze turkey principally a muscular structure and, therefore, different from that of many other animals used in studies on atherosclerosis?

2. Have you observed any ground substance changes preceeding these aortic dissections similar to those seen in Marfan's syndrome, pregnancy or protracted progestational hormone therapy?

G.A. GRESHAM (Cambridge): The subdiaphragmatic portion of the turkey aorta is muscular. No changes in ground substance preceeding the dissections were observed.

DARIA HAUST (Kingston, Ont.): With respect to Dr. MALMROS' suggestion to feed corn oil rather on fat free diet, I should like to mention that Dr. JAMES BEVERIDGE conducted the orignal experiments with corn oil on young men (university students) at Queen's University at Kingston. I believe that this is one of the few instances where human experiments aimed at lowering the serum cholesterol levels preceeded animal experiments.

H. MALMROS (Lund): In the literature you will find numerous experiments in humans with different kinds of fats. Dr. GROEN and Dr. KINSELL are the pioneers and I am sure they can tell you more about the beginning of the whole story.

J.J. GROEN (Jerusalem): Scientific congresses are not for the purpose of establishing priorities. In the present case, it may be interesting, however, to recall how the facts were discovered because it was through a piecing together of contributions of several workers, rather than by one single discovery. In 1949, we studied the effects of three types of diets on the serum cholesterol of 60 healthy volunteers. The diets were of a 'rich' american type of diet, containing a lot of butter, cream and meat, an almost purely vegetarian diet, containing skimmed milk but otherwise only vegetable foods, and oil as the source of fat, and an 'inbetween'. Dutch diet containing half butter-half margarine and milk. The quality of total fat was the same in the three diets. The 'american' diet increased the serum cholesterol, the vegetable diet decreased it. At that time, we could not conclude which factors in the diet were responsible. As you know, KINSELL, independently, found the same effect of vegetable fat in diabetic patients about the same time. We published these results in 1950 and 1951 and in 1957, or 1952. I do not remember exactly. I received a letter from Dr. BEVERIDGE, asking me what exactly was the oil which we had fed in the vegetable diet. It was sunflower seed oil. It remained then for BEVERIDGE and especially AHRENS and his co-workers to demonstrate that it was the chemical nature, especially the degree of saturation, of the fatty acids, which together with the difference in cholesterol content, was responsible for our results. This was soon futher elaborated by Professor MALMROS, AHRENS and KINSELL as far as chain length was involved. My excuses to those colleagues whom I may have forgotten in this summary of the events as I remember them.

R. WISSLER (Chicago, Ill.): In concluding this session I would like to thank each of the participants and discussants. The major problem still before us is to learn more about the mechanisms or pathogenesis by which atherosclerosis develops. We should, as Dr. CONSTANTINIDES suggests, study this as a biological and pathological problem *in depth*, using every modern tool and technique possible. In particular we need to devote more time and energy in understanding the factors which contribute to the development of a clinically important complicated atheroma from a simple fatty streak or fatty plaque. But I most emphasize that it is not 19th century pathology to try to produce the best possible model of the human disease process. This task should deserve the effort of as many talented and modern experimental pathologists as possible. In fact there may be no more important task in modern medicine.

Platelets

Department of Pathology, McMaster University, Hamilton. Blood and Vascular Disease Research Unit, Departments of Medicine and Pathology, University of Toronto

Recent Advances in Platelets, Blood Coagulation Factors and Thrombosis*

J. F. Mustard, M. F. Glynn, L. Jørgensen, E. E. Nishizawa, M. A. Packham and H. C. Rowsell, Hamilton and Toronto, Ont.

Introduction

The exact relationship of thrombosis to the development of atherosclerosis and its complications still requires more careful definition. There is no doubt that thrombi can occlude diseased arteries, producing infarcts, and that this is a major factor in the development of the complications of atherosclerosis [47]. There is also good evidence that mural thrombi become incorporated into the vessel wall, producing thickening of the intima. These intimal lesions range from ones rich in lipid to those rich in smooth muscle cells and connective tissue [17, 50, 49]. The frequency with which thrombosis contributes to the development of intimal lesions and the stage at which it becomes an important factor is still not well defined. However, the recent advances in our knowledge of the mechanisms involved in the formation of arterial thrombi make it possible to establish more clearly the relationships between thrombosis and atherosclerosis. It was recognized in the early investigations [3, 76, 18] that platelets are the primary constituent of thrombi formed from flowing blood. Recent studies with the electron micro-

---

* This work was supported in part by grants from the Medical Research Council of Canada, MT 1309, the Ontario Heart Foundation, Banting Research Foundation, and Geigy Pharmaceuticals, Division of Geigy (Canada) Ltd.

scope [22, 21, 41] have re-emphasized this point and shown that the platelets remain discrete rather than becoming a fused mass. Coagulation, or the formation of fibrin, appears to be a secondary event dependent on the accumulation of platelet materials. In any attempt to understand the relationship between atherosclerosis and thrombosis it is important to explore the mechanisms involved in the adherence of platelets to surfaces and to each other.

Platelet Adherence and Aggregation

It has been shown that a number of different stimuli cause platelets to adhere to each other. These include collagen, antigen-antibody complexes, gamma-globulin coated surfaces (such as latex particles, uric acid crystals, bacteria, and viruses) [77, 51, 28] and other factors such as endotoxin, thrombin and trypsin [33, 30, 66]. Of the vessel wall constituents, only collagen has been found to have a property capable of causing platelets to adhere to each other [37, 74]. *In vitro* studies have failed to demonstrate a ready interaction of platelets with normal endothelium. In studies of thrombosis in injured pig arteries, platelets have been found adherent to the endothelium [40]. Also, it has been observed that endothelial cells can phagocytose platelets [42]. Clearly, under some circumstances, the platelets do react with endothelium *in vivo*. The details of this are unknown.

When platelets adhere to an appropriate surface such as collagen, they undergo marked biochemical and ultrastructural changes (Figure 1). These include release of serotonin, nucleotides (including ADP), amino-acids, and other constituents, swelling of the platelets and the appearance of breaks or gaps in the platelet membranes [77, 74, 34]. The adenosine diphosphate (ADP) which is released from the platelets during this process is considered to cause the platelets to adhere to each other [74, 34, 23, 4]. Much of the evidence suggests that ADP may be the common pathway for most of the factors which cause platelet aggregation. ADP, in addition to causing the platelets to adhere to each other, causes swelling [9], formation of pseudopods, release of some constituents [34] and makes available the platelet lipid [58] for coagulation. The initial stimulus provided by surface contact is therefore sufficient to produce a platelet mass without any participation of blood coagulation. When platelets interact with small particles, the par-

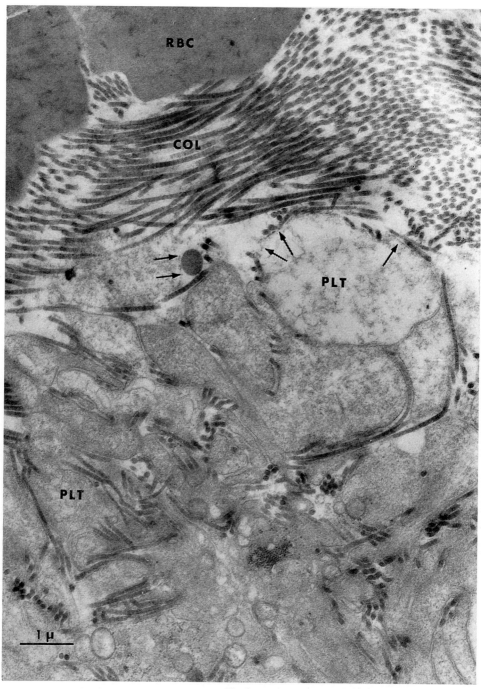

Fig. 1.

ticles may be phagocytosed [51, 14, 71, 27, 53]. Even collagen fibres, if they are ground finely enough, can be incorporated within the platelets [56]. Therefore, stimulation of the platelet surface by small particles can not only cause release of platelet ADP and platelet aggregation, but can also lead to phagocytosis of the particles by the platelets.

Another characteristic of the interaction of platelets with surfaces is that the platelets release factors which increase vessel permeability and injure cells [60, 57]. Apparently the platelets, like the leucocytes, can, when appropriately stimulated, release factors which influence vessel and tissue characteristics. In addition to surface stimuli, thrombin and ADP cause the release of permeability factors from the platelets [60].

In view of the similarity between the platelet surface reaction and such processes as inflammation, phagocytosis, and antigen-antibody reactions, we examined the effect on platelets of compounds known to affect these processes. Diisopropylfluorophosphate (DFP) inhibits the release of nucleotides from platelets induced by thrombin, collagen, antigen-antibody complexes, and gamma-globulin coated latex [57]. Salicylaldoxime, a compound known to interfere with the action of complement [15] and possibly other enzyme systems, also inhibits the release of platelet nucleotides induced by these stimuli [57]. The pyrazole drugs (such as phenylbutazone and sulphinpyrazone), some of which are known to have anti-inflammatory effects [10], inhibit the release of platelet nucleotides induced by collagen, antigen-antibody complexes and gamma-globulin coated latex [57, 62]. They do not block the release of platelet constituents induced by thrombin. Although the effect of the pyrazole compounds may not be on platelet adherence to surfaces, this aspect of the process can be modified with similar effects. It has been found that platelets do not adhere to fibrin in which thrombin has been neutralized [35]. If collagen is coated with this fibrin, platelet adherence to the collagen is prevented and the subsequent release of ADP and platelet aggregation do not occur [36].

Fig. 1. Platelets in contact with collagen in an injured vessel wall. The characteristic periodicity of the collagen fibres (COL) is evident. Platelets (PLT) in contact with the collagen are swollen and most of them have lost their internal structure. In several of these platelets there appear to be breaks (arrows) in the platelet membranes. Red blood cells (RBC) are seen at the top of the picture. ×14,500.

These observations show that failure of platelets to adhere to collagen prevents the subsequent changes in the platelets. Furthermore, contrary to what has been thought, fibrin does not always provide a surface to which platelets readily adhere. It should be pointed out, however, that platelets may adhere to freshly formed fibrin, particularly if the thrombin has not been neutralized.

There is some evidence that lipid, in the form of fine droplets, interacts with platelets. Schulz [72] observed that the addition of a lipid suspension to platelets in plasma led to phagocytosis of the lipid droplets by the platelets. Conner and associates [13], Hoak and colleagues [32], and Haslam [30] have reported that fatty acids can cause platelets to aggregate. This effect appears to be due to release of ADP from the platelets and it is most pronounced with the long chain saturated fatty acids. It is most likely that the action of these fatty acids is related to the fact that they are in the form of fine particles or droplets. The relationship between these observations and the possible effect of chylomicrons on platelets is not known. However, Husom [38] has reported that when lipemic plasma is centrifuged, some of the platelets go to the top with the chylomicrons.

Thrombus Formation

The knowledge of the mechanisms involved in the formation of platelet aggregates allows a reasonable explanation for the development of thrombi from flowing blood. Exposure of sub-endothelial tissues, such as collagen, provides a surface which allows platelets to adhere to the vessel wall. The collagen also stimulates the release of ADP from the platelets, which causes them to adhere to each other. In *in vitro* experiments, ADP will not cause platelets to aggregate unless the medium in which the platelets are suspended is stirred; *in vivo* it seems likely that the extent of platelet aggregation will be in part determined by the characteristics of blood flow. For example, in regions where flow is laminar, the size of the platelet aggregates

---

Fig.2. This figure shows platelets adherent to each other in the centre of a thrombus. Their granules (gr), mitochondria (mit) and glycogen (gly) are evident. Many of the granules show a lamellar type of structure. Pseudopods (pspd) are apparent and some of them have a fibrillar structure. ×30,500.

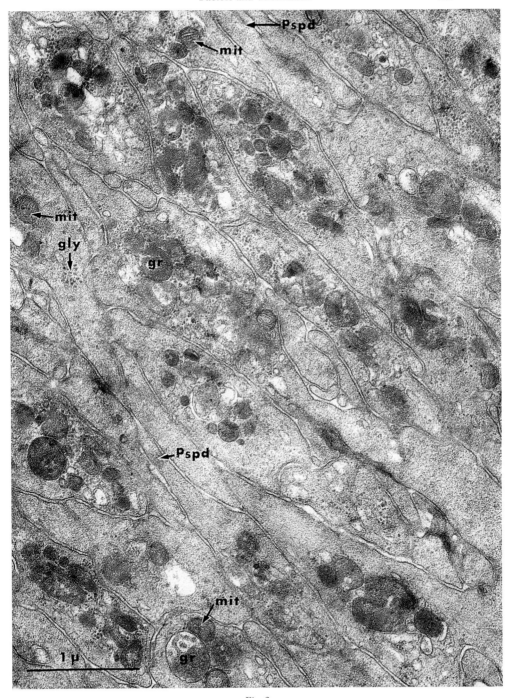

Fig. 2.

formed will be small, but where flow is disturbed, larger platelet aggregates will accumulate [55, 61]. In addition, the amount of ADP released, and the rate at which it is degraded by the plasma enzymes, will also influence the size and stability of the platelet aggregate which forms.

Collagen is known to activate Factor XII [63] and through this to initiate and accelerate coagulation. Furthermore, ADP, as well as aggregating platelets, makes available the platelet membrane phospholipid, thereby further accelerating the coagulation process [58, 11]. It thus appears that the mechanisms involved in the formation of the initial platelet aggregate also serve to initiate or accelerate coagulation. The thrombin which forms causes swelling and degranulation of platelets with release of ADP producing further platelet aggregation [30, 69]. In addition, when sufficient thrombin is formed, fibrinogen is converted to fibrin. It is fibrin formation which probably is the main factor in stabilizing the initial mass of platelets. The extent of the thrombin effect on the thrombus will be related to the amount of thrombin which is formed and the subsequent rate of loss of its activity by dilution, adsorption by fibrin, and neutralization by the blood anti-thrombins.

The ultrastructure of thrombi formed from flowing blood is in keeping with this concept (Figures 1, 2, 3).

Blood Flow and Thrombosis

The localization and structure of thrombi are recognized to be influenced by blood flow [55, 61]. In an arterial system the primary sites of thrombi are regions of disturbed flow. Even when the endothelium is injured, the thrombi forming in an area of laminar flow consist of thin layers of platelets and leucocytes, while at vessel bifurcations where flow is disturbed, the thrombi consist of masses of aggregated platelets, red cells and fibrin. In diseased arteries the

---

Fig. 3. This shows the peripheral region of a thrombus. The platelets toward the centre of the thrombus (right hand side) are more intact than those close to the lumen (left hand side). The platelets (PLT) at the left are swollen and some have breaks in their membranes (arrows). Some of the platelets contain a fibrillar material (fil). Strands of fibrin (fib) are interspersed amongst the platelets. A polymorphonuclear leucocyte (PMN) and red blood cells (RBC) are also present. ×8,000.

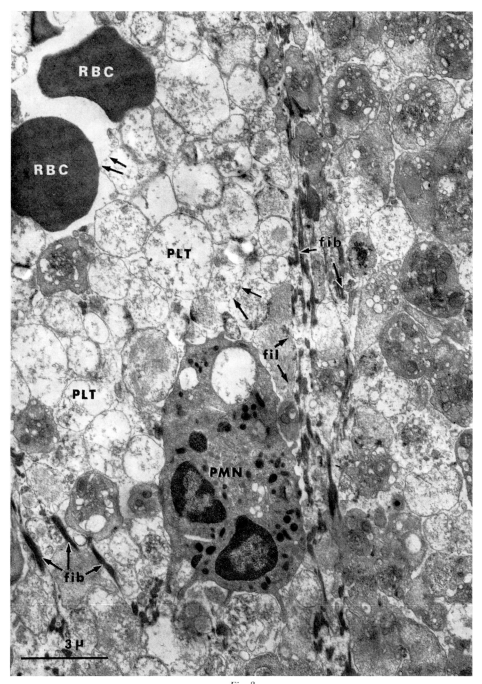

Fig. 3.

occlusive thrombi tend to occur in regions where there is marked disturbance in flow such as about a stenosis [47, 46]. It is unlikely that occlusive thrombi will occur in an arterial system unless there is a marked disturbance in blood flow.

It must be recognized that the formed elements themselves may interact with each other and the vessel wall in areas of disturbed blood flow. The red cells contain large amounts of ATP and ADP and when they are injured these nucleotides may be released. Thus, interaction of the red cells, the platelets, and the endothelial cells in these regions may result in the release of sufficient ADP to cause platelet aggregates to form. There is direct evidence that platelets adhere to collagen and only indirect evidence that they adhere to endothelium. Examination of sites where the formed elements interact with the vessel wall has shown platelet material on the surface in contact with the vessel wall and evidence of sub-endothelial edema [67] (Figure 4).

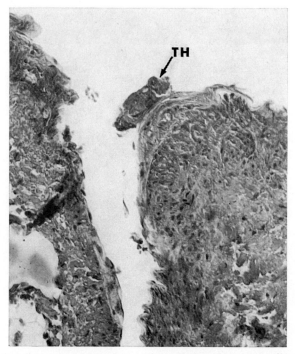

Fig. 4. This is a light micrograph of a small mural thrombus on the distal lip of the orifice of an intercostal vessel from a rabbit. The thrombus (TH) is adherent to a slightly thickened intima. There is some looseness of the intimal tissue, suggesting edema. MSB ×180.

The Pattern of Protein Accumulation in the Wall of the Aorta

We have shown that atherosclerosis in its early stages has a pattern about vessel orifices and bifurcations similar to the pattern with which the formed elements interact with surfaces at similar vessel configurations in model systems [55, 16]. Deposits of formed elements (platelets, white cells, and red blood cells) can be found on the surface of normal endothelium in young animals in these regions [24] (Figure 5). These small mural thrombi are rich in platelets

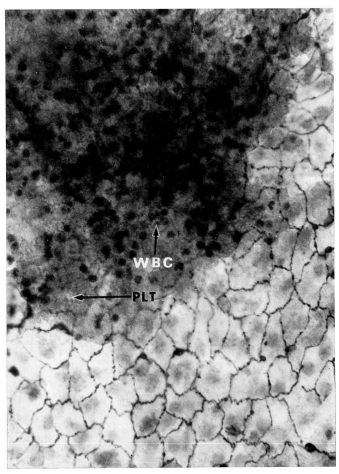

Fig. 5. Aortic endothelium (Hautchen preparation) showing a mass of formed elements including white blood cells (arrow), some red cells and an amorphous mass of material, probably platelets (PLT). The individual endothelial cells are outlined by the silver nitrate staining of their margins. ×360.

and, therefore, could be expected to have released factors which would increase vessel permeability. Even though these mural thrombi are probably transient, it seemed possible that they might have influenced the pattern of permeability in the aorta in these regions. In order to study this, I^{131} labelled albumin and Evans blue were infused into pigs or rabbits. The pattern of protein accumulation in the wall of the aorta was remarkably similar to the pattern of early atherosclerosis (Figures 6 and 7) and the pattern of accumulation of formed elements on the vessel wall [67, 68]. This in no way implies that the focal areas of increased protein accumulation are due to increased permeability. It suggests, however, that the interaction of formed elements with the vessel wall in these areas could be a factor in facilitating increased accumulation of

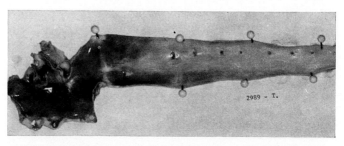

Fig.6. This figure shows the pattern of Evan's blue accumulation in the aortic arch and the thoracic aorta of a pig. The dark regions are the areas in which there is extensive deposition of Evan's blue. The dye is bound to albumin and therefore its accumulation probably represents accumulation of protein in the sub-endothelial tissues.

plasma proteins in the vessel wall. It may be that there is a close relationship between the focal accumulation of lipo-proteins in the vessel wall and areas of increased permeability induced by small mural thrombi.

The platelet permeability factor or factors have many of the characteristics of those which have been reported for leucocytes [52]. The factor which has been studied in some detail does not appear to be simply serotonin or histamine (particularly since it has been found in all species examined, including platelets which do not contain histamine, from pigs and humans) but is mainly a substance or substances of molecular weight less than 10,000 which is heat stable and has both short and long acting effects on vessel permeability [57].

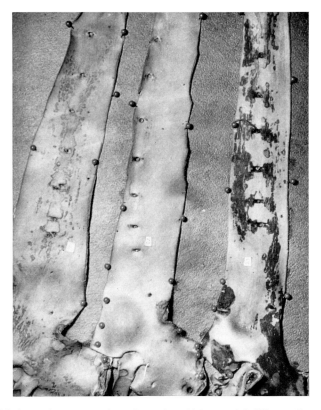

Fig. 7. This figure shows aortas from three pigs which were fed different diets. (The left is from a pig which received an egg yolk enriched diet, the middle is from a control animal, and the right is from a pig fed a butter enriched diet.) The aortas were stained with Sudan IV. The lipid staining areas, representing early atherosclerosis, show a pattern which is similar to that of figure 6, particularly for the aorta of the animal fed egg yolk.

The Fate of Mural Thrombi

One point, the importance of which is not fully understood, is the relationship between persistent mural thrombi and the development of the atherosclerotic lesions. DUGUID [17] in his revival of ROKITANSKY'S [75] hypothesis, suggested that fibrin thrombi became incorporated into the vessel wall and thereby produced intimal thickening. Although earlier experimentalists and pathologists clearly demonstrated that arterial thrombi begin as masses of platelets [3, 76, 18], the concept developed that it is fibrin thrombi which are important in atherosclerosis [17, 50]. Many investigators assumed that since arterial thrombi are fibrin clots, all one need study is blood coagulation to gain an understanding about the relationship between thrombosis and atherosclerosis [2], but it is now apparent that it is best to study the complete inter-relationship between platelets, surface stimuli, and blood coagulation.

Earlier pathologists such as WELCH [76] and ASCHOFF [1] showed that arterial thrombi began as platelet-rich thrombi which then were transformed into masses rich in fibrin. The thrombi were subsequently organized into the vessel wall. We have re-examined the fate of platelet-rich arterial thrombi induced by either a silk suture or collagen fibres in the carotid artery of pigs [40]. One hour after placing the suture of fibres in the vessel, a platelet-rich thrombus with its perimetric bands of fibrin was found. Study of the thrombus at this stage by electronmicroscopy showed the mass to be composed principally of adherent platelets. The platelets in the region of the collagen fibres were adherent to them, swollen, degranulated, and in many cases had breaks in their membranes (Figure 1). Platelets in the centre of the thrombus were adherent to each other and had developed pseudopods but had not lost their internal structure (Figure 2). Fibrin strands were interspersed among the platelets in the periphery of the thrombus (Figure 3). The platelets in this region were swollen, had lost their granules and there were breaks in the platelet membranes.

Three or four hours after these thrombi form, the platelets in the centre are often no longer adherent to each other and some appear to have undergone further morphological changes. At this stage it is unlikely that platelet adherence is responsible for holding the mass together. The amount of fibrin present around the platelets in the periphery of the thrombus is increased. It is probably this fibrin which now holds the thrombus together. By 24 h most of the platelets have disintegrated and an extensive network of fibrin has formed amongst the disintegrating platelets (Figure 8). Numerous

---

Fig. 8. This is an electronmicrograph of a thrombus in a carotid artery of a pig 44 h after a collagen fibre was inserted in the lumen through the wall of the vessel. There is considerable cellular debris (cd) interspersed amongst the fibrin (fib) strands. The cellular debris is probably disintegrated platelets. A mononuclear cell can be seen at the bottom of the picture with many vacuoles (VAC) containing cellular debris (cd). The nucleus (nuc) of this macrophage is apparent. Red blood cells (RBC) are also present. × 9,000.

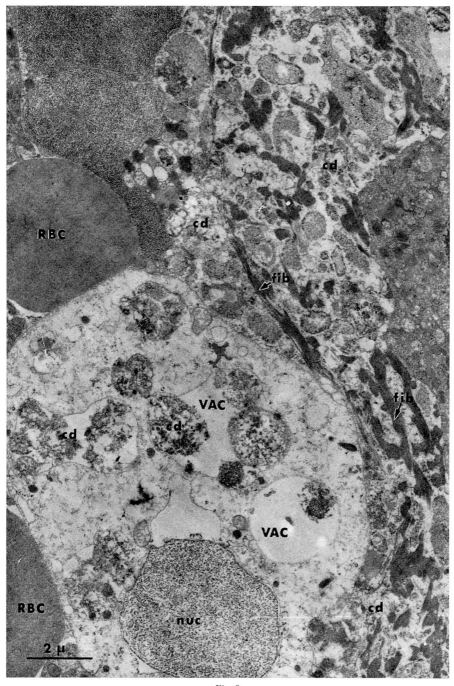

Fig. 8.

leucocytes, some of which appear to be disintegrating, and macrophages are now present. The macrophages have phagocytosed much of the thrombus material including platelets and fibrin. CHANDLER and HAND [12, 29] originally described phagocytosis of platelets by mononuclear cells and showed that such macrophages could, if they ingested large amounts of platelet material, become transformed into foam cells. In our study, although phagocytosis of the platelet material was clearly evident, the macrophages did not become converted into foam cells. By seven days the thrombus was covered with endothelial-like cells and was filled with fibroblasts. Collagen was evident by this stage and residual fibrin, cellular debris, and red cells were still present. From the tenth day on, cells with the characteristics of smooth muscle cells were present (Figure 9), along with a marked increase in collagen and some elastic tissue. Traces of fibrin and residual cellular debris were still evident.

Clearly, a 10–24 h old arterial thrombus may have changed sufficiently so as to make interpretation of the initial events difficult. Thus, although a thrombus may have the histological appearance of a fibrin mass when examined by light microscopy, the initial deposit will most likely have been composed of platelets. It is obviously important to understand the mechanisms whereby the original platelet thrombus undergoes disintegration and is replaced by the fibrin mass. It was apparent from these studies that reduction in the size of the thrombus which has been attributed, by some investigators, to clot retraction, was chiefly due to disintegration and phagocytosis of the platelets, which were replaced by a more compact fibrin mass. As mentioned earlier, since platelets do not readily adhere to fibrin in which thrombin has been neutralized, it may well be that development of a fibrin surface restricts further accumulation of new platelet material and thereby retards growth of the thrombus.

It has been found in some experiments that platelet thromboemboli can become transformed into lipid-rich intimal thickenings containing foam cells. HAND and CHANDLER [29] were able to

---

Fig. 9. This is an electronmicrograph of a thrombus 14 days after injury to the wall of a carotid artery. The myofilaments (MF) of the cells and their basement membranes (bm) are evident. There is some collagen (COL) present and dark staining material which may be residual fibrin and/or elastin. Some of the cells have an extensive endoplasmic reticulum (er). Their nuclei (nuc) are apparent. ×9,200.

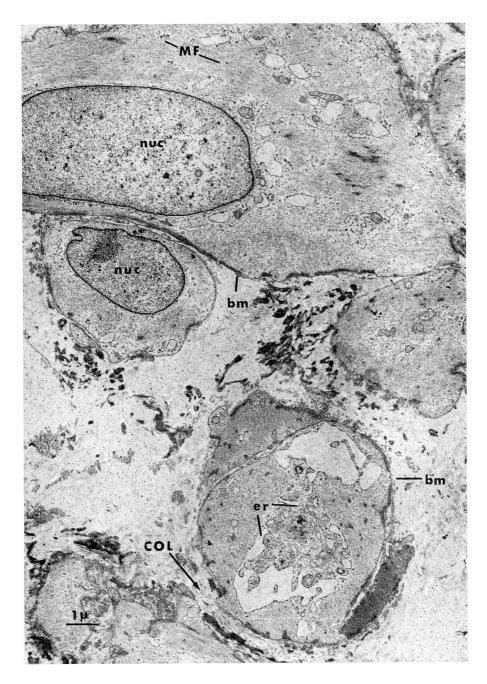

Fig. 9.

demonstrate this in their experiments in which they infused platelet emboli into the pulmonary circulation of rabbits. It remains to be seen what the difference is between their experiments and those reported here. It may be that if the platelets fail to disintegrate, phagocytosis of platelets becomes the main mechanism for their removal. At present it would appear that platelet-rich thrombi can produce intimal thickenings ranging from those rich in lipid and foam cells to those rich in smooth muscle cells and collagen.

Platelet Thrombo-Emboli

It has been found that a considerable number of myocardial infarcts are not associated with occlusive arterial thrombi [49, 19]. Some have suggested that in the pathogenesis of complications of arterial disease transient thrombo-emboli may be important [39, 48]. Examples of this are some forms of intermittent cerebral vascular or retinal ischemia, usually assciated with thrombotic disease at the bifurcation of the carotid artery [20, 43]. It has been shown that attacks of blindness or cerebral dysfunction are precipitated by showers of platelet-rich embolic material into the cerebral arterial bed.

It is possible that similar mechanisms to those in the cerebral circulation could be operating in the circulation of other organs. We have studied the effect of producing transient platelet thrombo-emboli in the myocardial and renal circulations. Infusion of ADP into the coronary artery of a pig produces a collapse of cardiac function with accompanying changes in blood pressure and electrocardiogram [39]. If the infusion is maintained for a period of 3–5 min in doses sufficient to keep the platelet count depressed during the infusion, the majority of animals so treated develop gross myocardial infarcts within 5 h of the experiment. Infusions of other compounds (e.g. AMP, adenosine) which do not cause platelet aggregation do not produce the severe effects. Furthermore, infusion of ADP into the coronary artery of thrombocytopenic pigs does not produce the severe changes in cardiac function and subsequent development of myocardial infarcts. This evidence indicates that transient platelet thrombo-emboli occurring in the myocardial circulation can produce cardiac dysfunction and subsequent tissue injury. It is not clear whether in the injured areas some of the platelet aggregates persist and maintain a permanent block of the microcirculation or whether the infarct is produced by a transient ischemia coupled with tissue injury induced by the platelet aggregates. Since the platelets are known to release, during aggregation, compounds which can influence vessel permeability and injure tissue [60, 57], this latter possibility must be considered as potentially important. Whatever the mechanisms, these experiments indicate that reversible platelet thrombo-emboli can produce myocardial infarcts. Since the diseased coronary arteries with their stenotic lesions are sites of predilection for the formation of such platelet thrombo-emboli, it seems likely that such a mechanism may be operating in some cases of myocardial infarction in man.

In similar studies it was found that transient platelet thromboemboli can produce acute injury to the kidney, giving rise to proteinuria, hematuria and casts in the urine [26]. These acute changes usually disappeared within 3–4 h. However, histological studies showed evidence of a more permanent injury to the kidney. The characteristic changes in the kidney at 1–2 days were increased cellularity of the glomerular tufts, and degeneration and atrophy of the collecting and proximal tubules. In certain peripheral or mid-zonal cortical areas the tubules showed marked degeneration or necrosis. By three days there was often evidence of adhesion between the glomerular tufts and Bowman's capsule. At 2–4 months, localized scars in the cortex were evident and there were also thickening of Bowman's capsules, thickening of the arterioles, and collapse of some glomerular tufts.

The rabbits allowed to live 2–4 months after the ADP infusion consistently showed an elevated systolic blood pressure. These changes were not found in animals given AMP nor in thrombocytopenic animals given ADP. It, therefore, appears that transient platelet thrombo-emboli in the renal circulation can produce renal dysfunction and injury with subsequent effects on the blood pressure.

In view of the number of stimuli which can cause platelets to form aggregates, the arterial and tissue lesions seen in a number of conditions (particularly those often referred to as disseminated intravascular coagulation [44]) may be related to platelet thromboembolism. As was stated in the introduction to this paper, antigen-antibody complexes and micro-organisms can, under appropriate circumstances, induce platelet aggregation. It is important to recognize that gamma-globulin is involved in these reactions and that its effect is modified by the other plasma proteins [28].

In the experiments in which platelet thrombo-emboli were induced by ADP infusion, those which persisted in arteries or veins could be seen adherent to the vessel wall at 24 h. Endothelium beneath them had disappeared and there was some tissue reaction at the site. This further emphasizes the importance of the platelet aggregates in inducing vessel wall injury.

The Effect of Factors which Influence the Platelet-Surface Reaction on Thrombosis and Hemostasis

Consideration of the nature of the mechanisms whereby platelet-rich thrombo-emboli are induced and their subsequent fate makes it seem likely that a number of factors other than anticoagulants could influence thrombosis. As discussed in the earlier part of this paper, the platelet surface reaction is influenced by the pyrazole drugs which do not have any effect on blood coagulation. Since they do not inhibit the action of ADP or thrombin, their primary action would appear to be on the surface stimuli. We have found that when sufficient amounts of these drugs are given to experimental animals, the formation of hemostatic plugs at the ends of transected vessels is impaired [57]. In the formation of the haemostatic plug there are at least three important factors: collagen, thrombin and ADP released from the injured tissues and from the platelets. These are also involved in the formation of thrombi. The impairment of haemostasis produced by the pyrazole drugs is due to their effect on the platelet-collagen reaction. These observations suggest that these pyrazole compounds would influence thrombosis in which surface stimuli were of major importance.

Another class of compound which affects the platelet surface reaction is phospholipids. We have found that phosphatidyl serine, a known inhibitor of the extrinsic and intrinsic pathway of coagulation [73, 59], also suppresses the interaction of platelets with surfaces [64]. The action of phosphatidyl serine appears to be similar to that of the pyrazole compounds in that nucleotide release, but not adherence to surfaces, is suppressed. In studies with synthetic phosphatidyl serines containing fatty acids of various chain lengths, it was found that phosphatidyl serine containing C-14 or C-16 fatty acids had the greatest inhibiting effect on coagulation; phosphatidyl serine containing C-10 fatty acids was the most effective inhibitor of the platelet-collagen reaction [65]. Study of haemostatic plug formation in animals infused with the natural phosphatidyl serine material or the synthetic phosphatidyl serines showed that formation of the haemostatic plug was impaired. With the synthetic material it was observed that the C-10 fatty acid phosphatidyl serine molecule (the one with the predominant effect on the platelet-collagen reaction and little effect on coagulation) produced as great an effect on hemostasis as the synthetic phosphatidyl serine which had anticoagulant properties. Furthermore, it has been found that the natural phosphatidyl serine fractions suppress thrombus formation in extracorporeal shunts [64].

Other phospholipid preparations (inositol, phosphatidyl ethanolamine, phosphatidyl choline) seem to have either no effect or potentiate the platelet surface reaction. Thus, the balance of phospholipids in the plasma may influence the response of the platelets to surface stimuli.

In contrast to the effect of phosphatidyl serine, a drug such as heparin in the usual therapeutic doses does not influence the action of collagen or ADP on platelets [34, 5]. It does, however, suppress the effect of thrombin on fibrinogen [8]. Thus, heparin could be anticipated to influence the growth and possibly the stability of the initial thrombus but certainly would not inhibit its initiation. It must be pointed out, however, that heparin in much higher doses does influence some membrane functions and, in our experience, can inhibit the platelet-surface reaction [70].

Dicumarol, a drug which predominantly influences the formation of blood clotting proteins, does not prevent platelet aggregation induced by collagen, ADP or thrombin. Dicumarol may, under certain circumstances, actually potentiate the platelet response to surface stimuli [25]. Thus, one would not anticipate that dicumarol would have much effect on the initial development of arterial thrombi. In very high doses it would depress the coagulation component of the response and, therefore, perhaps the stability and size of the thrombus which forms. In studies with extracorporeal shunts we found it impossible to suppress thrombus formation even when very high doses of dicumarol were given [54].

Since fibrin is important in the stabilization and subsequent transformation of arterial thrombi, fibrinolytic therapy could interfere with the later stages of arterial thrombosis. It has been shown that fibrinolytic agents can lyse the fibrin which forms around platelet aggregates and thereby allow them to break up [45]. Such platelet aggregates might well form platelet thrombo-emboli which would shower the organ downstream from the site of the thrombus. This might have a disasterous effect on an organ such as the brain in carotid artery thrombosis. There seems little doubt that fibrinolytic therapy is important in the dissolution of thrombi which are 8 or 9 h old and have formed a large fibrin component. In our studies, 24 h old carotid artery thrombi could be virtually completly lysed by fibrinolytic therapy [31].

Born and associates [6] have shown that a variety of nucleotides and nucleosides, such as AMP, adenosine, and 2-chloroadenosine will inhibit the action of ADP on platelets *in vitro* and *in vivo*. In experimental studies of thrombus formation in injured cerebral arteries and veins, it was found that the infusion of adenosine or 2-chloroadenosine would inhibit formation of the platelet mass [7].

Summary

There have been several major developments in our understanding of the mechanisms involved in the formation of thrombi from flowing blood:

1. Adenosine diphosphate causes platelet aggregation and appears to be a final common pathway for many stimuli which induce platelet aggregation.

2. Surfaces such as collagen, and particulate matter such as antigen-antibody complexes, viruses and bacteria can induce platelet aggregation.

3. The interaction of platelets with these stimuli can be inhibited by some compounds with anti-inflammatory properties.

4. Blood coagulation appears to be important for the growth and stabilization of the initial platelet mass.

5. Platelets release factors which increase vessel permeability when they are stimulated by materials which induce platelet aggregation.

Transient platelet thrombo-emboli can produce organ dysfunction and tissue injury. Platelet masses lodging on a vessel wall may give rise to a focus of injury and increased vessel permeability. This may be of some importance in determining the sites of focal accumulation of lipids in the early stages of atherosclerosis. The results of recent studies have re-emphasized that the platelet-rich thrombi which initially form from flowing blood are rapidly transformed into fibrin masses and organization of these thrombi produces a fibrous intimal thickening. It is likely that it will be possible to manage thrombo-embolic disease with drugs which influence platelet function, as well as with anticoagulants.

References

1. ASCHOFF, L.: Über den Aufbau der menschlichen Thromben und das Vorkommen von Plättchen in den blutbildenden Organen. Virchows Arch. path. Anat. *130:* 93–144 (1892).
2. ASTRUP, T.: The biological significance of fibrinolysis. Lancet *271:* 565–568 (1956).
3. BIZZOZERO, G.: Über einen neuen Formbestandteil des Blutes und dessen Rolle bei der Thrombose und der Blutgerinnung. Virchows Arch. path. Anat. *90:* 261–332 (1882).
4. BORN, G.V.R.: Aggregation of blood platelets by adenosine diphosphate and its reversal. Nature *194:* 927–929 (1962).
5. BORN, G.V.R. and CROSS, M.J.: The aggregation of blood platelets. J. Physiol. *168:* 178–195 (1963).
6. BORN, G.V.R.; HASLAM, R.J.; GOLDMAN, M. and LOWE, R.D.: Comparative effectiveness of adenosine analogues as inhibitors of blood platelet aggregation and as vasodilators in man. Nature *205:* 678–680 (1965).
7. BORN, G.V.R.; HONOUR, A.J. and MITCHELL, J.R.A.: Inhibition by adenosine and by 2-chloroadenosine of the formation and embolization of platelet thrombi. Nature *202:* 761–765 (1964).
8. BIGGS, R. and MACFARLANE, R.G.: Human blood coagulation. 3rd ed., p. 103 (Blackwell, Oxford 1962).
9. BULL, B.S. and ZUCKER, M.B.: Changes in platelet volume produced by temperature, metabolic inhibitors, and aggregating agents. Proc. Soc. exp. Biol., N.Y. *120:* 296–301 (1965).
10. BURNS, J.J.; YÜ, T.F.; DAYTON, P.G.; GUTMAN, A.B. and BRODIE, B.B.: Biochemical, Pharmacological considerations of phenylbutazone and its analogues. Ann. N.Y. Acad. Sci. *86:* 253–291 (1960).
11. CASTALDI, P.A.; LARRIEU, M.J. and CAEN, J.: Availability of platelet factor 3 and activation of factor XII in thrombasthenia. Nature *207:* 422–424 (1965).
12. CHANDLER, A.B. and HAND, R.A.: Phagocytized platelets: a source of lipids in human thrombi and atherosclerotic plaques. Science *134:* 946–947 (1961).
13. CONNOR, W.E.: The acceleration of thrombus formation by certain fatty acids. J. clin. Invest. *41:* 1199–1205 (1962).
14. DAVID-FERREIRA, J.F.: Sur la structure et le pouvoir phagocytaire des plaquettes sanguines. Z. Zellforsch. *55:* 89–103 (1961).
15. DIAS DA SILVA, W. and LEPOW, I.H.: Anaphylatoxin formation by purified human C^1 1 esterase. J. Immunol. *95:* 1080–1089 (1965).
16. DOWNIE, H.G.; MUSTARD, J.F. and ROWSELL, H.C.: Atherosclerosis: The relationship of lipids and blood coagulation to its development. Ann. N.Y. Acad. Sci. *104:* 539–562 (1963).
17. DUGUID, J.B.: Thrombosis as a factor in the pathogenesis of aortic atherosclerosis. J. Path. Bact. *60:* 57–61 (1948).
18. EBERTH, J.C. und SCHIMMELBUSCH, C.: Die Thrombose nach Versuchen und Leichenbefunden (Ferdinand Enke, Stuttgart 1888).
19. EHRLICH, J.C. and SHINOHARA, J.: Low incidence of coronary thrombosis in myocardial infarction. A restudy by serial block technique. Arch. Path. *78:* 432–445 (1964).
20. FISHER, C.M.: Observations of the fundas oculi in transient monocular blindness. Neurology (Minneap.) *9:* 333–347 (1959).
21. FRENCH, J.E.; MCFARLANE, R.G. and SANDERS, A.G.: The structure of haemostatic plugs and experimental thrombi in small arteries. Brit. J. exp. Path. *45:* 467–474 (1964).

22. FRENCH, J.E. and POOLE, J.C.F.: Electron microscopy of the platelets in artificial thrombi. Proc.roy.Soc. (Biol.) *157:* 170–176 (1963).
23. GAARDER, A.; JONSEN, J.; LALAND, S.; HELLEM, A. and OWREN, P.A.: Adenosine diphosphate in red cells as a factor in the adhesiveness of human platelets. Nature *192:* 531–532 (1961).
24. GEISSINGER, H.D.; MUSTARD, J.F. and ROWSELL, H.C.: The occurrence of microthrombi on the aortic endothelium of swine. Canad.med.Ass.J. *87:* 405–408 (1962).
25. GLYNN, M.F.: Unpublished observations.
26. GLYNN, M.F.; JØRGENSEN, L. and BUCHANAN, M.R.: Platelet aggregation, renal function and hypertension. Fed.Proc. *25:* 554 (1966).
27. GLYNN, M.F.; MOVAT, H.Z.; MURPHY, E.A. and MUSTARD, J.F.: Study of platelet adhesiveness and aggregation, with latex particles. J.Lab.clin.Med. *65:* 179–201 (1965).
28. GLYNN, M.F.; PACKHAM, M.A. and MUSTARD, J.F.: Protein requirement for platelet aggregation (in preparation).
29. HAND, R.A. and CHANDLER, A.B.: Atherosclerotic metamorphosis of autologous pulmonary thromboemboli in the rabbit. Amer.J.Path. *40:* 469–486 (1962).
30. HASLAM, R.J.: Role of adenosine diphosphate in the aggregation of human blood platelets by thrombin and by fatty acids. Nature *202:* 765–768 (1964).
31. HIRSH, J.; JØRGENSEN, L.; GLYNN, M.F. and MUSTARD, J.F.: Unpublished observations.
32. HOAK, J.C.; CONNOR, W.E. and WARNER, E.D.: Thrombogenic effects of albumin-bound fatty acids. Arch.Path. *81:* 136–139 (1966).
33. HOROWITZ, H.I.; DES PREZ, R.M. and HOOK, E.W.: Effects of bacterial endotoxin on rabbit platelets. II. Enhancement of platelet factor 3 activity *in vitro* and *in vivo*. J.exp.Med. *116:* 619–633 (1962).
34. HOVIG, T.: Release of a platelet-aggregating substance (adenosine diphosphate) from rabbit blood platelets induced by saline 'extract' of tendons. Thromb.Diath.haemorrh. *9:* 264–278 (1963).
35. HOVIG, T.: The effect of fibrin on platelet adherence to collagen and the subsequent aggregation reaction. Fed.Proc. *24:* 155 (1965).
36. HOVIG, T.; JØRGENSEN, L.; PACKHAM, M.A. and MUSTARD, J.F.: Platelet adherence to fibrin and collagen. J. Lab. clin. Med. (in press).
37. HUGUES, J.: Accolement des plaquettes aux structures conjonctives perivasculaires. Thromb.Diath.haemorrh. *8:* 241–255 (1962).
38. HUSOM, O.: The effect of human chylomicron suspensions on the recalcification time of platelet-poor plasma: an effect due to blood platelets. Scand.J.clin.Lab.Invest. *13:* 619–624 (1961).
39. JØRGENSEN, L.; ROWSELL, H.C.; HOVIG, T.; GLYNN, M.F. and MUSTARD, J.F.: Adenosine diphosphate induced platelet aggregation and myocardial infarction in swine. Lab. Invest. 1967 (in press).
40. JØRGENSEN, L.; ROWSELL, H.C.; HOVIG, T. and MUSTARD, J.F.: Fate of platelet mural thrombi in swine arteries. Circulation *32:* (Suppl.II), 19 (1965).
41. KJAERHEIM, A. and HOVIG, T.: The ultrastructure of haemostatic blood platelet plugs in rabbit mesenterium. Thromb.Diath.haemorrh. *7:* 1–15 (1962).
42. MARCHESI, V.T.: Some electron microscopic observations on interactions between leukocytes, platelets, and endothelial cells in acute inflammation. Ann.N.Y.Acad.Sci. *116:* 774–788 (1964).
43. MCBRIEN, D.J.; BRADLEY, R.D. and ASHTON, N.: The nature of retinal emboli in stenosis of the internal carotid artery. Lancet *i:* 697–699 (1963).
44. MCKAY, D.G.: Disseminated intravascular coagulation. Hoeber Medical Division (Harper and Row, New York 1965).

45. McNicol, G.P.; Bain, W.H.; Walker, F.; Rifkind, B.M. and Douglas, A.S.: Thrombolysis studied in an artificial circulation. Lancet i: 838–843 (1965).
46. Mitchell, J.R.A. and Schwartz, C.J.: Arterial disease (Blackwell, Oxford 1965).
47. Mitchell, J.R.A. and Schwartz, C.J.: The relationship between myocardial lesions and coronary artery disease II. A selected group of patients with massive cardiac necrosis or scarring. Brit. Heart J. 25: 1–24 (1963).
48. Moore, S. and Mersereau, W.A.: Platelet embolism and renal ischemia. J. Path. Bact. 90: 579–588 (1965).
49. More, R.H. and Haust, M.D.: The role of thrombosis in occlusive disease of coronary arteries. In R.L. MacMillan and J.F. Mustard (eds.) International Symposium: Anticoagulants and fibrinolysins, pp. 143–153 (Lea and Febiger, Philadelphia, Pa. 1961).
50. Morgan, A.D.: The pathogenesis of coronary occlusion. London: Oxford, 1956.
51. Movat, H.Z.; Mustard, J.F.; Taichman, N.S. and Uriuhara, T.: Platelet aggregation and release of ADP, serotonin and histamine associated with phagocytosis of antigen-antibody complexes. Proc. Soc. exp. Biol., N.Y. 120: 232–237 (1965).
52. Movat, H.Z.; Uriuhara, T.; MacMorine, D.R.L. and Burke, J.S.: A permeability factor released from leukocytes after phagocytosis of immune complexes and its possible role in the Arthus reaction. Life Sci. 3: 1025–1032 (1964).
53. Movat, H.Z.; Weiser, W.J.; Glynn, M.F. and Mustard, J.F.: Platelet phagocytosis and aggregation. J. Cell Biol. 27: 531–543 (1965).
54. Murphy, E.A.; Mustard, J.F.; Rowsell, H.C. and Downie, H.G.: Quantitative studies on the effect of dicumarol on experimental thrombosis. J. Lab. clin. Med. 61: 935–943 (1963).
55. Murphy, E.A.; Rowsell, H.C.; Downie, H.G.; Robinson, G.A. and Mustard, J.F.: Encrustations and atherosclerosis: The analogy between early in vivo lesions and deposits which occur in extracorporeal circulations. Canad. med. Ass. J. 87: 259–274 (1962).
56. Mustard, J.F.; Glynn, M.F.; Nishizawa, E.E. and Packham, M.A.: Platelet surface interactions: relationship to thrombosis and hemostasis. Fed. Proc. 26: 106–114 (1967).
57. Mustard, J.F.; Glynn, M.F.; Packham, M.A. and Nishizawa, E.E.: Unpublished observations.
58. Mustard, J.F.; Hegardt, B.; Rowsell, H.C. and MacMillan, R.L.: The effect of adenosine nucleotides on platelet aggregation and clotting time. J. Lab. clin. Med. 64: 548–559 (1964).
59. Mustard, J.F.; Medway, W.; Downie, H.G. and Rowsell, H.C.: Effects of intravenous phospholipid containing phosphatidyl serine on blood clotting with particular reference to the Russell's viper venom time. Nature 196: 1063–1065 (1962).
60. Mustard, J.F.; Movat, H.Z.; MacMorine, D.R.L. and Senyi, A.: Release of permeability factors from the blood platelet. Proc. Soc. exp. Biol., N.Y. 119: 988–991 (1965).
61. Mustard, J.F.; Murphy, E.A.; Rowsell, H.C. and Downie, H.G.: Factors influencing thrombus formation in vivo. Amer. J. Med. 33: 621–647 (1962).
62. Mustard, J.F.; Rowsell, H.C. and Murphy, E.A.: Platelet economy (platelet survival and turnover). Brit. J. Haemat. 12: 1–24 (1966).
63. Niewiarowski, St.; Bankowski, E. and Rogowicka, I.: Studies on the adsorption and activation of the Hageman Factor (factor XII) by collagen and elastin. Thromb. Diath. haemorrh. 14: 387–400 (1965).
64. Nishizawa, E.E.: Phospholipid, blood coagulation, platelet aggregation and thrombosis. Fed. Proc. 24: 154 (1965).
65. Nishizawa, E.E.: Unpublished observations.

66. O'BRIEN, J.R.: Platelet aggregation. 1. Some effects of the adenosine phosphates, thrombin and cocaine upon platelet adhesiveness. J. clin. Path. *15:* 446–452 (1962).
67. PACKHAM, M.A.; ROWSELL, H.C.; JØRGENSEN, L. and MUSTARD, J.F.: Localized protein accumulation in the wall of the aorta. Exptl. Mol. Pathol. (in press).
68. PACKHAM, M.A.; ROWSELL, H.C. and MUSTARD, J.F.: Localized accumulation of protein in the aorta. Fed. Proc. *25:* 619 (1966).
69. RODMAN, N.F.; PAINTER, J.C. and MCDEVITT, N.V.: Platelet disintegration during clotting. J. Cell Biol. *16:* 225–241 (1963).
70. ROWSELL, H.C.; GLYNN, M.F.; MUSTARD, J.F. and MURPHY, E.A.: Effect of heparin on platelet economy in dogs. Amer. J. Physiol. (in press).
71. SCHULZ, H. und LANDGRABER, E.: Elektronenmikroskopische Untersuchungen über die Adsorption und Phagozytose von Influenzaviren durch Thrombozyten. Information Exchange Group #2, Sci. Memo #103.
72. SCHULZ, H. and WEDELL, J.: Elektronenmikroskopische Untersuchungen zur Frage der Fettphagocytose und des Fett-Transportes durch Thrombozyten. Klin. Wschr. *40:* 1114–1120 (1962).
73. SILVER, M.J.; TURNER, D.L.; HOLBURN, R.R. and TOCANTINS, L.M.: Anticoagulant activity of phosphatidyl serine free of lysophosphatidyl serine. Proc. Soc. exp. Biol., N.Y. *100:* 692–695 (1959).
74. SPAET, T.H. and ZUCKER, M.B.: Mechanism of platelet plug formation and role of adenosine diphosphate. Amer. J. Physiol. *206:* 1267–1274 (1964).
75. VON ROKITANSKY, C.: Handbuch der pathologischen Anatomie (Braumuller u. Seidel, Vienna 1841–46).
76. WELCH, W.H.: The structure of white thrombi. Trans. Path. Soc., Philadelphia *13:* 281–300 (1887).
77. ZUCKER, M.B. and BORRELLI, J.: Platelet clumping produced by connective tissue suspensions and by collagen. Proc. Soc. expt. Biol., N.Y. *109:* 779–787 (1962).

Authors' address: Dr. J. F. Mustard, Dr. M. F. Glynn, Dr. E. E. Nishizawa, Dr. H. A. Packam and Dr. H. C. Rowsell, Dept. of Pathology, Faculty of Medicine, McMaster University, *Hamilton, Ont.* (Canada).

From the Department of Pathology, Medical School, University of British Columbia, Vancouver, B.C.

The Cause of Cerebral Artery Thrombosis in Man

P. Constantinides, Vancouver, B.C.

It has been known for a long time that breaks or ulcerations of atheromata will cause thrombosis in arteries, but since all histological studies in the past were done on only a few random or step-serial sections through occluded vessels (taken e.g. every 500 or every 1000 μ), such breaks could be demonstrated in only a small percentage of all cases examined, and the pathogenesis of coronary thrombosis remained obscure.

This phenomenon thus came to be regarded as polyetiologic (i.e. as induced by different causes in different cases). It was speculated that it might be caused by hypercoagulability of the blood, by increased platelet stickiness, by stasis of the blood, or by hemorrhages of plaque capillaries. It could certainly not be blamed primarily on plaque ulcerations, since ulcerations were found in only a few instances.

Two years ago, we made for the first time a *complete* serial section study of 20 consecutive cases of coronary artery thrombosis in St. Louis and found that (a) *all thrombi* had been caused by breaks in the atherosclerotic lining and (b) *most of the accompanying hemorrhages* had resulted from the entry of blood through the same breaks in the wall [1–4].

Similar results have since been reported with frequent or with step-serial sections by Sinapius in Göttingen [5], by Chapman in New York [6] and by Friedman and Bovenkamp in San Francisco [7].

To find out whether a similar pathogenesis applies to arterial thrombosis in the brain we made a complete serial section study of

10 consecutive cases of cerebral artery thrombosis that came to autopsy during 1965 in Vancouver.

As in our previous coronary study, the occluded arterial segments were decalcified and cut every 10 μ, from one end to the other, without skipping any sections. *All* of the more than 20,000 resulting sections were examined histologically after staining with Mallory.

We found that (a) all thrombi were accompanied by breaks in the atherosclerotic lining, and (b) nine of the ten thrombi were accompanied by plaque hemorrhages, all of which originated from the lumen through the same fractures (see Figs. 1–12 and their legends).

In conclusion, our results show that, as in coronary arteries, thrombosis in cerebral arteries is initiated primarily by breaks in the atherosclerotic or fibrosed arterial wall, even though the speed of

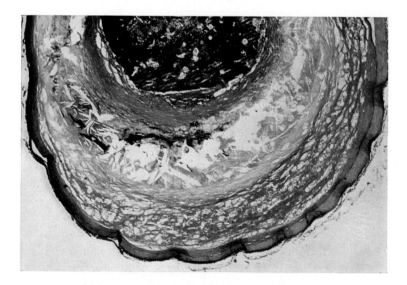

Fig. 1. Thrombus (the black material) occluding the narrowed lumen of a cerebral artery of case No. 1. A circular atheroma with a relatively thin collagenic capsule surrounds the thrombus; there is a hemorrhage in the fatty gruel underneath the capsule. 25×. All sections of this work were stained with Mallory (Fuchsin-Aniline Blue) which differentiates very well the various tissue components of interest in studies of the present type. Thus collagen stains blue, blood and red thrombus red, and white (platelet-rich) thrombus various hues of purple or gray. In the black and white photographs shown here, however, only marked structural changes show up.

development, the size, and the ultimate fate of the thrombi may well depend on humoral or other factors.

As in our previous coronary study, we are certain that the breaks were not postmortem or technical artefacts and that they did not develop after thrombosis had occurred, but that they preceded and caused the thrombi. This belief is based on the fact that in 8 of the 10 cases the breaks were overlaid and (or) filled by the white nuclei (i.e. the oldest portions) of the thrombi, and that hemorrhages into the arterial wall had clearly taken place through the fissures underlying the thrombi in almost all instances (in 9 of the 10 cases). It is further reinforced by the finding of fragments of the arterial wall buried in the thrombi in 7 of the 10 cases.

It seems that thrombosis is initiated by the exposure of blood to the raw surfaces of plaque trauma, regardless of the composition

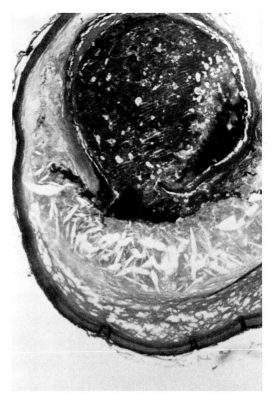

Fig.2. Section through another level of the artery of case No. 1 showing the break in the atheroma capsule that caused both the thrombus and the hemorrhage in the gruel. 25×.

of the plaque. Since more than half the broken plaques of this series were of the predominantly fibrous type, without any crystalline gruel or atheromatous abscess contents, it appears that massive release of plaque lipid is *not* indispensable for the induction of thrombosis. In fact, present evidence is quite compatible with the assumption that the exposed collagen in plaque trauma is just as thrombogenic as the exposed lipid, although it will be up to future work to identify the most thrombogenic principle(s) active at the site of plaque trauma.

Finally, we must once more point out that while it seems certain that atheroma breaks initiate thrombosis, we need more research to find out what causes these breaks in human atherosclerotic arteries.

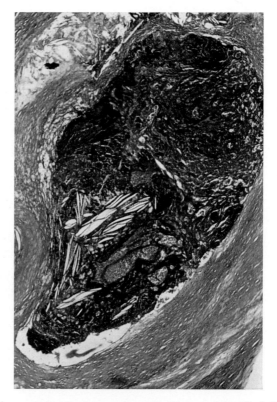

Fig. 3. Thrombus occluding the lumen of a cerebral artery of case No. 2. Here, too, the collagenic capsule of a circular atheroma surrounds the thrombus; a mass of cholesterol crystals is buried in the thrombus. 25 ×.

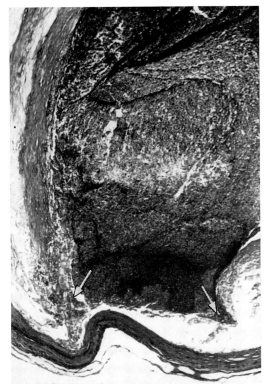

Fig. 4. Section through another level of the artery of case No. 2 showing the break in the atheroma capsule (lower field) over which the thrombus developed. The margins of the break are identified by arrows; the dark gray, wavy band at the bottom of the figure represents part of the fibrosed outer wall of the vessel (which surrounds the atheroma). 25×.

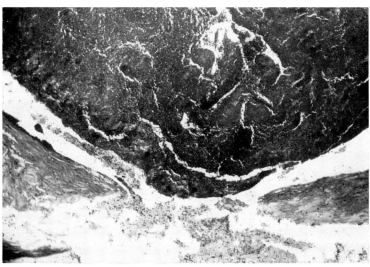

Fig. 5. Thrombus over a break in the collagenic capsule of an atheroma of case No. 3. Some fatty, hemorrhagic gruel has moved around the left margin of the break into the lumen between the thrombus and the capsule. 40×.

Fig. 6. Thrombus over a disintegrated atheroma in a cerebral artery of case No. 4. The capsule has broken into bits and pieces of collagen; there is intimate contact between the thrombus and the exposed, hemorrhagic gruel. 40×.

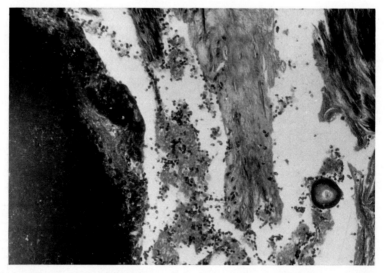

Fig. 7. Close-up of the right margin of the break shown in Fig. 6. Note two broken stumps of collagen jutting from above, as well as the hemorrhagic gruel right of the thrombus. The round, ball-like structure at right consists of a shell of elastic membrane around a core of collagen. We found several such structures in the cerebral arteries of this study and called them for convenience "spheroids"; they seem to be associated with the internal elastic membrane of arteries, developing perhaps from clefts in the latter followed by ingrowth of collagen; they appear to break off the arterial lining during the thrombogenic episode and travel sometimes into atheroma breaks. 100×.

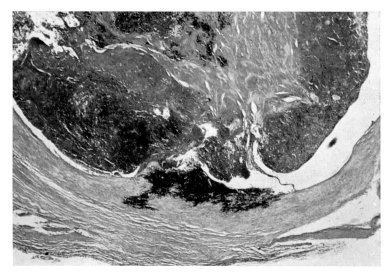

Fig. 8. A section through the thrombosed artery of case No. 5. The thrombus lies over and is attached to a fissure in the completely fibrosed arterial wall, in the lower field. The fissure, as well as the adjacent subendothelial portion of the wall is filled with blood (the black mass). Note the complete absence of fatty gruel from the cracked collagenic wall. 25×.

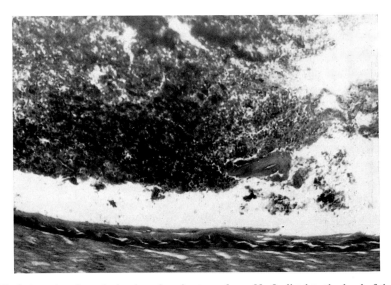

Fig. 9. A section through the thrombosed artery of case No. 8, distal to the level of the thrombogenic break. A collagen fragment (presumably originating from the break in the plaque surface) has travelled downstream and has been caught in the thrombus. 100×.

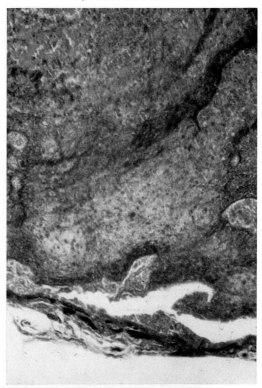

Fig. 10. Section through the thrombosed artery of case No. 9. This thrombus developed over a stretched, extremely thin part of the wall that has been transformed into a collagenic membrane. There are numerous tiny breaks in this membrane, plugged with platelets. All that is left of the wall is a few shreds of collagen (here appearing black) at the base of the thrombus, in the bottom of the figure. $100\times$.

Fig. 11. Thrombosed artery of case No. 10. The thrombus here developed over several breaks of the endothelium and the internal elastic membrane that are covering a largely fibrosed media with only negligible atherosclerosis. The free margin of one of the elastic membrane ruptures can be seen in this section; the broken end of the membrane has been lifted from the wall and has curled up, pointing to the right. $100\times$.

Fig. 12. A 'spheroid' structure under the thrombus attached to the internal elastic membrane of a non-atherosclerotic region of the wall of the artery of case No. 10. 100×.

Acknowledgements

This study was supported by the Medical Research Council of Canada and the British Columbia Heart Foundation. The author is very grateful to Dr. L. DOLMAN of the Vancouver General Hospital and to Dr. S. ENGLISH of St. Paul's Hospital for the supply of the thrombosed cerebral arteries from the autopsy services of the above institutions. Thanks are also due to Mr. B. COX for the preparation of the serial sections.

References

1. CONSTANTINIDES, P.: Plaque fissures in human coronary thrombosis. Fed. Proc. *23:* 443 (1964).
2. CONSTANTINIDES, P.: Plaque fissures in human coronary thrombosis. J. amer. med. Ass. *188* (6): 35–37, Medical News Section (1964).
3. CONSTANTINIDES, P.: The immediate cause of human coronary thrombosis. Progress in Biochemical Pharmacology, Vol. 2 (Karger, Basel/New York 1967).
4. CONSTANTINIDES, P.: Plaque fissures in human coronary thrombosis. J. Atheroscler. Res. *6:* 1–17 (1966).
5. SINAPIUS, D.: Über Wandveränderungen bei Coronarthrombose. Klin. Wschr. *43* (16): 875–880 (1965).
6. CHAPMAN, I.: Morphogenesis of occluding coronary artery thrombosis. Arch. Path. *80* (3): 256–261 (1965).
7. FRIEDMAN, M. and VAN DEN BOVENKAMP, G. T.: The pathogenesis of a coronary thrombus. Amer. J. Path. *48:* 19–39 (1966).

Author's address: Dr. P. Constantinides, Department of Pathology, University of British Columbia School of Medicine, *Vancouver, B.C.* (Canada).

From the Division of Medicine Cedars of Lebanon Hospital

The Effect of Catecholamines upon the *in vitro* Thrombotic Coagulation of Human Blood

H. ENGELBERG and M. FUTTERMAN, Beverly Hills, Cal.*

Introduction

It has recently been shown that cigarette smoking accelerates thrombotic coagulation of the blood *in vitro* in many individuals [1]. Since nicotine is one of the most powerful stimulants for the release of epinephrine from the adrenal medulla it seemed logical to investigate the effect of catecholamines upon the thrombosing tendency of human blood.

Methods

All subjects were either patients in the hospital or members of the house staff. Obviously tense or anxious individuals were not used in an attempt to avoid those who would be more likely to release larger amounts of endogenous catecholamines. Smoking was forbidden for three hours prior to the experimental procedure. Blood was drawn by clean venepuncture and an infusion of 250–500 ml of isotonic saline started. Blood pressure and pulse rate were recorded every few minutes, and blood was obtained from another vein approximately 10 min after the first sample. One mg of the catecholamine was then added to the infusion. (Although an attempt was made to conceal the exact time from the subject, the occurrence of palpitation or tingling soon revealed that the drug had been added.) The drug was given at rates varying from 10–30 drops per minute for 4–15 min depending upon the reaction. Blood samples were usually obtained every 5–10 min during the administration of the drug, and every 30 min for several hours after the intravenous had been discontinued. In several patients the catecholamine infusion was repeated another day, 30–60 min after the oral ingestion of 1 g of nicotinic acid. In one patient isoproterenol was given by nebulizer. Blood was drawn before, and several times after the patient had inhaled her usual dose of the medication.

Blood was drawn using plastic syringes and placed in citrated plastic or siliconed glass tubes. The thrombus formation time (TFT) was determined in a rotating loop

*Supported by the California Arteriosclerosis Research Foundation.

apparatus [2] as soon as the blood arrived at the laboratory, usually within 1–3 min after venepuncture. It was necessary to have 3 or 4 rotating loops available as the TFT rapidly decreases upon standing even if blood is kept in ice water. Free fatty acids were determined [3] on each blood sample.

When the effect of *in vitro* catecholamines was studied aliquots of blood were placed in tubes containing citrate. Varying quantities of the drugs were added and the TFT tests were run simultaneously with control blood samples.

Results

There were 16 *in vivo* experiments in 12 subjects. Epinephrine was administered in 9 studies, norepinephrine in 6 and isoproterenol once. The control value and the lowest TFT after intravenous catecholamine administration are shown in Table I. There was a

Table I. Control and lowest TFT values after I-V catecholamines. Time in minutes

Experiments	Control value	Lowest TFT after IV catecholamine
1	11.00	9.25
2	16.50	11.25
3	16.00	6.00
4	12.25	6.50
5	10.25	5.00
6	17.50	10.00
7	13.50	7.25
8	11.25	8.00
9	11.75	6.50
10	9.25	9.50
11	13.75	5.50
12	12.25	9.00
13	13.50	12.00
14	11.75	8.75
15	21.25	17.00
Average	13.50	8.75

reduction of the TFT in 14 of the 15 instances where epinephrine or norepinephrine was injected. Figures 1 and 2 are representative. The reduction in TFT persisted for a variable time ranging from 15 min – 1–2 h. In 2 instances, prior to the acceleration of thrombosis, a temporary prolongation of the TFT was observed, as shown

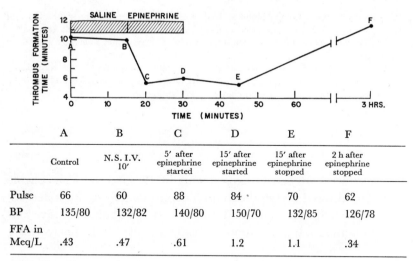

Fig. 1. The effect of intravenous epinephrine upon the thrombus formation time.

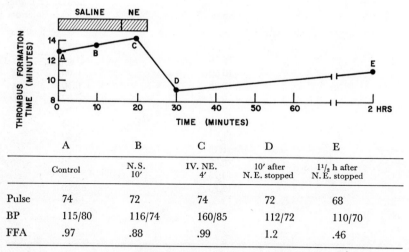

Fig. 2. The effect of intravenous norepinephrine upon the thrombus formation time.

in Figure 3. In three experiments 1 g of nicotinic acid was ingested 1 h prior to the catecholamine infusion, but it did not prevent the shortening of the TFT although the free fatty acid release was completely inhibited. The findings are shown in Figures 4 and 5. Figures 1 and 4 present the results of experiments in the same in-

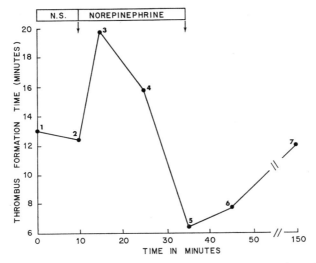

Fig. 3. The effect of intravenous norepinephrine upon the thrombus formation time.

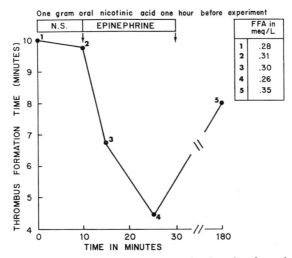

Fig. 4. The effect of intravenous epinephrine upon the thrombus formation time with prior oral nicotinic acid.

dividual on different days. Table II shows the data obtained in the one individual who received isoproterenol by inhalation.

There were 20 *in vitro* experiments; 9 each with epinephrine or norepinephrine and 2 using isoproterenol. The drugs were added in amounts varying from 0.001 µg/ml blood to 100 µg/ml. The TFT

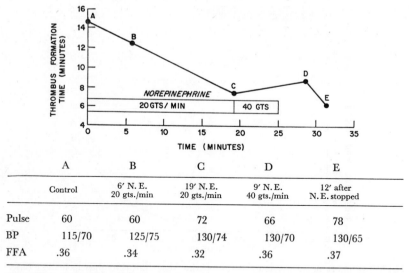

	A	B	C	D	E
	Control	6' N.E. 20 gts./min	19' N.E. 20 gts./min	9' N.E. 40 gts./min	12' after N.E. stopped
Pulse	60	60	72	66	78
BP	115/70	125/75	130/74	130/70	130/65
FFA	.36	.34	.32	.36	.37

Fig. 5. The effect of intravenous norepinephrine upon the thrombus formation time with prior oral nicotinic acid.

Table II. Thrombus formation time before and after the inhalation of isoproterenol (isuprel)

Blood sample	Thrombus formation time in minutes
Control	11
3' after isuprel	9.5
15' after isuprel	7.5

was decreased from 1–3 min in 3 of the 9 epinephrine studies, in 2 of the 9 where norepinephrine was used, and no decrease was observed in the 2 studies with isoproterenol.

Discussion

It was demonstrated long ago that epinephrine affected clotting in animals [4]. This observation was either unconfirmed or neglected for many years except for a few isolated reports. A rise in platelet adhesiveness was found after the infusion of adrenalin in rabbits [5], and in man a fall occurred in the clotting time in siliconed tubes, but not in glass tubes [6]. Despite these findings published reviews dealing with the effect of smoking on the cardiovascular system, or symposia on catecholamines, ignored the effect on blood clotting or platelets. Within the past few years however, the increase of platelet adhesiveness after adrenalin infusions in animals and man, and increased platelet aggregation when adrenalin and noradrenalin are added *in vitro*, has been reported [7–10], although the effect in rats is less definite [11]. In swine increased thrombus formation in an extracorporeal shunt was noted after moderate doses of adrenalin intravenously [12]. Adrenalin, but not noradrenalin, increased antihemophilic globulin activity in human plasma [13].

The results of the present study demonstrate that catecholamines administered *in vivo* in man accelerate the thrombotic coagulation of the blood as determined by the rotating loop technique. This occurred with epinephrine, norepinephrine and isoproterenol. The average TFT value was reduced 35% below control levels. The standard deviation of the method in our laboratory when duplicate tests on blood aliquots are started within 1 min of each other is 0.45 s [1] so that these catecholamine effects are significant. In several instances there was an initial prolongation of the TFT which was rapidly followed by a shortening. This is reminiscent of the occasionally prolonged clotting time after adrenalin observed by CANNON [4]. It is well known that catecholamines may produce opposing results depending upon the dose [14] and the state of the target tissue. The initially prolonged TFT after epinephrine parallels the similar finding in a few cases after smoking [1]. It is interesting that the drugs were far less effective in accelerating thrombosis when added *in vitro*. Again, this is in keeping with CANNON who found, in cats, that adrenalin did not accelerate clotting when added to blood *in vitro*, or even *in vivo* if the blood was prevented from circulating through the abdominal viscera [4].

It is not in agreement however with the results of WALDRON who reported that epinephrine and norepinephrine added *in vitro* decreased the clotting time of blood in collodion lined tubes [15]. Our finding is at variance with the observation that catecholamines cause platelet aggregation when added to platelet-rich plasma [8–10]. It must be remembered that plasma is not blood, and that the correlation of the various techniques for the study of platelet aggregation and adhesiveness with thrombosis is only assumed.

There is much to suggest a correlation but it has not, as yet, been proven. O'BRIEN stated that his observations of adrenalin enhanced platelet aggregation had no direct bearing on thrombosis [10]. This being so it is probable that techniques which use whole blood and produce a platelet-head thrombus are more valid for the study of thrombosis than determinations of platelet aggregation in plasma, or than studies of platelet adhesiveness to glass.

Finally the observation that large doses of nicotinic acid suppressed the relase of free fatty acids by catecholamines but did not prevent the acceleration of thrombosis indicates that these 2 actions of catecholamines are probably mediated via different mechanisms. It also proves that the effect on thrombosis does not result from the increase in free fatty acid levels, in agreement with other observations [1].

Summary

Smoking accelerates the thrombotic coagulation of human blood as determined *in vitro* by the rotating loop technique. Since this effect may be mediated via the epinephrine released by nicotine, the effect of catecholamines was investigated. The results show that catecholamines *in vivo* accelerate (decrease) the thrombus formation time of blood measured *in vitro*. This effect is not prevented by large doses of nicotinic acid which inhibit the free fatty acid increase after catecholamine administration. When catecholamines are added to blood *in vitro* the thrombus formation time is decreased in only a few instances and to a slight extent.

References

1. ENGELBERG, H.: Cigarette smoking and the *in vitro* thrombosis of human blood. J. amer. med. Ass. *193:* 1033 (1965).
2. CHANDLER, A.B.: *In vitro* thrombotic coagulation of the blood. Lab. Invest. *7:* 110 (1958).
3. DOLE, V.P.: A relation between non-esterified fatty acids in plasma and the metabolism of glucose. J. clin. Invest. *35:* 150 (1956).
4. CANNON, W.B. and GRAY, H.: Factors affecting the coagulation time of the blood. II. The hastening or retarding of coagulation by adrenaline injections. Amer. J. Physiol. *34:* 232 (1914).
5. WRIGHT, H.P.: The sources of blood platelets and their adhesiveness in experimental thrombocytosis. J. Path. Bact. *56:* 151 (1944).

6. FORWELL, D.G. and INGRAM, G.I.C.: The effects of adrenaline infusion in human blood coagulation. J. Physiol. *135:* 371 (1957).
7. MCCLURE, P.D.; INGRAM, G.I.C. and JONES, R.V.: Platelet changes after adrenaline infusion with and without adrenaline blockers. Thromb. Diath. haemorrh. *13:* 136 (1965).
8. MITCHELL, J.R.A. and SHARP, A.A.: Platelet clumping *in vitro*. Brit. J. Haemat. *10:* 78 (1964).
9. CLAYTON, S. and CROSS, M.J.: The aggregation of platelets by catecholamines and by thrombin. J. Physiol. *169:* 82 (1963).
10. O'BRIEN, J.R.: Some effects of adrenaline and anti-adrenaline compounds on platelets *in vitro* and *in vivo*. Nature *200:* 763 (1963).
11. NORDOY, A. and ROWIK, T.O.: Some effects of adrenaline on rat platelets *in vitro* and *in vivo*. Scand. J. clin. Lab. Invest. *17:* Suppl. *84,* 151 (1965).
12. OZGE, O.H.; MUSTARD, J.F.; HEGARDT, B.; ROWSELL, H.C. and DOWNIE, H.G.: Abst. The effect of adrenaline on blood coagulation, platelet economy and thrombus formation. Canad. med. Ass. J. *88:* 265 (1963).
13. INGRAM, G.I.C.: Increase in antihemophilic globulin activity following infusion of adrenaline. J. Physiol. *156:* 217 (1961).
14. SHIMAMOTO, T.: The relationship of edematous reaction in arteries to atherosclerosis and thrombosis. J. Atheroscl. Res. *3:* 87 (1963).
15. WALDRON, J.M.: Clot-accelerating property of *in vitro* epinephrine and norepinephrine in whole blood coagulation. J. appl. Physiol. *3:* 558 (1950–51).

Authors' address: Dr. H. Engelberg and Dr. M. Futterman, 465 N. Roxbury Drive, Suite 1003, *Beverly Hills Cal.* (USA).

Sir William Dunn School of Pathology, University of Oxford

The Fine Structure of Platelets and Platelet Aggregates *in vivo*

J. E. FRENCH and J. A. BARCAT*, Oxford

The study of blood platelets by means of the electron microscope has contributed much new information about their structure and properties. Since the subject has recently been reviewed by DAVID-FERREIRA (1964), only the main structural characteristics will be mentioned. Platelets are bounded by a distinct plasma membrane and contain a variety of organelles which include dense membrane bound granules, small mitochondria, a variety of vesicular elements with relatively translucent contents, and, less constantly, particles of glycogen; there is little or no rough-surfaced endoplasmic reticulum. Some of these features are illustrated in Figure 1. Platelets which have been fixed in glutaraldehyde show a system of microtubules arranged as a bundle which appears to circumscribe the cell. These microtubules are thought to preserve the discoid shape of circulating platelets and they perhaps represent the contractile protein which platelets are known to contain (see BEHNKE, 1965; 1966).

The most characteristic property of the platelets is their ability to adhere to foreign surfaces and to each other with the formation of aggregates. Within the lumen of the blood vessels platelets are normally dispersed in a random manner among the other formed elements of the blood, but aggregation readily occurs at sites of vascular injury and/or disturbed flow and can be induced by the intravascular injection of different agents which include ADP, thromboplastins, thrombin and various particulate preparations. The platelets in the aggregates undergo morphological changes with time and these may be associated with fibrin formation (see

* Supported by a fellowship from the Consejo Nacional de Investigaciones Científicas y Técnicas de la República Argentina.

FRENCH, 1966). The observations to be described here are concerned with the effect of injected particles on platelet behaviour *in vivo*.

Material and Methods

The experiments were carried out in mice, rats and guinea pigs. The following preparations were used for intravenous injection: (a) Colloidal carbon (Gunther Wagner-Preparation Cl1/1431a–0.2 ml/100 g); (b) Colloidal thorium dioxide (Thorotrast, Testagar–0.1 ml/100 g); (c) An artificial fat emulsion (Intralipid-Vitrum–0.4 ml/100 g); (d) Homologous chyle (obtained by thoracic duct cannulation from donor rats fed corn oil – 1 ml/100 g). After intervals of 10–30 min the animals were anaesthetized with ether and the lung, liver and, in some experiments, spleen were sampled for light and electron microscopy. The pieces for electron microscopy were fixed for 2 h in 2% Osmium tetroxide in 0.1 M phosphate buffer, pH 7.4; dehydrated in acetone and embedded in Araldite. Sections were stained with lead citrate or with uranyl acetate and lead citrate.

Results

Following the injection of carbon, small platelet aggregates were readily identified by light microscopy in small vessels in the lung. As seen by electron microscopy, most of the platelets in the aggregates had intact plasma membranes and had retained their organelles but were very closely packed together in a mosaic pattern. Occasionally, when an aggregate completely occluded the lumen of a vessel, platelets at its edge showed an apparent loss of granules or had formed empty sac-like projections of their cytoplasm. Where these changes occurred, there was some fibrin at the edge of the aggregate. Carbon particles were found among the platelets, either grouped together in masses or more widely dispersed in the narrow interspace between adjacent plasma membranes. Uptake of the particles and their enclosure within vacuoles, as has been shown to occur when platelets are mixed *in vitro* with carbon particles (MOVAT, WEISER, GLYNN and MUSTARD, 1965) was observed in some of the aggregates.

Essentially similar results were obtained following injection of thorium dioxide, although the aggregates were generally smaller, more loosely arranged, and seldom occluded a vessel. Where the platelets were tightly packed together the small thorium dioxide particles were fairly uniformly distributed between the adjacent plasma membranes. Uptake of thorium dioxide particles by the platelets, as has previously been described both *in vitro* and *in vivo* by DAVID-FERREIRA (1961) was also observed (Fig. 1).

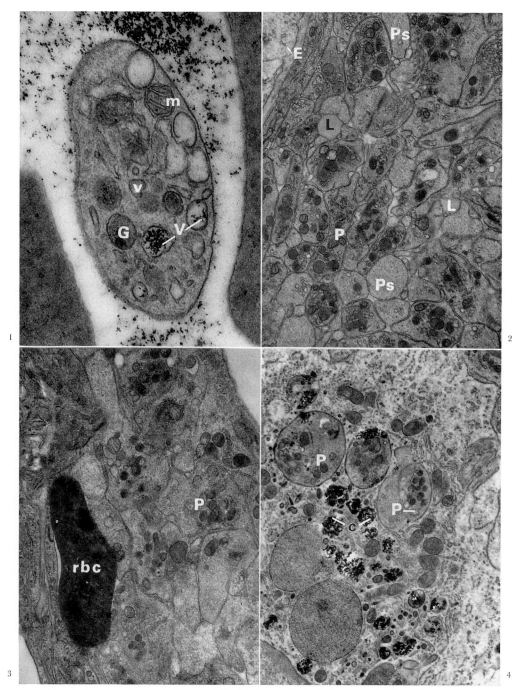

Figs. 1, 2, 3 and 4.

Injection of the two lipid preparations, Intralipid and homologous chyle, led to the formation of similar platelet aggregates in small vessels in the lung. In general the aggregates were more frequent with the Intralipid than with chyle and were sometimes large enough to occlude a small vessel completely. The platelets were tightly packed and most of them had retained their organelles, though in a few there was evidence of pseudopod formation.

Typical examples of the aggregates are illustrated in Figures 2 and 3. The lipid particles could usually be identified among the platelets but there was no clear evidence that they had actually been engulfed. It appears, however, from other studies that platelets can under certain circumstances take up lipid particles (SCHULZ and WEDELL, 1962).

Following injection of colloidal carbon, thorotrast or Intralipid, it was possible to find evidence of phagocytosis of platelets by cells of the reticulo-endothelial system in liver and spleen. Figure 4 illustrates this phenomenon in the spleen of a rat; it shows that in addition to taking up the carbon, a phagocyte has also engulfed several platelets.

Discussion and Conclusions

The aggregation of platelets which can be induced *in vitro* by particles of latex or carbon has recently been investigated by GLYNN, MOVAT, MURPHY and MUSTARD (1965). They have shown that the particles can be taken up by the platelets and at the same

---

Fig. 1. A platelet in a liver sinusoid from a mouse injected with Thorotrast. Vesicles containing Thorotrast (V), dense granules (G), mitochondria (m) and also empty vesicles (v) are seen. (Lead stain, ×35,000.)

Fig. 2. Platelet aggregate occluding a small vessel in the lung of a mouse injected with Intralipid. Most of the platelets (P) are unaltered but some show pseudopod formation (Ps). Lipid particles (L) can be seen among the platelets. The endothelial lining (E) of the vessel is at top left. (Lead stain, ×9,000.)

Fig. 3. Closely-packed platelets (P) and a red cell (rbc) in a capillary in the lung of a rat injected with chyle. No chylomicra are seen in this picture. (Lead stain, ×10,000.)

Fig. 4. Part of a macrophage in the spleen of a rat injected with carbon. Platelets (P) and carbon particles (c) are present within phagocytic vacuoles. (Lead stain, ×7,500.)

time ADP is released into the medium. They consider that phagocytosis of particles stimulates aggregation in the same way as does contact with other 'foreign' surfaces such as collagen fibres. It has been known for a long time that injection of various particles, including bacteria, will cause platelets to aggregate *in vivo* and that the aggregates may be trapped, at least temporarily, in small vessels of the lung and elsewhere. The present observations confirm that the aggregates induced in this way consist essentially of tightly-packed platelets and show that they correspond in structure to the aggregates formed in the initial stages of thrombosis and haemostasis (FRENCH, MACFARLANE and SANDERS, 1964) and to the aggregates induced by addition or release of ADP *in vitro* (HOVIG, 1962). The aggregation induced by carbon or thorium dioxide is accompanied by uptake of the particles by the platelets, which presumably become more adhesive during the process. In the case of the lipid particles it is less clear that actual phagocytosis of the particles is involved.

Since there are a number of indications that lipaemia and thrombosis are linked, particular interest attaches to the effect of lipid particles on platelet adhesiveness *in vivo*. SCHULZ and RABANUS (1965) have shown that platelet aggregates were present in small vessels when the artificial fat emulsion, *Lipofundin*, was given intravenously to rabbits. This emulsion apparently damaged the endothelium so it was not clear that the effect was a direct one on the platelets. In our experiments, where Intralipid or homologous chyle were used, there was no evidence of endothelial injury, and it seems probable that the presence of these particles in the bloodstream may itself have been a sufficient stimulus to induce some platelets to aggregate. This effect cannot apparently be attributed entirely to the artificial nature of the particles, since we have also observed occasional small platelet aggregates in the vessels of animals which had not been injected but had been given corn oil by mouth.

Platelets may normally be taken up by cells of the reticulo-endothelial system when they become effete, but evidence for this was rarely found in liver or spleen of control animals. Following injection of the particles, however, examples were found without difficulty. It appeared that when the surface properties of platelets had been changed in the circulation by contact or uptake of the injected particles they were more susceptible to ingestion by the tissue macrophages. Phagocytosis of platelets by monocytes has been observed *in vitro* in the experiments with latex particles already mentioned (MOVAT et al., 1965) and *in vivo* in organising thrombi (HAND and CHANDLER, 1962; POOLE, 1966).

Summary

The injection of particles or lipid emulsions into the circulation leads to the formation of closely-packed platelet aggregates which may be retained, at least temporarily, in small vessels in the lung. Fibrin may be formed in relation to the aggregates when they cause vascular occlusion. Particles of carbon or thorium dioxide are taken up into platelet vesicles. Some of the platelets, which presumably have altered surface properties, are ingested by cells of the reticulo-endothelial system.

References

BEHNKE, O.: Further studies on microtubules. A marginal bundle in human and rat thrombocytes. J. Ultrastruct. Res. *13*: 469–477 (1965). – Morphological changes in the hyalomere of rat blood platelets in experimental venous thrombi. Scand. J. Haemat. *3*: 136–148 (1966).

DAVID-FERREIRA, J. F.: Sur la structure et le pouvoir phagocytaire des plaquettes sanguines. Z. Zellforsch. *55*: 89–103 (1961). – The blood platelet: electron microscopic studies. Int. Rev. Cytol. *17*: 99–148 (1964).

FRENCH, J. E.: Electron microscopy of thrombus formation; in Modern trends in pathology, vol. 2. Ed. T. CRAWFORD (Butterworth, London).

FRENCH, J. E.; MACFARLANE, R. G. and SANDERS, A. G.: The structure of haemostatic plugs and experimental thrombi in small arteries. Brit. J. exp. Path. *45*: 467–474 (1964).

GLYNN, M. F.; MOVAT, H. Z.; MURPHY, E. A. and MUSTARD, J. F.: Study of platelet adhesiveness and aggregation, with latex particles. J. lab. clin. Med. *65*: 179–201 (1965).

HAND, R. A. and CHANDLER, A. B.: Atherosclerotic metamorphosis of autologous pulmonary thrombo-emboli in the rabbit. Amer. J. Path. *40*: 469–486 (1962).

HOVIG, T.: The ultrastructure of rabbit blood platelet aggregates. Thromb. Diath. haemorrh. *8*: 455–471 (1962).

MOVAT, H. Z.; WEISER, W. J.; GLYNN, M. F. and MUSTARD, J. F.: Platelet phagocytosis and aggregation. J. Cell Biol. *27*: 531–543 (1965).

POOLE, J. C. F.: Phagocytosis of platelets by monocytes in organizing arterial thrombi. An electron microscopical study. Quart. J. exp. Physiol. *51*: 54–59 (1966).

SCHULZ, H. and RABANUS, B.: Die kapilläre Plättchenthrombose im elektronenmikroskopischen Bild. Beitr. path. Anat. *131*: 290–311 (1965).

SCHULZ, H. and WEDELL, J.: Elektronenmikroskopische Untersuchungen zur Frage der Fettphagozytose und des Fetttransportes durch Thrombozyten. Klin. Wschr. *40*: 1114–1120 (1962).

Authors' address: Dr. J. E. French and Dr. J. A. Barcat, Sir William Dunn School of Pathology, University of Oxford, *Oxford* (England).

Nuffield Institute of Comparative Medicine, The Zoological Society of London

The Breakdown of Artificial Platelet Thrombi *in vitro*

Christine Hawkey[*], London

If it is accepted that platelet masses play a part in atherogenesis and thrombotic occlusion, then mechanisms capable of reversing platelet aggregation are of major importance in preventing these conditions. Platelet aggregation *in vitro* is reversible [1, 2] and one set of factors controlling whether or not platelet aggregates persist in the circulation would seem to be the relative amounts of aggregating substance and its inhibitor present at a given time. However, circumstances favouring platelet aggregation – such as cellular breakdown, tissue damage and the presence of thrombin, also favour activation of the blood coagulation mechanism with consequent fibrin formation. A second way in which platelet masses can be made permanent is by becoming bound on the outside by fibrin fibres, and, if this happens, even if the bonds linking the platelets within the mass are broken, platelets will not be set free from the mass unless this fibrin is removed. Therefore if fibrin is formed before disaggregation is complete, the effectiveness of the disaggregating mechanism will be lost unless the fibrin is subsequently lysed.

Release of platelets from an artificially formed thrombus *in vitro* can be demonstrated by Chandlers technique [3]. When platelet-rich plasma is rotated in the plastic loop, three well defined stages in formation of the artificial thrombus can be observed [1]. These are aggregation of platelets into clumps; the coming together of

[*] Research Fellow.

the clumps to form a white 'platelet thrombus' and, finally, the appearance of fibrin. By including an activator of the fibrinolytic mechanism in the system, these stages can be to some extent reversed. Active fibrinolysis results in removal of the fibrin, leaving the 'platelet thrombus', and further observations can then be made on the fate of this platelet mass.

Artificial platelet thrombi have been prepared from platelet-rich plasma of 15 normal humans. The fibrinolytic system was activated by addition to the plasma of an optimum concentration of streptokinase. This brought about lysis of the fibrin comprising the thrombus in 10–20 min, leaving a white, rounded structure, composed of platelets and histologically similar to an intravascular platelet thrombus. With continued rotation this platelet mass became progressively smaller until it was no longer distinguishable. Concurrent microscopical examination of the surrounding serum revealed the presence of an increasing number of free platelets and some small aggregates. After 24 h the mean number of free platelets in the serum was 52% ($\pm 6\%$) of the original platelet count, the remainder of the platelets being present as small aggregates.

Examination of the free platelets by phase contrast microscopy did not reveal any gross changes in morphology. Platelets released from the thrombus within the first few hours of the experiment possessed the ability to bring about normal clot retraction when added to recalcified platelet-poor plasma. However, no aggregation occurred when adenosine diphosphate (ADP) was added to a suspension of these platelets in normal plasma.

This experiment demonstrates that, *in vitro* at least, platelet thrombi are not permanent structures. It seems unlikely that the fibrinolytic system is directly responsible for release of platelets from the thrombus, but that by removal of fibrin binding the aggregates together, disaggregation is revealed. Breakdown of ADP-induced platelet aggregates has been studied by BORN and associates [2] and normal plasma probably contains enzymes capable of bringing this about. However, disaggregation of platelets in clumps bound by fibrin will only be effective in the presence of an active fibrinolytic mechanism. This provides a further example of the interdependence of the enzyme systems governing platelet aggregation, blood coagulation and fibrinolysis in maintaining the patency of blood vessels (Figure 1).

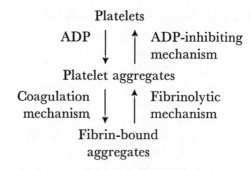

Fig. 1. Stages in formation and dispersal of a 'platelet thrombus'.

Acknowledgements

This work was supported by research grants from St. George's Hospital, London, and the British Heart Foundation.

References

1. Sharp, A.A.: Henry Ford Symposium on platelets. S.A. Johnson, Ed. (Churchill, London 1960).
2. Born, G.V.R. and Cross, M.J.: J. Physiol. *166*: 29 (1963).
3. Chandler, A.B.: Lab. Invest. *7:* 110 (1958).

Author's address: Dr. Christine Hawkey, Nuffield Institute of Comparative Medicine, The Zoological Society of London, Regent's Park, *London N.W. 1* (England).

Department of Experimental Pathology, University of Birmingham

The Nature and Significance of Platelet-Products in Plasma

K.W. Walton and P. Wolf, Birmingham

It has been found that fresh citrated plasma, freed from intact platelets, contains minute particulate material which originates from the osmiophilic granules of platelets. This material has therefore been called 'platelet dust' [1].

Platelet dust is extruded from platelets on storage. Larger yields are recovered from citrated plasma, in which platelets are labile, than from plasma kept fluid with sequestrene (ethylene diamine tetracetate) in which platelets are stable. Platelet dust is also released when blood coagulation occurs and can be demonstrated in serum.

Platelet dust contains proteins which can be shown to be identical, immunologically, with the proteins present in extracts of intact platelets, but distinct from the protein moieties of the soluble lipoproteins (alpha and beta lipoproteins) of plasma. Platelet dust is also rich in phosphatidyl ethanolamine and phosphatidyl serine which are known to be actively involved in the early stage of blood coagulation. The presence of platelet dust in serum may account for the 'platelet like activity' previously described in serum [2].

Platelet dust can be distinguished from chylomicra both by differences in physico-chemical characteristics and by biological activity in that platelet dust appears to provide the lipid requirements for the early stages of the intrinsic system of thrombin generation whereas the coagulation activity of chylomicra washed free of platelet dust is only manifested in the stypven test.

Preliminary evidence suggests that release of platelet dust occurs *in vivo* in association with intravascular thrombosis.

References

1. WOLF, P.: Brit.J.Haemat. *13:* 269 (1967).
2. O'BRIEN, J.R.: Brit.J.Haemat. *1:* 223 (1955).

Authors' address: Dr. K. W. Walton and Dr. P. Wolf, Department of Experimental Pathology, University of Birmingham, *Birmingham* (England).

Discussion Session 10

R.J. HASLAM (Macclesfield): I would like to comment on what Dr. ENGELBERG said about the mechanisms of action of adrenaline on platelets. Our own experiments, using both inhibitors of ADP-mediated platelet aggregation of the adenosine type and enzymes which remove ADP liberated from platelets, show that the effect of adrenaline is due to the release of ADP from the platelets. Thus, not only could the effect of adrenaline be almost completely inhibited by addition of snake-venom ADPase before adrenaline, but marked reversal of aggregation was produced when the enzyme was added after adrenaline. Our work also emphasizes the importance of the naturally occurring ADPase present in plasma.

A.L. ROBERTSON (Cleveland, Ohio): I would like to ask Dr. CONSTANTINIDES whether in his very laborious study of serial sections of cerebral artery thrombosis he could identify any relation between breaks on the surface of the arterial intima with thrombus formation and/or lipid deposition. His observations as well as those of FRIEDMAN ET AL. in coronary vessels need yet to show whether the atheromatous plaque or the thrombus are the major contributors to rupture of the continuity of the arterial wall.

P. CONSTANTINIDES (Vancouver, B.C.): There was no constant relation between the lipid content of the plaques and thrombosis. In about half the cases, the thrombi developed over plaques with a lot of fatty gruel in the area of the break, and in the remaining cases they developed over plaques with no visible gruel in the region of the fracture (i.e. over purely fibrous plaques). Thus, I do not know which is the most thrombogenic principle that acts on blood at the site of plaque trauma. It may be collagen, lipid, broken muscle cells, ADP, or any combination of the above. I also think it will take some years before we discover what causes breaks of plaque surfaces.

E.H. AHRENS, Jr. (New York, N.Y.): Dr. CONSTANTINIDES how does the size of the thrombus (length, width, or area) relate to the size of the 'crack' in the vessel wall so nicely demonstrated in your studies?

P. CONSTANTINIDES (Vancouver, B.C.): The length, width, shape and number of the 'cracks' vary greatly. The ones I encountered in the present series ranged from about 200 to 1500 μ in length, and from about 200 to 400 μ in width, as a very approximate estimate of order of magnitude. Thus, the ratio of crack length to thrombus length would be of the order of 1:10, 1:20, 1:50, etc.

O.J. POLLAK (Dover, Del.): I stained whole arterial sections with the Von Kossa stain (silver impregnation) prior to decalcification. I could see a very thin calcium phosphate shell in practically all lesions. Such 'egg-shell' lesion cracks and the crack provides the communication between lesion and vascular lumen.

J.L. BEAUMONT (Paris): Dr. MUSTARD, what kind of γ globulin did you use in your experiments?

J.F. MUSTARD (Toronto, Ont.): The gamma globulin used in our experiments was prepared by means of DEAE cellulose column from a crude fraction of human gamma globulin. The material gave a single band on immunoelectrophoresis against a human anti-serum antibody or an antibody against immune globulin (IGG). The material gave a single peak in the analytical ultracentrifuge.

J.L. BEAUMONT (Paris): Dr. WALTON, does this 'dust' material also occur in serum? What is the relation of chylomicra to the coagulative activity of plasma?

K.W. WALTON (Birmingham): (1) Yes, 'platelet dust' can also be recovered by high speed centrifugation from serum and may account for the 'platelet-like' activity of serum described by O'BRIEN [1]. (2) Citrated plasma freed from chylomicra retains its coagulative properties on recalcification. On the other hand, removal of platelet dust from citrated plasma renders it virtually incoagulable on recalcification. As mentioned in my reply to Dr. AHREN's question (q.v.) chylomicra show activity in the stypven test but not in the thrombin generation test of PITNEY and DACIE [2] so that the relation of chylomicra to the coagulative activity of plasma is dependent upon the method used to test it [3].

References

1. O'BRIEN, J.R.: Brit.J.Haemat. *1:* p.223 (1955).
2. PITNEY, W.R. and DACIE, J.V.: J.clin.Path. *6:* p.9 (1953).
3. POOLE, J.C.F.: Brit.J.Haemat. *1:* p.229 (1955).

E.H. AHRENS (New York, N.Y.): Dr. WALTON have you compared the possibly different performance of chylomicrons gathered from chyle versus chylomicrons derived from plasma? There is evidence that the protein coating of the particle is different in these two cases.

K.W. WALTON (Birmingham): We have not personally examined the behaviour of chylomicra from chyle. It was reported by POOLE and ROBINSON [1] that chylomicra from this source showed coagulative activity. When my colleague, Dr. WOLF, examined serum chylomicra prepared in the conventional fashion, these initially showed activity in the thrombin generation test and the stypven test. But when the chylomicra were washed to free them from 'platelet dust' they lost all activity in the thrombin generation test although activity was retained in the stypven test, suggesting that the lipid requirements for these two tests differed [2]. This difference in behaviour according to the coagulation test system used must also be kept in mind in comparing chylomicra from different sources.

References

1. POOLE, J.C.F. and ROBINSON, D.S.: Quart.J.exp.Physiol. *41:* p.31 (1956).
2. WOLF, P.: Brit.J.Haemat., in press (1966).

L. JØRGENSON (Oslo): Platelets and other formed elements of the blood are found along the lining of the aorta of normal rabbits and pigs. They are particularly associated with the branching sites, forming small mural thrombi. If permanent, these thrombi may be organized to fibrous plaques. If transient, the platelets or other formed elements may increase the permeability locally, allowing lipoproteins to infiltrate at these sites. These are two mechanisms whereby platelets may contribute to the development of atherosclerotic plaques.

R.W. WISSLER (Chicago, Ill.): I would like to ask Dr. FRENCH whether he believes that platelets making their way into the vessel wall could be responsible for the localized deposits of fibrin demonstrated by Dr. KAO and myself (J. Mol. Exp. Path., Vol. 4, No. 5, October 1966) as well as by Dr. HAUST's beautiful studies (M.D. HAUST and R.H. MORE, Progr. biochem. Pharmacol., Vol. 4. Karger, Basel/New York 1967, in preparation). These deposits of fibrin which are almost uniformly present in atheromatous lesions in man are (in our experience) not related closely to the lipid deposits and are much more frequent than any deposit of fibrin on the intima. In fact, we have difficulty in finding any substantial evidence of intimal thrombosis during the development of atherosclerosis in man, monkey or rat, but we wonder if this frequent deposit of fibrin among the normal vessel components may not contribute significantly to the pathogenesis of the fully developed disease.

J. FRENCH (Oxford): Platelet aggregates which adhere to the walls of blood vessels will undergo a series of changes with time if they are not dislodged. The changes in the platelets are associated with the appearance of increasing amounts of fibrin. The fibrin in atherosclerotic plaques could be the result of such surface encrustations, but I would hesitate to say that this is the only way in which fibrin can accumulate in the intima.

J. F. Mustard (Toronto, Ont.): I would like to comment on Dr. Wissler's question. In a study with Dr. Jørgensen and Dr. Rowsell we examined the fate of platelet-rich thrombi induced in the carotid artery of pigs. Initially these thrombi were rich in tightly packed platelets with a small thin layer of fibrin around the periphery. By 12 hours the platelets in the centre had separated from each other, some had disintegrated and a large amount of fibrin was beginning to form between the platelets. This was still most marked around the periphery. By 24 hours the original platelet-rich thrombus had become a mass of fibrin. Platelets which had disintegrated were evident between the fibrin strands. If one had not known the sequence of events it would have been very difficult to identify some of the constituents. This simply re-emphasizes what Welch and Aschoff in earlier investigations had shown, that is, platelet-rich thrombi rapidly transform to a fibrin mass. The question of whether platelets pass through the vessel wall and then undergo such transformation is a more difficult question to answer. We have had no difficulty in finding surface deposits of platelet material on the vessel of normal animals. Because of the factors which bind the platelet material together great care must be taken in the preparation of these specimens, since the thrombi can be lost. I think it would be extremely difficult to find such surface thrombi in routine *post mortem* material. However, they probably do occur and it seems likely that a proportion of these may undergo such a transformation and become incorporated into the vessel wall. It would be my opinion that fibrin forming within the vessel wall which was not due to platelet precipitation on the vessel wall was probably related to the transformation of fibrinogen in the intimal space itself.

Another interesting point in respect to the experiment which I am referring to is that the size of the thrombus is greatly reduced by 24 hours. Many people have attributed this to clot retraction. However, it would appear from our experiments that a bulk of this reduction is due to the disintegration of the platelets.

T. Zemplényi (Prague): My question to Dr. French is whether he has no experience concerning the role of histamine in the plagocytosis of platelet aggregates as referred by him. In experiments connected with indian-ink intake, we had, some years ago, some evidence that histamine might play a role under such conditions (Quart. J. exp. Physiol., *45:* p. 50, 1960).

J. French (Oxford): Observations with the electron microscope have shown that the effect of histamine on the localisation of carbon particles in the walls of small blood vessels can be explained by the formation of intercellular gaps in the endothelium which allow the carbon to escape (Majno, G. and Palade, G. E.: J. biophys. biochem. Cytol. *11:* pp. 571–605, 1961). Platelets will readily adhere to such gaps and appear stuck in the wall. These changes do not necessarily involve uptake of the carbon or platelets by endothelial cells. Some examples of platelets *within* endothelial cells at sites of minor injury in peripheral vessels have been published (Marchesi, V. T.: Ann. N.Y. Acad. Sci. *116:* pp. 774-788, 1964), but it is not clear that this is an effect of histamine.

H. Engelberg (Beverly Hills, Cal.): Dr. Born how long can platelet aggregates exist before fibrin forms?

G.V.R. Born (London): In rat cerebral veins *in vivo* platelet aggregates can exist for up to at least two hours before fibrin forms, other than a small white clot on the exact site of injury. The platelet-rich plasma used for the *in vitro* experiments contained anticoagulant.

G. A. Gresham (Cambridge): *In vivo* observations of platelet aggregation and their subsequent fate are, I think, an important aspect of research in atherosclerosis. One problem, however, is to administer a measurable stimulus to the vessel wall, if quantitative measurements are to be of value. Is Prof. BORN satisfied with the method that he employed to promote the formation of 'white bodies'?

G.V.R. BORN (London): I am only satisfied with the method insofar as it is the simplest so far devised that can be made to give quantitative and reproducible results. However, better methods will certainly be devised.

T. ZEMPLÉNYI (Prague): May I make a comment to the paper of Dr. BORN? In experiments with the 'thrombogenic' diet (J. Atheroscler. Res. 5: 540, 1965), we consistently found a higher ATPase activity of the vessel wall. This refers to the diet consisting of butter-fat, while in the diet containing arachid oil there was no such an increase of ATPase activity (and no intravascular thrombosis). Since it is reasonable to expect a higher accumulation of ADP in the vascular wall under such conditions, it is possible that this vascular factor could also play a role in platelet aggregation and initial phases of intravascular thrombosis.

H. PAYLING-WRIGHT (London): As a collolary to the work of BORN and HILP on white-body formation *in vivo* in rats with and without lipaemia, PHILIP and I made a series of observations on platelet stickiness in man. For these observations we used the rotating bulb technique. Blood was taken from fasting volonteers and a further sample was taken about 2 h after a fatty meal, when the lipaemia had increased the plasma optical density about four-fold. In lipaemia, the platelet stickiness was enhanced over that formed in fasting state. In fasting blood the *in vitro* addition of adenosine reduced the stickiness of platelets considerably. In lipaemia, however, the effect of added adenosine was significantly diminished. So, it seems that the effect of adenosine on platelet stickiness is antagonized by lipaemia under the conditions of our experiments.

G.A. GARCIA (Buenos Aires, Arg.): Dr. POLLAK made some remarks on my work about vascular cerebral failure. Those remarks were left without answer and they can be related to this discussion. – Dr. POLLAK has said that 'thrombosis' and 'hypertension' are both a manifestation of a general condition. This may be right or wrong. It is generally wrong in the thrombosis process, because when we find vascular disease in the legs or in the arms, there are frequently noticeable difference in blood pressure taking simultaneously blood in both legs or between the legs and arms.

Considering Dr. BORNS' work, I think it is extremely interesting and has possible application in preventive medicine though not in curative medicine. This is so, because Dr. BORN speaks about the inhibition of increased platelets aggregation.

Both the adenosine and heparine prevent 'thrombus' formation, but do not interfere in the formation of 'hemostatic plugs'. This fact points out a remarkable difference between thrombosis and hemostasis.

T. SHIMAMOTO (Tokyo): Since we found a highly potent platelet-agglutinating substance appearing in the plasma of rabbits treated by endotoxin, we also noticed that endotoxins, given as one dose, induce a shortening of blood clotting times and a reduction of platelet count (OLEF). Later, we also found that an intravenous injection of 0.1 to 1.0 mg/kg of adrenaline induces an immediate reduction in adhesive platelet count and a shortening in one stage prothrombosis time, in rabbits. These hematological phenomena were also found in man.

We also found that the pretreatment of rabbits or men with pyridinolcarbamate (20 to 30 mg/kg) prevents the effect of adrenaline in man and rabbits. The master's two step test in patients suffering from angina pectoris induces an immediate shortening in one stage prothrombin time and calcium clotting time and, at the same time, an immediate reduction in adhesive count. Using a double blind test, the preventive effect of pyridinolcarbamate given by mouth in a dose of one gram has been demonstrated.

Clinical

Department of Medicine (Ludolf-Krehl-Klinik), University of Heidelberg, Germany
(Director: Prof. Dr. G. Schettler)

Current Trends in the Treatment of Atherosclerosis from the Clinical View Point

G. Schettler, Heidelberg

There is no single cause of human atherosclerosis and its pathogenesis today is as controversial as ever. Although it is beyond the scope of this presentation to discuss the evidence for various theories of atherogenesis, it appears worthwhile to mention them briefly, since attempts of treatment are usually aimed at influencing various parameters thought to be related to atherogenesis.

The main current theories of atherogenesis include the lipid-filtration theory (Page, 1954), the thrombogenic theory (Rokitansky, 1852; Duguid, 1949), the lipophage migration theory (Leary, 1941), and the lipid synthesis theory. Metabolic abnormalities of the intimal lining of blood vessels and its study by means of biochemical examination and organ cultures are recently receiving increased attention (Böttcher, 1965; Page et al., 1966).

With the exception of fatty streaks in infants, the reversibility of atherosclerotic lesions in the human has never been proved and seems highly unlikely in the case of ulcerated and calcified advanced lesions. Dissolution of atherosclerotic plaques by local perfusion with lipase and surface-active agents as described by Mansfield and Howard (1966) in cadaver vessels, and with proteolytic enzymes and chelating agents as proposed by the same authors, appears problematic, since it may result in injury to the vessel wall with rupture, bleeding or aneurism formation. It is therefore reasonable that such changes should be prevented by mitigation or elimination of risk factors.

Combating Recognized Risk Factors

With regard to modification of *hyperlipidemia* the dietary approach is still held to be the basis of management. Both the Council on Foods and Nutrition of the American Medical Association and the Board of Directors of the American Heart Association have recently (1965 and 1966) extended previous recommendations and recommended dietary modification not only in hyperlipidemic adults, but also in children and young men at risk.

Of *hypocholesterolemic agents, nicotinic acid* in the doses required has subjective and objective side effects, of which transient impairment of liver function and glucose tolerance, and even manifestation of latent diabetes mellitus seems to be the most serious. Since the drug increases skin circulation without concomitant rise in cardiac output, perfusion of tissues such as muscle including heart muscle, is diminished. As a result angina pectoris and intermittent claudication may be caused or aggravated by nicotinic acid in arteriosclerotic subjects. Recently, a good hypocholesterolemic effect was observed with the use of *beta-pyridylcarbinol*, a nicotinic acid derivative. Of this compound only about $1/4$ of the amount of nicotinic acid has to be given for a comparable hypocholesterolemic effect (ZÖLLNER, 1965; SCHOEN et al., 1965).

Cholestyramine, a nonabsorbable anion exchange resin, binds bile acids in the intestine and seems to lower serum cholesterol by both promoting degradation of cholesterol to bile acids and diminishing reabsorption of cholesterol. With this drug HASHIM and VAN ITALLIE (1965) have reported disappearance of xanthomas so that the possibility of favorable effects upon vascular lesions cannot be excluded in such cases.

Comparing the hypolipoproteinemic action of *dextrothyroxine and chlorophenoxyisobutyrate* (CPIB, available as Atromid-S), STRISOWER and STRISOWER (1964) found that the former lowered cholesterol-rich Sf 0–20 lipoproteins more than the triglyceride-rich Sf 20–400 fraction, while the opposite was true for chlorophenoxyisobutyrate. With dextrothyroxine, aggravation of angina pectoris is considerably less than with the levorotatory isomers. Nevertheless, I do not use thyroxine homologues in patients with coronary heart disease. Even treatment of hypothyroidism with thyroid hormones requires the utmost caution since myocardial infarctions have been seen after institution of therapy in such patients. While thyroid hormone analogues seem to lower plasma cholesterol levels by increasing the catabolism and excretion of cholesterol, the mechanism of action of CPIB is not yet completely understood. Some relevant aspects will be discussed by Dr. GOULD and Dr. THORP during this symposium. In addition to its hypolipidemic effect, chlorophenoxyisobutyrate has been found to lower fibrinogen levels in subjects with atherosclerosis (COTTON and WADE, 1965) and to decrease the anticoagulant requirement (HELLMAN et al., 1963, and others).

With regard to the use of *heparin* in the management of subjects with coronary heart disease, a cooperative study at several German universities was initiated last year to evaluate its effectiveness in reducing mortality from myocardial infarction and reinfarction rate. In a blind study patients are randomized and treated with either 200 mg of heparin subcutaneously twice weekly or placebo for periods of 2–3 years.

Dr. PICK, during this symposium, will review experiences with *estrogens* in animal atherosclerosis. Although these hormones have a significant hypocholesterolemic effect, long-term clinical trials involving survivors of myocardial infarction have not yielded clear cut effects on mortality, and most men are unwilling to accept the feminizing effects of adequate doses.

Forthcoming information on the mechanism of free fatty acid mobilisation from adipose tissue has facilitated the development of *drugs* which, by *inhibiting free fatty acid liberation*, lower plasma free fatty acid levels. These aspects are discussed in a separate presentation by Dr. CARLSON, their clinical usefulness remains to be determined.

While the propensity of diabetics for the development of early and severe atherosclerosis is well documented, the recognition of an increased incidence of carbohydrate intolerance in patients with atherosclerosis (SCHRADE et al., 1960; ALBRINK, 1965; HERMAN and GORLIN, 1965), and elevation of blood sugar as a risk factor (EFSTEIN, 1965) is of more recent origin. It remains to be shown in future studies whether *drug-induced improvement of carbohydrate tolerance* in such patients will improve prognosis.

Evidence of the hypolipidemic effect of 'essential' *phospholipids* (EPL) (a purified soy bean phosphatide fraction rich in polyunsaturated fatty acids) is still controversial and has been summarized recently by HILD (1967). Other pertinent effects reported with the use of this drug include increased lipid clearing after fat infusions, improvement of blood vessel elasticity and reduction of the hypercoagulability state occurring after fat loading. The Shwartzman-Sanarelli phenomenon with its clotting defects is prevented by EPL (MÜLLER-BERGHAUS et al., 1964).

Platelet Stickiness and Thrombosis

Various findings obtained during the last few years support the thesis that increased platelet turnover and increased platelet stickiness may have some relation to the pathogenesis of thrombosis and arteriosclerosis. Therefore, an approach to the treatment of arteriosclerosis may consist in an *attempt to lower platelet stickiness* on a long-term basis. While a number of drugs such as adenosine, antihistaminics, phenylbutazone, sulfinpyrazone (Anturan), heparin, aspergillus enzyme preparations, atromid, monoaminooxidase inhibitors, 'essential' phospholipids, etc. have been found to exert this effect *in vitro*, these drugs either lack activity in clinical trials or have not been sufficiently tested in the human. In addition, some of these substances have considerable side effects. Nevertheless, the search for a suitable substance for lowering platelet stickiness *in vivo* on a long-term basis might be of great importance and I wonder whether Dr. MUSTARD and Dr. BORN will have something to say on this subject.

Fat phagocytosis by platelets has been recently studied by Dr. PFLEIDERER and coworkers (1966) of our group. With the use of electronmicroscopy, true phagocytosis of fat by platelets was established. During fat phagocytosis platelet aggregation increases and intracellular ATP and glycogen concentration decreases, similar to the changes described by MUSTARD et al. (1966) during phagocytosis of other particles. It is of interest that fat phagocytosis by platelets is inhibited when fat particles are modified by addition of surfaceactive agents. BÖHLE et al. (1965) were able to normalize abnormal platelet agglutination for periods up to 8 h by the oral administration of oils containing large proportions of unsaturated fatty acids to patients with atherosclerosis. It remains to be seen whether findings like these will improve our understanding of the role of platelets in arteriosclerosis and thrombosis.

The place of *anticoagulants* in the treatment of acute myocardial infarction seems firmly established. According to most recent figures

(KOLLER, 1966) the acute mortality from myocardial infarction with anticoagulants is 10% as compared to 22% without anticoagulants.

Long-term administration of anticoagulants results in reduction of mortality only during the first year after the infarction and only in patients below age 55. During the second year and later and in patients older than 55 years there is no difference in mortality between treated and untreated subjects (CLAUSEN et al., 1961, British Medical Research Council, 1964).

Fibrinolysis or thrombolytic therapy represents a new principle in the management of atherosclerotic complications such as arterial thrombosis or embolism. Details of the mechanism and execution of this form of treatment have been summarized in a number of recent reviews (DOUGLAS and McNICOL, 1964; PECHET, 1965; PIPER, 1965). Its effectiveness seems established in peripheral thrombosis and embolism, where results can be documented by angiography (GROSS et al., 1960). A beneficial effect in myocardial infarction is difficult to substantiate and has not been uniformly accepted. A cooperative study of medical departments of six Swiss and German universities, comparing 297 patients with myocardial infarction treated by fibrinolysis with 261 matched patients who were treated with anticoagulants showed a significant improvement of mortality in the former group as compared to the latter, in the absence of significant complications (SCHMUTZLER et al., 1966). While the activator of the fibrinolytic system used in this trial was streptokinase, urokinase was chosen by the Committee on Thrombolytic Agents of the National Heart Institute as most worthy of investigation and further development. In contrast to some other investigators, we feel that fibrinolytic therapy is not indicated in cerebrovascular occlusion because of the danger of hemorrhage.

Ischemia of Extremities

Ischemia produced by atherosclerotic narrowing or occlusion of peripheral arteries results in a discrepancy between substrate requirement and supply, and consecutively in inhibition of aerobic and increase of anaerobic metabolism. An ergometric work load during intraarterial administration of adenyl phosphate will result in significant improvement of hypoxia as judged by the arteriovenous balance of lactate and pyruvate. Results obtained with this method in a great number of patients were markedly better than those obtained by intraarterial administration of vasodilating agents alone, due to improvement of perfusion of ischemic tissues occurring with the work load (HILD et al., 1966).

In conclusion it may be worthwhile to emphasize that, with regard to hypolipidemic regimens, definite proof of a beneficial effect is still lacking, although preliminary results are promising (CHRISTAKIS et al., 1966). Time will also show the effect of elimination of other risk factors, such as smoking, hypertension, or obesity on the incidence and prevalence of coronary heart disease. With regard to drug therapy, double-blind, long-term studies are urgently needed and should not be endangered by premature reports on uncontrolled trials. Herein lies an important field of international research.

Summary

Current trends in the management of arteriosclerosis aim at mitigation on risk factors of the one hand and prevention and treatment of complications on the other.

The present paper reviews some drugs used in combating hyperlipidemia, deals briefly with some aspects of platelet stickiness and thrombosis, and reports our method for management of ischemia of the lower extremities.

References

ALBRINK, M.J.: Carbohydrate metabolism in cardiovascular disease. Ann.int.Med. *62:* 1330 (1965).
AMERICAN HEART ASSOCIATION: News release. Dec. 31, 1965; Jan. 1, 1966.
AMERICAN MEDICAL ASSOCIATION, COUNCIL ON FOODS AND NUTRITION: Diet and the possible prevention of coronary atheroma. J.amer.med.Ass. *194:* 1149 (1965).
BÖHLE, E.; BAUKE, J.; HARTMUTH, E. and BREDDIN, K.: Untersuchungen über die Agglutination der Blutplättchen nach Zufuhr verschiedener Nahrungsfette. Klin. Wschr. *43:* 555 (1965).
BÖTTCHER, C.J.F.: Chemical aspects of atherosclerosis in different types of arteries. In: Symp.path.klin.Aspekte Fettstoffwechsels (Thieme, Stuttgart 1966).
BRIT. MED. RES. COUNCIL: An assessment of long-term anticoagulant administration after cardiac infarction. Brit.med.J. *ii:* 837 (1964).
CHRISTAKIS, G.; RINZLER, S.H.; ARCHER, M.; WINSLOW, G.; JAMPEL, S.; STEPHENSON, J.; FRIEDMAN, G.; FEIN, H.; KRAUS, A. and JAMES, G.: The Anti-Coronary Club. A dietary approach to the prevention of coronary heart disease – a seven-year report. Amer.J.Publ.Hlth. *56:* 299 (1966).

Clausen, J.; Andersen, P.E.; Andresen, P.; Gruelund, S.; Harslof, E.; Andersen, U.H.; Jorgensen, J. and Mose, C.: Langtids-antikoagulans behandling efter akut Hjerteinfarkt. Ugeskr. Laeg. *123:* 987 (1961).

Cotton, R.C. and Wade, E.G.: Further observations on the effect of ethyl-a-p-chlorophenoxyisobutyrate + androsterone (Atromid) on plasma fibrinogen and serum cholesterol in patients with ischemic heart disease. J. Atheroscl. Res. *6:* 98 (1966).

Douglas, A.S. and McNicol, G.P.: Thrombolytic therapy. Brit. med. Bull. *20:* 228 (1964).

Duguid, J.B.: Pathogenesis of atherosclerosis. Lancet *ii:* 925 (1949).

Epstein, F.H.; Ostrander, Jr., L.D.; Johnson, B.C.; Payne, M.W.; Hayner, N.S.; Keller, J.B. and Francis, Jr., T.: Epidemiological studies of cardiovascular disease in a total community – Tecumseh, Michigan. Ann. int. Med. *62:* 1170 (1965).

Gross, R.; Hartl, W.; Kloss, G. and Rahn, B.: Thrombolyse durch Infusionen hochgereinigter Streptokinase. Dtsch. med. Wschr. *85:* 2129 (1960).

Hashim, S.A. and Van Itallie, T.B.: Cholestyramine resin therapy for hyper-cholesterolemia. J. amer. med. Ass. *192:* 289 (1965).

Hellman, L.; Zumoff, B.; Kessler, G.; Kara, E.; Rubin, I.L. and Rosenfeld, R.S.: Reduction of cholesterol and lipids in man by ethyl p-chlorophenoxyisobutyrate. Ann. Int. Med. *59:* 477 (1963).

Herman, M.V. and Gorlin, R.: Premature coronary artery disease and the preclinical diabetic state. Amer. J. Med. *38:* 481 (1965).

Hild, R.: Surface-active substances in the treatment of arteriosclerosis. In: Arteriosclerosis (Elsevier, Amsterdam 1966).

Hild, R.; Brecht, Th. and Zolg, H.: Die Wirkung eines Nucleotid-Nucleosid-Gemisches auf die hypoxische Stoffwechseländerung bei ergometrischer Belastung durchblutungsgestörter Gliedmaßen. Klin. Wschr. *44:* 388 (1966).

Koller, F.: Unpublished data. 72th Congr. German Soc. Intern. Med. (Wiesbaden 1966).

Leary, T.: Genesis of atherosclerosis. Arch. Path. *32:* 507 (1941).

Mansfield, A.O. and Howard, J.M.: *In vitro* mobilization of atherosclerotic plaque lipids. II. Perfusion. Arch. Surg. Chicago *92:* 414 (1966).

Müller-Berghaus, G.; Huth, K.; Krecke, H.J. and Lasch, H.G.: Blutlipide und intravasale Gerinnung in der Pathogenese des Sanarelli-Shwartzman-Phänomens. Schweiz. med. Wschr. *94:* 1519 (1964).

Mustard, J.F.; Rowsell, H.C. and Murphy, E.A.: Platelet economy (Platelet survival and turnover). Brit. J. Haemat. *12:* 1 (1966).

Page, I.H.: The Lewis A. Conner Memorial Lecture. Atherosclerosis. An Introduction. Circulation *10:* 1 (1954).

Page, I.H.; Green, J.G. and Lazzarini-Robertson, A.: The physicians incompleat guide to atherosclerosis. Ann. int. Med. *64:* 189 (1966).

Pechet, L.: Fibrinolysis. New Engl. J. Med. *273:* 966 (1965).

Pfleiderer, Th. and Morgenstern, E.: Zusammenhänge zwischen Phagozytose und Klebrigkeitsphänomenen bei Thrombozyten. Verh. dtsch. Ges. inn. Med. (1966) In press.

Pfleiderer, Th.; Morgenstern, E. and Weber, E.: Funktionelle, morphologische und biochemische Veränderungen in Blutplättchen während der Fettphagozytose *in vitro*. Klin. Wschr. (In press).

Piper, W.: Fibrinolyse als therapeutisches Prinzip. Mkurse ärztl. Fortbild. *15:* 542 (1965).

Rokitansky, C.: A manual of pathological anatomy. Sydenham Society (London 1852).

Schmutzler, R.; Heckner, F.; Körtge, P.; van de Loo, J.; Pezold, A.; Poliwoda, H.; Praetorius, F. and Zekorn D.: Zur thrombolytischen Therapie des frischen Herzinfarktes. Dtsch. Med. Wschr. *91:* 581 (1966).

Schön, H.; Zeller, W. and Henning, H.: Untersuchungen zur Behandlung der endogenen Hyperlipoproteidämie mit β-pyridyl-carbinol. Klin. Wschr. *41:* 1108 (1963).

Schrade, W.; Boehle, E. and Biegler, R.: Humoral changes in arteriosclerosis. Investigations on lipids, fatty acids, ketone bodies, pyruvic acid, lactic acid, and glucose in the blood. Lancet *ii:* 1409 (1960).

Strisower, E. H. and Strisower, B.: The separate hypolipoproteinemic effects of dextrothyroxine and ethyl chlorophenoxyisobutyrate. J. clin. Endocrin. *24:* 139 (1964).

Zöllner, N. and Wernecke, G.: A comparative study of the hypocholesterolemic effect of nicotinic acid and some compounds related to it. 6th Int. Congr. Geront. (Copenhagen 1963).

Author's address: G. Schettler, M. D., Med. Universitätsklinik Bergheimer Strasse 58, *69 Heidelberg* (Germany).

Istituto di Ricerche Farmacologiche 'Mario Negri', Milano

Inhibition of Fatty Acid Release by Pyrazole Derivatives

A. Bizzi, Milan

In these last years considerable attention has been focused on the problems of free fatty acid (FFA) mobilization from adipose tissue.

A large number of studies has been performed in order to establish the nature of the factors controlling the mobilization of FFA. Our interest has been concentrated on drugs able to decrease the level of plasma FFA with the aim of classifying these drugs on the bases of their specificity of action in blocking FFA mobilization [1] induced by various experimental procedures. This report is intended to summarize some data obtained with 3.5 dimethylpyrazole (3.5 DMP), originally reported by Gerritsen and Dulin mostly as hypoglycemic agent [2], and with two derivatives, namely 5 carboxyl-3-methylpyrazole (5C 3MP) and 5 carbinol-3-methylpyrazole (5A 3MP) (see for chemical structures Fig. 1).

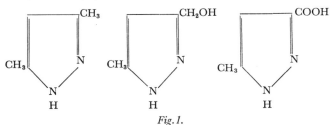

Fig. 1.

5 carboxyl-3-methylpyrazole has been identified by Smith and others [3] in the urine of humans and rats treated with 3.5 DMP.

5 carbinol-3-methylpyrazole [4] has been tested because it could be an intermediate in the formation of the carboxyl derivative.

This research has been partially financed by grant No. 1RO1 HEO 9971-01 of the U.S. Department of Health Education and Welfare P.H.S., N.I.H., N.H.I.

Previous work [1–5] showed that 3.5 DMP is quite active, when administered *in vivo*, in decreasing FFA and glycerol in adipose tissue as well as in plasma. These effects can be obtained with doses as low as 50 μg/kg i. p. This drug is quite active in fed and in fasted animals as well as in animals exposed to cold. The ablation of the adrenal gland does not modify the activity of 3.5 DMP. Furthermore a lowering of plasma FFA can be obtained in animals showing an high level of FFA as a result of pharmacological treatments, such as noradrenaline, ACTH, chlorpromazine, amphetamine and theophylline [1].

The effect of 3.5 DMP and of its derivatives on plasma FFA of fasted rats is reported in Table I. The compounds are equally active when the determinations of plasma FFA were performed 30 min after the administration. However, 2 hours later, 3.5 DMP and its carboxyl derivative appear to be more active than the alcohol derivative.

When the level of plasma glycerol was measured, similar results were obtained.

The activity *in vitro* of these compounds was also investigated. When the drugs are added to the adipose tissue during a stimulation induced by noradrenaline, ACTH or theophylline, different results have been obtained according to the compound under study.

Table I. Effect of pyrazole derivatives on the level of plasma FFA and glycerol

Treatment mg/kg i. p.		Plasma			
		FFA		Glycerol	
		30 min	120 min	30 min	120 min
Saline		100	100	100	100
3.5 DMP	0.1	32 ± 2	82 ± 15	33 ± 4	71 ± 4
	1	36 ± 1	45 ± 4	36 ± 3	59 ± 8
	10	54 ± 8	32 ± 2	52 ± 3	36 ± 4
5C 3MP	0.1	31 ± 4	64 ± 3	33 ± 8	71 ± 8
	1	31 ± 2	49 ± 3	53 ± 14	56 ± 4
	10	45 ± 2	37 ± 2	54 ± 5	43 ± 2
5A 3MP	0.1	31 ± 2	110 ± 6	35 ± 4	114 ± 2
	1	38 ± 5	72 ± 9	49 ± 7	72 ± 9
	10	35 ± 5	32 ± 4	49 ± 5	40 ± 6

Rats fasted overnight. Determinations, carried out 30 min and 120 min after drug administration, were done according to TROUT [7] with minor modification for FFA and to WIELAND [8] for glycerol.

Level of plasma FFA in fasted rats was 650 ± 32 μEq/1., glycerol was 144 ± 16 μM/1.

The results reported in Table II show that 3.5 DMP and its alcohol derivative was inactive even at a dose of 60 μg/ml while the carboxyl derivative was already effective at 1.5 μg/ml with a relationship between concentration and intensity of activity.

Table II. Effect of pyrazole derivatives on the release of FFA from adipose tissue incubated *in vitro*

Drug	μg/ml	FFA μEq/g/h after		
		norepinephrine 0.056 μg/ml	ACTH 0.05 μg/ml	theophylline 75 μg/ml
Saline		10.7 ± 0.1	7.3 ± 0.5	8.3 ± 0.4
3.5 DMP	60	10.1 ± 0.1	7.6 ± 0.2	7.9 ± 0.2
5A 3MP	60	11.0 ± 0.3	8.1 ± 0.3	7.6 ± 0.1
5C 3MP	1.5	8.5 ± 0.1	2.9 ± 0.3	3.8 ± 0.6
	3	7.6 ± 0.6	2.7 ± 0.2	3.6 ± 0.4
	6	7.1 ± 0.1	2.9 ± 0.1	2.4 ± 0.5

Pooled epididymal tissue 200 ± 10 mg was incubated in 5 ml Krebs Ringer bicarbonate + 2% albumin and 0.1% glucose pH 7.4 at 37° for 1 h.

Similar results were obtained when glycerol was measured in the medium, as an index of the lipolysis (Table III).

Having available drugs which are powerful inhibitors of lipolysis and therefore powerful agents in decreasing the level of plasma FFA, it was of interest to investigate their effect on the level of plasma triglycerides. It is in fact well known that a considerable percentage of the circulating FFA are taken up by the liver to form triglycerides which are then excreted in plasma and carried out by lipoproteins.

3.5 DMP has been found effective in decreasing liver triglycerides in normal conditions and in conditions of increased triglycerides accumulation [6].

Table III. Effect of pyrazole derivatives on the release of glycerol from adipose tissue incubated *in vitro*

Drug	μg/ml	Glycerol μM/g/h after		
		norepinephrine 0.056 μg/ml	ACTH 0.05 μg/ml	theophylline 75 μg/ml
Saline		9.0 ± 0.2	11.0 ± 0.2	5.7 ± 0.2
3.5 DMP	60	8.7 ± 0.3	9.8 ± 0.4	5.7 ± 0.1
5A 3MP	60	7.7 ± 0.5	10.9 ± 0.5	4.9 ± 0.1
5C 3MP	1.5	8.0 ± 0.1	7.8 ± 0.1	2.9 ± 0.3
	3	6.7 ± 0.3	6.6 ± 0.1	2.7 ± 0.2
	6	5.7 ± 0.1	6.4 ± 0.1	2.9 ± 0.1

Furthermore 3.5 DMP and its derivatives given by oral route are able to decrease plasma triglycerides in a manner which is proportional to the dose used. Data reported in Table IV show that after this treatment it is possible to reach even a 50% decrease of plasma triglycerides. This fall can be hardly achieved with any one of the known hypolipemic drugs.

Table IV. Effect of pyrazole derivates on plasma triglycerides concentration

Dose mg/kg os	Plasma Triglycerides (controls = 100) after		
	3.5 DMP	5C 3MP	5A 3MP
0.75	100 ± 4	93 ± 5	68 ± 5
1.50	56 ± 11	74 ± 7	61 ± 6
3.70	65 ± 11	73 ± 4	47 ± 6
7.50	55 ± 3	50 ± 3	61 ± 3
15.00	46 ± 2	53 ± 4	48 ± 2
30.00	47 ± 2	66 ± 5	59 ± 4

Rats fasted overnight were killed 4 h after administration of the drugs. Triglycerides determinations were performed according to VAN HANDEL and ZILVERSMIT [9].

Finally other studies have been carried out to see whether this fall of plasma FFA and triglycerides could be obtained after a subacute treatment. 3.5 DMP was given daily during a period of 10 days. One hour before the animals were sacrificed, half of the controls and treated animals received a dose of 3.5 DMP, the animals were killed one hour later.

Results in Table V show that 3.5 DMP exerts the expected effect on plasma FFA and triglycerides even after a subacute treatment.

Table V. Effect of 3.5 DMP following a subacute treatment

Treatment		Plasma	
Chronic	Acute	FFA μEq/l	Triglycerides mg/ml
–	–	668 ± 30	92 ± 4
–	3.5 DMP (60 min)	108 ± 12	60 ± 7
3.5 DMP	–	735 ± 27	95 ± 5
3.5 DMP	3.5 DMP (60 min)	164 ± 7	50 ± 3

3.5 DMP was injected i. p. at the dose of 15 mg/kg daily for 10 days. After the last administration, rats were fasted overnight and the following morning received one dose of 3.5 DMP (7.5 mg/kg i. p.). Rats were killed 60 min later.

However, in these conditions the rebound effect (increase of plasma FFA following the period of decrease) was more rapid and evident.

If the level of plasma triglycerides is somewhat involved with the presence or the development of atherosclerosis, these new drugs may be offered as a tool to understand the significance of lipid metabolism in the lesions of the arterial walls.

References

1. Garattini, S. and Bizzi, A.: Effect of drug on mobilization of FFA. Pharmacol. Rev. *18:* 1, 243–245 (1966).
2. Gerritsen, G. C. and Dulin, W. E.: Effect of a new hypoglycemic agent, 3,5 dimethylpyrazole, on carbohydrate and free fatty acid metabolism. Diabetes *14:* 507–515 (1965).
3. Smith, D. L.; Forist, A. A. and Dulin, W. E.: 5-Methylpyrazole-3-carboxylic acid. The potent hypoglycemic metabolite of 3.5 dimethylpyrazole in the rat. J. med. Chem. *8:* 350 (1965).
4. Rubessa, F.: Sintesi di nuovi derivati pirazolici. Il Farmaco (in press) (1967).
5. Bizzi, A.; Jori, A.; Veneroni, E. and Garattini, S.: Effect of 3.5 dimethylpyrazole on blood free fatty acids and glucose. Life Sciences *3:* 1371–1375 (1964).
6. Bizzi, A.; Tacconi, M. T.; Veneroni, E. and Garattini, S.: Triglyceride accumulation in liver. Nature *209:* 1025–1026 (1966).
7. Trout, D. L.; Estes, E. H. and Friedberg, S. J.: Titration of free fatty acids of plasma; A study of current methods and a new modification. J. Lipid Res. *1:* pp. 199–202 (1960).
8. Wieland, O.: Eine enzymatische Methode zur Bestimmung von Glyzerin, Biochem. Zschr. *329:* 313–319 (1957).
9. Handel, E. Van and Zilversmit, D. B.: Micromethod for the direct determination of serum triglycerides. J. lab. clin. Med. *50:* 152–157 (1957).

Author's address: Dr. A. Bizzi, Istituto di Ricerche Farmacologiche 'Mario Negri', Via Eritrea 62, *Milano* (Italy).

Department of Pharmacology (Director: Prof. T. L. Chruściel) and Department of Pathological Anatomy (Director: Prof. W. Niepolomski), Silesian Medical Academy, Katowice

Influence of Clofibrate and Vatensol on the Development of Experimental Atherosclerosis in White Rats*

T. L. Chruściel, Z. S. Herman, R. Brus, A. Plech,
D. Habczyńska and C. Ważna-Boguńska, Zabrze

Continuing the research on the influence of 'antiatherosclerotic' substances upon the development of experimental atherosclerosis we have investigated the effect of ethyl ester of p-chlorphenoxy-isobutyric acid (Clofibrate, Atromid S, Imp. Chem. Ind.) as well as of ([2-]2,6-bichlorphenoxy[ethyl]amino) guanidine sulphate (Vatensol, Comp. 1029, Pfizer Ltd.).

Clofibrate (CPIB) causes a marked decrease of the level of triglycerides and cholesterol in rabbits [9] and does not influence the degree of development of experimental atherosclerosis in rabbits [9, 17]. Thorp and Waring [17] have shown that CPIB increases the level of cholesterol in the serum of monkeys and chickens fed atherosclerogenic diet. In men with coronary disease and with increased levels of cholesterol and triglycerides in serum CPIB decreases hypercholesterolemia and hyperlipemia [1, 7, 8, 11, 16, 18]. Vatensol (V) decreases the activity of dopa-decarboxylase and β-dopamine hydroxylase, thereby inhibiting the synthesis of catecholamines [10]. This effect prompted us to apply V in experimental atherosclerosis.

Material and Methods

Experiments were performed on 30 albino, male rats of the Wistar strain, weighing 100–200 g. The animals were divided in 3 groups: Group I was fed during 90 days an

* Partly supported by a grant of the Polish Academy of Sciences.

atherogenic diet (AD prepared after HOOGERWERF [6] and HARTROFT and THOMAS [4]) in a daily dose of 20 g/kg. Diet AD contained cholesterol 20%, hydrogenated fat 40%, rape-seed oil 11%, wheat flour 29%, carbimazole 0,0025% and vitamin D_3 0,0025%. Group II was fed diet AD, containing CPIB in the dose of 40 mg/kg. Group III was fed diet AD, containing V in the dose of 10 mg/kg. All animals, besides AD diet, received in the evening usual food, water or milk. Samples of blood were taken by heart puncture. We estimated the level of total lipids in serum [SWAHN, 15], esterified fatty acids [STERN and SHAPIRO, 14], lipid-phosphor after ZILVERSMIT and DAVIES [19] and free and total cholesterol after SPERRY and WEBB [13]. The animals were killed by decapitation and macroscopic examinations of the heart, great arteries, liver and kidneys were made. These organs were put into 4% neutralised formaldehyde solution. The frozen sections were stained with Sudan III. The paraffin sections were stained with haematoxylin and eosin, BEST carmine, resorcine and fuchsin. The statistical analysis of results was performed with Student's t-test.

Results

In rats fed diet AD with CPIB we found a statistically significant decrease of serum levels of totals fats, esterified fatty acids, lipid phosphor and total cholesterol as compared with rats fed diet AD only (Table I). In rats fed diet AD with V a statistically significant decrease of esterified fatty acids serum level was observed whereas the level of other considered substances did not differ significantly from their content in the serum of the animals of group I.

In six rats fed by AD among nine animals numerous small, white-yellowish atherosclerotic nodules in the intima of main artery were observed; in one of these rats considerable changes were seen (Fig. 1).

All animals have shown similar histological changes; however the degree of their intensity was different in individual experimental groups (Table II, Fig. 2).

The animals treated with diet AD only, have shown the greatest degree of the histological changes.

Discussion

We had shown that CPIB caused a significant decrease of the serum levels of total lipids, esterified fatty acids, lipid phosphorus and total cholesterol in rats fed an atherogenic diet. A similar effect of CPIB was described in rabbits by JONES et al. [9]. JONES et al. [9] and THORP and WARING [17] did not find an antiatherosclerotic effect of CPIB added to the diet.

Table I. Biochemical changes in the serum of rats fed atherogenic diet (AD) and of rats fed diet AD with clofibrate (Atromid S) or Vatensol added

Group	Diet	Substance added	Number of animals	Total fats	p	Esterified fatty acids	p	Serum level in mg % Lipid phosphor	p	total	p	Cholesterol free	p
I	AD	–	8	297±40.0		108.3±9.5		4.62±0.34		129.2±12.3		44.0±10.8	
II	AD	Clofibrate	7	145.3±25.4	<0.05	59.9±7.1	<0.01	2.94±0.016	<0.0025	91.7±9.2	0.05	28.6±5.3	>0.2
III	AD	Vatensol	10	266.2±24.3	>0.6	80.7±5.9	<0.05	4.94±0.36	>0.5	113.9±4.3	0.2	40.7±5.7	>0.7

Development of Experimental Atherosclerosis in White Rats 581

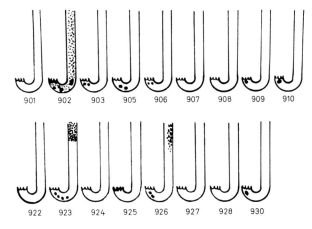

Fig. 1. Macroscopic changes in the main artery of rats treated with atherogenic diet (AD) with or without CPIB added. Top: group I, diet AD. Bottom: group II, diet AD containing CPIB. Figures under schemes indicate number of animal.

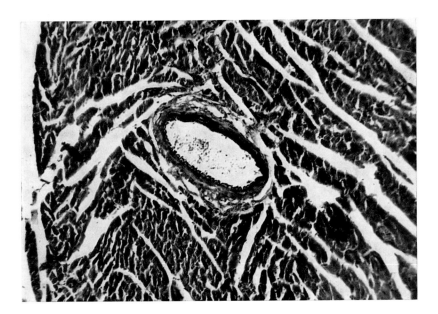

Fig. 2. Thickening and calcification in the media of coronary artery. H.E. Magn. ×400.

Table II. Histological changes in rats. The figures in the table indicate number of animals

Changes in organs	Atherogenic diet (AD) 9 rats	Diet AD + clofibrate 8 rats	Diet AD + Vatensol 10 rats
Heart:			
Hyperaemia	3	1	–
Haemorhage in myocardium	1	–	–
Fatty degeneration	–	1	–
Fragmentation	9	7	9
Granular disintegration	6	1	3
Fibrosis	–	1	1
Inflammatory granulation	–	1	–
Coronary vessels:			
Proliferation of endothelium	7	3	2
Thickening of vessels	9	4	8
Calcification	1	1	–
Thrombosis	–	1	–
Main artery:			
Proliferation of endothelium	5	2	–
Disintegration of elastic fibers	2	3	–
Connective tissue proliferation	3	2	–
Calcification	2	1	1
Kidneys:			
Hyperaemia	–	6	1
Changes in parenchyma	2	1	2
Fatty degeneration	8	6	10
Inflammatory infiltration	8	2	1
Necrosis	–	1	–
Walls thickening	5	3	1
Calcium deposits	2	3	–
Liver:			
Hyperaemia	–	–	–
Fatty degeneration	9	7	8
Glycogenic degeneration	–	–	3
Inflammatory infiltrations	1	1	–

According to the above data it seems that CPIB does not protect the walls of blood vessels from early injuries caused by atherogenic diet or blood dyscrasia.

Vatensol decreased considerably the serum level of fatty acids and it prevented the appearance of macroscopic atherosclerotic changes, although it did not prevent the development of the histological changes in the heart and parenchymatous organs. A decrease of the intensity of histological changes in the main artery was shown. Since we did not observe in animals fed diet AD with Vatensol any macroscopic changes, we may suggest that Vatensol has some preventive effect upon the development of atherosclerotic changes. We should like to point out the fact that all animals fed diet AD with Vatensol survived the experiment, whereas in the group fed with AD one of the rats died in the course of experiment and in the group fed AD and CPIB 2 animals died after 10 weeks of administration of AD. If we take into consideration some role of catecholamines in the pathogenesis of atherosclerosis [3, 5], we may relate the activity of Vatensol with its inhibitory effect on the synthesis of those substances.

Conclusions

(1) Ethyl-2-(p-chlorophenoxy) isobutyrate administered for 3 months in the diet to white rats treated concomitantly with an atherogenic diet produced a decrease of hyperlipidemia and hypercholesterolemia as compared with control rats with experimental dietary atherosclerosis. However, no decrease in the degree of development of athcrosclerotic changes in the circulatory system was observed in this experiment.

(2) Dichlorophenoxyethylamino-guanidine sulfate (Vatensol) administered for 3 months in the diet to white rats treated concomitantly with an atherogenic diet causes a decrease of increased fatty acids level in the serum and diminishes the degree of development of the atherosclerotic changes in the circulatory system. Vatensol decreases the toxicity of chronically given atherogenic diet.

Summary

Ethyl-p-chlorophenoxyisobutyrate (clofibrate) in a dose of 40 mg/kg p.o. decreased hypercholesterolemia and hyperlipidemia in white rats treated for 3 months with an atherogenic diet. Dichlorophenoxyethylamino-guanidine sulphate (Vatensol) in a dose of 10 mg/kg p.o. decreased an increased serum level of fatty acids and diminished the intensity of development of atherosclerotic changes in the heart and great arteries of rats treated with the same atherogenic diet.

References

1. Askanas, Z.; Malanowicz, W. i Zambrowicz, K.: Wpływ atromidu na lipidy osoczowe i fibrynolize u chorych z przebytym zawałem serca. Pol. Arch. Med. Wewn. *35:* 5–11 (1965).
2. Baccino, F.M. e Congin, L.: Atero-arteriosclerosi sperimentale da adrenalina e colesterolo. Lo Sperimentale *114:* 129–139 (1964).
3. Brux, J. De; Calvarin, R. et Kaufmann, H.: Lésions artérielles provoquées expérimentalement par l'adrénaline. Presse Med. *72:* 1539–1544 (1964).
4. Hartroft, W.S. and Thomas, W.A.: Pathological lesions related for disturbances of fat and cholesterol metabolismus in man. J. amer. med. Ass. *164:* 1899 (1957).
5. Heimberg, M. and Fizette, N.B.: The action of norepinephrine on the transport of fatty acids and triglycerides by the isolated perfused rat liver. Biochem. Pharmacol. *12:* 392–394 (1963).
6. Hoogerwerf, S.: Der Einfluß von Vasolastine auf künstliche Sklerose bei Ratten und Atherosklerose des Menschen. Ärztl. Forsch. *9:* 540–546 (1955).
7. Howard, R.P.; Alaupovic, P.; Brusco, O.J. and Furman, R.H.: Effects of ethyl chlorophenoxyisobutyrate, alone or with androsterone (atromid) on serum lipids, lipoproteins and related metabolic parameters in normal and hyperlipidemic subjects. J. Atheroscler. Res. *3:* 482–499 (1963).
8. Jurand, J. and Oliver, M.F.: The effects of ethyl chlorophenoxyisobutyrate on serum cholesterol, triglyceride and phospholipid fatty acids. J. Atheroscler. Res. *3:* 547–553 (1963).
9. Jones, D.; Gresham, G.A. and Howard, A.N.: The effect of atromid on arterial disease induced by semi-synthetic diet in the rabbit. J. Atheroscler. Res. *3:* 716–724 (1963).
10. Lawrie, T.D.W.; Lorimer, A.R.; Alpine, Mc. S.G. and Reinert, H.: Clinical trial and pharmacological study of compound 1029 ('Vatensol'). Brit. med. J. *i:* 402–406 (1964).
11. Oliver, M.F.: Further observations on the effects of atromid and of ethyl chlorophenoxyisobutyrate on serum lipid levels. J. Atheroscler. Res. *3:* 427–444 (1963).
12. Robinson, R.W.: Changes in serum lipids with atromid and chlorophenoxyisobutyric ester. J. Atheroscler. Res. *3:* 566–570 (1963).
13. Sperry, W.M. and Webb, M.: A revision on the Schoenheimer-Sperry method for cholesterol determination. J. Biol. Chem. *187:* 97–106 (1950).
14. Stern, J. and Shapiro, B.: A rapid and simple method for determination of esterified fatty acids and total acids in blood. J. clin. Pathol. *6:* 158–160 (1963).
15. Swahn, B.: A micromethod for the determination of total lipids in serum. Scand. J. Clin. Lab. Invest. *5:* 1–5 (1953).
16. Sznajderman, M.; Ciświcka-Sznajderman, M.; Ignatowska, H.; Gajewski, J. i Skorykow, A.: Badania kliniczne nad atromidem. Pol. Arch. Med. Wewn. *34:* 1641–1652 (1964).
17. Thorp, J.M. and Waring, W.S.: Modification of metabolism and distribution of lipids by ethyl chlorophenoxyisobutyrate. Nature *194:* 948–949 (1962).
18. Thorp, J.M.: An experimental approach to the problem of disordered lipid metabolism. J. Atheroscler. Res. *3:* 351–360 (1963).
19. Zilversmit, D.B. and Davis, A.K.: Microdetermination of plasma phospholipids by trichloroacetic acid precipitation. J. lab. clin. Med. *35:* 155–160 (1950).

Authors' addresses: Dr. T. L. Chruściel, Dr. Z. S. Herman, Dr. R. Brus, Dr. A. Plech, Department of Pharmacology, Silesian Medical Academy, Marksa 38, *Zabrze 8*; Dr. D. Habczyńska, Dr. C. Ważna-Boguńska, Department of Pathological Anatomy, Silesian Medical Academy, 3-go Maja 15, *Zabrze 1* (Poland).

Department of Experimental Medicine, F. Hoffmann-La Roche & Co. Ltd., Basel

Comparison of Nicotinic Acid and β-Pyridylcarbinol with Respect to Lipid Metabolism in the Rat *in vivo*

K. F. Gey and E. Lorch, Basel

Nicotinic acid [2, 3] and the corresponding alcohol, i.e. β-pyridylcarbinol (Ronicol®, Roniacol®) [1, 8, 9, 10] are known to inhibit the lipolysis in adipose tissue and to decrease the free fatty acids in the blood plasma. Both compounds have been investigated almost independently. Therefore, this paper presents a comparison of both pyridine derivatives in female albino rats of about 150 g body weight. The drugs were given as aqueous suspensions in doses of 500 mg/kg by stomach tube. Both compounds were suspended with the powdered galenic constituents of the long-acting preparation of Ronicol®. The galenic constituents alone were used as placebo; they do not react chemically with the pyridine derivatives and lack pharmacological effects of their own.

When normal rats were fasted for 2½ days and given the pyridine derivatives 5× within this time period the drugs did not substantially change the total lipids (measured gravimetrically [5]) of the liver, heart and muscle of the upper thigh (Table I).

Table I. Normal fasted rats treated 5 times with nicotinic acid or β-pyridylcarbinol (Ronicol®) within 2½ days (Groups of 16 animals)

Treatment (per stomach tube twice daily)	Tissue		
	liver total lipids in %	heart of dry weight	muscle
Placebo	22.2±0.4	23.0±0.4	15.2±1.2
β-Pyridylcarbinol (500 mg/kg)	21.1±0.9	23.2±0.3	17.2±1.4
Nicotinic acid (500 mg/kg)	23.9±0.7	26.3±1.5	16.2±0.5

Since triglycerides represent only a minor fraction of total lipids in the liver, measurement of the latter does, of course, not exclude a relative decrease of triglycerides [4]. – When together with the last administration of the pyridine derivatives 1-^{14}C-acetate was injected s.c. (Table II) (in further experiments 1-^{14}C-acetate was

Table II. Normal fasted rats injected with 1-^{14}C-acetate. 2 Groups of 2 rats received 1.5 mC/kg (4.7 mg/kg) s.c. simultaneously with the last administration of pyridine derivatives

Treatment (per stomach tube 5× within 2 ½ days)	$^{14}CO_2$ Expiration % of injected ^{14}C-acetate	
	1 ½ h	4 h
Placebo	51±3	70±1
β-Pyridylcarbinol (500 mg/kg)	61±2	72±5
Nicotinic acid (500 mg/kg)	62±8	77±8

given before or after the last drug administration), the expiration of $^{14}CO_2$ was not significantly changed either. This suggests that the acetate pool and thus probably also the caloric conditions of the rat were not modified by both pyridine derivatives.

In normal rats having access to food *ad libitum* treatment with nicotinic acid or pyridylcarbinol for 4 weeks decreased the total liver lipids only to a minimal degree even though the difference was statistically significant. In the 5 hours' fasting period subsequent to the last drug administration the plasma free fatty acids (FFA) as measured by photometric titration [7] were depressed by both pyridine derivatives to the same extent (Table III).

For further experiments with a pathological triglyceride accumulation the fatty liver induced by either CCl_4 or choline deficiency was chosen as a model. When CCl_4-treated rats (8×770 mg/kg CCl_4 in olive oil s.c. within 4 weeks) thereafter were fasted and given the pyridine derivatives for 2 ½ days, pyridylcarbinol – in contrast to nicotinic acid – caused a marked (32%) regression of the liver lipids which according to detailed measurements was mainly due to a decrease of triglycerides. There was a fair correlation between the depression of liver lipids and plasma FFA (Table IV). – In choline-deficient rats (choline-free diet including 38% fat for 4 weeks) again subsequent treatment with pyridylcarbinol

Table III. Normal rats treated with β-pyridylcarbinol (Ronicol®) or nicotinic acid for 4 weeks (Groups of 12 animals each)

Treatment per stomach tube (twice daily)	Liver lipids %	Plasma-FFA (5 h after the last admin.) %
No	96±1	–
Placebo	100±2[a]	100±7[b]
β-Pyridylcarbinol (500 mg/kg)	90±1*	43±5*
Nicotinic acid (500 mg/kg)	87±2*	40±6*

[a] absolute value of total liver lipids: $19.5 \pm 0.4\%$ of dry weight.
[b] absolute value of plasma-FFA: 177 ± 13 μEq./l.
* $p < 0.01$ in comparison to placebo.

Table IV. Regression of the CCl$_4$-induced fatty liver of the rat within 2½ days (Groups of 52 animals)

Treatment (per stomach tube 2 × daily) after CCl$_4$	Liver lipids %	Plasma-FFA (5 h after the last admin.) %
No	98±3	–
Placebo	100±2[a]	100±6[b]
β-Pyridylcarbinol (500 mg/kg)	68±3*	63±3*
Nicotinic acid (500 mg/kg)	95±2	93±3

[a] absolute value of total liver lipids: 53.2 ± 1.2 of dry weight.
[b] absolute value of plasma-FFA: 822 ± 43 μEq./l.
* $p < 0.01$ in comparison to placebo.

markedly reduced both the lipid infiltration of the liver and the plasma FFA whereas nicotinic acid had only a weak effect which was not statistically significant (Table V). – Pyridylcarbinol did not only induce the regression of a fatty liver but counteracted also, in part, its development. Thus, in previous experiments [6] in which rats had access to food ad lib. and received both CCl$_4$ and the pyridine derivatives for 4 weeks, pyridylcarbinol – but not nicotinic acid – reduced the CCl$_4$-induced fatty liver significantly.

Table V. Fatty liver induced by choline-deficiency in the rat – regression within 2 ½ days (Groups of 15 animals)

Treatment per stomach tube (twice daily)	Liver lipids %	Plasma-FFA (5 h after the last admin.) %
No	90±8	–
Placebo	100±6[a]	100±5[b]
β-Pyridylcarbinol (500 mg/kg)	63±8*	75±4*
Nicotinic acid (500 mg/kg)	88±7	84±4

[a] absolute value of total liver lipids: 50.5±3.0% of dry weight.
[b] absolute value of plasma-FFA: 953±53 µEq./l.
* $p < 0.01$ in comparison to placebo.

The differences between nicotinic acid and pyridylcarbinol with regard to the diminution of a fatty liver may be explained mainly by a longer duration of action of pyridylcarbinol. Thus, in normal fasting rats the last of five administrations of either nicotinic acid or pyridylcarbinol caused about the same decrease of plasma FFA within about 5 h but thereafter nicotinic acid showed no further action whereas the FFA of pyridylcarbinol-treated rats remained depressed for at least 24 h (Fig. 1). In rats with a CCl_4-induced

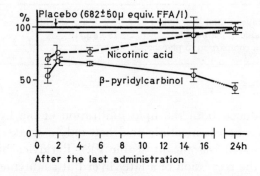

Fig. 1. Behaviour of free fatty acids of blood plasma (FFA) in the normal fasted rat after five administrations of nicotinic acid and β-pyridylcarbinol resp. Groups of 5–16 rats were fasted and given the pyridine derivatives (500 mg/kg twice daily) per stomach tube for 2 ½ days. Abscissa: hours after the last administration. Ordinate: % change of FFA in comparison to placebo-treated rats decapitated after the same time interval.

accumulation of triglycerides in the liver this difference was even more pronounced since under these conditions the last of 5 doses of nicotinic acid decreased the plasma FFA substantially only for about 2 h but pyridylcarbinol for at least 24 h (Fig. 2).

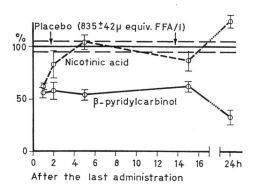

Fig. 2. Behaviour of FFA in CCl$_4$-treated rats subsequently treated with five doses of nicotinic acid or β-pyridylcarbinol. Groups of 5–16 rats received CCl$_4$ (8×770 mg/kg s.c. within 4 weeks). Thereafter the animals were fasted and given the pyridine derivatives (500 mg/kg twice daily) per stomach tube for 2 ½ days). Abscissa: hours after the last administration of pyridine derivatives. Ordinate: % change of FFA in comparison to placebo-treated animals decapitated after the same time interval.

Summary and Conclusion

In the normal rat large doses of nicotinic acid and β-pyridylcarbinol resp. given by stomach tube caused no substantial change in the total lipids of the liver, heart and skeletal muscle and in the expiratory $^{14}CO_2$ derived from exogenous 1-^{14}C-acetate. β-Pyridylcarbinol, however, in contrast to nicotinic acid, diminished markedly the pathological lipid accumulation as induced in the rat liver by CCl$_4$ or choline deficiency. This difference seems mainly to be due to a longer-lasting decrease of plasma FFA by β-pyridylcarbinol.

It may therefore be concluded that inhibition of lipolysis in the adipose tissue is a useful approach to the counteraction of a pathological triglyceride accumulation, at least in the rat liver – provided that the plasma FFA are decreased not only transiently (as by nicotinic acid) but rather permanently (as after β-pyridylcarbinol).

References

1. Ammon, H.P.T. und Zeller, W.: Der Einfluss von β-Pyridylcarbinol auf die durch Alkohol erzeugte Fettleber der Ratte. Arzneimittelforsch. *15*: 1369–1371 (1965).
2. Carlson, L.A.: Inhibition of the mobilization of free fatty acids from adipose tissue. Ann.N.Y.Sci. *131*: 119–142 (1965).
3. Carlson, L.A.: Consequences of inhibition of normal and excessive lipid mobilization. Progr. biochem. Pharmacol. *3:* 151–166 (Karger, Basel/New York 1967).
4. Carlson, L.A. and Nye, E.R.: Acute effects of nicotinic acid in the rat. I. Plasma and liver lipids and blood glucose. Acta med.scand. *179*: 453–461 (1966).
5. Folch, J.; Lees, M. and Stanley, G.H.S.: A simple method for the isolation and purification of total lipids from animal tissues. J.biol.Chem. *226*: 497–509 (1957).
6. Gey, K.F.: Diminution of the CCl_4-induced fatty liver by β-pyridylcarbinol. Progr. biochem. Pharmacol. *3:* 445–449 (Karger, Basel/New York 1967).
7. Lorch, E. and Gey, K.F.: Photometric titration of free fatty acids with the Technicon Autoanalyzer. Analyt. Biochem. *16:* 244–252 (1966).
8. Crastes de Paulet, A.; Barjon, P. and Descomp, B.: Effects of β-pyridylcarbinol on plasma free fatty acids in normal men. Progr. biochem. Pharmacol. *2:* 256–269 (Karger, Basel/New York 1967).
9. Schön, H.; Zeller, W. und Henning, N.: Untersuchungen zur Behandlung der endogenen Hyperlipoproteidämie mit β-Pyridylcarbinol. Klin.Wschr. *41*: 1108–1109 (1963).
10. Zöllner, N. and Gudenzi, Malda.: Treatment of hypercholesterolemia with beta-pyridylcarbinol. Progr. biochem. Pharmacol. *2:* 406–411 (Karger, Basel/New York 1967).

Authors' address: Dr. K. F. Gey and Dr. E. Lorch, Department of Experimental Medicine, F. Hoffmann-La Roche & Co. Ltd., *4002 Basel* (Switzerland).

Blindness Caused by Atherosclerosis of the Choroid and Retina Treatment with Enzymes of Prostatic Origin

J.J. Lijó-Pavía, Buenos Aires, Arg.*

Atherosclerosis is the result of a biochemical disturbance superimposed on an anatomical substrate. It is not purely an alteration of arteries but rather the result of a close interplay of the organic and the functional phase, with the latter initiating the morbid anatomy.

If the concept that atherosclerosis is primarily a metabolic disorder is correct, the process should be amiable to control. The prerequisite is knowledge of the exogenous and endogenous causative blood factors. On that basis, attempts should be made to modify these factors, with particular stress on solubility.

Bianchi divides the morphologic picture into five phases. (1) Discreet and generalized thickening of the 'intima' caused by 'lipid' loading and swelling of endothelial cells. (2) Obliteration of cellular borders, fusion of cytoplasm, nuclear pyknosis and production of mucoids. (3) Disappearance of the cell membrane. (4) Proliferation of subendothelial fibroblasts and degeneration. (5) Accumulated 'lipids' are surrounded by fibrous proliferation and hyalinization leading to sclerotization.

Experimental *in vivo* study correlating derangement of biochemical blood constituents with vascular alterations is extremely difficult. However, lesions of the eye lend themselves to such study [Lijó-Pavía 1953] since gradual increase in serum polysaccharides, cholesterol esters and triglycerides is paralleled by *fibrosis* and progressive structural disorganization of the wall of the retinal vessel.

Geer *et al.* describe 'lipids' in arterial smooth muscle cells in the first stage of atherosclerosis. Takahashi demonstrates mucoid degeneration and an increase in polysaccharides and mucopolysaccharides in the vessel wall. Electron microscopic studies reveal

---

* Chairman of the Argentine Association of Atherosclerosis.

moderate thickening of the elastica, increase in arterial ground substance and slight thickening of the adventitia. At the same stage there exists humoral changes evidenced by enzymatic variations of 'lipids', proteins, carbohydrates, etc.

Certainly constitutional, hereditary, physicomechanical and nutritional factors, endocrinopathies, climate and psychic behavior play a role in atherogenesis [RINTELEN]. With regard to psychic stress one must consider it as a 'sine qua non' condition since tension causes not only metabolic derangement initiating hyperlipaemia but also causes terminal thrombosis [BAGUENA].

The proliferative changes involve the connective tissue—elastic and collagen fibers, endothelial cells, monocytes, histiocytes, granulocytes, plasmacytes and fibroblasts, water, enzymes, proteins, and polysaccharides of the ground substance [LINNER and TENGROTH]. The latter form a gel in which *hyaluronic acid* [MAYER and PALMER, 1930] *predominates* and which also contains chondroitin-sulfuric acid and mucoitin-sulfonic acid. These trigger the proliferative reaction leading to replacement of atrophied muscle fibers by *fibrous tissue* and resulting in vascular obstruction.

Collagen fibers are part of the plastic matrix which undergoes mucoid degeneration proven by metachromasia. Elastic fibers lose their tinctorial characteristics and disappear. As the end result, the obliterated arterioles undergo hyaline degeneration. This process has been observed in renal, cerebral and retinal arterioles [BAILLIART]. In the fundus of the eye this process gives rise to an *alarm signal* [LIJÓ-PAVÍA, 1956].

Retinal Atherosclerosis

Atherosclerosis of the retinal vessels interferes with the nutrition of the retina and leads to secondary atrophy of the optic nerve [TADROS]. The retina is an extension of the central nervous system. Structurally it is essentially nervous tissue. It derives its blood supply from one of the principal cerebral arteries.

The retina is instantaneously vulnerable to sudden reduction of the blood pressure. All blood entering the eyes has to overcome an intraocular pressure of about 20 mm. This is documented by the loss of vision during or after an episode of rapid general hemorrhage, though it is impossible to verify ocular vascular occlusion at necropsy. The demand of oxygen of the retina is similar to that of the cerebral cortex. A blood arrest lasting from 4 to 5 minutes has the same consequence as a cardiac arrest. May we also stress that the retina, similar to cerebral tissue, possesses a notable thromboplastic activity and hyperactivity of hemocoagulant factors. This hemostatic balance explains the trend toward organization of hemorrhages and the frequency of arteriolar and venular thrombosis.

The circulation, though controlled by nerves, depends ultimately on the pressure gradient in the ventricle and on the distensibility and elasticity of the vessels. Any loss of reactibility will change the conditions of the distal parts as the ophthalmic artery and its branches.

Fibrous replacement is a common cause for loss of reactibility of arterioles. It diminishes the contractibility of the vessel wall since cellular elements are replaced by inert collagen. The *fibrosis* produces diffuse narrowing of retinal arterioles and stretching in the vessel's lines.

Treatment

Two factors provide the basis for therapeutic considerations: the composition of the circulating blood, and the state of the small retinal vessels.

It is necessary to combat the *fibrous replacement* of the vessel wall and of the stroma which surrounds the arterioles and capillaries. Furthermore, the hyaline constriction of arterioles and precapillaries leading to obstructive sclerosis must be counteracted [ASHTON].

The concept of cellular pathology has pretty well run its course. We have to consider most events on the subcellular level and we could speak of mitochondrial pathology. Search for therapeutic measures has to be based on the study of cellular synthesis. Each tissue has its own, more or less specific, cellular enzyme constellation [ZÖLLNER]. In cells, the concentration of enzymes is many thousandfold greater than in serum. For example, the enzyme concentration gradient in the prostate is between 1:200,000 and 1:1,000,000. This may be held proof that cell membranes are impervious to molecules the size of enzymes. Organ extracts may act as biologic catalysts which accelerate synthesis block the desintegration of functional groups of protein molecules and promote multiple chemical reactions in a fraction of a second.

We initiated experiments with *prostatic hyaluronidase-acid phosphatase* at the Veterinary Faculty of the University of Buenos Aires in 1953, using dogs who had experimentally induced atherosclerosis of one year's duration. We evaluated our results by study of the eye fundi by ophthalmidynamometry. The maximum dose of our enzyme preparation used was 1 ml per kg (1 ml of injected product corresponding to 1 gm of prostate extract).

On the basis of animal experiments, we applied in 1954 this treatment to human beings using the intravenous, carotid [BRAGE], intrarachidean [DUARTE] and our own original retrobulbar [LIJÓ-PAVÍA] way of enzyme administration.

The bibliography of hyaluronidase (this enzyme was first defined in testes and semen by GLEAN and DUTHIE, 1943) exceeds the limit of this report. The same is also true for acid phosphatase. A few selected references should be cited: The fibrinolytic components in the human prostate have been defined by RASMUSSEN and ALBRECHTSEN; changes in fibrinolysis have been reported after administration of acid phosphatase by KNIPPER, YING *et al.* and by RIGOTTI and MAIRANO.

The enzymatic complex *hyaluronidase-acid phosphatase* causes degradation of hyaluronic acid and chondroitin sulfuric acid. The two acids co-exist in many tissues, namely in the collagen. Depolimerization of the mucopolysaccharides causes impairment of permeability, inhibition of cellular growth; fibroblastic proliferation forms a barrier hindering the permeation of tissue and tissue oxygenation. Addition of testicular and ovarian extracts containing lipoids with hormonal action enhances the enzymatic action of our preparation through humoral and neural excitation. Multiple common characteristics of androgens and estrogens contribute to maintenance of a hormonal balance.

Casuistics

Very good results were obtained with this enzymatic complex preparation over a ten years' observation period. The improvement obtained with a large dose during the first year was satisfactorily maintained with a reduced dose applied every two weeks.

Ten men and ten women, all with normal blood pressure, all with retinal and choroid atherosclerosis, referred by cardiologists, internists and psychiatrists were treated. All twenty patients had an intense and regular concentric reduction of the field of vision. All patients improved after one year of treatment (Table I).

Table I

Number	Age	Year	Vision RE	LE	After a year RE	LE
Women						
1	43	1960	0.3	light	0.4	0.1
2	55	1962	0.1	0.2	0.2	0.5
3	36	1958	0.2	0.5	0.6	0.7
4	63	1960	0.2	0.3	0.4	0.6
5	57	1957	0.1	0.5	0.2	0.5
6	58	1959	0.6	0.4	0.6	0.5
7	61	1961	0.2	0.4	0.4	0.5
8	57	1954	0.5	0.4	0.8	0.7
9	59	1962	0.2	0.4	0.5	0.4
10	70	1962	H*	0.3	0.1	0.5
Men						
11	64	1960	0.4	H*	0.6	0.1
12	39	1959	0.1	0.2	0.3	0.5
13	53	1962	H*	0.3	0.1	0.5
14	66	1963	0.4	0.3	0.4	0.3
15	34	1962	H*	0.1	0.1	0.3
16	61	1960	0.4	0.1	0.5	0.3
17	55	1959	0.1	H*	0.1	0.1
18	49	1960	0.1	0.3	0.3	0.4
19	59	1959	H*	0.4	0.2	0.6
20	53	1955	0.2	0.6	0.5	0.7

H* = intraocular hemorrhage and vision bulks

Increase of vision in the eyes of 20 patients (40 eyes)

Amount of the vision: 0.1 = 14 eyes
0.2 = 11 eyes
0.3 = 7 eyes
0.4 = 1 eye
0.6 = 1 eye

Without change: 6 eyes

Summary

The improvement obtained in our patients is reminiscent of the results obtained in other morbid processes which have *fibrosis* as a common denominator. The effect is attributed to the combination of two enzymes. Hyaluronidase is used by ZIMMERMAN and OLIVERA who treated myocardial edema with intravenous doses of up to 100,000 IU and obtained normalization of the electrocardiographic patterns within a few hours. This is ascribed to reduction of edema. The fact that acid phosphatase is present in considerable quantity in seven of the ten layers of the retina gives rise to speculation of the mode of therapeutic action of acid phosphatase administration.

We hope that we have contributed in a small way to a positive solution of an important problem.

Author's address: J. J. Lijó-Pavía, M. D., Asociación Argentina de la Aterosclerosis, Av. Quintana 104-2° T. E. 44-2927, *Buenos Aires* (Argentina).

Institute for Cardiovascular Diseases, Tokyo Medical and Dental University, Medical School, 1-Yushima, Bunkyo, Tokyo

Treatment of Atherosclerosis with Pyridinolcarbamate

T. Shimamoto, T. Atsumi, F. Numano and T. Fujita, Tokyo

Pyridinolcarbamate is a potent substance capable of preventing and abolishing edema of arterial wall [1]. The substance has been tested in prevention of experimental atherosclerosis of cholesterol-fed rabbits and its preventive effect without lowering of blood cholesterol level has been repetitiously confirmed by the authors [4].

In this report the treatment with pyridinolcarbamate [5] of established atherosclerosis of cholesterol-fed rabbits and also the treatment with this compound of patients suffering from arteriosclerosis obliterans have been subjected to the publication.

Part I

Treatment of Atherosclerosis of Cholesterol-Fed Rabbits

This report is based on the observation of 65 rabbits received 10 mg and 20 mg per kg of pyridinolcarbamate regimen and placebo regimen.

Material and Method

Sixty-five albino rabbits from Takeda firm (6 females and 59 males, 8 months old) with starting body weight of 2.4±0.5 kg were utilized. In order to produce atherosclerosis all animals were kept on pellets produced by Oriental East Co. containing 1% of cholesterol for 15 weeks. These 65 atheromatous rabbits were divided into the placebo control group and the treated group. The treated group received daily pyridinolcarbamate in a dose of 10 mg/kg or 20 mg/kg placed in a gelatin capsule by mouth and the placebo control group received potato starch placed in a gelatin capsule.

Each 4 to 6 animals of the placebo control group and of 10 mg/kg pyridinolcarbamate group were sacrificed on the 4th and 7th day of the treatment respectively under the continuous cholesterol feeding.

In the remaining animals the cholesterol pellets were withdrawn after the initiation of the treatment and they were kept on RC-5 pellets which contain no cholesterol. After 6, 10, and 20 weeks of the treatment with 10 mg/kg or 20 mg/kg of pyridinolcarbamate or with placebo each 2 to 6 animal of pyridinolcarbamate regimens and each 3 to 9 animals of placebo regimens respectively were sacrificed.

In order to see the morphological changes including edematous changes of arterial walls, the rapid sampling and fixation of aortic specimen in ice-cold glutaraldehyde solution (5%) or in 1% osmic tetraoxide were used and the procedure was finished within 30 s after giving a stunning blow on the head of animals. For the embedding paraffin and also celloidin were used and hematoxylin-eosin stain, Elastica Van Gieson stain, Heidenhain's Azan variant, Sudan III and Sudan IV stain were used. For the identification of myoblast an electron microscopic technique was used.

The total cholesterol content of aortic wall was measured by gaschromatography. Pyridinolcarbamate (Anginin) was supplied by Banyu Co., Ltd.

Results

(1) Short term treatment. Aortas of the animals of the placebo control groups exhibited grossly a severe atheromatous changes occupying over 50% of the whole internal surface.

The histological findings. In the placebo control group a typical feature of well known atheroma of cholesterol-fed rabbits was observed. The lesions exhibited some fibrous tendency in the superficial portion accompanied by marked edematous features as shown in Fig. 1. The aortas of animals treated with pyridinolcarbamate for 4 to 7 days under the cholesterol diet, exhibited grossly almost the same features as in the placebo control group. However, the histological analysis revealed the disappearance of the edematous features of atheroma and a definite flattening of foam cells. An appearance of myoblasts, fibroblasts, and of fibrocytes inside the atheromatous mass (Fig. 1) and an extracellular deposition of collagenous substances was already found in some atheromatous lesions though slight. In animals received the compound for one week a definite deposition of collagen fibres was found in many atheromatous lesions, in which young muscle cells with their extremely fine and short fibres and the elastic fibres appeared already in the middle part of atheromatous lesions.

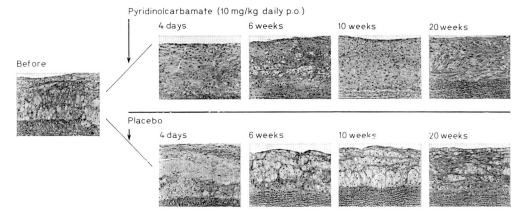

Fig. 1.

(2) Long term treatment. In animals of the placebo control group kept on cholesterol-free diet and placebo for 6 to 20 weeks fibrous changes were found inside the atheroma and smooth muscle fibres appeared on the surface area of atheroma. However the amount of the fibrous and muscle fibres was small and the main parts of the atheroma were occupied either by atheromatous mass showing fatty degeneration and necrotic foci or by an abundant accumulation of foam cells and the Sudan stain exhibited also an abundant accumulation of sudanophilic substance occupying the whole atheromatous lesions.

In animals received the compound for 6 to 20 weeks, the most striking histological finding was the appearance of abundant mature smooth muscle fibres inside the atheroma, invading into the atheromatous mass occupying the hyalinous foci, and dividing the hyalinous substances and the mass of degenerated foam cells into smaller portions as shown in Fig. 1. The appearance of smooth muscle fibres was also accompanied by elastic fibres. In animals received the substance for 10 to 20 weeks the foam cells exhibited further absorption and almost disappeared in the majority of the lesions except the atheroma locating on the point of bifurcation although showing definite degenerative features (Fig. 1). The sudanophilic substance in the majority of the atheromatous lesions exhibited a definite reduction especially in the areas adjacent to medial layers.

In animals received the compound in a dose of 10 mg/kg for 20 weeks or the compound in a dose of 20 mg/kg for 10 weeks the regenerated smooth muscle fibres seem to occupy almost fully the lesions in almost over 50% of atheromatous lesions and the elastic fibres were also found among the smooth muscle fibres. In such lesions no foam cell was seen and the Sudan positive substance became minimal and scattered as small particles. In the remaining lesions a few degenerated foam cells were seen divided into small parts surrounded entirely by regenerated smooth muscle fibres. A sudanophilic substance was still found in these lesions, however, it exhibited a striking further reduction as compared with that of the treated animals for shorter period. There were no necrotic or hyalinous foci.

At the beginning of this experiment the mean serum cholesterol level of test animals was 1651.3 ± 68.5 mg%. At 2, 4, 6, 10 and 15 weeks after the withdrawal of cholesterol diet it was still high amounting 1101, 651, 428, 191 and 113 mg% respectively. There was no significant difference between the placebo control and treated group.

The total cholesterol content of aortic wall of the rabbits received the compound exhibited a significant decrease as compared with the placebo control group as shown in Fig. 2.

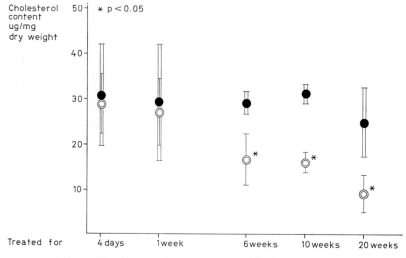

Fig. 2. ● Placebo control group. ◉ Pyridinolcarbamate group.

Discussion

The prompt disappearance of edematous feature from the atheromatous lesions and the early appearance of repair process in the atheroma were the most striking evidence in this treatment of experimental atherosclerosis with pyridinolcarbamate. It is well known that the repair process of common aseptic wound is fast and used to be finished within 8 to 10 days. In the repair process there are two processes; the production of fibrous material filling the dead spaces and the regeneration of local tissues. In the repair of atheromatous lesions the rapid and striking regeneration of smooth muscle fibres replacing the atheromatous mass is the most important finding in the treatment of atheroma with pyridinolcarbamate.

Since VIRCHOW [6] the appearance of edema, small round cells and of polymorphonuclear leucocytes has been described by many morphologists [7] in atheromatous lesions and RÖSSLE [8] proposed 'the serous inflammation of artery' as a causative mechanism involved in atherogenesis. The authors [3] found that the traumatic stress given to test animals or the one shot treatment of animals with so-called atherogenic substances such as cholesterol or adrenaline or high molecular weight substances induces the edematous arterial changes as an immediate reaction and the pretreatment of animals with bradykinin-forming enzyme inhibitor; trasylol or with bradykinin-antagonists such as pyridinolcarbamate or cyproheptadine or acetylsalicylate prevents the edematous reaction induced by these procedures. These evidences have indicated that kinin-induced veno-constriction [9] and the increase in vascular permeability in the vasa vasorum may be the causative factor in the production of exudation in the arterial wall. Namely this compound may inhibit the increase in permeability and the excessive veno-constriction of vasa vasorum by bradykinin or by its active homologues which produce the exudation of arterial wall leading to the production, maintenance and progression of atheromatous changes and prevents the healing of atheromatous lesions.

The antagonistic effect of pyridinolcarbamate [4] against the increased permeability by bradykinin and other active agents may prevent the congestion and exudation and establish a smooth circulation in vasa vasorum which contributes to the healing process of the hitherto incurable atheroma by regeneration of local structural elements even under the continuous high blood cholesterol level. Needless to say the above mentioned explanation is merely a tentative hypothesis.

Summary

Pyridinolcarbamate, a bradykinin-antagonist, exhibited definitely a curative effect in atheroma of cholesterol-fed rabbits. The appearance of its curative effect was prompt and the leading process was the replacement of atheromatous mass with regenerated smooth muscle fibres which begins early within the first week of the treatment and progresses gradually in the course of over 20 weeks. The 20 mg/kg pyridinolcarbamate regimen seems much more effective as compared with the 10 mg/kg pyridinolcarbamate regimen.

Part II

Treatment of Human Atherosclerosis

The first therapeutic attempt with pyridinolcarbamate in human atherosclerosis has been successfully applied in patients suffering from atherosclerotic diseases of extremity, renal, coronary and creebral arteries [5, 10, 11].

This report is limited to the results obtained in 67 patients suffering from atherosclerotic diseases of extremities, i.e. from arterisclerosis obliterans, in which the spontaneous improvement is rare and the morbid condition is always progressive and the improvement of the atheromatous lesions reflects directly to the easily observable increase in the peripheral circulation; namely to the reappearance of the absent arterial pulsation or to the increase in the amplitude of peripheral arterial pulsation, or a disappearance of cyanosis or an increase in skin temperature etc.

Material and Method

67 cases of 61 male and 6 female patients with age of 24–72 years (46.6 ± 1.8 years) were hospitalized in our university hospital during 1965 and have been subjecte to the

present clinical trial with pyridinolcarbamate. For the diagnosis of this morbid condition the criteria of McPherson et al. (1963) [12] was used.

The average case history amounted 3.7 ± 0.7 years and the number of patients with case history less than one year amounted only 13 and the majority of cases, i.e. the remaining 54 cases, had longer histories over one to 20 years. In order to accumulate the cases the authors attempted a type of advertisement of this therapeutic trial on a weekly, so that 34 cases participated to this trial had had a long history with experiences of various surgical and medical treatments.

(1) After the preliminary trial [10, 11] 34 patients who visited our hospital during October of 1965 were divided into two groups; the placebo control group and pyridinolcarbamate group. The placebo control group was composed by 16 patients 15 male and 1 female patients hospitalized on odd days and their ages were 24 to 68 years (average 49.8 ± 3.7 years) and the duration of case history amounted from 1 month to 20 years (3.6 ± 1.2 years).

The pyridinolcarbamate group was composed of patients (18 male patients hospitalized on even day and their ages were 39 to 72 years (average 51.7 ± 2.4 years) and the duration of case history amounted from 1 month to 20 years (4.3 ± 1.2 years).

Under the double blind technique these 34 patients were treated either with placebo or with pyridinolcarbamate in a daily dose of 1 g, for 2 weeks.

For the evaluation of clinical improvement the arteriography, the plethysmography, the electrical registration of arterial pulsation and the measurement of the amplitude of arterial pulsation under the controlled pressure (20 mm Hg) given on the affected arteries through the skin and after keeping patients at least over 30 min under the constant room temperature of $25°$C, the measurement of skin temperature with electrothermometer under the constant room temperature of $25°$C, and the measurement of claudication time under the standard rate of 120 steps per minute were performed before and in the course after the initiation of pyridinolcarbamate administration beside the routine clinical and laboratory examination of the university hospital. The clinical findings were recorded every day by attending doctors.

(2) The retrospective comparison between the clinical efficacy of various treatment which had been performed on all 34 patients aged 24 to 66 years (41.6 ± 1.8 years) before the present trial with pyridinolcarbamate and the present pyridinolcarbamate treatment. The analysis depended on the inquiries of patients and of their responsible physicians and it was inevitable that the change of subjective symptoms was the main subject of inquiries. The forgoing treatments to be compared with the present treatment were sympathectomy in 13 cases, mesoinositol hexanicotinate administration in 14 cases, kallikrein administration in 14 cases, 2-Benzyl imidazoline hydrochloride in 11 cases and Procain blocking of sympathetic ganglions in 2 cases.

(3) All 67 patients, including patients received placebo during the above mentioned double blind trail, were given 1g of pyridinolcarbamate daily for 1 month to one and half year and the clinical course was subjected to the present observation.

The whole patients were classified into three categories I, II, and III according to their symptoms. 17 patients (4 female and 13 male; age 24 to 66 years, average 52.3 years) with an absolutely absent or minimal pedal arterial pulsation not accompanied by intermittent claudication or by skin changes were classified into category I. 22 patients (one female and 21 male, age 24 to 72 years, average 53.1 years) with intermittent claudication without skin changes were classified in category II. 28 patients (one female 27 male, age 31 to 64 years average 41.1 years) with skin changes such as ulcer or cyanosis or red coloration of affected feet accompanied either by absent peripheral arterial pulsations or by intermittent claudication were classified in category III.

Pyridinolcarbamate was supplied by Banyu Co. Ltd. as a tablet form containing 250 mg of pyridinolcarbamate in one tablet of lactose. The placebo is also similar tablet with lactose. The patients received 2 tablets after each breakfast and supper.

Results

(1) Double blind test in 34 cases (Fig. 3).

None of patients received placebo exhibited an improvement in any of clinical and laboratory examination during 2 weeks of hospitalization. Namely all 16 cases had an impairment or absent arterial pulsation in the dorsal pedal artery, 11 had a definite short claudication time amounting 2 to 8 min, one had pain in affected feet and 2 had ulcer and 4 had rubor and one had cyanosed toes and none exhibited an improvement.

On the other hand 8 of all 18 patients exhibited an unmistakable favorable response during 2 weeks of hospitalization with the administration of pyridinolcarbamate as detailed in Fig. 3 ($P < 0.05$). In these patients no change was noted in cyanosed toes in one case and in ulcer of affected feet in all 3 cases, however, an appearance or a definite increase in the amplitude of the previously absent or minimal arterial pulsation of dorsal pedal artery was noted in 3 of all 18 cases and a definite prolongation in claudica-

Clinical Improvement in Arteriosclerosis Obliterans by Pyridinolcarbamate Treatment

Placebo Control Group (16 Cases)

Age + Sex	History	Category	Rubor Cyanosis Ulcer	Cholesterol Level (mg/dl)	Symptoms 1 — 2 Weeks
56 M	1Y7M	II		165	
39 M	1Y10M	III	Rubor		(Impr.)
47 M	3Y2M	III	Rubor	232	(Impr.)
60 M	1Y6M	I		236	
64 M	1M	III	Ulcer	174	(Impr.)
24 M	6Y	I		214	
63 M	1Y	II		183	
27 M	4Y	I		254	(Impr.)
72 M	1Y10M	II		231	(Impr.)
37 M	20Y	III	Rubor	164	(Impr.)
33 M	5Y	III	Rubor Cyanosis Ulcer	172	(Impr.)
68 M	2M	I		208	
49 F	1M	I			
40 M	6Y	II		156	
61 M	4Y	II		276	
57 M	1Y	II		190	(Impr.)
49.8±3.7Y	3.6±1.2Y	I : 5 II : 6 III : 5		203.9±10.0	

Fig. 3.

Pyridinolcarbamate Treated Group (18 Cases)

Age + Sex	History	Category	Rubor Cyanosis Ulcer	Cholesterol Level (mg %)	Symptoms 1 2 Weeks
62 M	5Y	I		208	
40 M	1Y	I		206	
39 M	4Y	III	Rubor Ulcer	206	
39 M	6M	III	Rubor	220	
66 M	1Y	I		246	
58 M	1M	II		246	
65 M	1M	I	Cyanosis	200	
49 M	12Y	III	Rubor Ulcer	194	
47 M	3Y2M	III	Rubor	232	
46 M	1Y1M	II		219	
60 M	3Y	II		184	
54 M	2Y	III	Rubor	192	
44 M	1Y6M	II			
72 M	20Y	II		231	
40 M	7Y	III	Rubor	203	(Impr.)
54 M	5Y	II		228	
47 M	8Y	III	Rubor		
49 M	3Y4M	III	Rubor Ulcer	151	(Impr.)
51.7±2.4Y	4.3±1.2Y	I :4 II :6 III: 8		210.4±6.1	

Fig. 3. (continued)   (Impr.): Improvement after the Long Term Treatment

	Symptoms	Sympt. Improved
Impaired Arterial Pulsation		
Claudication		
Sensory Disturbances		

tion time was noted in 5 of all 11 cases and a definite decrease or disappearance of pain of affected feet was noted in 2 of all 3 cases and a complete disappearance of numbess of the affected foot was noted in one of all one case and a definite reduction in red coloration was noted in 2 of all 8 cases.

(2) The results obtained in the comparison done between the treatment of this drug and of other treatment performed previously (Fig. 4).

The improvement ratio obtained by sympathectomy was 38.5% and that by the other treatments with drugs such as Kallikrein, meso-inositol hexanicotinate, benzyl-immidazoline and isoxsuprine were less than 10% and in total 34 cases, improvement was seen in 13.1%. On the countrary, pyridinolcarbamate exhibited signi-

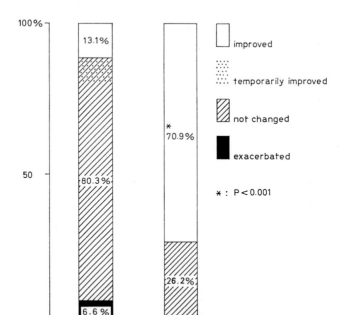

Fig. 4.

ficantly larger improvement ratio amounting 70.9% was more effective and clinical improvement was shown. Namely 24 of all 34 cases exhibied a definite clinical improvement.

(3) The results of pyridinolcarbamate treatment in 67 cases (Fig. 5).

(a) In patients of category I: Among all 17 patients of category I, 5 cases remained unchanged during 16 to 25 weeks of pyridinolcarbamate treatment, while 12 cases exhibited a definite clinical improvement starting at the first to 12th week of the treatment during 4 to 36 weeks of the whole observation periods under pyridinolcarbamate treatment. The return of previously absent arterial pulsation took place in 1 of all 4 cases and a definite increase over at least twice in the amplitude of pedal arterial pulsation was found in 7 of all 13 patients. The paraesthesia disappeared in 6 of all 6 patients already in the first 5 weeks of pyridinolcarbamate treatment.

(b) In patients of category II: Among all 22 patients of category II, 9 patients remained unchanged during 4 to 20 weeks of treat-

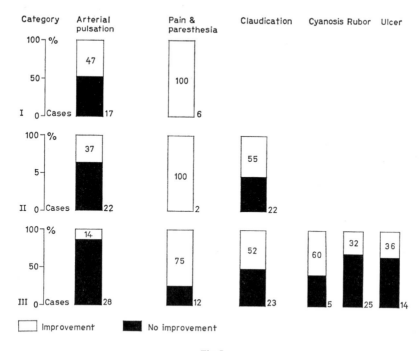

Fig. 5.

ment, while 13 cases showed a definite clinical improvement starting at the first to 5th week during 4 to 54 weeks of the whole periods under the treatment. The reappearance of previous absent arterial pulsation took place in 2 cases of all 8 cases and a definite increase in the amplitude of arterial pulsation was found in 6 cases of all 14 patients. Intermittent claudication was improved in 12 cases out of total 22 cases starting at the first to 5th week. The sensory disturbances disappeared in two cases within two weeks.

(c) In patients of category III: Among all 28 cases of category III, 9 patients remained unchanged during 9 to 24 weeks of treatment, while 19 patients showed a definite clinical improvement starting at the first to 12th week during 5 to 42 weeks of the whole periods of the treatment. The reappearance of absent arterial pulsation took place in 2 cases of all 14 cases and a definite increase in the amplitude of arterial pulsation was found in 2 cases of all 14 cases. Cyanosis disappeared within one week in three cases of all 5 patients. Red discoloration in the foot was grown dull within 4 weeks in 8 cases of 25 patients. And ulcer was cured in 5 cases of

14 patients. Intermittent claudication was improved in 12 cases of all 23 cases and sensory disturbances was improved definitively in 9 cases of all 12 patients.

The number of patients, whose absent arterial pulsations were reappeared, was 1 out of 4, 2 out of 8 and 2 out of 14 in patients of category I, II and III respectively and the number of patients, whose decreased arterial pulsations were definitively augmented, was 7 out of 13, 6 out of 14 and 2 out of 14 respectively. The improvement ratio is higher in patients of category I than in patients of category III ($P < 0.02$). Namely the reappearance and augmentation of arterial pulsation was revealed in 47% (8/17), 37% (8/22) and 14% (4/28) in patients of category I, II and III respectively. There was a significant difference in the improvement ratio between the patients of category I and the patients of category III.

Sensory disturbances was improved in 6/6, 2/2 and 9/12 in patients of the patients of category I, II and III respectively and the improvement of claudication was showed in 12/22 (55%) in patients of category II and 12/23 (52%) in those of category III. But these improvement ratio was not statistically significant.

Discussion

The prompt improvement of local circulatory disturbance such as the disappearance of local cyanosis or pain of affected feet taking place on the 3rd to 20th day corresponded with the rapid disappearance of edema and the appearance of repair process in the experimental atherosclerosis by pyridinolcarbamate shown in the previous chapter. It also corresponds with the prompt disappearance or definite amelioration of anginal attack by pyridinolcarbamate treatment taking place often on the 3rd to 5th day of the treatment in patients suffering from coronary sclerosis by the authors [5].

It has been known to be uncommon for the collateral circulation to develop to the extent that absent pedal pulsations again will become detectable. Morris [13] reported the return of absent pedal pulsation in her 12 cases out of all 300 cases suffering from chronic obliterative arterial disease of the lower extremities during 10 years of her observation. However the shortest time interval for the return of a readily perceptible pedal pulse was 6 months and the longest was 51 months.

However, it was noted that the return took place significantly earlier and more frequent in our patients treated with pyridinolcarbamate than Morris's [13] cases. Namely the return of absent arterial pulsation took place within 6 months of the treatment in 7 of all 27 cases and it took place within the first 1, 2, 3, 10, 13, 19 and 21 weeks of the treatment respectively. The definite augmentation of affected arterial pulsation seen in 11 of all 67 cases took place also at the short time intervals of 2 to 5 weeks of the pyridinolcarbamate treatment in the majority of cases. Such characteristic features of the course of clinical improvement by pyridinolcarbamate and especially the arteriographical finding seen also to correspond with the pathological course of improvement of experimental atherosclerosis shown in the previous paper.

Summary

Pyridinolcarbamate treatment with a daily dose of one gm exhibited a definite curative effect in 67 patients suffering from arteriosclerosis obliterans.

References

1. Shimamoto, T. and Sunaga, T.: Edematous arterial reaction by adrenaline and cholesterol and its prevention by MAO inhibitor observed by electron microscopic technique. Jap.Heart J. *3:* 581–601 (1962).
2. Shimamoto, T.: The relationship of edematous reaction in arteries to atherosclerosis and thrombosis. J.Atheroscler.Res. *3:* 87–102 (1963).
3. Shimamoto, T.; Maezawa, H.; Yamazaki, H.; Ishioka, T.; Sunaga, T. and Fujita, T.: Edematous arterial reaction and its relationship to atherosclerosis and thrombosis. Meth.achievm.exp.Path. *1:* 337–354 (Karger, Basel/New York 1966).
4. Shimamoto, T.; Numano, F. and Fujita, T.: Atherosclerosis-inhibiting effect of an antibradykinin agent, pyridinolcarbamate. Amer.Heart J. *71:* 216–227 (1966).
5. Shimamoto, T.; Maezawa, H.; Yamazaki, H.; Atsumi, T.; Fujita, T.; Ishioka, T. and Sunaga, T.: Pyridinolcarbamate, a bradykinin antagonist in veins. Amer. Heart J. *71:* 297–312 (1966).
6. Virchow, R.: Die Zellular-Pathologie in ihrer Begründung auf physiologische und pathologische Gewebelehre (Hirschwald, Berlin 1871).
7. McGill, H.C.; Geer, J.C. and Strong, J.P.: Natural history of human atherosclerotic lesion; in Sander and Bourne's atherosclerosis and its origin. pp.39–65 (Academic Press, New York/London 1963).
8. Rössle, R.: Über wenig beachtete Formen der Entzündung von Parenchymen und ihre Beziehung zu Organsklerosen. Verh.dtsch.ges.Path. *27:* 152–164 (1943).

9. Rowley, D.A.: Venous constriction as the cause of increased vascular permeability produced by 5-hydroxytryptamine, histamine, bradykinin and 48/80 in the rat. J. exp. Path. 45: 56–67 (1964).
10. Shimamoto, T. and Atsumi, T.: Pyridinolcarbamate; bradykinin antagonist, in patients suffering from arteriosclerosis obliterans and thromboangiitis obliterans (Buerger's disease), a preliminary report on the treatment of human atherosclerosis. Jap. Heart J. 6: 407–415 (1965).
11. Shimamoto, T.; Numano, F.; Fujita, T.; Ishioka, T. and Atsumi, T.: Treatment of experimental and human atherosclerosis with pyridinolcarbamate. A preliminary report. Asian med. J. 8: 825–838 (1965).
12. McPherson, J.R.; Jürgens, J.L. and Griford, R.W.: Thromboangiitis obliterans and arteriosclerosis obliterans, clinical and prognostic differences. Ann. intern. Med. 59: 288–296 (1963).
13. Morris, L.E.: The return of absent pedal pulsations without surgical intervention. Angiology, Baltimore 17: 104–108 (1966).

Authors' address: Dr. T. Shimamoto, Dr. T. Atsumi, Dr. F. Numano and Dr. T. Fujita, Institute for Cardiovascular Diseases, Tokyo Medical and Dental University, Medical School, 1-Yushima, Bunkyo-ku, *Tokyo* (Japan).

Imperial Chemical Industries Limited, Pharmaceuticals Division, Macclesfield, and
Royal Infirmary, Edinburgh

Role of the Endocrine System in the Regulation of Plasma Lipids and Fibrinogen, with Particular Reference to the Effects of 'Atromid'-S

J. M. Thorp, R. C. Cotton, Macclesfield, and
M. F. Oliver, Edinburgh

Rather than implicate one of the components of blood as being the 'prime mover' in the evolution of atherosclerosis or thrombosis, it may be potentially more useful to consider the nature of their inter-related responses to various stimuli or treatments. Such an analysis may serve to indicate the requirements for comprehensive, rather than selective, methods of preventing, or reversing, blood and vascular changes associated with the progression of arterial disease. However, the comparison of data derived from different species and subjects, under the influence of stimuli of varying intensity and duration, can hardly be undertaken without some degree of selectivity. It is the necessarily limited object of this paper to outline some interactions involving the thyroid hormones, and the plasma lipoproteins, free fatty acids (FFA), and fibrinogen. These are correlated with previous observations of the effects of 'Atromid' * and 'Atromid'-S * [ethyl-a-(4-chlorophenoxy)-a-methylpropionate] on serum cholesterol and triglycerides [1], free fatty acids [2], plasma fibrinogen [3, 4], and on the role of the endocrine system in the action of the compound [5, 6].

It is generally recognized that a variety of endocrine-related factors (age, sex, pregnancy, thyroid or adrenal dysfunction) can lead to a 'chronic' alteration in the levels of blood lipids, FFA and fibrinogen; but that these may differ from the changes following

---

* Trade mark of Imperial Chemical Industries Limited.

an 'acute' stimulus (e.g. surgery, myocardial infarction, inflammation), which gives rise to marked changes in endocrine function. The rise in free fatty acids associated with acute myocardial infarction [7], is accompanied or succeeded by a fall in cholesterol and triglyceride [8], and by a rise in fibrinogen [9]. A similar postoperative rise in fibrinogen [10] may contribute to the increased tendency to thrombosis following these acute stimuli.

These acute and chronic changes are compared with the responses to an excess or deficiency of thyroxine, or to treatment with 'Atromid'-S, in Table I. If these generalizations are valid, they

Table I. Comparison of changes in plasma fibrinogen, serum cholesterol and triglycerides, and the serum free fatty acid (FFA): albumin ratio in response to 'acute' and 'chronic' stimuli

Condition		Change from normal of:		
	fibrinogen	cholesterol	triglyceride	FFA/albumin
'Acute' (e.g. myocardial infarction, surgery)	+	−	−	+
Increased thyroxine	+	−	±	+
Decreased thyroxine	−	+	±	−
'Chronic'	+	+	+	+
'Atromid-S' treatment	−	−	−	−

+ = increased; − = decreased; ± = variable.

suggest that increases of fibrinogen and the FFA/albumin ratio are associated with an increase of thyroxine (which may itself be secondary to e.g. an acute release of catecholamine); but that a dissociation occurs in the case of the chronic cholesterol and triglyceride changes, and in response to 'Atromid'-S.

What is the nature of this dissociation? The complex contributory factors involved in chronic changes cannot here be analysed in detail, but may include the influence of diet on adrenocortical hormones, and of these and androgens on the plasma proteins, which either directly or indirectly cause alterations in the distribution and intracellular fate of thyroxine. In the case of the female the additional cyclical effects of oestrogens and progestagens [11], and, for example, a relative deficiency of pyridoxine during pregnancy [12], may ultimately affect thyroxine equilibria.

It has been observed that 'Atromid'-S causes an increased localization of labelled thyroxine (or its metabolites) in the liver of the rat, in a manner different from that of salicylate [13]. In man, labelled thyroxine is rapidly taken up by liver, and re-excreted into the hepatic vein as a compound behaving with the properties of tetraiodothyroacetic acid (TA_4) [14]. These findings suggest that alterations in the distribution of thyroxine, or of its deaminated metabolites, between liver, plasma and extra-hepatic tissues, may account for the differing patterns of thyroxine-linked changes shown in Table I.

One factor involved in the distribution of thyroid hormones between liver and plasma is their binding to plasma proteins. We have therefore examined the effect of p-chlorophenoxyisobutyrate (CPIB, the anion arising from 'Atromid'-S *in vivo*), and related aryloxy-isobutyrates (with similar patterns of biological activity [5]), on the binding to serum proteins of L-thyroxine (LT_4) and tetraiodothyropropionic acid (TP_4, of similar distribution and biological activity to TA_4). Free hormone in the presence of binding protein was determined by a modification of the method of TRITSCH et al. [15]. This method is not affected by iodide.

Addition of an excess of CPIB (10 mol/mol albumin) *in vitro* strikingly suppressed the primary binding of LT_4 to human albumin (Fraction V) in phosphate buffer (pH 7.4) at 21°C (Figure 1). Little effect is evident on the secondary binding sites for LT_4. Similar results were obtained in the case of TP_4.

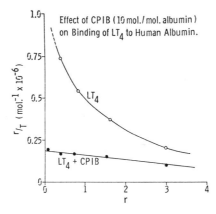

Fig. 1. Reduction of binding of LT_4 to albumin in presence of chlorophenoxyisobutyrate (CPIB).

To determine whether a displacing effect was present at the concentration of competing anion obtainable *in vivo*, rats were dosed orally with an aryloxyisobutyrate, and blood collected 3, 5½, and 24 h later. The binding of TP_4 and LT_4 to these serum samples was inhibited when compared with that to control serum. One result is illustrated in Figure 2, and in Table II it is shown that

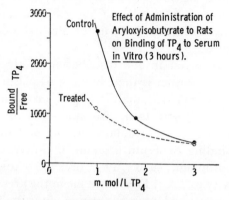

Fig. 2. Effect of aryloxyisobutyrate in serum of dosed rats on binding of TP_4 to serum proteins. Serum sample taken at 3 h after dosing (cf. Table II).

Table II. Effects of an aryloxyisobutyrate administered orally to rats on binding of LT_4 and TP_4 added *in vitro* (at 1 m mol/l) to serum collected at different intervals after dosing

Time after dosing (hours)	Concentration of compound in serum (m mol/l)	% Free LT_4	% Free TP_4	% Increase in free fraction due to treatment LT_4	% Increase in free fraction due to treatment TP_4
0	0	0.31	0.038	–	–
3	0.47	0.40	0.091	29	140
5½	0.44	0.37	0.058	19	52
24	0.24	0.33	0.042	6	10

at a particular concentration of thyroid hormone, the reduction in binding is related to the changing concentration of the compound in serum. The effects on the binding of TP_4 (which has an affinity for serum proteins approximately $10\times$ that of LT_4) are greater than those on LT_4.

Reduction of the availability of plasma protein binding sites for TA_4, occurring in the presence of aryloxyisobutyrates, would tend to reduce the rate of release of the hormone from liver. This is the same physico-chemical mechanism, occurring extracellularly, by which CPIB tends to decrease the release of FFA from adipose tissue and their uptake by albumin [16]. It is an interesting possibility that competition between FFA and thyroid hormones for plasma protein binding sites may act as a form of autoregulation of the balance between extrahepatic and hepatic metabolism.

In patients with untreated myxoedema, 'Atromid'-S is devoid of hypocholesterolaemic activity, but 'potentiates' the action of exogenous thyroxine in such patients [17]. The major component of the plasma protein bound iodine (PBI) in man is globulin-bound thyroxine. Levels of PBI are not significantly affected by 'Atromid'-S treatment [18]. The effectiveness of the compound in the rat (in which thyroxine-binding globulin is absent), and the present *in vitro* observations, therefore suggest that the major point of action of CPIB, in both man and rat, is at anionic binding sites of albumin and pre-albumin. Alterations in the amounts of these proteins, or in their affinity for thyroid hormones [19, 20], may serve to account for the differences between 'acute' and 'chronic' conditions, and the effects of 'Atromid'-S, shown in Table I.

The concerted effects of 'Atromid'-S on both the atherogenic and the thrombogenic components of blood are likely, in terms of this mechanism of action, to be evinced in conditions where these components are chronically altered. In the response to an 'acute' stimulus, the occurrence of an abnormal increase in free thyroxine [19], and a decrease in thyroxine-binding capacity of pre-albumin [20], may be sufficient to obscure the action of CPIB. Understanding of the events occurring during this 'acute' response is a prerequisite to the design of methods or treatments for the prevention or reversal of their consequences.

Summary

Analysis of the plasma lipoprotein, free fatty acid, and fibrinogen changes, in response to acute or chronic stimuli, suggests that thyroxine may play a major role in the regulation of these potentially atherogenic or thrombogenic components of blood.

Differences in the peripheral distribution and metabolism of thyroxine may arise because of changes in the anion-binding capacity of albumin or pre-albumin. It is demonstrated that the anion CPIB (chlorophenoxyisobutyrate), arising from 'Atromid'-S *in vivo*, reduces the binding affinity of L-thyroxine and tetraiodothyropropionic acid (TP_4) for serum protein. Preferential localization of TP_4 or TA_4 in the liver, with a reduction of FFA transport, may account for the observed effects of 'Atromid'-S.

References

1. OLIVER, M.F.: Further observations on the effects of 'Atromid' and of ethylchlorophenoxyisobutyrate on serum lipid levels. J. Atheroscl. Res. *3:* 427–444 (1963).
2. MACMILLAN, D.C.; OLIVER, M.F.; SIMPSON, J.D. and TOTHILL, P.: Effect of ethylchlorophenoxyisobutyrate on weight, plasma volume, total body water, and free fatty acids. Lancet *ii:* 924–926 (1965).
3. COTTON, R.C.; WADE, E.G. and SPILLER, G.W.: The effect of 'Atromid' on plasma fibrinogen and heparin resistance. J. Atheroscl. Res. *3:* 648–652 (1963).
4. COTTON, R.C. and WADE, E.G.: Further observations on the effect of ethyl-*a*-*p*-chlorophenoxyisobutyrate+androsterone ('Atromid') on plasma fibrinogen and serum cholesterol in patients with ischaemic heart disease. J. Atheroscl. Res. *6:* 98–102 (1966).
5. THORP, J.M.: An experimental approach to the problem of disordered lipid metabolism. J. Atheroscl. Res. *3:* 351–360 (1963).
6. THORP, J.M. and BARRETT, A.M.: Studies on the mode of action of 'Atromid'-S (chlorophenoxyisobutyrate, CPIB). Progr. biochem. Pharmacol. In press (Karger, Basel/New York 1966).
7. KURIEN, V.A. and OLIVER, M.F.: Serum-free-fatty-acids after acute myocardial infarction and cerebral vascular occlusion. Lancet *ii:* 122–127 (1966).
8. DEEGAN, T. and HAYWARD, P.J.: Serum lipid changes following myocardial infarction. J. Atheroscl. Res. *5:* 267–275 (1965).
9. LOSNER, S.; VOLK, B.W. and WILENSKY, N.D.: Fibrinogen concentration in acute myocardial infarction. Arch. intern. Med. *93:* 231–245 (1954).
10. GODAL, H.C.: Quantitative and qualitative changes in fibrinogen following major surgery. Acta. med. scand. *171:* 687–694 (1962).
11. OLIVER, M.F. and BOYD, G.S.: Changes in the plasma lipids during the menstrual cycle. Clin. Sci. *12:* 217–222 (1953).
12. BROWN, R.R.; THORNTON, M.J. and PRICE, J.M.: The effect of vitamin supplementation on the urinary excretion of tryptophan metabolites by pregnant women. J. clin. Invest. *40:* 617–623 (1961).
13. OSORIO, C.; WALTON, K.W.; BROWNE, C.H.W.; WEST, D. and WHYSTOCK, P.: The effect of chlorophenoxyisobutyrate ('Atromid'-S) on the biliary excretion and distribution of thyroxine in the rat. Biochem. Pharmacol. *14:* 1479–1481 (1965).
14. APPLETON, W.G. and DOWLING, J.T.: Dynamics of hepatic thyroxine metabolism in man. Physiologist *8:* 102 (1965).
15. TRITSCH, G.L.; RATHKE, C.R.; TRISTCH, N.E. and WEISS, C.M.: Thyroxine binding by human serum albumin. J. biol. Chem. *236:* 3163–3167 (1961).
16. BARRETT, A.M.: The effect of chlorophenoxyisobutyric acid on the release of free fatty acids from isolated adipose tissue *in vitro*. Brit. J. Pharmacol. *26:* 363–371 (1966).

17. HARRISON, M.T. and HARDEN, R.McG.: Some effects of clofibrate in hypothyroidism and on the metabolism of thyroxine. Scot.med.J. *11:* 213–217 (1966).
18. MOGENSEN, E.F.: The effect of ethyl chlorophenoxyisobutyrate on the binding between tri-iodo thyronine and protein. J.Atheroscl.Res. *3:* 415–417 (1963).
19. INGBAR, S.H.; BRAVERMAN, L.E.; DAWBER, N.A. and LEE, G.Y.: A new method for measuring the free thyroid hormone in human serum and an analysis of the factors that influence its concentration. J.clin.Invest. *44:* 1679–1689 (1965).
20. SURKS, M.I. and OPPENHEIMER, J.H.: Postoperative changes in the concentration of thyroxine-binding pre albumin and serum free thyroxine. J.clin.Endocrin. *24:* 794–802 (1964).

Authors' addresses: Dr. J. M. Thorp and Dr. R. C. Cotton, Imperial Chemical Industries Limited, Pharmaceuticals Division, *Macclesfield* (England) and Dr. M. F. Oliver, Royal Infirmary, *Edinburgh* (Scotland).

Discussion Session 11

S. GERÖ (Budapest): Dr. BIZZI, the enhancing effect of theophylline on FFA levels in animals would be rather unfavourable in coronary patients. Are there any data concerning the effects of theophylline on the serum FFA level in man? What was the dosage per kg/wt used in the experimental animals compared to that used generally in human patients?

L. BIZZI and S. GARATTINI (Milan): Our rats received 150 mg/kg theophylline. This dosage is certainly much higher than that one used in man. We have no knowledge of similar type of determinations in man.

A. N. HOWARD (Cambridge): Is there a species difference between the effect of drugs which mobilize free fatty acids? Recently we have been doing some work, in collaboration with Dr. HYAMS, on the effect of 3-methyl salicylic acid, which is very similar we believe to salicylic acid except it is much longer acting. We find that in the rabbit this drug causes an increase in the fasting level of FFA because of a rebound effect and if one gives the drug to rabbits fed a semisynthetic diet there is an increase in atherosclerosis after 4 months. However, when we test this drug, in patients, there are decreased fasting FFA levels overnight and also two hours after a 17 hours fast.

L. BIZZI (Milan): We have not studied species other than the rat but there may be species differences as you point out.

J. SKOŘEPA (Prague): At what pH have you examined the lipolytic activity?

L. BIZZI (Milan): We used Krebs Ringer bicarbonate medium with bovine serum albumin and 0.1% plasma at pH 7.4.

D. KRITCHEVSKY (Philadelphia, Pa.): We have tested W-398 (Benzyl N-Benzylcarbethoxyhydroxamate) in rats, feeding it at the levels indicated by the original authors (BERGER et al., Proc. Soc. exp. Biol. N.Y. *114:* 337, 1963). At a level of 2% in the diet, W-398 causes a marked weight loss and liver enlargement. When fed as 0.5% of the diet the effects are less marked, but still evident. (KRITCHEVSKY and TEPPER; Proc. Soc. exper. Biol. N.Y. *121:* 1162, 1966).

D. KRITCHEVSKY (Philadelphia, Pa.): Do you have any data on reduced mortality in patients being treated with your very interesting drug?

T. SHIMAMOTO (Tokyo): Prof. MASUYA of Kyushu University Medical School reported the reduced mortality rate in atherosclerotic patients being treated with pyridinolcarbamate. In 1964 he had 15 deaths out of all 63 hospitalized patients suffering from various atherosclerotic diseases. In 1965 all patients with atherosclerotic diseases received pyridinolcarbamate in daily dose of one gram and he had just 2 deaths out of all 59 hospitalized patients.

F. K. GEY (Basel): Compounds similar to pyridinol carbamate can inhibit cholinesterase. Is this also true for pyridinolcarbamate?

T. SHIMAMOTO (Tokyo): No cholinesterase inhibiting effect was observed in pyridinolcarbamate at least at the present stage of experiment.

E. H. AHRENS (New York, N.Y.): Dr. THORP's ideas never fail to excite the imagination. May I ask whether you have obtained evidence that CPIB-binding is different in the plasma proteins of patients who respond to this drug, as contrasted to those who do not? For instance, you have mentioned that patients with myxedema or with panhypo-pituitarism fail to show a plasma cholesterol decrease with CPIB dosage, as do a significant number of patients with simple familial hypercholesteraemia. Are these groups of patients typically different in their binding capacities for CPIB, and for thyroid globulin?

J. M. THORP (Macclesfield): A type II hypercholesterolaemia may be due to one or more of several possible alterations in thyroxine secretion, transport (distribution), metabolism and excretion. In the simplest case, where thyroxine secretion is low or absent, it is recognized that CPIB is ineffective, but potentiates the effect of orally administered thyroxine, that is, as measured by a reduction of the cholesterol-rich S_f 0–20 lipoprotein fraction. We suggest that this fraction *rises* in the *absence* of hepatic thyroxine, in contrast to the more usual statement that it *falls* in response to an *increase* in thyroxine.

Emphasis on the consequences of a thyroxine deficit enables one to envisage more readily the consequences of other changes in thyroxine transport or metabolism, leading to reduced liver levels. Thus an increased ratio of thyroxine bound to TBG, compared to that bound to pre-albumin and albumin, would tend to reduce the free thyroxine available for uptake by the liver, even though PBI and thyroid function were apparently normal. The specificity of binding and high affinity of thyroxine for TBG is such that this binding is not likely to be significantly affected by CPIB (whereas binding to albumin and probably pre-albumin is). This therefore could be a second cause of relative thyroxine deficiency in the liver, leading to a type II hypercholesterolaemia, and which, for these reasons, may prove unresponsive to treatment with 'Atromid'-S. There remains the case of altered metabolism and excretion of thyroxine by the liver. This is seen, for example, in vitamin E deficient animals, where deiodination of thyroxine is markedly increased, plasma cholesterol rises, and the response to CPIB is absent, but can be restored by administration of vitamin E. It is possible that the ratio of unsaturated to saturated fatty acids in the diet influences plasma cholesterol by a similar mechanism, i.e. altering the redox balance of liver, and hence hepatic metabolism of thyroxine. Another metabolic defect may be concerned with the role of pyridoxal phosphate as a co-factor for a transaminase involved in the formation of an 'active' deaminated metabolite of thyroxine in liver. As with vitamin E, pyridoxine deficiency in the rat leads to increased plasma cholesterol, which does not respond to CPIB. Admittedly, the relevance of these mechanisms to the case of the patient who does not respond to CPIB is not yet known, but they suggest that measurements of thyroxine-binding by the plasma proteins of these subjects, which should undoubtedly be undertaken, may not necessarily provide the answer. It would be of interest to determine the effects of additional treatment with tocopherol, pyridoxine or unsaturated fat in such 'non-responding' subjects with apparently normal thyroid function.

G. BOYD (Edinburgh): In the *in vitro* studies you showed bindings of tetraiodothyropropionic acid to albumin whereas of course for many species the thyroid hormone is not carried by albumin but it is carried on an other protein. In those animals treated with CPIB where there is a marked rise of liver triglycerides, have you any information on whether the liver endoplasmic reticulum is modified in any way?

J. M. THORP (Macclesfield): Both tetraiodothyropropionic acid and tetraiodothyroacetic acid have considerably higher association constants for binding to albumin than has thyroxine. We used tetraiodothyropropionic acid because it was the only tetraiodothyroalkanoic acid commercially available in amounts sufficient to enable us to undertake the large number of measurements required, and ancillary animal experiments. We wished to examine a general physico-chemical principle which applies to competition between any two anions. Albumin was used for the *in vitro* studies for the same reasons. It is probable that the magnitude of the effect will differ in the case of pre-albumin, and particularly so for thyroxine-binding globulin, which, like other globulins, has a relatively low affinity for anions other than thyroxine. This suggests that the major contribution to TBG-binding of thyroxine is the ionized phenolic group. When the zwitterion of the aminoacid moiety of thyroxine is replaced by an alkanoic acid, affinity for TBG is decreased, and that for albumin and pre-albumin increased. By measuring the binding to total serum proteins of samples from dosed animals or patients in comparison with undosed controls, we assumed that the effects of CPIB on all relevant protein-binding sites would be shown. In both rat and human serum, the results obtained paralleled those found when using albumin alone. However, the amounts of added thyroid hormone required to provide sufficient free (unbound) hormone, for analysis by the technique used, exceeded the very limited capacity of non-albumin binding sites. Therefore the significance of these findings is to be further tested by a study of the effect of CPIB on the hepatic clearance of labelled thyroxine, and the excretion into plasma of thyroxine metabolites, in man.

We have not encountered a marked rise in liver triglycerides in any animals treated with CPIB, and so cannot say whether this is associated with changes in the sub-cellular morphology of the liver. Perhaps Dr. Hess could comment on this?

S. Garattini (Milan): Dr. Thorp, how do you reconcile the decrease of plasma FFA with the increase of liver triglycerides? If there is a decrease of FFA transport, there should be a decrease in the synthesis of liver triglycerides; at least there is not a decrease in the triglyceride excretion.

J. M. Thorp (Macclesfield): We have not been able to confirm that any significant increase in liver triglycerides occurs in our strain of rats, with our techniques of feeding, dosing, killing, sampling, and analysis. One or more of these variables may contribute to the small increase observed by other workers. The explanation for this almost certainly involves the effect of CPIB on the uptake of FFA by the liver. As shown by my colleague Dr. Speake, incubation of liver slices in the presence of CPIB enhances the uptake of labelled palmitic acid from plasma, and its incorporation into triglyceride is increased. This happens at all ratios of FFA to albumin. The situation is precisely the converse of the effect of CPIB in reducing release of FFA from adipose tissue, although lipolysis persists at the normal rate, with the result that FFA accumulate in the tissue, and triglycerides decrease. The same principle applies in both cases, and depends upon the fact that CPIB reduces the association constants of the FFA for albumin binding sites. Thus for a given total concentration of FFA, there will be a higher concentration of non-protein bound FFA in the presence of CPIB than in its absence, for a given concentration of albumin. Since the non-protein bound extra-cellular fraction of the FFA determines the gradient between the extracellular and intracellular FFA, CPIB causes a decrease in the gradient in the case of a cell producing FFA (adipose tissue), with decreased flux of FFA, whereas there is an increase in the tradient in the case of a cell (liver) utilizing FFA, with increased flux, and accumulation in the cell of the products of FFA metabolism (triglycerides in liver). This is exactly analogous to the postulated physico-chemical effects of CPIB on the flux of thyroxine out of the thyroid (decreased) and into the liver (increased), and of the flux of thyroalkanoic acids out of liver into plasma (decreased).

It has been shown that CPIB only partially suppresses the plasma FFA rise induced by catecholamines, and does not reduce the rise in FFA induced by fasting. Since the half-life of the compound in the plasma of rats is about 6–7 h, the ratios of CPIB and of FFA to albumin will be continuously changing with time, and will be modified by the patterns of feeding, activity, fasting, and dosing of the animals, as well as their age, sex, and strain. Diurnal and seasonal variations in liver composition and plasma FFA are superimposed on this, and the effects of environmental 'stress' (noise, handling, anaesthetic, etc.) can cause marked acute variations in FFA. Most of these factors influencing FFA, and hence presumably liver triglycerides, will act in the direction of increasing FFA, and the presence of CPIB will therefore tend to cause an increase in liver triglyceride. An adequately controlled study of this point could take 2–3 years. As I have stated, under the conditions of our experiments, no increase in liver triglycerides in CPIB-treated rats has been observed, nor has there been any histological evidence of increase of liver fat in animals treated chronically, for periods of up to two years.

In the normal situation, the question to be answered is whether, over 24 h of every day, CPIB causes a net increase of flow of FFA into liver, or a net decrease of flow of FFA out of adipose tissue? Our observations in animal experiments and in man suggest that the latter is the case. A consequence of this is the possibility that subjects in positive caloric balance will show an increase of adipose tissue, which is the probable explanation of the weight gain reported to occur in a proportion of patients treated with 'Atromid'-S, whose diet has been uncontrolled. This lends additional force to the arguments for controlling caloric intake in such patients. As far as tissues other than adipose tissue or liver are concerned, reduced uptake of FFA as a result of their lowered plasma concentration would be partially balanced by the stimulation of uptake in the presence of CPIB.

L.W. KINSELL (Oakland, Cal.): We have had occasion to reexamine our ideas on so-called 'pure' hyper-β-lipoproteinemia. Some patients with high cholesterol, high cholesterol/phospholipid ratios have responded to CPIB or have had augmentation of effects of cholestyramine, when CPIB was added. Re-evaluation of these patients indicates that some, at least, have slight but significant elevation of plasma triglycerides and material in the pre-β-region of the Lee-Hatch strip. Some may be in the general type III category (FREDRICKSON). These observations may help to make sense of some of the apparent discrepancies in response or lack of response to CPIB (and diet as well).

R.G. GOULD (Palo Alto, Cal.): The hypothesis proposed by THORP is most provocative and intriguing but I have difficulty in reconciling it with some published work. STRISOWER et al. have reported that CPIB acts primarily on S_f 20–400 lipoproteins and thyroid hormones primarily on S_f 0–20. This seems to indicate that the two act by different mechanisms. Dr. GEORGE BOYD first found evidence that CPIB decreases hepatic cholesterol synthesis, and we have found this to be a definite and marked effect. Since it is generally agreed that thyroid hormones increase cholesterol synthesis, how can an increased thyroid hormone activity in liver be reconciled with a decreased rate of cholesterol synthesis?

Does Dr. THORP consider that the effect of CPIB on FFA release from adipose tissue is sufficiently long lasting to account for the decrease in serum triglyceride and cholesterol levels or is it a transient effect like that of nicotinic acid?

Does he consider that the effect of the drug on cholesterol metabolism is independent of its effect on triglyceride metabolism? The suggestion that familial hypercholesterolemia may be a form of hypothyroidism raises the question whether patients with this syndrome show clinical evidence of myxedema.

J.M. THORP (Macclesfield): It is perhaps not surprising that Dr. GOULD experiences difficulty in reconciling results based (a) on the lipoprotein responses of what are initially different groups of patients, and (b) the *rate* of cholesterol biosynthesis by the liver of rats which are, prior to treatment, normal; following treatment with either thyroid hormones or CPIB. Apart from the marked differences between the two species in the levels and patterns of plasma lipoproteins and thyroxine-binding proteins, an obvious inconsistency in arguing from these two sets of results is that, whilst a deficiency of thyroxine in man certainly leads to an increase of plasma cholesterol, which is correctable by bringing thyroxine levels up to normal; in the case of the euthyroid subject or of a normal rat, orally administered thyroxine only becomes effective at doses giving a plasma level of free thyroxine which is higher than that resulting from the normal feed-back regulation via thyrotrophin. This is seen to be so by the changes in whole body metabolic rate. When B.M.R. is above normal there occurs an increased mobilisation of fat, and an increase in plasma FFA, the availability of which, as a source of energy and as a substrate, is probably essential to the observed increase *net* rate of cholesterol biosynthesis observed in the liver of the thyroxine-treated rat. I stress that the increase is in the *net* rate of biosynthesis, since there also occurs an increase in the rate of cholesterol degradation and excretion by liver in such animals. By contrast, in the rat treated with CPIB, there is a small but consistent decrease in whole body metabolic rate, a decrease in plasma FFA, an increase in liver weight and metabolic rate, and an increase in the level of thyroxine (or its metabolites) in liver. Reduced availability of FFA, as a substrate and source of energy for hepatic cholesterol metabolism, would account for the observation that the rate of cholesterol biosynthesis is decreased by CPIB, by implication because thyroxine-induced cholesterol degradation is still increased. It would be of interest to measure both synthesis and degradation rates of cholesterol in the livers of CPIB or thyroxine treated animals, perfused with their own plasma, normal plasma, or FFA-enriched plasma.

The persistence of CPIB in plasma is at least six times longer than that of nicotinic acid, and the suppression of plasma FFA, though less marked than that by nicotinic acid, is not followed by the rebound seen with the latter. Suppression of FFA is probably not sufficient to account for the reduction of serum cholesterol by CPIB, since this is not seen in thyroidectomized animals, but may play a part in the reduction of triglyceride levels. In this sense, then, we consider that the effect of the drug on cholesterol metabolism is partly independent of its effect on triglyceride metabolism. Using other organic anions, which are highly protein-bound, we have been able to demonstrate relatively selective effects on either serum cholesterol or triglyceride in the rat, but as yet we do not know the practical usefulness of these compounds in man.

I think it unlikely that patients with familial hypercholesterolaemia will show clinical evidence of myxoedema to any marked degree, since this generally involves a gross reduction of thyroxine secretion. It is more probable that they manifest other abnormalities of thyroxine transport, or of its hepatic metabolism, as discussed in reply to Dr. AHRENS.

R. PAOLETTI (Milan): Have you ever measured coenzyme A levels in the liver – because we know that thyroxine induces an increase of coenzyme A concentration and of course the same is true for the adenine cyclase level in the liver which could also be measured. Have you shown any effect for CPIB on free fatty acid in plasma during prolonged fasting?

J.M. THORP (Macclesfield): We have not measured CoA or adenylcyclase in liver. As we reported at the Milan meeting, CPIB does not inhibit the rise of FFA in fasted rats, when given 18 h prior to fasting. The acute effect of a single dose on fasting levels has not been measured.

K.W. WALTON (Birmingham): While we have confirmed [1] Dr. THORP's observation that selective retention of isotopically labelled thyroxins occurs in the liver in animals treated with CPIB, I have difficulty in accepting his general hypothesis concerning its mode of action. Thyroxine exerts an overall effect on protein metabolism generally. In excess dosage catabolism is increased over synthesis. This might account for the alteration of fibrinogen levels in serum which THORP has found. But in reality on to β-lipoprotein, it is now well established that CPIB is most effective in hyperlipidaemias characterized by elevation of S_f 20–400 lipoproteins [2] and ineffective in those characterized by elevation of S_f 0–20 lipoproteins (including the hyper-β-lipoproteinaemia of this pattern encountered in hypothyroidism). Since the β-lipoproteins share a common protein moiety [3] it is difficult to account for this differential behaviour in terms of alteration of intrahepatic metabolism of lipoprotein by a localized action of thyroxine at this site.

References

1. OSORIO, C.S.; WALTON, K.W.; BROWNE, C.H.W. and WHYSTOCK, P.: Biochem. Pharmacol. *14:* p.1479 and *15:* p.213 (1965 and 1966).
2. WALTON, K.W.; SCOTT, P.J.; VERNIER JONES, J.; FLETCHER, R.F. and WHITEHEAD, T.: J. Atheroscler. Res. *3:* p.396 (1963).
3. WALTON, K.W. and DARKE, S.J.: Immunochemistry *1:* p.267 (1964).

J. M. Thorp (Macclesfield): We agree that there are differences in the responses of the S_f 0–20 and 20–400 lipoproteins, and of fibrinogen, which cannot be accounted for by a single generalized effect of thyroxine on protein metabolism. The fibrinogen changes in plasma may be primarily dependent upon changes in fibrinolysin activity, as suggested by the work of Hume (Brit. Med. J. i: p. 686, 1965), who found that administration of L-thyroxine diminished the fibrinolytic activity of blood.

S. Garattini (Milan): From a theoretical point of view, if the main mechanism of CPIB is to impose bonding at the albumin level, two consequences should be expected:
1. FFA metabolism may be decreased in conditions of emergency in which they are needed.
2. Drugs which are bound to albumin, when given to patients on CPIB treatment, may become more active and, perhaps, more toxic.

J. M. Thorp (Macclesfield): The answer to your next question leads on from the previous discussion. Increased FFA metabolism in conditions of emergency is generally accepted to be initiated by catecholamine release, the effects of which are only partially suppressed by CPIB. The most important tissue utilizing FFA in emergency is heart muscle. Only when protein-carbohydrate reserves become depleted, which may take days, does the liver switch predominantly to FFA metabolism. The requirement of heart muscle for FFA as an energy source means that there will be a large gradient between extra-cellular non-protein bound FFA and intra-cellular FFA. As in the case of liver, therefore, the presence of CPIB will stimulate the uptake of FFA by heart muscle. However, there is probably a case to be made for limiting the 'acute' rise in FFA which follows an episode such as myocardial infarction or surgery. If this rise is primarily due to catecholamines, and is subsequently reinforced by thyroxine-mediated effects, attention should be directed to the nature and origin of the initial stimulus, and to ways in which it can be limited. This is a problem which is presently occupying our attention.

The question of the possible consequences of CPIB treatment on the activity and toxicity of albumin-bound drugs has, of course, been in our minds for some years. Such a mechanism probably accounts for the increased prothrombin-lowering activity of anticoagulant drugs, whose activity is localized in the liver. Increased uptake by the liver of compounds acting at that site would be comparable to the effect of CPIB on FFA uptake by liver. However, most other acidic drugs act at extra-hepatic sites, and are generally metabolized to less active and less toxic compounds by the liver. Thus the effects of CPIB would tend to offset each other, and in practice no significant increase in activity or toxicity of commonly-used acidic drugs has been observed. It is perhaps relevant to this topic that my colleague Dr. Platt has not been able to demonstrate any significant increase in microsomal protein or its associated drug-metabolizing enzymes in the liver of CPIB-treated rats. This contrasts with the effects of a large number of compounds which undergo hydroxylation by the liver, and lends additional support to the view that the effects of CPIB on liver structure, metabolism, and composition are indirect.

D. Fredrickson (Bethesda, Md.): Since elucidation of the mechanism of action of Atomid-S may be aided by better understanding of the abnormalities in the patients to whom it is given, I would like to appeal again for the use of lipoprotein patterns in characterizing the hyperlipidemia under therapeutic trial. From the report in the J. Artheroscler. Res., 1965, on the Conference on Atromid one can find two different kinds of 'hyperbetalipoproteinemics' to whom the drug was given. Several of these appear to us to have been Type II in our classification. They did not respond. One example of Type III appears to be present and he apparently responded. In this syndrome, the increased beta lipoprotein concentration seems to be related to an increase in triglycerides. If CPIB effects triglycerides primarily, then the beta lipoprotein concentration will fall with the decrease in glycerides. Type II patterns may be associated with modest increases in glycerides, but these appear to be in different relationship to the beta lipoprotein. Type III is a quite different clinical and genetic entity from Type II. In answer to Dr. GOULD, Type II or III may be associated with '*hepatic*' myxedema, but there is not generalized evidence of thyroid deficiency.

J. M. THORP (Macclesfield): Dr. FREDRICKSON's observations indicate the rapid progress that has been made, largely through his efforts, towards wider recognition of the different varieties of lipoprotein abnormality that may occur. May I suggest that one possible role of 'Atromid'-S might be to differentiate between apparently similar abnormalities of lipoproteins of differing aetiology. Dr. STONE and I have been developing a simple ultrafiltration and nephelometric method for the fractionation and quantitation of the very low density lipoproteins (including chylomicra), which should prove helpful in the elucidation of these problems.

A. SOLYOM (Milan): Dr. THORP, if the plasma FFA lowering effect of CPIB is a consequence of a defect in the FFA transport mechanism in normally fed animals, what happens to adipose tissue metabolism? Does lipolysis proceed normally? In this case, do FFA accumulate in the adipose tissue or are they increasingly metabolized, e.g. by increased re-esterification? Have you investigated triglyceride synthesis in adipose tissue?

J. M. THORP (Macclesfield): These questions have been only partially answered by studies of adipose tissue *in vitro*. Here one sees an accumulation of FFA in the tissue, suggesting that lipolysis proceeds normally. *In vivo* it is probable that a different equilibrium between lipolysis and re-esterification will be reached. This would be worth mesuring.

R. H. FURMAN (Oklahoma City, Okla.): I should like to comment further on what Dr. FREDRICKSON has said relative to so-called 'pure' or familial hypercholesterolemia. In this disorder there are high serum cholesterol concentrations due to excessive amounts of β-lipoproteins without excessive amounts of serum triglycerides. We have had a large number of these subjects under study and so far have failed to find any evidence of thyroid abnormality. I think that most clinicians and students of the problem would agree that there is no evidence of abnormal thyroid function in this disorder. As a matter of fact, we have no evidence that the primary difficulty here is with sterol metabolism. We could very well be dealing with an abnormality of the metabolism of the protein carrier unit or 'apolipoprotein'.

In respect to the use of CPIB in subjects with familial hypercholesterolemia, it has been our experience that a patient will respond occasionally with a lowering of serum cholesterol concentration. This is not common, however. On the other hand, in some subjects with hypercholesterolemia associated with slight hypertriglyceridemia, we have achieved reduction of serum cholesterol and triglyceride levels with CPIB, but after a period of several weeks have been surprised to see serum cholesterol levels beginning to rise. I would be most interested to know if others have observed this phenomenon.

In respect to Dr. KRITCHEVSKY's remarks about STRISOWER's report, I would simply point out that if one employs a pharmacological agent which diminishes the amount of triglyceride requiring transport in the serum, the low density β-lipoproteins will obviously contain less triglyceride and thus there will be a shift in the lipoprotein spectrum in the direction of smaller, denser lipoprotein particles.

Finally, Dr. THORP is very keen to relate the action of CPIB to the phenomenon of protein binding. We recall that his original thesis was that CPIB in combination with androsterone was effective because the CPIB displaced andosterone from a protein binding site, thus releasing the androsterone which was presumed to be responsible for the lipid-lowering effect. This suggestion was proved false when it was demonstrated by our laboratory and others that CPIB alone was equally as effective as CPIB-androsterone combination in lowering lipids. In respect to his suggestion that CPIB induces a state of hepatic 'hyperthyroidism', I find this unpalatable in view of our published metabolic balance studies of subjects administered CPIB. It would seem almost certain that, if hepatic hyperthyroidism existed, a protein catabolic effect would have been apparent in the nitrogen balance data. We have seen no evidence of protein catabolism in subjects treated with CPIB under metabolic balance conditions.

J. M. THORP (Macclesfield): I agree with Dr. FURMAN's comments which imply that one should consider the total spectrum of the lipoproteins, regarded as discrete groups of particles, differing in amount, but each with relatively constant contents of protein, phospholipid, cholesterol and triglyceride; rather than any one lipoprotein or component. Analysis of only one component, common to all classes of particle, gives only a very poor indication of what is happening to the spectrum. Since there are discernible differences only in the composition and properties of the protein fraction of the lipoproteins, this is clearly the component which will most readily provide an insight into the mechanisms of their synthesis and disposal. Whether, under normal circumstances, apolipoprotein availability limits the secretion of lipoproteins by the liver is not yet clear, but it seems probable that the synthesis of the apolipoproteins themselves is effected by the liver. Dr. WALTON's group have demonstrated the effects of thyroxine on turnover of the protein-labelled moiety of reinjected lipoproteins. These results seem to me to indicate the care needed in deciding the extent to which apparent catabolic or anabolic effects are determined by differential changes in biosynthesis and degradation. If the effective concentration of thyroxine in liver, either directly or indirectly (e.g. by alteration of corticosteroid and 17-ketosteroid metabolism), is concerned with the rate of synthesis and degradation both of proteins (apolipoproteins, albumin, fibrinogen) and of lipids (triglyceride, cholesterol, phospholipid), it is the *net* balance, as well as the availability of suitable carriers, which will determine the rate of secretion of any lipoprotein or protein by the liver. I doubt whether adequate hepatic balance measurements to enable this to be studied in man are feasible. One has therefore to consider results obtained in experimental animals, in which physiological levels of thyroxine have an anabolic effect on liver. In hyperthyroidism the net effect is, of course, catabolic. Perhaps we have not stressed sufficiently that the effect of CPIB in the rat is to induce a *relative* increase in liver thyroxine, and that this requires that thyroxine secretion, transport, and hepatic metabolism are not grossly abnormal. Since the action on liver and the rest of the body are in opposite directions, a gross effect on nitrogen balance in man would not be expected.

We may appear keen to relate the effects of CPIB to its influence on the protein-binding of physiologically active anions, but this is primarily because we know of no other explanation which will accommodate all the known observations, and not merely those in any one experimental system or clinical situation. Our original observations in the rat, that CPIB alone was effective, but was potentiated by androsterone, are still valid. This led us to examine the effects of thyroxine, our interpretation of which was made in terms of the type of observation, clinical and experimental, which we have now been discussing, and of the reported effects of thyroxine on steroid metabolism, but without the insight acquired from subsequent experiments. On this limited basis, we concluded, perhaps erroneously, that the effects of thyroxine on liver metabolism could be ascribed to androsterone. Correction of a lipid defect in man appeared therefore to require the addition of androsterone to CPIB. This seemed to be true in the first clinical experiments, and was compatible with the observation that CPIB displaced protein-bound androsterone sulphate. The demonstration that CPIB alone was effective in man does not alter the fact that we originally observed it to be active alone in the rat, or that it will compete with androsterone sulphate for protein binding sites. The more interesting problem is why it is not effective in all cases. Surely this is because the origin and nature of the hyperlipidaemia differs. Equally so, we would not maintain inflexibly that the effects on binding of thyroxine and its metabolites are the only explanation for the action of CPIB. We think that its effects on the binding to plasma proteins of FFA, androsterone and other steroid sulphates, and tryptophan (amongst the anions known to be affected) may independently, or in concert with thyroxine, contribute to its observed activity. Moreover, this mode of action also serves to account for the finding, to which you have referred, that an increase in serum cholesterol may occur during treatment. We have also described this, at high plasma levels of CPIB, in experimental animals, and ascribed this to a decreased secretion of thyroxine from the thyroid, because the necessary transport capacity of plasma was decreased by the protein-bound CPIB.

R. PAOLETTI (Milan): Nicotinic is known to be concentrated in adipose tissue. Is this also true for pyridylcarbinol?

F. K. GEY (Basel): We have no data on this subject at the present time.

R. PAOLETTI (Milan): Dr. GEY, what is the rate of transformation of Ronicol® (β-Pyridylcarbinol) into nicotinic acid?

F. K. GEY (Basel): In the intact rat Ronicol® is converted into nicotinic acid rapidly and to a large extent. Therefore, three reasons may be considered for the longer effect of Ronicol on plasma FFA:
  1. Ronicol is absorbed from the intestinal tract to a larger extent or more continuously than nicotinic acid;
  2. Ronicol has an effect of its own either by the remainder of the unchanged compound or by metabolic alterations due to the oxidation of Ronicol to nicotinic acid;
  3. Ronicol is transformed into nicotinic acid within intracellular compartments which cannot be entered by nicotinic acid.

J. STAMLER (Chicago, Ill.): In response to several inquiries, may I briefly describe the Coronary Drug Project recently launched as a national cooperative undertaking in the United States. This investigation has been several years in the planning. Its fundamental objective is to assess the efficacy of several lipid-influencing drugs in the longterm care of men age 30–64 with a history of one or more previous myocardial infarctions. The patients are accepted into the study after recovery from their most recent infarction, provided they do not have marked impairment of myocardial reserve, or diabetes mellitus requiring insulin therapy, or other major life limiting diseases. The drugs being rested are: nicotinic acid at a dosage of 3.0 g/day, dextro-thyroxine at a dosage of 6.0 mg/day, mixed conjugated equine estrogens at dosages of 2.5 and 5.0 mg respectively (two separate groups), CPIB at a dosage of 1.8 g/day, and placebo. The cited final dosage is arrived at over a two month period, with monthly stepwise increments, beginning with one-third the final dosage. At this juncture, 28 research groups throughout the United States are participating, and the plan is to expand this to 50 or 60, in order to recruit 8,000 patients estimated as necessary for the study. Multiple measurements are being accomplished on these men throughout their five year period on drug, including biochemical evaluations of toxic effects, adherence, and changes in serum cholesterol, triglycerides and lipoproteins. The primary end-point being evaluated is survival, although several other end-points (e.g. recurrence of non-fatal myocardial infarction) are being studied. This cooperative investigation is being supported by the National Heart Institute, National Institutes of Health, United States Public Health Service.

Authors' Index

Abdulla, Y.H., 218
Adachi, Minoru, 294
Adams, C.W.M., 218, 260, 261, 262, 263
Ahrens, E.H., Jr., 54, 88, 90, 165, 166, 169, 561, 562, 618
Alaupovic, O., 91, 163, 164, 165, 210, 260, 261, 332, 334
Andelman, S.L., 30
Anderson, J.A., 71
Andrus, S.B., 393, 457, 458
Aravanis, C., 12
Atsumi, T., 597

Barcat, J.A., 550
Bates, D., 235
Bayliss, O.B., 218
Beaumont, J.L., 110, 165, 166, 169, 454, 561
Begg, T., 206
Berkson, D.M., 30
Bhose, A., 352
Blaton, V., 122, 144
Bihari-Varga, M., 183
Billimoria, J.D., 225, 264
Bizzi, A., 573, 618
Boberg, J., 126, 165
Böttcher, C.J.F., 231, 260
Borgström, A., 166
Born, G.V.R., 563, 564
Bowyer, D.E., 89, 235, 260, 263, 287, 438, 454, 455
Boyd, G.S., 132, 166, 619
Bradford, R.H., 334
Brunner, D., 19, 52, 140, 168, 375, 457
Brus, R., 578
Bullock, B.C., 420
Burkey, Frances, 30

Carlson, L.A., 167, 168, 170, 209, 210, 373
Christensen, S., 244
Chruściel, T.L., 578
Cohen, L., 141, 166
Colombo, M.A., 270

Constantinides, P., 52, 262, 330, 331, 457, 501, 533, 561
Cotton, R.C., 611
Clarke, G.B., 354
Clarkson, T.B., 420, 455, 501

Djordjevich, Juliana, 141
Dontas, A., 12
Dreyfuss, F., 20

Eisenberg, S., 253
Engelberg, H., 50, 330, 542, 563
Epstein, M.B., 30

Farquhar, J.W., 197
Forman, D.T., 71
Fredrickson, D., 164, 168, 169, 623
French, J.E., 550, 562, 563
Frosch, B., 179
Fujita, T., 597
Funmilayo, D., 132
Furman, R.H., 168, 334, 374, 452, 455, 458, 624
Futterman, M., 542

Gale, M., 206
Garattini, S., 209, 377, 618, 620, 623
Garcia, G.A., 503, 564
Gerö, S., 52, 183, 618
Gerschenson, L.E., 282
Gey, K.F., 585, 618, 626
Glynn, M.F., 508
Gonet, A.E., 363
Gould, R.G., 88, 89, 166, 191, 209, 375, 621
Granda, J.L., 153
Gresham, G.A., 122, 235, 287, 438, 460, 501, 503, 506, 563
Groen, J.J., 1, 20, 50, 53, 210, 375, 459, 506
Gross, R.C., 197
Gupta, S., 352
Gustafson, A., 143
Gutmann, L., 20

Habczyńska, D., 578
Haggerty, D.F., 282
Hall, Yolanda, 30
Harary, I., 282
Hartzell, R., 317
Haslam, R.J., 561
Haust, M. Daria, 429, 452, 506
Hawkey, Christine, 556
Herman, Z.S., 578
Hess, R., 470, 501
Hirsch, R.L., 351, 502
Hollander, W., 270, 332, 333
Howard, A.N., 122, 209, 235, 287, 334, 438, 454, 455, 456, 457, 460, 503, 618
Hrabák, P., 161

Inoue, G., 270

Jackson, I., 206
Jeanrenaud, B., 363
Jørgensen, L., 408, 562

Kalant, N., 333
Katz, L.N., 354
Keys, A., 12
Kim, D.N., 445, 488
Kinsell, L.W., 59, 90, 166, 375, 455, 458, 621
Kohout, M., 325
Kramsch, D.M., 270
Krcilková, M., 503
Kritchevski, D., 317, 330, 457, 459, 474, 480, 501, 502, 503, 618
Krut, L.H., 249, 454
Kurien, V.A., 204

Lee, Y.L., 59, 445, 488
Lehner, N.D.M., 420
Lekos, D., 12
Levinson, M., 30
Lijó-Pavía, J.J., 591
Lindberg, H.A., 30
Lindgren, F.T., 438
Loebl, K., 140

Lofland, H.B., 211, 260
Lorch, E., 585

Mahler, R.F., 218
Mancini, S., 53, 166
Malmros, H., 50, 88, 452, 456, 482, 501, 502, 503, 506
MacMillan, G.C., 280
Marčan, K., 161
McGill, H.C., Jr., 26
McGill, H., 455, 456
McMillan, G., 331
Mead, J.F., 282, 331
Mialhe, H., 317
Michaelides, G., 12
Miettinen, T.A., 68, 90, 208
Mikkelson, B., 71
Miller, Wilda, 30
Mojonnier, Louise, 30
Mrhová, O., 325
Mukherjee, S., 352
Mustard, J.F., 508, 561, 563

Nikkilä, E.A., 208
Nishizawa, E.E., 508
Novak, Š., 161
Numano, F., 597
Nye, E.R., 126

Oliver, M.F., 204, 611
Onajobi, Funmilayo, D., 132

Packham, M.A., 508
Palmer, B.V., 235
Paoletti, R., 90, 377, 456, 622, 626
Pascaud, M., 89
Patelski, J., 287, 458

Payling-Wright, H., 564
Peeters, H., 122, 144, 165, 459
Pelkonen, R., 208
Percy-Robb, I.W., 132
Pick, Ruth, 166, 354, 501, 502
Plech, A., 578
Pollak, O.J., 50, 261, 265, 294, 377, 502, 561
Portman, O.W., 393
Puglisi, L., 371

Reaven, G.M., 197
Renold, A.E., 363, 373
Rifkind, B.M., 53, 168, 206
Riopelle, A.J., 393
Robertson, A.L., 261, 305, 330, 331, 332, 561
Robinson, R.W., 370
Root, M.A., 218
Rothblat, G.H., 317
Rothwell, T.J., 225
Rowsell, H.C., 508

Scanu, A.M., 153, 164, 166, 331, 332
Scott, R.F., 488
Shimamoto, T., 564, 597, 618
Shioda, R., 59
Sisson, J.A., 488
Skořepa, J., 161, 210, 263, 618
Smith, E.B., 164, 260, 455
Sólyom, A., 371, 624
Sullivan, M.F., 270
Swyryd, E.A., 191
Székely, J., 183

Szulc, S., 287
Schettler, G., 565
Schlierf, G., 59
Schwartzkopff, W., 81
Stamler, J., 30, 50, 51, 52, 53, 263, 373, 375, 626
Stary, H.C., 280
Stäubli, W., 470
Stauffacher, W., 363
Stein, O., 253
Stein, Y., 253
Sternby, N.H., 482

Taskinen, M.-R., 208
Taylor, C.B., 71, 88, 89, 90, 455, 456, 457, 459, 506
Tepper, Shirley A., 474, 480
Thomas, W.A., 333, 445, 488
Thorp, J.M., 165, 611, 618, 619, 620, 621, 622, 623, 624, 625
Todorovičová, H., 161

Urbanová, D., 325

Wagener, H., 81, 89, 179
Waligóra, Z., 287
Walton, K.W., 159, 163, 262, 559, 561, 562, 622
Ważna-Boguńska, C., 578
Wilkens, J.A., 249
Wissler, R.W., 165, 262, 378, 452, 457, 502, 506, 562
Wolf, P., 559
Wood, P.D.S., 59

Zemplényi, T., 51, 209, 262, 325, 563, 564

Subject Index

Adenosine Diphosphate
  effect on platelet aggregation 526
Adiposity
  and blood lipid levels 206
Age
  and atherosclerosis 1
  and ischemic heart disease 337
Androgens
  and serum lipids 343
Aneurysms, cerebral
  in primates 393
Anti- a_2 lipoprotein serums 164
Aorta (see also Arterial Wall)
  arterial lipids 236
  cholesterol esters in
    tissue cultures 317
  enzymes 218, 326
  mode of lipid transfer 331
  mucopolysaccharide changes 333
  perfusion 214, 235
  phospholipid synthesis 225
  uptake of cholesterol 273
  uptake of cholesterol esters 240
  uptake of FFA 239
Apolipoproteins (see also Lipoproteins, Plasma)
  delipidization procedure 94
  distribution 100
  properties 92
  role of low density lipoproteins 159
  type A 95
  type B 97
  type C 163
Arachidonic Acid
  and hyperlipemia 84
  metabolism in man 81
Arterial Wall
  chemical composition 266
  cholesterol esterase 287
  chondroitin sulphate C 233
  enzymes in birds 325
  enzymes in man 325
  esterase activity 222, 261
  flux of cholesterol 212
  lipid accumulation 517
  lipid synthesis 215
  lipolytic enzymes 218, 220, 261, 287
  mast cells 266
  metabolism 211
  metabolism of lysolecithin 253
  metabolism of mucopolysaccharides 270
  mucopolysaccharide gel 232
  oxidative enzymes 218
  oxygen measurements 305
  oxygen uptake 268
  penetration of lipids 212
  phospholipid synthesis 225
  plasma constituents in 249
  production of mucopolysaccharides 268
  protein synthesis 445, 497
  protein synthesis in 267
  tissue cells 294
  tissue cultures 265
  transacylating enzyme 260
  uptake of phospholipids 258
Atheromata (see also atherosclerotic lesions)
  presence of lipoproteins 212
Atherosclerosis
  age 1
  and estrogens 354
  blindness 581
  cerebral complications 7
  comparative pathology 460
  diabetes 6, 363
  dicumarol 358
  diet 4, 35
  dietary, in dogs 482
  effect of clofibrate 578
  effect of ephedrine 381
  effect of linolexamid 480
  effect of meprobamate 381
  effect of special fats 474
  effect of vatensol 578
  epidemiology 1, 26, 30
  ethnic groups 2
  geographic pathology 26
  hypercholesteremia 3
  hypertension 6
  in baboons 438
  in birds 464
  in chimpanzees 393, 401
  in monkey 420
  in parakeet 470
  in pig 466
  in primates 378
  in rabbit 463
  in rat 460
  international project 27
  obesity 5, 363
  physical activity 5
  prostatic enzyme treatment 581
  protein metabolism 488
  relation to thrombosis 508

sex 2
smoking 5
thiouracil 357
treatment 565
treatment of human 602
Atherosclerotic Lesions (see also *atheromata*)
  acetate utilisation 310
  and lipoprotein uptake 244
  cell hypoxia 314
  cholesterol synthesis 274
  DNA synthesis 280
  fatty streak 236
  metabolism of cholesterol 270
  metabolism of polysaccharides 270
  origin 231
  oxygen measurement 305
  release of lipids 314
  uptake of cholesterol 245
ATP -ase
  in human aorta 328
Atromid (Atromids, Clofibrate, CPIB)
  effect on plasma lipids 192
Baboons
  aortic atherosclerosis 438
  lipoprotein composition 122
  plasma lipoproteins 443
Bile Acid
  excretion in man 62, 63
Bile Acids
  and hyperlipemia 179
  and regulation of cholesterol synthesis 165
  effect of estrogens 360
Biliary Drainage
  and rat lipoproteins 136
Birds
  pathology of atherosclerosis 464
Bread
  fatty acids in, 50
Chilomicrons
  function 170
  protein composition 100
  transformation in lipoproteins, 143
Chimpanzees (see also Primates)
  spontaneous atherosclerosis 393
2- Chloro Adenosine
  effect on platelet aggregation 526
Clofibrate (*Atromid*)
  see CPIB
Cholesterol
  degradation in man 90
  dietary, and rat lipoproteins, 134
  effect of gonadal hormones 352
  ester metabolism 317
  esterase 321
  esterase in aorta 287
  esters in tissue culture 317
  esters, uptake in aorta 240
  feeding in monkeys 424
  flux in arterial wall 212, 272, 273
  in baboons 442
  in human liver 89
  in lipoproteins in aorta 276
  in obesity 206
  in primates 403
  metabolism in atheromata 270
  subcellular distribution in aorta 278
  synthesis in aorta 215, 274
  turnover, effect of CPIB 195
  uptake in atheromata 245, 309, 260
Cholesterol, Metabolism
  absorption and gastric mucoid 185
  absorption in man 57
  and unsaturated dietary fat 66
  in blood and atherosclerosis 42
  in blood and coronary heart disease 50
  in man 54
  in rabbits fed hard fat diets 58
Cholesterol Synthesis
  absence of 'feed back' in man 78, 89
  effect of cholic acid 68
  effect of dietary cholesterol 68
  in man 71
Cholestyramine
  and cholesteremia 165
  and rat lipoproteins 136
Cholic Acid (see also Bile Acids)
  and cholesterol synthesis 68
  and rat lipoproteins 137
Chondroitin Sulphate C
  in arterial wall 233
Coronary Drug Project 626
Coronary Heart Disease (see also Ischemic Heart Disease)
  and cholesterol in alphalipoproteins 140
  and diabetes 6
  and plasma FFA 176
  epidemiology 1, 10, 20, 30
  ethnic groups 22
  in Israel 20
  lipid transport 171
  mortality rate 46
  pathogenesis 2

prevention 10
psycho-social factors 9
'risk factors' 34
smoking 50
Coronary Prevention
  evaluation program 35
CPIB (Atromid, Clofibrate)
  block of acetate incorporation 193
  effect on cholesterol synthesis 192
  liver TGL synthesis 194
CPIB
  and endocrine regulation of lipids 611
  in atherosclerosis treatment 566
  in rat atherosclerosis 578
Dextrothyroxine
  effect on plasma lipids 192
Diabetes
  and atherosclerosis 6, 52, 363
  and coronary heart disease 51
  lipoprotein turnover 351
Dicumarol
  and atherosclerosis 358
Diets
  and atherosclerosis 35
  and protein metabolism 488
  aortic lipid synthesis 228
  atherogenic and protein synthesis 445
  atherogenic in baboons 439
  atherogenic in dog 482
  in primates 386, 397
DNA Synthesis
  in arteriosclerotic lesions 280
  in monkey aortas 448
Dog
  atherogenic diet 482
Electrochromatography of
  lipoproteins 144
Ephedrine
  and primate atherosclerosis 381
Epinephrine
  thrombus formation 545
Estrogens
  and atherosclerosis 354
  and bile lipids 360
  and plasma lipids 360
  and polyunsaturated fatty acids 361
  effect on lipoproteins 340
Estrone
  and lipoprotein composition 342
Fat
  and atherosclerosis 38
  special, and atherosclerosis 474
  unsaturated and blood

cholesterol 66, 88
unsaturated and steroid excretion 59
Fatty Acids, Polyunsaturated
  and atherosclerosis 37
  effect of estrogens 361
  in diet 4
  metabolism 282
  uptake in heart 284
  uptake in HeLa cells 285
Fatty Streaks
  in human aorta 429
  in man, age 1
  in perfused aorta 236
Free Fatty Acids
  after vascular occlusion 204
  and nicotinic acid 175, 371
  and norepinephrine 371
  function 170
  inhibition 174
  in obese mice 373
  metabolism 172
  mobilization 173
Glycerol
  and TGL Turnover studies 200
Glycocholic Acid
  in human serum 182
Gonadal Hormones
  and coronary heart disease 339
  effect on cholesterol metabolism 352
Heart Lipids
  and FFA mobilization 209
Heparin
  and aorta lipase 290
  and aorta lipids 330
Hypercholesteremia (see also *Cholesterol*)
  and human atherosclerosis 3, 30
Hyperlipidemia
  and atherosclerosis 6, 31
  and serum bile acids 179
  arachidonic acid
  Turnover 84
  auto immune 110
  insulin 208
  with anti-béta-lipoprotein 110
Hypertension
  and atherosclerosis 6, 31
Immuno Histochemistry
  of aorta fatty streaks 429
Insulin
  in hyperglyceridemia 208
Ischemic Heart Disease (see also
  Coronary Heart Disease)
  and auto immune hyperlipidemia 117

and blood lipids 334
and gonadal hormones 334, 339
EKG findings 16
in Americans 337
in Greece 12
in primates 393
mortality 17
mortality ratio 337
physical activity 19
Isoproterenol
  thrombus formation 546
Lactate Dehydrogenase
  in human aorta 328
Lecithin
  synthesis from lysolecithin
  in arterial wall 256
Linolexamid
  and experimental atherosclerosis 480
Lipase
  effect of heparin 290
  in aorta 287
Lipids (see also Cholesterol,
  Triglycerides, Phospholipides)
  in aortic wall 236
  plasma, endocrine regulation 611
  serum, effect of androgens 343
  serum, effect of estrogens 341, 360
  serum, in parakeet 470
Lipids, Plasma
  absorption 183
  metabolism 170, 191
Lipoprotein Lipase
  in human plasma, 161
Lipoproteins
  alpha, variation in man 141
  anti beta lipoprotein 110
  apolipoproteins 91
  delipidization 94, 153
  effect of estrogens 340, 341, 343
  effect of methyltestosterone 345
  fatty acid composition 122, 144, 148
  in atheromata 212
  in baboons 122, 442
  in diabetic rats 351
  in human aorta 275
  in lipid transport 170
  in primates 398
  in rat 132
  metabolism 91
  new classification 103
  optical properties 153
  penetration in the arterial walls 212
  protein moiety of VHDL 98

protein moiety of VLDL 97
recirculation of H^3 palmitate 126
role in atherosclerosis 159
spectrum 92
synthesis in aorta 276
uptake in atheromata 244
Lysolecithin
  conversion to lecithin 256
  metabolism in arteries 253
  uptake 254
Malate Dehydrogenase
  in human aorta 328
Mast cells
  in arterial walls 266
Meprobamate
  and primate atherosclerosis 381
Methyltestosterone
  effect on lipoproteins 345
Mevalonate
  in cholesterol synthesis 69
Monkeys
  atherosclerosis 420
Mucopolysaccharides
  and lipid absorption 183
  gel in arterial wall 232
  in experimental atherosclerosis 272
  in rabbit aorta 333
  metabolism in atheromata 270
Myocardial Infarction (see also
  Coronary Heart Disease, Ischemic
  Heart Disease)
  ethnic groups 23
  physical activity 19
Neomycin
  effect on plasma lipids 192
Nicotinic Acid
  and arterial Triglycerides 174
  and FFA transport 371
  and plasma FFA 175
  effect on lipid metabolism 585
Norepinephrine
  effect on FFA transport 371
5'-Nucleotidase
  in human aorta 328
Obesity
  and atherosclerosis 5, 363
  and blood lipids 206
Oxygen
  and lipid clearance 330
  tension in atheromata 314
Parakeet
  atherogenesis 470
Phosphatidyl Ethanolamine

in aorta 228
Phospholipase A
　in aortic wall 287
Phospholipids 'essential'
　in atherosclerosis treatment 567
Phospholipid, Synthesis
　at subcellular level 262
　effect of diet 228
　in aorta 225
　in perfused aorta 227
　in rat tissues 226
　synthesis in atheromata 247
　uptake by artery 258
　uptake in atheromata 245
Physical activity
　and atherosclerosis 5
Pig
　pathology of atherosclerosis 466
Pigeons
　atherosclerosis in 213
Platelets
　adherence and aggregation 509, 568
　breakdown of platelet thrombi 556
　fine structure 550
　products in plasma 558
　relation to thrombosis 508, 568
　thrombo-emboli 524
Primates
　atherosclerosis 378
　diets 386
Protamine Sulphate
　and aortic lipase 290
Protein Metabolism
　and atherogenic diets 484
Protein Synthesis
　in arterial wall 267
(3,5 Dimethyl) Pyrazole
　and FFA mobilization 209
Pyrazole Derivates
　inhibition of FFA release 573
Pyridinol Carbamate
　treatment of atherosclerosis 587
Beta-Pyridylcarbinol
　and atherosclerosis treatment 566
　and lipid metabolism 585
Rabbit
　arterial tissue culture 294
　pathology of atherosclerosis 463
　special fat and atherosclerosis 474
Race
　and coronary heart disease 338
Rat
　pathology of atherosclerosis 460

Retinal atherosclerosis 582
Salicylic Acid
　and FFA mobilization 209
Sex (see also Gonadal Hormones)
　and atherosclerosis 2
　and ischemic heart disease 337
(Beta)-Sitosterol
　effect on plasma lipids 192
Smoking
　and atherosclerosis 5, 32
　and serum cholesterol 33
　and thrombi formation 542
Smooth Muscle Cells
　in monkey aorta 450
　lipids 389
Steroid
　excretion 56
　fecal, effect of diet 59
　feed, measurement 54, 61
　gas liquid chromatography 55
　in plasma 59
Taurocholic Acid
　in human serum 182
Thiouracil
　and atherosclerosis 357
Thrombosis
　and atherosclerosis 508
　and blood flow 514
　and catecholamines 542
　cerebral 7, 8, 370
　cerebral, in man 533
　platelet-thrombi 556
　relation to platelets 508, 568
Tissue Cultures
　and cholesterol esters 317
　cholesterol esterase 321
　subcellular distribution of
　　cholesterol 323
Triglycerides
　in blood and atherosclerosis 43
　in blood, obesity 206
　in lipoproteins of diabetic rats 351
　synthesis and diet 126
　synthesis, effect of CPIB 194
　turnover in liver and plasma 197, 200
Triparanol
　site of action 194
Vascular Occlusion
　and plasma FFA 204
Vatensol
　in rat atherosclerosis 578
Weight Reduction
　and blood lipids 53